CHEMICAL ANALYSIS

Other volumes in preparation

CHEMICAL ANALYSIS

A SERIES OF MONOGRAPHS ON
ANALYTICAL CHEMISTRY AND ITS APPLICATIONS

Editors

P. J. ELVING • I. M. KOLTHOFF

VOLUME 27

INTERSCIENCE PUBLISHERS

A division of John Wiley & Sons, New York/London/Sydney/Toronto

Rock and Mineral Analysis

JOHN A. MAXWELL

Geochemistry, Mineralogy and Economic Geology Division
Geological Survey of Canada
Ottawa

1968

INTERSCIENCE PUBLISHERS

a division of John Wiley & Sons, New York/London/Sydney/Toronto

Library of Congress Catalog Card Number 68-29396

SBN 470   57900X

PRINTED IN THE UNITED STATES OF AMERICA

PREFACE

The methods and techniques of rock and mineral analysis have undergone many changes during their long history of service but never so rapidly and extensively as in recent years. To a considerable extent this is a reflection of the advances made in the development and analytical application of instrumental methods in general, particularly those which lessen the time required to make an analysis and which reduce the high degree of personal skill that has heretofore been a requirement of the rock and mineral analyst. No small credit, however, must be given to the salutary, although painful, lesson of G-1 and W-1 which initiated a searching reappraisal of the [then] current methods and techniques. The winds of change have been blowing with telling effect, and seem likely to continue to do so for some time to come.

These statements, however, necessitate an explanation of the reasoning behind the preparation of yet another book on rock and mineral analysis in which the majority of the methods described are still of the classical or conventional type. In the first place, rocks and minerals are being analyzed in many laboratories which do not have access to the necessary instrumentation required for the newer methods, and which must perforce continue to use the older ones; for these laboratories an attempt has been made to collect and evaluate a significant part, at least, of the many pertinent literature references that have accumulated since 1950. Secondly, the skillful application of the so-called classical methods in their present refined and modified form is still the best means for providing those accurate reference standards by which relative instrumental readings are converted into meaningful compositional values, and a detailed restatement of these methods and techniques is overdue. The writer has endeavored to provide for these two needs, while at the same time drawing attention to the advantages to be gained from an exploitation of the newer instrumental methods.

The preanalysis steps, which include sampling, sample preparation and sample decomposition, are presented in Part I, together with a discussion of the accuracy and precision to be expected from the use of different methods, and information on laboratory reagents and equipment. Part II is both a review of the recent literature on methods for the determination of each constituent and a discussion of the analytical problems associated with the determination. The review is not an

v

9 2335

exhaustive one but, cumulatively, the cited references cover a significant
portion of the new developments in rock and mineral analysis over the
past fifteen to twenty years. Part III is a detailed description of
methods, most of which have been used and modified to some degree
by the writer. Part IV touches briefly upon X-ray emission spectrography and atomic absorption spectroscopy, two recent developments
which have had, and continue to have, a considerable effect upon the
state of the art. Finally, three schemes for the analysis of silicate rocks
are outlined in the appendixes, with each step being referred to the
appropriate section in the text. The emphasis throughout has been
placed upon the analysis of silicates and carbonates, both because of
the writer's familiarity with this type of analysis and because these
methods, with minor modification, are generally applicable to most
rocks and minerals. It is felt that the analysis of ores requires a
separate presentation.

The monumental treatise of W. F. Hillebrand and G. E. F. Lundell
(1929) has set a pattern for the treatment of the subject that is certainly
reflected in the following pages. The writer was privileged to take his
formal training in the modified classical methods of rock and mineral
analysis in the Rock Analysis Laboratory at the University of Minnesota
and these methods, further modified in the writer's laboratories at the
Geological Survey of Canada, constitute a large portion of Part III.
The debt that I owe to the Rock Analysis Laboratory in general, and to
S. S. Goldich, Eileen K. Oslund, R. B. Ellestad and E. B. Sandell in
particular, is sincerely and gratefully acknowledged.

My thanks are due also to my colleagues, Mr. Sydney Abbey and
Mr. G. R. Lachance, for their critical reading of Chapters 11 and 12,
respectively, and for their helpful suggestions. The Geological Survey
of Canada encouraged me to undertake the task of preparing this monograph, and provided the necessary secretarial assistance; the contents
are, however, my responsibility alone.

Figures 4.5 and 4.6 are reproduced with the kind permission of
The Canadian Mineralogist, Figure 9.1 and Table 9.4 with that of
Chemistry in Canada; all other illustrations were provided by the
Photographic Section of the Geological Survey of Canada.

Finally, I must express my deep appreciation of the patience and
forbearance of my wife, Helen, whose timely prodding did much to
bring this manuscript to completion.

J. A. MAXWELL

Ottawa, Canada
August 1968

CONTENTS

PART I Preliminary Considerations

Part IV Methods of Analysis—Special

APPENDIXES

PART I

Preliminary Considerations

CHAPTER 1

THE NATURE OF THE ANALYSIS

1.1. General

The importance of chemical analyses of rocks and minerals to geological and related studies is now so well established that there is no need to begin here by justifying the effort, time and expense required to provide them; the writer is well content to let the arguments presented by advocates having greater skill and authority, such as the late Professor Arthur Holmes (Ref. 7, pp. xi–xviii), speak for him.[8,25] The quantity of such analyses appearing in print since about 1950 has increased remarkably over that of previous years, and fully as many more must lie buried in filing cabinets and card files.* The relatively recent trend toward the use of computer programs for the processing of geological data has emphasized the need to have a sufficient amount of data for statistical evaluation and the demand for rock and mineral analyses shows no signs of abating.

The chemical analysis of such naturally occurring substances as minerals and ores was very much in vogue as far back as the eighteenth century† and, according to Hillebrand (Ref. 8, p. 793), "the composition of the ultimate ingredients of the earth's crust—the different mineral species that are found there . . . —was the favorite theme of the great workers in chemistry of the earlier half of the nineteenth century. . . . As an outgrowth of the analysis of minerals and closely associated with it came the analysis of the more or less complex mixtures of them—the rocks." This resulted in the early development of a systematic ana-

---

* A recent[13] publication of the Geological Survey of Canada lists some 1,300 chemical analyses of Canadian rocks, minerals and ores compiled from the publications and files of the Survey for the 110-year period 1846–1955. By contrast, the number of such analyses added to the Survey Data File during the 10-year period 1956–1965 is almost 6,200, and this does not include the very large number made in the course of geochemical studies. Chemical analysis was made an integral part of Survey activity at a very early date by [Sir] William Logan, its founder and first Director; in an official letter written in December, 1843, in which he outlined the necessary staff requirements for the proposed work, he stated that "it is, however, with regard to the analysis of the minerals that I am most anxious."

† The close association between the analysis of rocks and minerals and the discovery of the elements has been presented in fascinating detail by Weeks.[26]

lytical scheme which underwent continuous modification in the light of the increasing knowledge of the principles of analytical chemistry and the development of new and improved laboratory facilities. Such famous names as Berzelius, Wöhler, Rose, Klaproth and Scheele lead in direct succession to Washington, Clarke, Hillebrand, Lundell and Groves, to whom we owe the "classical" methods of rock and mineral analysis that are still widely used. But these methods are too time-consuming to provide the quantity of analyses needed today and, starting about 1947, various "rapid" or "novel" analytical schemes have been proposed as an alternative.[4,11,16,19,21] For the most part these offer increased speed of determination at the expense of decreased accuracy; mention is made frequently in the succeeding pages of the methods proposed by Hedin, Shapiro and Brannock, Corey and Jackson, J. P. Riley and others, in which gravimetry is minimized and emphasis is placed upon spectrophotometric, titrimetric and flame photometric methods. It must be realized at once, however, that these proposed analytical schemes are only adaptations of the classical scheme and still retain many of the characteristics of the latter, including the need to make certain separations. Chirnside, in his review of silicate analysis (Ref. 4, p. 18T), refers to the frequent claim that less-skilled analysts are readily able to perform these analyses and states that this latter "is not borne out by our experience. Indeed, we have found that the methods as described could not be made to work at first by analysts of the very highest skill. I am not always impressed by claims for the ability to turn out larger and larger numbers of analyses by less and less skilled analysts. There is, I think, more often a case for fewer analyses by more highly skilled workers." With these statements the writer is completely in accord.

Since about 1960 much work has been done on methods involving the simultaneous or sequential determination of many constituents by physicochemical instrumental methods, such as those of optical emission, X-ray fluorescence and, most recently, atomic absorption spectroscopy, as well as chromatographic and activation methods. Analytical schemes which combine these various approaches with some still necessary "wet" chemical methods have been proposed for general use in rock analysis (mineral analysis is, for the most part, too "specialized" to permit the application of these methods, unless the analysis is treated as a problem in applied research) and, within the limitations mentioned previously for the rapid chemical schemes, are capable of being truly rapid. The application of these newer methods and techniques is discussed at appropriate places in the following pages* but, as a means of illustrating the changing nature of rock and mineral analysis, some of the newer and different analytical schemes will be briefly described here.

Maynes[14] has described a procedure for the analysis of silicate rocks in which nine of the regular constituents are determined in a 1-g sample, following a two-column ion-exchange separation of Fe, Ti, Al and P from Ca, Mg, and Mn; Fe^{2+}, Na, K and total water are determined on separate portions of the sample by conventional methods. The 1-g portion, after the determination of H_2O^-, is fused with Na_2CO_3 and a single separation of SiO_2 is made as usual, following evaporation with hydrochloric acid; the unrecovered Si is determined colorimetrically in the filtrate.[10] To the latter, evaporated to a small volume, is added sulfosalicylic acid, the pH is adjusted to 7.3 $\pm$ 0.2 with aqueous NH_3, and the solution is passed through two ion-exchange columns, the first containing an anion-exchange resin, Dowex 1-X8 (treated with aqueous NH_3 to pH 7.3 $\pm$ 0.2), and the second a cation-exchange resin, Amberlite IRC-50 converted to the ammonium form, to effect the separation mentioned previously. The Dowex-1 column is washed first with hydrochloric acid (concentrated at first, then 1:1) to remove Ti, Al and P, then with sulfuric acid (1:19) to recover the iron. Similarly, Mg and Mn are removed first from the Amberlite column with a 1% solution of 8-hydroxyquinoline-5-sulfonic acid adjusted to pH 10.0 with aqueous NH_3, then the Ca is removed with hydrochloric acid (1:19). The various constituents are determined separately in an aliquot of the appropriate eluates by essentially conventional procedures (Al_2O_3, for example, is determined gravimetrically with 8-hydroxyquinoline). The accuracy and precision of the proposed analytical scheme is demonstrated by replicate analyses of G-1 and W-1 (see Sec. 2.2, Table 2.4) and compare favorably with those obtained by the conventional procedures; it is stated that six complete analyses may be carried out in about 8 days. Maynes[15] has also adapted the procedure to the semimicro analysis of minerals as well as rocks. The possibilities of a similar scheme, in which ion-exchange separations followed an acid decomposition of the sample but with solvent extraction steps as well, were investigated by Ahrens, Edge and Brooks[1] for the determination of major, minor and trace constituents in common silicate rocks; final determination of the constituents is done spectrochemically on the dried residues of the eluates and extractions.

---

* May and Cuttitta have recently reviewed in some detail a variety of new techniques now being used in geochemical analysis (Supp. Ref.).

A detailed study by Langmyhr and Graff (Sec. 5.2.1) of the use of hydrofluoric acid as a decomposing agent for silicate rocks demonstrated that there is no, or only a negligible, loss of Si when a siliceous material is dissolved in an excess of concentrated (38–40%) hydrofluoric acid, and upon this fact the authors have built a scheme of analysis for the determination of 11 main constituents, using 3 separate samples.[11] Sample A (0.2000 g) is decomposed by hydrofluoric acid in a special closed Teflon vessel (or in a special Teflon-lined bomb) and the excess HF is complexed and the precipitated fluorides dissolved, by the addition of aluminum trichloride solution. Two aliquots are taken for the colorimetric determination of Si as the yellow α-12-molybdosilicic acid, using as the blank solution an aliquot of solution B. The latter is prepared by decomposing 0.4000 g of sample in an open Teflon dish with hydrofluoric and sulfuric acids, and all F and Si are removed by a double evaporation to dryness; aliquots of the sulfate solution are used for the Si blank, the colorimetric determination of Fe, Ti, P and Mn, and for the titrimetric determination of aluminum (back-titration with standard zinc solution after complexation with excess EDTA) and of calcium and magnesium (EDTA titration of Ca and of Ca plus Mg), and the flame photometric determination of sodium and potassium. A third decomposition (sample C) is made in a covered crucible of platinum or recrystallized alumina by a modification of the Pratt method (Sec. 9.22.1) for the determination of ferrous iron. Established methods are used for the determination of H_2O^-, total H_2O and carbon dioxide. The precision and accuracy of the results obtained by this analytical scheme are shown in Sec. 2.1, Table 2.2 (Part II) and Sec. 2.2, Table 2.4 (column VII).

Emission spectrography has, for some years, been used to provide a rapid determination of the major constituents and, by means of close operating control and replicate determinations, a satisfactory accuracy has been achieved in the analysis of the simpler silicates and related material. An automated scheme based upon the use of a direct-reading optical emission spectrometer for the determination of both the major and trace elements in silicate rocks has been developed at the Centre de Recherches Pétrographiques et Géochimiques, Nancy, France. The trace elements[6] are determined by interrupted arcing of a briquette prepared by mixing the sample with graphite; for the major constituents[18] the sample is fused with a mixture of boric acid, lithium and strontium carbonates and cobalt oxide (as internal standard), powdered and introduced into the arc by means of a tape machine. Ingamells and Suhr have recently[9,20] described a rapid chemical and spectrochemical scheme of analysis, designed primarily for silicates but adaptable to other classes of materials, that is simple in operation but capable of producing results having a high degree of accuracy. The proposed combination of techniques is particularly useful with small samples and it is possible to make a complete analysis with as little as 50 mg of material, the only difference in sensitivity or accuracy being occasioned by the potentially greater sampling error. The sample, usually 100–200 mg, is fused with lithium metaborate ($LiBO_2$) in a graphite crucible at 950°C and the liquid melt is dissolved in dilute HNO_3 containing cobalt, the latter for use as an internal standard in the spectrochemical part of the scheme. Aliquots of the solution are used for the flame photometric determination of Na, K, Rb and Cs, the spectrophotometric determination of Si (silicomolybdenum blue), P (phosphomolybdenum blue) and Al (aluminon) and, if necessary, for the determination of Fe, Mn, Cr, Ni, Ti, Ca and Mg. These latter are usually determined spectrometrically, and Ingamells and Suhr recommend the use of a rotating disk electrode with a high-voltage ac spark in conjunction with a direct-reading spectrometer. The solution procedure overcomes the major difficulty encountered in the use of powders for spectrochemical analysis, that of the usual lack of adequate standard samples for calibration and comparison. Standard solutions are easily and accurately prepared at any desired concentration by adding a known weight of the desired element (in sulfate or nitrate solution) to a base solution containing $LiBO_2$, HNO_3 and cobalt nitrate (internal standard), and the problems of inhomogeneity and differences in physical condition are avoided. Langmyhr and Graff[11] have also emphasized the usefulness of synthetic "rock" solutions in testing the precision and accuracy of a new analytical scheme. The authors of the new scheme stress the fact that the accuracy attainable in the determination of the major constituents is not comparable to that achieved by the "classical" methods, which indeed are used to provide the necessary standards, but point out that these direct secondary methods are capable of giving results which are entirely adequate for the routine analysis of large numbers of samples and can be carried out by semi-skilled personnel after a short training period. This scheme, like that of Langmyhr and Graff, has been presented with a welcome abundance of detail.

If the writer has seemed to dwell overlong on these new schemes, it is because they are indicative of the radical changes that are currently taking place in the methods for the analysis of rocks and minerals. It is important that those who are embarking upon major programs which require large numbers of analyses, and who may seek here for some guidance, should be made aware at the beginning that there are other approaches very different from those outlined in the following pages and that these, depending upon the particular needs and resources, may be worth a closer examination.

1.2. The constituents to be determined

This is a subject that defies a simple answer because, earth scientists being what they are, the rock and mineral analyst will eventually be called upon to determine about three-quarters of the elements in the Periodic Table, in concentrations ranging from those of major constituents to less (often much less!) than 1 ppm. This is a recent development, in part arising out of the increased awareness of the possibilities of the optical emission spectrograph and accentuated now by the growing application of the electron microprobe to the study of individual minerals grains, and to the emphasis being placed upon the distribution of trace elements among the various physical components of the lithosphere. Table 1.1, which is reprinted here from *Outlines of Methods of Chemical Analysis* (Ref. 12, Table 8, p. 15), shows the elements most often encountered in rock analysis. Among the additional minor constituents frequently determined now are Rb, Cs, Be, Y, La and the rare earths, Th, Nb, U, Ag, B, Sn, Pb, As, Sb and Bi; less frequent requests are received for the Pt group and Au. Many of these additional elements are determined routinely by optical emission spectrography, and at the Geological Survey of Canada it is the practice to obtain such an analysis before proceeding with the chemical analysis of those samples which are not strictly routine in nature; it alerts the analyst to possible sources of difficulty in the analysis, quantitative correction of some major constituents can be made for the presence of such minor trace constituents as Sr, Ba, Cr and Ni, and the photographic plate or film serves as a useful reference for possible future questions about the trace element composition of the sample.

There is no problem about the choice of constituents in a mineral analysis. The use that is to be made of the data, such as the calculation of the formula from the analytical results, or perhaps a demonstration of the degree of solid solution of one mineral in another, together with the wide variations in composition that exist among minerals, will dictate the elements to be determined; this also applies to such specialized geological materials as meteorites. With regard to rocks, however, it has been the long-standing practice to determine between 13 and 18 constituents for a "complete" rock analysis, of which SiO_2,

TABLE 1.1

Elements Most Often Encountered in the Chemical Analysis of Rocks

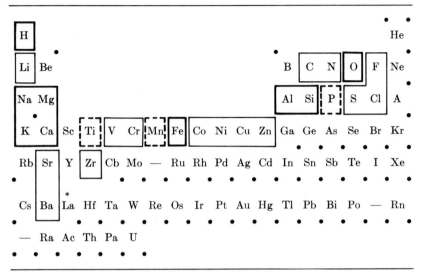

* Also elements 57–71.
Heavy blocks inclose major constituents.
Heavy broken blocks inclose common minor constitutents.
Light blocks inclose minor constituents that may occur in weighable or easily discoverable quantities.

Al_2O_3, Fe_2O_3, FeO, CaO, MgO, Na_2O, K_2O, H_2O^+, H_2O^-, TiO_2, MnO, P_2O_5 and CO_2 are the most commonly determined, and total S, Cl, F and C less frequently so. The chief reason for this is that most of these constituents are required for normative calculations, when such are made, and for convenience in comparing one analysis with another.[7,8,25] A less important reason is the fact that the percentages of the above constituents will total close to 100% for most rocks, thus giving confidence to both the submitter and analyst that the values obtained are correct. The fallaciousness of this reasoning is discussed at some length in Chapter 2 and it can no longer be considered as a valid excuse for requesting a "complete" analysis. Because of the availability of reliable methods for the individual determination of the constituents of the "main portion," i.e., Si, Al, Fe, Ti, Ca and Mg, the choice of constituents to be determined should now be governed by the particular petrographical, mineralogical or geochemical goal, rather than by long-established but outdated custom. The following types of analyses are, for example, routinely made in the Analytical Chemistry Laboratories of the Geological Survey of Canada.

1. Mineral Analysis. The number of constituents to be determined will depend upon the complexity of the mineral, the purpose of the analysis and the quantity of sample available—each sample must be considered individually.

2. Conventional ("Classical") Rock Analysis. As in (*1*) but the majority of routine requests are for the 14 constituents listed previously, to be used in petrographic calculations or to serve as monitors of the accuracy of data obtained by other procedures. In the latter connection, the determination of only a few constituents may be required.

3. Rapid Rock Analysis. The increased interest in the statistical treatment of data has resulted in a very considerable increase in the number of analyses needed, particularly for reconnaissance and major petrological studies. The individual constituents are determined as follows:

- (*a*) CO_2, Total H_2O. By conventional methods.
- (*b*) FeO. Done in lots of 30 by a modified version of Wilson's ammonium vanadate method (Sec. 9.22.3).
- (*c*) Na_2O. By flame photometry.
- (*d*) P_2O_5. Colorimetrically, as phosphomolybdenum blue, using an aliquot of the solution prepared for the determination of sodium.
- (*e*) Si, Al, Total Fe, Ti, Ca, Mg, K and Mn. By X-ray fluorescence spectroscopy, with an automatic 10-channel spectrometer. The sample is either simply mixed with a little boric acid and pressed into a disk, or is fused with a special flux mixture, the fused bead crushed and ground, and the powder pressed into a disk.

It is thus possible to supply various combinations of constituents by methods which will best serve the particular need in terms of completeness, accuracy and speed. The relative importance of each constituent in the analysis is discussed briefly in the appropriate section of Chapter 6 and should be consulted when there is doubt as to the effect of the nondetermination of a particular constituent.

1.3. The statement of the analysis

This subject can be divided into three parts—the *order* in which the constituents are to be reported, in what *form* and the *number of figures* to be given—and all have been treated in some detail in the standard texts.[8,25] Washington's suggested order is that which is still generally followed in the presentation of data and the writer knows of no significant effort that has been made to report the data in a form other than that of the conventional oxides. Little progress has been made toward a more meaningful statement of results as regards the implied precision and accuracy of the latter.

Washington (Ref. 24, p. 61) based his conclusions upon a premise that is no longer tenable, i.e., that "a rock [or mineral] analysis . . . is primarily intended for, and almost exclusively used by, petrographers," the latter having been elbowed out of the way by the geochemists with their insatiable demand for new and more abundant data. This does not change his dictum, however, that "what one needs especially is an arrangement which shall bring the essential chemical features—both the percentage figures and the molecular ratios—prominently and compactly before the eye, so that the general chemical character and the relations of the various constituents may be seen at a glance. It is also of importance that the arrangement be such as to facilitate comparison of one analysis with another." With regard to the form in which the constituents should be presented, Washington (op. cit., p. 62) urged the use of oxides, on the grounds of convenience and practicability sanctioned by [then] long and universal usage. Although written at the turn of the present century, these ideas have continued to dominate the field, and not without justification.

The reporting of the constituents as oxides with appropriate application of oxygen equivalents, although subject to questioning from the viewpoint of the physical chemistry of rocks and minerals, remains the most useful system of reporting yet suggested. Because oxygen is the major constituent of most rocks and many minerals, this method of expression enables one to judge the completeness of the analysis at a glance, and the total, although it may be misleading and must be considered in the light of such variables as the analyst and the methods used, does give a reasonable suggestion about the probable accuracy of the analysis. It is only recently that the direct determination of oxygen by neutron activation[5,22,23] has been attempted and the writer cannot comment from experience upon its efficacy, but such a value would be very useful and would remove much that is empirical in nature from the present system of reporting. The attribute of convenience in the comparison of analyses cannot be ignored, particularly when one has had occasion to work with older analyses in which the constituents are expressed as a variety of compounds, but this is not so important at the present time because of the ready availability of computers prepared to convert the statement of the analysis into any form and order for which a program can be written.

The usual *order* in which the statement of the analysis is written is in two parts—the main portion containing the principal constituents, and a subordinate part containing the remainder. The main portion is as follows:

SiO_2	The list begins with the chief acid radical and the constituent usually
Al_2O_3	present in the largest amount. The trivalent elements follow, then
Fe_2O_3	the divalent ones, with ferrous iron and magnesium together be-
FeO	cause they are closely associated in ferromagnesian minerals and the
MgO	calcium last because it is common to these latter and to the feld-
CaO	spars, particularly the plagioclase feldspars. These are logically
Na_2O	followed by Na and K in that order. H_2O^+ and H_2O^-, both given
K_2O	because the latter is not an essential constituent and will vary with
H_2O^+	the sample environment, together with CO_2 serve as a measure of
H_2O^-	the freshness of the material and complete a list of constituents
CO_2	that will, for most rocks, constitute about 95% of the sample. This
	order is usually followed, the major discrepancy being that CaO is
	placed before MgO in keeping with the analytical order in which it
	is determined.

The order of the subordinate part follows the principle expressed in the main portion:

TiO_2	The acid radicals begin the list, followed by the subordinate metallic
ZrO_2	oxides, again in the order of R_2O_3, RO and R_2O. This order is by
P_2O_5	no means universally used: MnO is frequently placed after P_2O_5
V_2O_5	(and the two are often placed between H_2O^- and CO_2), C is usually
SO_3	found with Cl, F and S, and SrO is often placed after CaO. Many
Cl	of the subordinate metallic oxides listed here, plus many others not
F	listed, are often determined by emission spectroscopy, and by con-
S, FeS_2	vention these are usually reported in the elemental form; in order to
Cr_2O_3	avoid unnecessary conversion of elements to the oxide form, it is
NiO, CoO	customary to include in the oxide statement only those spectro-
MnO	graphically determined constituents needed to correct the values for
BaO	major constituents, and to list the others separately as the ele-
SrO	ments, usually in part per million rather than in weight percent.
Li_2O	
C	

The correction for the oxygen equivalent of S, Cl, and F, when one or all of these are present, is given after the summation and from which it is subtracted to give the final total.

Mineral analyses may be expressed in the format given above but it is often more convenient and illustrative to list the constituents in the order of their importance in the mineral structure. Such specialized geological materials as meteorites, which may contain silicate, sulfide and metallic phases, require a different format, although the silicate portion is usually expressed in the form outlined previously.

It is only rarely now that the constituents of a rock or mineral analysis are expressed to more than four significant figures (in many old analyses the use of three decimal places may be noted) and there is a strong sentiment in favor of retaining only one figure beyond the decimal point. Prior to the salutary experience of G-1 and W-1 (see Chapter 2), no quantitative interlaboratory expression of the accuracy of the individual determinations was available and both the analyst and customer judged the reliability of the analysis by the nearness of the summation to 100.00%, assuming a "complete" analysis was made. Constituents were quoted to two figures to the right of the decimal, even though the conscientious analyst knew that the first of these figures, let alone the second one, was often in doubt. This knowledge, unfortunately, was not always communicated to the customer, who tended to accept the second decimal figure as a "true" value and who was thus less prepared for the shock occasioned by comparison of the data obtained in the interlaboratory investigation of G-1 and W-1.

Although he at first advocated elimination of the second place after the decimal in the statement of conventional silicate analyses, Chayes[3] later argued for its retention, chiefly on the grounds that if the second decimal is eliminated by rounding off to the first decimal place, the user of the data is thus denied the opportunity to either use it or to discard it, as he sees fit. This argument assumes that the user is motivated by more than blind faith in his consideration of the data. Chalmers and Page[2] have discussed the problem in some detail and point out (p. 247) that "the errors in even the most painstaking work may be sufficiently large to make the reporting of four significant figures unjustifiable." They prefer, however, "to state the data as observed and the estimated standard deviation of the observational error, whether or not the former are truncated in a final statement of results. In such a summary statement it will usually be found that not more than three significant figures can be justified" (p. 248). Thus the user of the data, knowing its estimated accuracy, can better evaluate the significance of differences between one analysis and another. The analyst has presented his whole story and the user is protected against being misled by an inadequate statement of results. Unfortunately, as Chalmers and Page conclude, this sound practice is seldom followed.

The foregoing discussion applies to the conventional or "classical" type of rock and mineral analysis. For those determinations made by the "rapid" methods of analysis, there is seldom any justification for reporting more than one figure to the right of the decimal point for major constituents (i.e. greater than 1%), and often little justification for doing so for those constituents present to less than 1%. However, it has been shown that many of the "rapid" methods of analysis are capable of giving results equally as good as the conventional methods, and it again emphasizes the need to state the limit of accuracy for each determination and to give as complete a statement as is consistent with the nature of the sample and the type of analysis. It is recommended also that the type of analysis be stated if it is other than a conventional one in order to avoid misinterpretation of the validity of the data when they are lifted out of the context in which they were presented.

1.4. The Summation

It has already been suggested that too much reliance is placed upon the summation as an indicator of the relative accuracy of an analysis, when the analysis is detailed enough for the summation to approach close to 100%, but it is not without its usefulness, provided that its limitations are understood.

Hillebrand[8] set the permissible range of variation in the summation of a "complete" conventional analysis as 99.75–100.50%, and preferred a total greater than 100% to one that was less than 100% because of the inevitability of positive errors due to blanks accumulated during the course of the analysis. Washington[25] was more liberal and raised the upper limit to 100.75, a figure that Groves[7] considers to be too high, preferring instead the range 99.75–100.50. Peck,[17] on the other hand, would reduce the upper limit to 100.25 while maintaining the lower limit as 99.50, a not unreasonable suggestion in light of the decreased magnitude of the blank now to be expected from the use of present-day laboratory ware and reagents; he also perceives a trend toward lower totals and suggests that recent research into the methods and techniques of rock and mineral analysis has tended to reduce plus errors more than minus ones.

A very high or very low summation clearly indicates that something is wrong; either one value at least is appreciably in error or some constituent present in significant amount has not been determined. An acceptable summation does not prove that an analysis is accurate; it may be due to the compensation of unrelated errors in individual constituents or, more importantly, to the compensation of related errors, as in the "silica-alumina discrepancy" discussed in Chapter 2. The best criterion of careful work is that of consistently good summations by an analyst of known integrity.

A wider range of acceptable summation is allowed for those analyses made by the "rapid" chemical methods, generally 99.0–101.0%. Because most of the determinations are done independently of other constituents, the summation is a more reliable guide to the overall accuracy of the analysis than is that for a conventional analysis. It is necessary to accept a wide range, however, because of the potentially greater errors inherent in the methods used for the determination of the major constituents.

References

1. Ahrens, L. H., Edge, R. A., and Brooks, R. R. Investigations on the development of a scheme of silicate analysis based principally on spectrographic and ion exchange techniques. *Anal. Chim. Acta* **28**, 551–573 (1963).
2. Chalmers, R. A., and Page, E. S. The reporting of chemical analyses of silicate rocks. *Geochim. Cosmochim. Acta* **11**, 247–251 (1957).
3. Chayes, F. In defense of the second decimal. *Am. Mineralog.* **38**, 784–793 (1953).
4. Chirnside, R. C. Silicate analysis—a review. *J. Soc. Glass Technol.* **43**, No. 210, 5T–29T (Feb. 1959).
5. Dibbs, H. P. The determination of oxygen by fast neutron activation analysis. *Mines Branch Tech. Bull.* **TB 55**, Dept. of Mines and Tech. Surveys, Ottawa, Canada (1964), 22 pp.
6. Govindaraju, K. Dosage des éléments de trace dans les roches silicatées par spectrométrie photoélectrique, avec la quantomètre A. R. L. Publication de la Groupement pour l'Avancement de Méthodes Spectrographiques, Paris, 1963; pp. 319–326.
7. Groves, A. W. *Silicate Analysis*, 2nd ed., 1951. London; George Allen and Unwin, pp. xi–xviii, 28–37.
8. Hillebrand, W. F., Lundell, G. E. F., Bright, H. A., and Hoffman, J. I. *Applied Inorganic Analysis*, 2nd ed., 1953. New York: John Wiley and Sons, pp. 793–807.
9. Ingamells, C. O. Absorptiometric methods in rapid silicate analysis. *Anal. Chem.* **38**, 1,228–1,234 (1966).
10. Jeffery, P. G., and Wilson, A. D. A combined gravimetric and photometric procedure for determining silica in silicate rocks and minerals. *The Analyst* **85**, 478–486 (1960).
11. Langmyhr, F. J., and Graff, P. R. A contribution to the analytical chemistry of silicate rocks; a scheme of analysis for eleven main constituents based on decomposition by hydrofluoric acid. *Norges Geol. Undersökelse* **230**, (1965), 128 pp.
12. Lundell, G. E. F., and Hoffman, J. I. *Outlines of Methods of Chemical Analysis*, 1938. New York; John Wiley and Sons, Inc., p. 15.
13. Maxwell, J. A., Dawson, K. R., Tomilson, M. E., Pocock, D. M. E., and Tetreault, D. Chemical analyses of Canadian rocks, minerals and ores. *Geol. Surv. Canada Bull.* **115**, (1965), 476 pp.
14. Maynes, A. D. A procedure for silicate rock analysis based on ion exchange and complex-ion formation. *Anal. Chim. Acta* **32**, 211–220 (1965).

16   ROCK AND MINERAL ANALYSIS

15. Maynes, A. D. A semi-micro procedure for mineral analysis. *Chem. Geol.* **1,** 61–75 (1966).
16. Mercy, E. L. P. The accuracy and precision of "rapid methods" of silicate analysis. *Geochim. Cosmochim. Acta* **9,** 161–173 (1965).
17. Peck, L. C. Systematic analysis of silicates. *U. S. Geol. Surv. Bull.* **1170,** (1964), 89 pp.
18. Roubault, M., de la Roche, H., and Govindaraju, K. L'analyse des roches silicatées par spectrométrie photoélectrique au Quantomètre A. R. L. et son contrôle par des roches étalons. *Sciences de la Terre* **9,** 339–371 (1962–63); see also Govindaraju, K., Supp. Ref.
19. Smales, A. A., and Wager, L. R. (Ed.) *Methods in Geochemistry*, 1960. New York: Interscience Publishers; "Analysis by gravimetric and volumetric methods, flame photometry, colorimetry and related techniques," E. A. Vincent, pp. 71–78.
20. Suhr, N. H., and Ingamells, C. O. Solution technique for analysis of silicates. *Anal. Chem.* **8,** 3730–734 (1966).
21. Voinovitch, I. A., Debras-Guedon, J., and Louvrier, J. *L'analyse des silicates*, 1962. Paris: Hermann, pp. 19–22.
22. Volborth, A. Total instrumental analysis of rocks. Part A. X-ray spectrographic determination of all major oxides in igneous rocks and precision and accuracy of a direct pelletizing method. *Rept. No.* **6,** Nevada Bureau of Mines, 71 pp. 1963. *Part B.* Oxygen determination in rocks by neutron activation. *Ibid.,* 13 pp.
23. Volborth, A., and Banta, H. E. Oxygen determination in rocks, minerals and water by neutron activation. *Anal. Chem.* **35,** 2203–2205 (1963).
24. Washington, H. S. The statement of rock analyses. *Am. J. Sci. 4th Ser.* **10,** 59–63 (1900).
25. Washington, H. S. *The Chemical Analysis of Rocks*, 2nd ed., 1910. New York: John Wiley and Sons, pp. 1–31.
26. Weeks, M. E. *Discovery of the Elements*, 6th ed., revised 1960. Easton, Pa.: *Journal of Chemical Education*; 910 pp.

Supplementary References

Asklund, A.-M., Grundulis, V., and Rönnholm, B., Våtkemisk analys av silikatbergarter—en handledning. *Sveriges Geol. Undersökn.* **61,** (1966), 55 pp. A rapid analysis scheme used at the Geological Survey of Sweden which is similar to the methods of Shapiro and Brannock (Chapter 6.2, Ref. 38), but in which Al is determined in Solution B by the alizarin red-S procedure.
Borgen, B. I. Analytical procedures used in the Geochemical Laboratory of the Survey. *Rept. 10, The Geological Survey of Greenland,* (1967), 44 pp. *Group A* (Si, Ca, Mg, Fe (total), Ti, Al) constituents are determined after fusion of sample with Na_2O_2. Al is separated with R_2O_3 group, then complexed with KF at pH 10 and the released OH^- ions titrated with HCl. Those of *Group B* (Na, K, Fe (total), Ti, Mn and P) are determined after decomposition of sample with HF and H_2SO_4.
Carmichael, I. S. E., Hampel, J., and Jack, R. N. Analytical data on the U.S.G.S. standard rocks. *Chem. Geol.* **3,** 59–64 (1968). The results of 36 determinations of major, minor and trace constituents by chemical and X-ray fluorescence methods are given for each of G-1, W-1 and the 6 U.S.G.S. rocks..

Davoine, P. L'analyse chimique des roches silicatées. *Documents des laboratoires de Géol. de la Faculté des Sciences de Lyon* **21**, 1–60 (1967).
Gordon, G. E., Randle, K., Goles, G. E., Corliss, J. B., Beeson, M. H., and Oxley, S. S. Instrumental activation analysis of standard rocks with high-resolution γ-ray detectors. *Geochim. Cosmochim. Acta* **32**, 369–396 (1968). 29 elements can be determined on a 0.8 g sample after irradiating it for periods ranging from 5 min to 1 hr.
Govindaraju, K. Dosage des éléments majeurs des roches silicatées par spectromètre photoélectrique avec le quantomètre A. R. L. *Bull. Soc. Fr. Céram.* **67**, 25–43 (1965). A more detailed account of the spectrographic procedure described in Reference 18.
Henderson, C. M. B. Analysis of standard silicate rocks. *Earth and Planetary Sci. Letters* **3**, 1–7 (1967). This scheme combines both rapid and classical procedures for the determination of 13 constituents using 7 sample portions. Improved agreement between laboratories will only occur through the introduction of standardized chemical methods.
May, I., and Cuttitta, F. New techniques in geochemical analysis. *U.S. Geol. Surv. Astrogeologic Studies Ann. Progr. Rept.*, July 1, 1965–July 1, 1966; *Part C, Cosmic Chemistry and Petrology*, 67–112, October, 1966.

CHAPTER 2

PRECISION AND ACCURACY

In 1938, Esper S. Larsen[20] described the results of comparative analyses of six amphiboles made by "from two to four of the better chemists upon whom mineralogists and geologists depend." Even allowing for the fact that for some of the replicate analyses it was unfortunately necessary to prepare new samples of professedly similar material, the comparison was disconcerting; for every amphibole there was lack of agreement on one or more constituent by as much as $1\frac{1}{2}\%$ in the amount stated and the replicate values for some constituents varied by as much as 4%. The failure of four of the five analysts taking part to determine fluorine ($\approx 2\%$) doubtless contributed much to the discrepancies noted, but Larsen quite rightly pointed out that if this was a fair indication of the best that could be achieved by two analysts working with the same sample, then it was time that results were reported only to the nearest 0.1% (except for minor constituents) so that the geologist or mineralogist unfamiliar with these probable errors in the chemical analysis would not draw unwarranted conclusions from small differences between analyses. Groves[10] expanded Larsen's point to deplore the lack of communication between the customer and the analyst, and the failure of the latter to check the validity of analytical data by mineralogical calculations.

About a decade after Larsen's disturbing discovery, the United States Geological Survey, the Department of Geology at Massachusetts Institute of Technology and the Geophysical Laboratory of the Carnegie Institution of Washington joined forces to initiate the first reconnaissance survey of the [then] state of rock analysis insofar as accuracy and precision were concerned, with the aim of determining what variability in rock analyses might be expected under the conditions [then] existing in the rock analysis laboratories throughout the world. Carefully mixed and divided samples, each weighing about 70 g, of a granite (G-1) and a diabase (W-1) ,were sent to 25 of these laboratories where they were analyzed by a total of 34 trained rock analysts using the "classical" methods, with some modifications. The data were compiled and treated statistically; the first presentation of the results was made in 1951 by Schlecht,[26] who commented, with what must now be regarded

18

as massive understatement, that the "first results show that rock analyses are generally not as precise as has been assumed." This international collaborative project must occupy an honored place in the long history of rock and mineral analysis, and much credit is due to those who conceived, organized and implemented it. It lifted the problem of reliability of data out of the realm of speculation and gave it quantitative form.[4,5,7-9,15,30] Vincent has presented a brief but cogent evaluation of the project.[29]

2.1. Precision

By *precision* of a measurement is meant the reproducibility or extent of agreement among themselves of the individual values obtained by replicate determination of a particular constituent in a given sample; the smaller the difference between these individual values and the arithmetical mean of all the values, the greater is the precision of the measurement. However, precise values are not necessarily accurate values, and high precision is no guarantee of accuracy; the arithmetical mean mentioned above may not coincide with the true value because the individual measurements are subject to both determinate and indeterminate errors. The individual measurement is really an algebraic statement, the complexity of which depends upon the magnitude and sign, which need not be constant, of the errors associated with the various parts of the measuring process, such as the sampling, the preparation of the sample, the method of measurement, operative technique and comprehension, and instrumental variability. The subject of errors in chemical analysis is too involved to attempt even a brief discussion here; for a detailed treatment of this, and also of accuracy and precision, more specialized sources should be consulted.[18,19,27,32] In rock and mineral analysis there is much opportunity, because of the sequential nature of the work, for determinate and indeterminate errors to occur, and the individual sources of potential error have been discussed in the appropriate sections of Parts II and III.

Statements of the reproducibility, or agreement, to be expected of duplicate analyses made by the conventional methods have been given by Washington[33] and Groves;[10] the allowable variations in duplicate determinations of the same constituents were also considered by Hillebrand,[12,13] who, while citing the limits set by Washington, believed that no rigid rules could be laid down. Part I of Table 2.1 lists the allowable variations cited by Groves as a reasonable expectation for first class work, an expectation presumably based upon experience in replicate analyses in the author's own laboratory but implying that these stated limits of variation should apply to work done in other laboratories as well. That these variations are reasonable for duplicate determinations made at the same time by the same analyst is shown by the duplicate analyses given in Part II, columns 1 and 2 of Table 2.1 (see also Part III, columns A, B and C of Table 2.3). But all this latter says is that the analyst was competent and careful, and that the unavoidable errors in his work had the same magnitude and sign in both analyses. Unfortunately this difference between precision and accuracy has not always been appreciated by the "customers" and the consumer of analytical data has tended to equate the one with the other. Schlecht[26] has discussed the probable development of this tradition of reproducibility and has emphasized that, contrary to popular belief, small differences between determinations carried out in parallel may give a very misleading idea of the accuracy of the determinations.* What, then, *should* be expected of analyses made of the same material by different analysts in different laboratories?

The determination of this actual variability was the aim of the cooperative project involving G-1 and W-1, by comparing data obtained on an international scale. It must be realized that each analysis submitted was probably the average of precise replicate determinations by the submitting laboratory. The variability in the individual statements, in terms of the standard deviation,† found when all of the analyses of G-1 submitted by the 24 laboratories were considered, is given in Part III of Table 2.1 and is seen to be a far cry from the limits proposed, or implied, by Groves; the variability in the data for W-1, not

---

* Schlecht[26] suggests the term "duplicity" for an unreliable estimate of error based upon the agreement of duplicate determinations made at the same time by the same analyst.

† Analysts may find useful the following simple statistical terms for expressing quantitatively the precision of their results. They are also encountered in considerations of the precision of determinations of constituents in G–1, W–1 and other international standards.

n = number of observations
$\bar{x}$ = arithmetical mean
d = deviation of an observation from mean
s = standard deviation (uncertainty of a single observation) = $[\Sigma d^2/(n-1)]^{1/2}$
s^2 = variance
$s_{\bar{x}}$ = standard error (the error of the arithmetic mean) = $s/n^{1/2}$
C = relative deviation (also called coefficient of variation) = $(s/\bar{x}) \times 100$
E = relative error = $C/n^{1/2}$

Note: C and E permit the comparison of data on a percentage basis.

TABLE 2.1

Precision (in wt. %) of Replicate Determinations by Conventional Methods

	I	II Basalt		III G–1			IV G–1			V Haplogranite			VI Tonalite (T–1)		
		1	2	$\bar{x}$	s	C	$\bar{x}$	s	C	$\bar{x}$	s	C	$\bar{x}$	s	C
SiO_2	±.20	48.12	48.05	72.22	0.43	0.60	72.64	0.23	0.31	72.23	0.34	0.5	62.69	0.26	0.42
Al_2O_3	.10	19.49	19.57	14.44	.541	3.75	14.13	.17	1.22	16.19	.26	1.6	16.55	.29	1.72
Fe_2O_3	.03	2.21	(2.21)	.94	.34	36.2	.86	.17	19.3				2.86	.29	10.1
FeO	.02	5.40	(5.40)	1.00	.135	13.5	1.06	.06	5.72	1.94	.17	8.7	2.85	.20	7.12
CaO	.02	11.33	11.33	1.42	.152	10.7	1.34	.09	6.53	.85	.05	5.4	5.19	.25	4.76
MgO	.03	10.29	10.28	.39	.135	34.6	.44	.04	9.90				1.85	.10	5.62
Na_2O	.03	2.23	2.11	3.26	.284	8.71	3.43	.21	6.23	3.24	.21	6.5	4.35	.19	4.32
K_2O	.02	.03	.03	5.51	.549	9.96	5.43	.21	3.83	5.69	.12	2.2	1.24	.09	7.04
H_2O^+	.02	.11	.12	.37	.104	28.1	.31	.08	26.7				1.57	.24	15.2
TiO_2	.01	.70	.70	.26	.067	25.7	.25	.03	11.2				.59	.04	6.90
P_2O_5	.01	.01	.01	.10	.045	45.0	.10	.02	23.4				.13	.03	19.9
MnO	.01	.25	.23	.03	.0122	40.7	.03	.01	30.5				.11	.02	18.8
Total		100.17	100.04	99.94			100.02						99.93		

I. Groves[10] (p. 228); allowable variations in duplicate determinations, expressed as percent of the whole rock.

II. Duplicate analyses of basalt (1091–64) by J. L. Bouvier, Analytical Chemistry Section, Geological Survey of Canada.

III. Fairbairn and others[5] (p. 37, Table 14); arithmetical mean ($\bar{x}$) and standard deviation (s), with added calculated relative deviation (C), of determinations of constituents in G–1 by 24 laboratories.

IV. Fairbairn and others[5] (p. 38, Table 17); as in III but by one laboratory (7 analysts).

V. Fairbairn and Schairer[6] (p. 752, Table 6); as in III but for the determination of constituents in synthetic granite glass by 11 selected laboratories.

VI. As in III but for the determination of constituents in Standard Geochemical Sample (T–1), Msusule tonalite, by 14 laboratories, as given in Supplement No. 1, Geological Survey Division, United Republic of Tanzania, Dodoma, Tanganyika, 1963.

quoted here, is of similar magnitude. Considerable improvement in precision is seen when replicate determinations (7) of G-1 made by different analysts from *one* laboratory are considered (Part IV, Table 2.1), and some improvement in precision resulted when 11 of the laboratories taking part in the G-1 project (those whose determinations were nearest to the mean values) cooperated further in the analysis of a synthetic granite glass (Haplogranite),[6] as shown in Part V of Table 2.1. The data also emphasize the fact that the same precision of measurement does *not* hold for any constituent regardless of its relative concentration,[1] as is evident from the values of C in Table 2.1.

Parts II–VI in Table 2.1 answer the question asked previously, i.e., what should be expected in the way of precision for analyses made of the same material by different analysts in different laboratories. It must be conceded, however, that the s values in Part III do represent a somewhat extreme case; as a result of the self-examination stimulated in most rock analysis laboratories by the G-1 and W-1 projects, the interlaboratory precision has generally improved, as witness the s values in Part VI of Table 2.1 for a somewhat similar cooperative venture which took place about ten years later. With regard to the variability to be expected between replicate determinations made routinely by a competent, trained rock analyst in a given laboratory, certainly the s values of Part IV, with the exception of those for Na_2O and K_2O, make up a more reasonable statement than that proposed by Groves (Part I). For alkalis in the amounts shown, a standard deviation of ± 0.10 would be more reasonable.

Let us now consider briefly the precision to be expected of some methods, other than the conventional ones, in general use. The *rapid methods* of analysis were not used to provide results for the first test of G-1 and W-1 but data were included in later summations.[7,30] The procedures used in these rapid methods are described in detail in Parts II and III, and suffice it to say here that they are largely photometric, are free of the involved separations necessary in some stages of the conventional scheme and are generally independent of each other although the same sample solution is used for the determination of several constituents. They are less subject to accumulated errors but, because of the smaller sample portion taken, sampling errors may be greater. They are also very subject to operational error and, contrary to the popular impression, consistently good results are obtained only by trained personnel who adhere rigidly to the procedural details. Shapiro and Brannock[28] and Riley[25] give replicate analyses for G-1 and W-1 which indicate a precision within their respective laboratories that is close to

the limits proposed by Groves (Part I, Table 2.1). Mercy has discussed
the precision of his methods (a modified Shapiro and Brannock scheme)
in much detail;[22] this is shown in Table 2.2 (Part I) and the values are,
with the exception of SiO_2 and Al_2O_3, generally better than those given
for the conventional results of 7 analysts from one laboratory (Part IV,
Table 2.1). He also gave additional precision data for different ranges
of concentration for each constituent which showed clearly how the
relative deviation (coefficient of variation) increased with decreasing
concentration of the constituent.[23]

Spectrographic data for G-1 and W-1 have been presented and evalu-
ated by Filby and Leininger (Ref. 30), but they emphasize their lack of
homogeneous populations of results for each constituent; they also have
discussed the factors which affect the precision (and accuracy) of the
method. Their values for G-1 are given in Part III of Table 2.2. Contrary to chemical precision, spectrochemical precision is independent of
the concentration of the constituent.

TABLE 2.2

Precision (in wt. %) of Replicate Determinations by "Rapid Chemical"
and Emission Spectrographic Methods

	I M149-G1			II G-1			III G-1		
	$\bar{x}$	s	C	$\bar{x}$	s	C	$\bar{x}$	n	s	C
SiO_2	73.4	0.33	0.46	72.73	0.14	0.20	72.2	6	1.92	2.7
Al_2O_3	14.2	.49	3.5	13.81	0.42	3.07	14.5	7	.52	3.6
Fe_2O_3	.55	.04	6.8	0.66	0.09	14.12	2.00[a]	6	.059	2.95
FeO	.94	.03	2.7	0.96	0.05	4.81				
CaO	.94	.009	.95	1.27	0.01	1.17	1.38	8	.18	13.0
MgO	.80	.01	1.5	0.39	0.02	5.06	.38	8	.025	6.6
Na_2O	4.1	.08	1.9	3.27	0.04	1.18	3.38	7	.29	8.6
K_2O	4.2	.12	2.8	5.05	0.06	1.24	5.37	6	.63	11.7
H_2O^+	1.0	.12	11.8	—	—	—	—	—	—	—
TiO_2	.21	.01	6.0	0.237	0.004	1.65	.24	10	.031	12.9
P_2O_5	.09	.008	8.6	0.071	0.008	11.66	—	—	—	—
MnO	.04	.004	11.2	0.034	0.003	9.93	.026	14	.0086	33.1

I. Mercy[22] (p. 168, Table 3); arithmetical mean ($\bar{x}$), standard deviation (s) and
relative deviation (C) of replicate (6) determinations of the constituents of a
granite (M 149–G1) by "rapid methods."

II. Langmyhr and Graff (Chap. 1, Ref. 11, Table 18, p. 109); as in I, but for 10
replicate determinations of G-1 by HF–decomposition scheme.

III. Filby and Leininger (Ref. 30, pp. 66–68, Tables 12 and 13); as in I but for the
determination of the constituents of G-1 by optical emission spectroscopy in
different laboratories (n). [a] Total Fe as Fe_2O_3.

There has been a notable increase in the use of both X-ray fluorescence and atomic absorption spectroscopy for the determination of major and minor constituents in rocks and minerals. There are, however, not sufficient data available on the precision of determinations made by atomic absorption spectroscopy for presentation here, and the precision (and accuracy) of those made by X-ray fluorescence spectroscopy will be discussed in a later section.

Emphasis has been placed upon the data accumulated for the two U.S. reference samples G-1 and W-1 in the discussion and in the tables. Similar, although less extensive, interlaboratory studies have been made on other samples, including the Msusule tonalite (Part VI of Table 2.1), the Canadian syenite and sulfide ore,[34] the French granites GR, GA, GH and basalt BR (Part VI of Table 2.4), a series of 10 basic silicate rocks of complex composition analyzed by three laboratories in the U.S.S.R. (Bykova *et al.*, Supp. Ref.), and four East German samples— granite (GM), basalt (BM), clay (TB) and carbonate (KH)—analyzed in more than 20 laboratories (Grassman, Supp. Ref.). Herzog[11] gives data for the precision of rubidium and strontium determinations in five lepidolites by four laboratories using a variety of methods.

It is also of interest to note another international comparison of precision, this time of different analytical methods for the determination of trace constituents in nuclear materials.[3] An aqueous solution containing Cu, Cr, Mn and Hg in the 0–20 ppm range was used for the comparison and, because the interference effect from the matrix was thus minimized, the variation in results could be attributed to the methods and techniques used. Determinations were made by emission spectrography, polarography, neutron activation and spectrophotometry; at the 95% confidence level, the precision ranged from about 40% for emission spectrography to about 10% for spectrophotometry, with an overall precision of about 25% for a single determination.

2.2. Accuracy

We may define the *accuracy* of a determination simply as the concordance between it and the true or most probable value; the numerical difference between the two values is the *error*.

Only sketchy statements of the probable precision of the separate determinations in a rock analysis were available prior to 1951; no similar statement of the probable accuracy of these determinations is known to the writer. Estimates of the accuracy of an analysis (when considered!) were evidently based largely upon the degree of success with which mineralogical formulas could be derived from the analytical data. The

initial stage of the G-1 and W-1 project was an exercise in the determination of the precision of rock analysis because no true values for the two
rock samples were known. This was, of course, recognized at the start
of the project and, partly because of the shock occasioned by the
disturbingly large variations in the results for some constituents, it was
decided to test both precision and accuracy by the use of a material of
known composition, in this case a six-component synthetic silicate glass
("haplogranite") approximating to the composition of G-1 (Part I of
Table 2.3).[6] Because the supply of material was limited, it was
distributed to only 12 selected laboratories as mentioned in Sec. 2.1, 11
of which had participated in the G-1 and W-1 project. Part II of
Table 2.3 gives the arithmetical mean and range of the results supplied
by these laboratories; for comparison, three analyses made in one
laboratory (and not included in the previous mean) are listed in Part III,
together with their arithmetical mean. The latter are remarkably
close to the true values (the precision is also excellent) but the extent of

TABLE 2.3

Accuracy of Determinations of Constituents in Synthetic Granite Glass
(in wt. %)

	I	II		III			
		$\bar{x}$	Range	A	B	C	$\bar{x}$
SiO_2	72.64	72.23	71.38– 72.75	72.69	72.75	72.51	72.65
Al_2O_3	15.78	16.19	15.85– 16.78	15.64	15.75	15.97	15.79
Fe_2O_3							
FeO							
CaO	1.82	1.94	1.67– 2.35	1.79	1.81	1.85	1.82
MgO	.80	.85	.75– .90	.84	.88	.87	.86
Na_2O	3.19	4.24	3.07– 3.82	3.23	3.05	3.16	3.15
K_2O	5.76	5.69	5.40– 5.84	5.61	5.64	5.66	5.64
H_2O^+							
TiO_2							
P_2O_5							
MnO							
$SiO_2 + Al_2O_3$	88.42	88.42	87.52– 88.87	88.83	88.50	88.48	88.44
Total	99.99	100.14	99.26–100.65	99.80	99.88	100.02	99.91

  I. Fairbairn and Schairer[6] (p. 752, Table 6); standard composition of the syn-
      thetic granite glass.
 II. *Idem.*; arithmetical mean ($\bar{x}$) and range for the determination of constituents
      by 11 selected laboratories.
III. *Idem.*; for triplicate determinations in one laboratory (data not included in
      calculations given in II.

TABLE 2.4

Preferred Values (wt. %) for Major and
Minor Constituents of Some Reference Samples

	I	II	III	IV	V	VI
				CAAS	CAAS	
	G-1	W-1	T-1	syenite	sulfide	GR
SiO_2	72.52	52.58	62.65	59.45[a]	34.52[a]	65.85
Al_2O_3	14.08	14.94	16.52	9.58[a]	9.46[a]	14.54
Fe_2O_3	0.85	1.38	2.81	2.27	32.70[b]	1.68
FeO	0.94	8.71	2.90	5.44		2.12
CaO	1.36	10.92	5.19	10.32[a]	3.96	2.47
MgO	0.35	6.52	1.89	4.07	3.93	2.34
Na_2O	3.29	2.15	4.39	3.24[a]	1.03	3.73
K_2O	5.52	0.63	1.23	2.75[a]	0.63	4.50
H_2O^+	0.25	0.45	1.53	0.42	2.91	0.87
H_2O^-	0.02	0.08	—	0.18	0.08	0.06
TiO_2	0.26	1.08	0.59	0.49	0.81	0.62
P_2O_5	0.09	0.14	0.14	0.20	0.10	0.34
MnO	0.026	0.17	0.11	0.40	0.11	0.056
CO_2	0.08	0.07	—	0.36	—	0.37[a]
F	0.07	0.03	—	—	—	0.10
S	—	—	—	0.04	12.04	—
Other	0.222[a]	0.166[a]	—	—	—	0.31[b]
Total	99.93	100.02	99.95	(99.21[b])	—	99.95
Less O≡F	0.03	0.01	—			0.04
Total	99.90	100.01	99.95	(99.21[b])	—	99.91

I.   Ingamells and Suhr[14] (Table 9, p. 908).   [a] Trace elements, including ZrO_2 0.028, Rb_2O 0.0236, BaO 0.12, SrO 0.031, SnO_2 0.0006, V_2O_5 0.0025, Cr_2O_3 0.0016, CuO 0.0014, BeO 0.0008, CoO 0.0003, PbO 0.0053, ZnO 0.0068.

II.   *Idem.*   [a] Trace elements, including ZrO_2 0.022, SnO_2 0.0003, V_2O_5 0.045, Cr_2O_3 0.016, NiO 0.0097, CuO 0.0138, BeO 0.0003, CoO 0.0056, PbO 0.0006, ZnO 0.014, Rb_2O 0.0022, Cs_2O 0.00006, BaO 0.015, SrO 0.021.

III.   Msusule tonalite (see Part VI, Table 2.1); values given here are for arithmetical mean of preferred values. See also Ingamells and Suhr[14] for their preferred values, which are in general agreement with above except for SiO_2 63.08, Al_2O_3 16.31 and H_2O^+ 1.31; they also give trace constituents, including ZnO 0.020, BaO 0.055, SrO 0.033, CO_2 0.08 and F 0.04.

IV.   Webber[34] (Table 3, pp. 233–235); arithmetical mean of results received from 32 laboratories.   [a] These means appear low for SiO_2 and Na_2O, and high for Al_2O_3, CaO and K_2O; the values preferred by Ingamells and Suhr[14] (Table 9, p. 908) and the Geological Survey of Canada (Ref. 34, Table 1, #9, p. 230) are respectively: SiO_2 59.78 and 59.76, Al_2O_3 9.01 and 8.86, CaO 10.09 and 10.05, Na_2O 3.38 and 3.49, K_2O 2.60 and 2.59.   [b] The syenite also contains relatively large amounts of Zr, Th, Rare earths and U (see Table 2.5, Column IV) not given here but some of which are probably responsible for the high the range of values obtained from the 11 laboratories is somewhat disconcerting (see Part V of Table 2.1, for the precision data), although the means for four of the constituents differ from the true values by only about ±0.1%. An important consideration arising out of the investigation was the recognition of a bias in the determinations of SiO_2 and Al_2O_3; the mean value for the combined $SiO_2 + Al_2O_3$ is in excellent agreement with the true value but that for SiO_2 is 0.41% *low* and that for Al_2O_3 is 0.41% *high*. The probable sources of error giving rise to this bias have been discussed in detail[4,9,15,30] and it has been shown that it can be minimized by careful attention to the techniques of dehydration and washing of precipitates when the conventional method is used, or by the use of other methods, such as the combined gravimetric–colorimetric method for silica described by Jeffery and Wilson.[16]

The haplogranite project did not provide a clear-cut answer to the question of the accuracy of a single analysis but it did spark a small controversy over the so-called silica–alumina "discrepancy" that resulted in a careful examination of the relevant procedures. It also emphasized the importance of G-1 and W-1 as yardsticks against which an analyst can measure his analytical performance. Although the true values for these reference materials will probably never be known,*

* It should be mentioned here that there is not complete agreement that the G-1 and W-1 powders were sufficiently homogeneous to eliminate heterogeneity as a contributing source of some of the variations found. One school of thought believes that only a glass can be sufficiently homogenous to permit a valid comparison such as was attempted with G-1 and W-1.

mean values for Al_2O_3 mentioned previously; the following values have been given for these constituents and others:

	Ingamells and Suhr[14]	GSC
ZrO_2	0.47%	0.59%
U_3O_8	0.28	0.28
ThO_2	} 0.43	0.14
Rare earths		0.16
F	0.17	—
S	—	0.022

V. Webber[34] (Table 4, pp. 236–237); arithmetical mean of results obtained from 32 laboratories. ᵃ For these constituents Ingamells and Suhr[14] (Table 9, p. 908) give SiO_2 34.84, Al_2O_3 9.73. ᵇ Total Fe calculated to Fe_2O_3.

VI. Preferred values given by Ingamells and Suhr[14] (Table 9, p. 908). ᵃ Total C as CO_2; ᵇ includes BaO 0.11, CuO 0.055, SrO 0.067, V_2O_5 0.010, Cr_2O_3 0.015 Rb_2O 0.0196, ZnO 0.012 and others; a report on interlaboratory comparative values for GR (and for 3 other standards, GA, GH and BR) has recently been published by Roubault et al. (Supp. Ref.).

TABLE 2.5

Preferred Values (in ppm) for Trace Constituents of Some Reference Samples

	I	II	III	IV CAAS syenite	V CAAS sulfide	VI
	G-1	W-1	T-1			GR
Ag	0.04[a]	0.06[a]		1.6	3.9	
As	0.8[a]	2.2[a]		10.4	424	
Au	0.005[a]	0.005[a]			0.11	
B	1.5–[m]	17[m]		107	23	
Ba	1220[m]	180[m]	680	273	227	1000
Be	3[m]	0.8[m]	1.1*	25	1	5.6
Bi	0.2[m]	—		0.25*	5	0.15
Cd	0.06[m]	0.3[m]				
Ce	200[m]	25[m]		593		
Co	2.4[v]	50[v]	13	19	546	10
Cr	22[v]	120[v]	24	58	368	100
Cs	1.5[m]	1.1[m]		0.9		
Cu	13[v]	110[v]	47	25	8291	440
Dy	2[m]	4[m]				
Er	2[m]	3[m]				
Eu	1.0[m]	1.1[m]				
Ga	18[v]	16[v]	21	18	13	
Gd	5[m]	4[m]				
Ge	1.0[m]	1.6[m]				
In	0.03[m]	0.08[m]				
La	120[m]	20[m]	15*	243		76
Li	24[v]	12[v]		126		
Mo	7[v]	0.5[v]		4.1	8.3	
Nb	20[m]	10[m]	6	146		
Nd	55[m]	17[m]		325		
Ni	1–2[v]	78[v]	13	42	13103	68
Os	0.0001[a]	0.0004[a]				
Pb	49[v]	8[v]	37	516	248	26
Pd	—	—	—	—	0.18	
Pt	—	—	—	—	0.17	
Rb	220[v]	22[v]	32*	204	38*	179
S	—	130[m]				
Sb	0.4[v]	1.1[v]				
Sc	3[v]	34[v]	11*	14	22	7
Sm	11[m]	5[m]		12		
Sn	4[v]	3[v]	43	5.6*	11	11
Sr	250[v]	180[v]	410	320	110	570
Ta	1.6[a]	0.7[a]				
Tb	0.6[m]	0.8[m]				
Th	52[v]	2.4[v]		1324		

(continued)

TABLE 2.5 (continued)

	I	II	III	IV	V	VI
				CAAS	CAAS	
	G-1	W-1	T-1	syenite	sulfide	GR
Tl	1.3[a]	0.13[a]				
U	3.7[v]	0.52[v]		2472		
V	16[v]	240[v]	96	87	192	55
W	0.4[a]	0.45[a]				
Y	13[v]	25[v]	25	447	21	15
Yb	1[v]	3[v]	2.3*	73	8.6	1.5
Zn	45[v]	82[v]	190	200	298	100
Zr	210[v]	100[v]	170	3048	108	62

I. Fleischer[7] (Table 5, p. 1268–1277), Fleischer and Stevens[8] (pp. 531–539); [v]recommended *value*, [a]recommended *average*, [m]*magnitude* only. The references given should be consulted for comments on the validity of the data. *See also* note on work of Brunfelt and Steinnes following.

II. *Idem.*

III. Msusule tonalite (see Part VI, Table 2.1); average of results from several laboratories (see Bowden and Luena[2] on validity of trace element data). * Preferred values from Ingamells and Suhr[14] (Table 10, p. 909) who differ considerably with some values given here, including Ba 490, Sr 280, Cr 8 and Zr 120.

IV. Arithmetical means (Webber[34], Table 5, pp. 238–243); *preferred values given by Ingamells and Suhr[14] (Table 10, p. 909).

V. Arithmetical means (Webber[34], Table 6, pp. 243–247); *preferred values given by Ingamells and Suhr[14] (Table 10, p. 909).

VI. Preferred values given by Ingamells and Suhr[14] (Table 10, p. 909); see also Roubault *et al.* (Supp. Ref.).

Note: Brunfelt and Steinnes (Supp. Ref.) have given values for some trace constituents in G-1 and W-1, as determined by an instrumental neutron activation method:

	G-1	W-1
		ppm
Co	2.2	41.6
La	97	8.8
Sc	2.83	33.3
Sm	7.7	3.2

they have been approached asymptotically by a consideration and statistical evaluation of the accumulated data.[7,8] It is of interest to note two recent demonstrations of the use to which G-1, W-1 and other standard materials may be put; when 110 post-1961 analyses of Australasian tektites, monitored with the above standards, are compared with 55 older analyses of similar tektites, "the deviations of the older analyses

TABLE 2.6

Availability of Some Existing and Potential Reference Samples[a]

Nature	Designation	Available from[b]
Granite	G–1	No longer available (U.S. Geol. Survey).
Diabase	W–1	Mr. Francis J. Flanagan, Liaison Officer, Analytical Laboratories Branch, U.S. Geological Survey, Washington, D.C. 20242.
Syenite	Syenite rock–1	No longer available (C.A.A.S.).
Sulfide ore	Sulfide ore–1	Dr. G. R. Webber, Department of Geological Sciences, McGill University, Montreal, Quebec, Canada. Five dollars per bottle ($\approx$ 100 g), payment on order, payable to: Nonmetallic Standards Committee, Canadian Association for Applied Spectroscopy.
Tonalite	Msusule tonalite, T-1	Mineral Resources Division, P. O. Box 903, Dodoma, Tanzania, East Africa. Issued in approximately 100 g lots.
Granite	GR	Centre de Recherches Pétrographiques et Géochimiques, Nancy-Vandoeuvre, B.P. No. 682 Nancy, France. GR no longer available. GA, GH and BR in limited quantity; not available for general distribution.
Granite	GA	
Granite	GH	
Basalt	BR	
Granite	Shetland granophyric granite No. 1771	Limited quantity; not available for general distribution (Geol. Survey of Great Britain.
Granite	G–2	Mr. Francis J. Flanagan, Liaison Officer, Analytical Laboratories Branch, U.S. Geological Survey, Washington, D. C. 20242.
Granodiorite	GSP–1	
Andesite	AGV–1	
Peridotite	PCC–1	
Dunite	DTS–1	
Basalt	BCR–1	

(continued)

TABLE 2.6 (*continued*)

Nature	Designation	Available from[b]
Dolomite	GFS 400	G. Frederick Smith Chemical
Limestone	GFS 401–402	Co., P. O. Box 23344, Colum-
Limestone–dolomite blends	GFS 403–419	bus, Ohio; $12/100 g; $25/250 g.
Magnetite	GFS 450–452	As above but $4/100 g and
Hematite	GFS 453–455	$7.50/250 g.
Magnetite–hematite— dolomite blends	GFS 456–490	
Cement	NBS 1011, 1013–1016	National Bureau of Standards, Washington, D. C. 20242; $6/unit, each unit consisting of 3 sealed vials each containing approximately 5 g of sample.
Ores (iron, bauxite, lithium, manganese, phosphate, tin and zinc)	See Descriptive List, NBS Miscellaneous Publication 241 (1962).	As above; $3–6/bottle (50–110 g).
Chrome, alumina and silica refractories	As above	As above; $6/bottle (45–60 g).
Glasses and glass sand	See Descriptive List, NBS Miscellaneous Publication 241 (1962).	National Bureau of Standards, Washington, D. C. 20242; $6/bottle (45–60 g).
Limestone	NBS 1a	As above; $6/per bottle (50 g).
Portland cement	NBS 177	As above; $6/bottle (15 g).
Silica brick	NBS 102	As above; $6/bottle (60 g).
Burned magnesite	NBS 104	As above.
Ores (iron, manganese, chrome)	See Descriptive List No. 408, British Chemical Standards and Photographic Standards, January, 1964.	Bureau of Analysed Samples, Ltd., Newham Hall, Newby, Middlebrough, England. Approximately $2.50/25 g; available also in 50-, 100- and 500-g lots.
Sillimanite, silica brick and firebrick	As above	As above.

[a] See also Flanagan and Gwyn (Supp. Ref.).
[b] Unless otherwise indicated, samples are issued free of charge.

TABLE 2.7

Accuracy of Determinations of Constituents (in wt. %) of G-1 by
"Rapid Chemical" and Optical Emission Spectroscopic Methods

	I	II	III	IV	V	VI			VII
SiO_2	72.5	72.4	72.5	72.5	72.40	71	73	74	72.73
Al_2O_3	14.1	14.5	14.0	14.3	14.15	14.0	15.0	14.5	13.81
Fe_2O_3	0.85	0.86	0.76	0.97	0.95	[a]1.93	2.00	2.03	0.66
FeO	0.94	0.96	0.94	0.91	0.97				0.96
CaO	1.36	1.4	1.5	1.38	1.38	1.50	1.35	1.32	1.27
MgO	0.35	0.31	0.46	0.39	0.39	0.39	0.40	0.38	0.39
Na_2O	3.29	3.3	3.8	3.41	3.32				3.27
K_2O	5.52	5.4	5.4	5.33	5.36				5.05
H_2O^+	0.25	—	0.4	0.33	0.35				—
TiO_2	0.26	0.24	0.26	0.22	0.25				0.237
P_2O_5	0.09	0.08	0.08	0.08	0.10				0.071
MnO	0.03	0.02	0.02	0.040	0.02				0.034
Other	0.39[b]	0.42[c]	—	—	—				—
Total	99.9	99.9	100.1	99.86	99.64				98.48

I. Preferred values for G-1, based upon Ingamells and Suhr[14] (Table 9, p. 908); [b] see Column I, Table 2.4, for individual values of constituents included in "Other."

II. Shapiro and Brannock[28] (Table 5, p. 56); average of 2 analyses by rapid methods.   [c] Total H_2O 0.31 and CO_2 0.11 included in "Other."

III. Mercy[22] (Table 1, p. 164); *single* analysis by modified rapid chemical methods.

IV. Riley[25] (Table II, p. 427); *single* analysis by modified rapid chemical methods.

V. Maynes[21] (Table I, p. 216, IV); *single* analysis by rapid chemical methods involving ion-exchange and complex-ion formation; Na + K, H_2O and Fe^{2+} determined on separate portions, remaining constituents (8) determined on a 1-g portion.

VI. Joensuu and Suhr[17] (Table IV, p. 103); emission spectrographic analysis following $LiBO_2$ or $Na_2B_4O_7$ fusion of sample.   [a] Total Fe calculated as Fe_2O_3.

VII. Langmyhr and Graff (Chap. 1, Ref. 11, Table 17, p. 107); average of 10 analyses by HF-decomposition method.   See Table 2.2 for precision data.

from the restricted values of the modern analyses are comparable to the imprecisions shown by early analyses of G-1 granite and W-1 diabase" (Ref. 31, p. 123). Similarly, Philpotts and Pinson[24] used replicate analyses of G-1 and W-1 to evaluate the accuracy of rapid chemical and X-ray fluorescence analyses of moldavites. The lesson of G-1 and W-1 would appear to have been learned and the inclusion among other analytical data of the results obtained for a recognized reference sample is an easy and usually reliable way to express the probable accuracy of the data. Tables 2.4 and 2.5 list the "preferred" major, minor and trace element values for a number of well-known reference samples; information about the availability of these and other potential reference samples is given in Table 2.6.

The accuracy attainable by a selection of "rapid chemical" schemes and by optical emission spectroscopy is illustrated in Table 2.7. Similar data for the X-ray fluorescence spectroscopy method are given in a later section.

References

1. Ahrens, L. H. A note on the relationship between the precision of classical methods of analysis and the concentration of each constituent. *Mineralog. Mag.* **30**, 467–470 (1954).
2. Bowden, P., and Luena, G. The use of T-1 as a geochemical standard. *Geochim. Cosmochim. Acta* **30**, 361 (1966).
3. Cook, G. B., Crespi, M. B. A., and Minczewski, J. International comparison of analytical methods for nuclear materials. I. Accuracy and precision of some techniques in routine trace analysis. *Talanta* **10**, 917–929 (1963).
4. Fairbairn, H. W. Precision and accuracy of chemical analysis of silicate rocks. *Geochim. Cosmochim. Acta* **4**, 143–156 (1953).
5. Fairbairn, H. W., and others. A cooperative investigation of precision and accuracy in chemical, spectrochemical and modal analysis of silicate rocks. *U.S. Geol. Survey Bull.* **980**, 1951, 71 pp.
6. Fairbairn, H. W., and Schairer, J. F. A test of the accuracy of chemical analysis of silicate rocks. *Am. Mineral.* **37**, 744–757 (1952).
7. Fleischer, M. Summary of new data on rock samples G-1 and W-1, 1962–1965. *Geochim. Cosmochim. Acta* **29**, 1263–1283 (1956).
8. Fleischer, M., and Stevens, R. E. Summary of new data on rock samples G-1 and W-1. *Geochim. Cosmochim. Acta* **26**, 525–543 (1962).
9. Goldich, S. S., and Oslund, E. H. Composition of Westerley Granite G-1 and Centerville diabase W-1. *Bull. Geol. Soc. Am.* **67**, 811–815 (1956).
10. Groves, A. W. *Silicate Analysis*, 2nd ed., 1951, London; George Allen and Unwin pp. 28–37, 224–236.
11. Herzog, L. F. Analyses of identical samples by more than one laboratory. *Ann. N. Y. Acad. Sci.* **91**, 207–220 (1961).
12. Hillebrand, W. F. The analysis of silicate and carbonate rocks. *U.S. Geol. Survey Bull.* **700**, 1919, 285 pp.
13. Hillebrand, W. F., Lundell, G. E. F., Bright, H. A., and Hoffman, J. I. *Applied Inorganic Analysis*, 2nd ed., 1953. New York: John Wiley, pp. 793–807.
14. Ingamells, C. O., and Suhr, N. H. Chemical and spectrochemical analysis of standard silicate samples. *Geochim. Cosmochim. Acta* **27**, 879–910 (1963).
15. Jeffery, P. G. The silica and alumina content of the standard rocks G-1 and W-1. *Geochim. Cosmochim. Acta* **19**, 127–133 (1960).
16. Jeffery, P. G., and Wilson, A. D. A combined gravimetric and photometric procedure for determining silica in silicate rocks and minerals. *The Analyst* **85**, 478–486 (1960).

17. Joensuu, O. I., and Suhr, N. H.   Spectrochemical analysis of rocks, minerals and related materials. *Appl. Spectr.* **16**, 101–104 (1962).
18. Kolthoff, I. M., and Elving, P. J., Eds.   *Treatise on Analytical Chemistry*, Part I, Sec. A, Vol. 1, 1959.   New York: Interscience; "Errors in Chemical Analysis," E. B. Sandell, pp. 19–46; "Accuracy and Precision: Evaluation and Interpretation of Analytical Data, W. J. Youden, pp. 47–66.
19. Laffitte, P.   Étude de la précision des analyses de roches. *Bull. Soc. Géol. France 6th Ser.* **3**, 723–745 (1953).
20. Larsen, E. S.   The accuracy of chemical analyses of amphiboles and other silicates. *Am. J. Sci. 5th Ser.* **35**, No. 206, 94–103 (1938).
21. Maynes, A. D.   A procedure for silicate rock analysis based on ion exchange and complex-ion formation. *Anal. Chim. Acta* **32**, 211–220 (1965).
22. Mercy, E. L. P.   The accuracy and precision of "rapid methods" of silicate analysis. *Geochim. Cosmochim. Acta* **9**, 161–173 (1956).
23. Mercy, E. L. P.   The geochemistry of the older granodiorite and of the Rosses granitic ring complex, Co. Donegal, Ireland. *Trans. Roy. Soc. Edinburgh* **64** (No. 5), 101–138 (1958–1959).
24. Philpotts, J. A., and Pinson, W. H., Jr.   New data on the chemical composition and origin of moldavites. *Geochim. Cosmochim. Acta* **30**, 253–266 (1966).
25. Riley, J. P.   The rapid analysis of silicate rocks and minerals. *Anal. Chim. Acta* **19**, 413–428 (1958).
26. Schlecht, W. G.   Cooperative investigation of precision and accuracy. *Anal. Chem.* **23**, 1568–1571 (1951).
27. Schlecht, W. G.   The probable error of a chemical analysis. *U.S. Geol. Survey Bull.* **992**, Part 6, 57–69 (1953).
28. Shapiro, L., and Brannock, W. W.   Rapid analysis of silicate rocks. *U.S. Geol. Survey Bull.* **1036C**, 1956, 56 pp.
29. Smales, A. A., and Wager, L. R., Eds., *Methods in Geochemistry*, 1960.   New York: Interscience; "Analysis by Gravimetric and Volumetric Methods, Flame Photometry, Colorimetry and Related Techniques," by E. A. Vincent, pp. 33–80.
30. Stevens, R. E., and others.   Second report on a cooperative investigation of the composition of two silicate rocks. *U.S. Geol. Survey Bull.* **1113**, 1960, 126 pp.
31. Tatlock, D. B.   Some alkali and titania analyses of tektites before and after G-1 precision monitoring. *Geochim. Cosmochim. Acta* **30**, 123–128 (1966).
32. Youden, W. J.   *Statistical Methods for Chemists*, 1951.   New York: John Wiley, 126 pp.
33. Washington, H. S.   *The Chemical Analysis of Rocks*, 2nd ed. 1910.   New York: John Wiley, pp. 22–27.
34. Webber, G. R.   Second report of analytical data for CAAS syenite and sulphide standards. *Geochim. Cosmochim. Acta* **29**, 229–248 (1965).

Supplementary References

Brunfelt, A. O., and Steinnes, E.   Instrumental neutron-activation analysis of "standard rocks." *Geochim. Cosmochim. Acta* **30**, 921–928 (1966).
Babko, A. K.   The accuracy and reproducibility of chemical analysis. *Zavodskaya Lab.* **21**, 269–277 (1955).   This was an early collaborative study of 5 silicate samples in 15 laboratories to determine principal sources of error.

Bykova, V. S., Knipovitch, Yu. N., and Stolyarova, I. A. Analysis of basic silicate rocks of complex composition, in *Methods of Analysis of Mineral Raw Materials*, Trudy VSEGEI–117 Leningrad, 1964, pp. 9–16.

Flanagan, F. J. U.S. Geological Survey silicate rock standards. *Geochim. Cosmochim. Acta* **31**, 289–308 (1967). Analytical data obtained in U.S.G.S. laboratories for the six new silicate samples are given.

Flanagan, F. J., and Gwyn, M. E. Sources of geochemical standards. *Geochim. Cosmochim. Acta* **31**, 1211–1213 (1967). 12 sources are given.

Friese, G., and Grassman, H. Die standardgesteinsproben des ZGI. 4. Mitteilung: Diskussion der Gehalt an einigen Hauptkomponenten auf Grund neuer Analysen. *Z. Angew. Geol.* **13**, 473–477 (1967). New determinations have been made of SiO_2, Al_2O_3, TiO_2, Fe (total) and MnO, with particular emphasis upon the determination of Al_2O_3.

Goldich, S. S., Ingamells, C. O., Suhr, N. H. and Anderson, D. H. Analyses of silicate rock and mineral standards. *Can. J. Earth Sci.* **4**, 747–755 (1967). Data are given for the new U.S.G.S. silicate samples and three silicate minerals (albite, adularia and pyroxene).

Grassmann, H. Die standardgesteinsproben des ZGI. *Z. Angew. Geol.* **12**, 368–378 (1966).

Ingamells, C. O., and Suhr, N. H. Chemical and spectrochemical analysis of standard carbonate rocks. *Geochim. Cosmochim. Acta* **31**, 1347–1350 (1967). Data are given for 4 limestone and dolomite samples from the G. Frederick Smith Chemical Co.

Nicholls, G. D., Graham, A. L., Williams, E., and Wood, M. Precision and accuracy in trace element analysis of geological materials using solid source spark mass spectrography. *Anal. Chem.* **39**, 584–590 (1967). A precision of better than ±5% is obtained, but a triple fusion of the sample with an internal standard (Re) is necessary before forming electrodes.

Roubault, M., de la Roche, H., and Govindaraju, K. Rapport sur quatre roches étalons géochimiques: Granite GR, GA, GH et Basalte BR. *Sciences de la Terre* **11**, 105–121 (1966).

CHAPTER 3

SAMPLING AND SAMPLE PREPARATION

3.1. General

The observation is often made that an analysis is no better than the sample that it represents. This relationship is, unfortunately, too often ignored, and much time and labour are expended upon samples not worthy of the effort. Conversely, conclusions are often drawn from these analyses that are not warranted by the nature of the samples that they represent, thus adding an often incorrect interpretation to unnecessary work. It is incumbent upon the analyst to insure that the sample taken is suitable for the analytical work requested; the nature of the sample is dictated in part by the expected degree of accuracy of the desired analytical information, and he must be prepared to interpret this to the geologist, and to give guidance where it is needed.

It is only recently that some attention has been paid to the importance in rock analysis of proper sampling and sample preparation other than qualitative discussions of inherent difficulties.[19,30] This is in contrast to the proper sampling and preparation of ore samples, about which much detailed discussion is available,[10,26,57] and which reflects the economic importance attached to the taking of a truly representative sample of an ore; such samples will decide whether or not mining is started in a particular area, what working areas are payable, and how long work will be continued. Such economic considerations do not usually apply to samples taken for rock analysis and, as a result, much less emphasis has been placed upon this initial step. Until recently, the number of rock analyses made was dictated largely by the time and expense involved, and the samples submitted for analysis were frugally selected. The advent of the more rapid methods of analysis has reduced the need for careful selection on the grounds of frugality. Instead of a single outcrop, large areas are subjected to detailed chemical study and the scope of interpretation of the analyses has broadened in consequence. With increased opportunity for the accumulation of data has come, however, an increased need to select proper samples and to apply the analytical data in a sound manner.

In recent years there has been a notable emphasis on the application of statistics to geology, in particular to geochemical studies. Sampling, analyzing and comparing results are considered to be three basic procedures that are governed fundamentally by statistical theory,[55,56] and the increased accumulation of data that now appears to characterize present-day studies is a response to this trend and its need for abundant data to satisfy statistical requirements. A plea for caution in this latter respect has, however, been made.[13]

It is neither possible nor desirable here to discuss sampling and sample preparation in an exhaustive fashion. The intent is rather to draw attention to the many problems involved in the process of securing a proper sample for analysis and to the means whereby these problems may be eliminated or minimized. As general reference books on statistical measurements, those of Youden,[68] and Dixon and Massey,[11] will be found useful. Discussions of sampling and sample preparation are included in most publications dealing with methods of rock and mineral analysis, such as those by Groves,[20] Milner[38] (sedimentary rocks), Oertel[44] (soils) and Hawkes and Webb[21] (geochemical prospecting), and they are treated in a general fashion by Wager and Brown in *Methods of Geochemistry*.[58] Wilson[66] has examined mathematically the errors introduced when samples are taken from heterogeneous powders for both major and trace element determination (see also Kleeman, Supp. Ref.).

The emphasis has so far been upon the taking of rock samples. The sampling of minerals presents a different problem, except for the few occasions when large deposits of such minerals as barite, gypsum and clays are to be examined. A mineral sample is usually obtained by concentration from a host rock and the quantity of material available for analysis is governed by both the percentage of the mineral in the rock and the ease with which it can be concentrated and purified (see Sec. 3.5).

3.2. Sampling

The process of sampling is often affected by factors which have nothing to do with the securing of a statistically sound sample. As mentioned previously, the availability of analytical services or facilities will influence the number of samples that are taken for analysis. If only a few analyses can be made, the geologist must select those samples which illustrate certain specific details, and local, rather than general, considerations will be paramount. If, on the other hand, the opportunities for analysis are favorable, the conditions for sampling may not be so, and the samples taken may have to be governed by such considerations as the availability of outcrops, the difficulty in securing them, and the distance and effort required to transport them. If we consider, however, a situation in which the influence of these factors is not significant, the problem becomes one of where to take the samples and how many and how much of each.

3.2.1. HOMOGENEITY AND GRAIN SIZE

The degree of homogeneity of a rock mass is an obvious consideration, but Laffitte[29] and Grillot[18] have also drawn attention to the importance of grain size as a factor in sample selection, a consideration discussed much earlier by Larsen.[31] Their conclusions as to the size of sample necessary to give significance to the analysis must be startling indeed to those who are content with a single hand specimen broken from an outcrop that is conveniently available and easily sampled.

Laffitte[29] considers the sampling of a gneiss containing crystals of microcline, from which a series of 100 samples of 1 kg each is taken from the top to the bottom of the mass. The samples are crushed to pieces approximately 1 cm^3 in size, mixed thoroughly and a representative sample of 1 kg is carefully prepared, the K_2O content of which is subsequently determined to be 6%. Assuming that all mineral grains have equal weight, it is shown that the possible error (twice the standard deviation, at the 95% confidence limit) is 0.58% and thus 5 of the 100 samples could have a K_2O content falling outside the limits 5.42–6.58%. This inhomogeneity is not revealed in the composite sample but requires a series of samples for its statistical evaluation. A similar situation has been described by Shaw and Bankier[65] for the occurrence of rubidium in 28 diabase samples from Ontario[15]; the calculated average Rb content (m) was found to be 144 ppm, with a standard deviation (s) of 86 ppm, and thus at the 95% confidence limit the range of Rb content ($m \pm 2s$) would be 0–316 ppm. A composite sample of the 28 diabases would probably have given an average Rb content close to 144 ppm, but would have given no indication of the range that existed. Burks and Harpum[7] found that bulk samples of granitic rocks may give analyses which are not representative of the area sampled and which may not even fall in the field of composition of any natural rock in the area. It is suggested[56] that for geochemical work the only safe rule is "the more samples the better!"

The importance of grain size and its influence upon the size of sample that should be taken has been considered by Laffitte[29] for the determination of both major and minor constituents. Consider a granodiorite in which all of the mineral grains are approximately the same size and which has an average CaO content of 2.4%; the CaO content of the

quartz, mica and potash feldspar grains can be taken as zero while that
of the plagioclase feldspar, which makes up 40% of the rock, is 6%.
If it is assumed that the weight of a grain having a volume of 1 cm^3 is
2.5 g, then the size of sample necessary to give desired limits of error in
the determination of CaO for a given grain size is shown in Table 3.1.

TABLE 3.1

The Influence of Grain Size on the Size of Sample Required
(After Laffitte,[29] p. 729)

Grain size (cm^3)	No. of grains	s (%)	$m \pm 2s$ (%)	Sample wt. (grams)
0.1	100	0.3	1.8 –3.0	25
0.1	2,500	0.06	2.28–2.52	625
0.1	10,000	0.03	2.34–2.46	2,500
0.01	10,000	0.03	2.34–2.46	250

If the size of the largest grain present is taken as the governing size,
then phenocrysts and porphyroblasts will greatly increase the size of
sample needed; the percentage of the constituent sought must also be
considered.

3.2.2. METHODS OF SAMPLING

There are four common types of sampling procedures. *Random*
sampling is self-explanatory, but there are many difficulties in the way
of obtaining a truly random sample.[8] *Stratified* sampling involves the
selection of samples from specific layers, as in the sampling of inter-
bedded shale and limestone, or banded gneiss. The taking of samples
at spaced intervals about a sample point is *cluster* sampling and is done
to give information about variability over a small distance. *Systematic*
sampling is probably the type most used, knowingly or unknowingly,
to give information about a specific phase or feature of a mass. All
of these types are best served by an organized sampling program which
will yield the desired information, and foresight is preferable to hindsight.

The proper sampling of a rock mass requires a geological knowledge
of the mass; this is usually confined to a plane surface but for some
layered intrusions the third dimension is exposed as well.[59] If the mass
is homogeneous, then as much randomness as possible should be
injected into the sampling procedure; if heterogeneous, the mass should
be dealt with as a series of rock types, and consideration must be given

to the effect of one type upon another (e.g. hybridization of surrounding material by xenoliths).   In the case of a banded gneiss, possibly the most difficult of any rock to sample meaningfully, stratified sampling should be used if the bands are sufficiently coarse to permit it; the alternative is to take a large sample and homogenize it by crushing, an obvious compromise.   Washington[64] preferred to take a carefully selected hand specimen from a homogeneous mass, but favored the sampling of individual facies in a heterogeneous one, and the combining of the resultant *analyses* in relative proportions to give an average composition.   Grout[19] obtained surprisingly good results when analyses of representative samples of banded gneiss were compared with those of larger chip samples.

There are several ways in which a sample may be obtained.   Single specimens, weighing about 6 lbs, are obtained by breaking them from an outcrop with a geological hammer (grab sampling); lichen and weathered material should be removed by washing and scrubbing, if necessary, and the sample thoroughly dried.   This should be done *before* the sample is submitted for analysis, for only the collector knows the extent to which trimming should be done on the specimen.   *Chip sampling* involves the taking of several pounds of chips, each weighing about 50–100 g, from various sites on the mass being sampled; *channel sampling*[22] is similar, except that this requires the cutting of a channel, 3–6 in. in width and 1–2 in. in depth, across the cleaned surface of the mass, using a hammer and chisel, or pneumatic drill.   All of the material cut out is included in the sample and generally the longer the channel, the better the sample.   *Core drilling* provides an excellent opportunity to obtain a series of samples, and is becoming increasingly popular; an elegant way to obtain an average sample from a diamond drill core is to grind the core lightly with a carborundum grinding wheel along one side and collect the powder.   A comparison of grab sampling versus channel sampling in limestones has been made by Galle.[17]

Mention has been made of the need to consider grain size in the selection of samples (see Sec. 3.2.1).   It must be emphasized, however, that the end purpose of the sampling must govern the sampling procedure. Most of the foregoing has been concerned with the sampling of a rock mass for the determination of the major constituents, a relatively simple job compared to that required for a study of the distribution of those constituents in low concentration which may occur in scattered accessory minerals, such as sulfides, zircon and apatite, or camouflaged in major mineral lattices.   The sampling plan needed for a study such as this latter one will be more complex and more extensive than that for the estimation of major constituents. Lamar and Thompson[30] have described the sampling of limestone and dolomite deposits for the spectrographic determination of trace and minor elements.

The sampling program used by Pitcher and Sinha[47] to obtain samples for a petrochemical study of the Ardara Aureole is instructive, although Flinn,[16] who subsequently analyzed their data statistically, has pointed out certain failings in their procedure. Pelitic schists make up about 1200 ft of the aureole surrounding the Ardara pluton, part of the Donegal Granite Complex. This horizon was sampled inside and outside by means of 8 cross-traverses, each traverse being allotted 5 rectangular areas measuring 50 × 150 yards each. Five collecting points were selected within each rectangular area, and samples of about 300 g each were taken at each point. They were washed, split and one half of each was broken into small pieces. The pieces from a single rectangular area were mixed to give an aggregate sample of about 750 g. In order to show that 5 collecting points were sufficient, chemical analyses were made on 5 samples taken from one rectangular area, and on a composite of the samples prepared as described above. Some variation occurred in the content of Si, Mg, Ca, Na and K between individual samples, but the average values of the five analyses were in good agreement with the values obtained for the composite sample.

3.3. Preliminary treatment of the sample

A preliminary examination of a rock or mineral sample received for analysis is a useful practice and may result in a saving of analytical time. When possible, the analytical work to be done should be discussed with the submitter; useful information can often be obtained about the nature of the sample, and the object of the analysis and the analytical work may be reduced or modified as a result. The shotgun approach of a "complete" analysis is often wasteful of analytical time, and the determination of 4 or 5 constituents may provide as much information as will the determination of 12 to 15 of them.

3.3.1. CLEANING AND TRIMMING

It is preferable that the removal of unwanted material from a specimen be done by the collector before it is submitted, so that no error will be introduced by the removal of pertinent material. If this is not done, then trimming instructions should be obtained before the sample is treated further, and lichen, encrusting dirt, and identification marks (including adhesive labels) should be noted for removal also. Wet samples should be dried at 100–105°C and allowed to come to equilibrium with their surroundings before being sampled. The disposition of the bulk of the residue after sampling should be determined, and arrangements made to save a piece for thin section preparation or other purposes *before* the whole sample is reduced to less than pea size!

3.3.2. PRELIMINARY EXAMINATION

The presence of sulfide minerals or accessory minerals containing such elements as fluorine and boron should be noted, because these will affect the choice of analytical methods to be used. When no other information is available, a modal analysis may suggest useful modifications in the analytical approach.

Mineral samples are usually submitted as material having a grain size between 60- and 80-mesh and are often obtained by concentration with heavy liquids. If the mineral has a noticeable odor of volatile organic compounds, the sample should be dried in an oven at 100–105°C in order to expel this organic material, then allowed to come to equilibrium in air before further treatment. The degree of purity of the sample is really the responsibility of the submitter, but a precautionary examination will do no harm and may reveal the presence of unwanted impurities which make further purification of the sample necessary; the analytical data may later be used to calculate the formula of the mineral and the analysis of an impure sample can be very misleading.

Such detailed examination of samples is not always possible nor practical, particularly where large numbers of samples are being submitted. The responsibility must then rest with the submitter but every effort should be made to acquaint him with the problems involved. Those engaged in sample preparation should be informed of the procedure to be followed and care taken to see that the usefulness of the sample is not impaired by careless preparatory work.

3.3.3. RECORDING OF SAMPLE DATA

It is a good practice to keep a record of the weight of the sample submitted for analysis, particularly of minerals, for such information is often useful at a later date. This again may not be practical when a large number of samples are submitted. All pertinent information about the sample should be recorded, such as the specimen number, where collected and by whom, the geographical and geological occurrence, a macroscopic and petrographic description (if the latter is available) and the purpose for which the analysis is required. Such information is more difficult to obtain after the analysis has been completed. At the Geological Survey of Canada a special form is filled out in code by the submitter to provide data of this nature when the samples are submitted; this information, together with the analytical data later obtained, are stored in the Survey Data File.

3.4. Preparation of rock samples for analysis

Sample preparation, which includes the crushing, splitting, grinding, sieving, mixing and, if necessary, the drying of the sample, is an important first step and much has been written about the proper preparation procedures for all types of materials. It is not possible nor desirable to discuss these in detail here and, for a fuller treatment of this subject, reference should be made to more general texts.[27,28,61,65,67] Because the efforts to provide a proper sample are easily undone by improper sample preparation, the latter must not be left to chance. There are many ways to prepare samples and it is necessary to choose the best procedure for the job at hand.[25]

Contamination of the sample during sample preparation is an ever-present and unavoidable danger. One cannot prepare a rock or mineral sample for analysis without changing its composition in some respect, either through the addition of foreign material, the preferential loss of material, or the oxidation of constituents. The goal is one of minimizing these undesirable effects, either by preventing their ready occurrence or by choosing a procedure that will favor a type of contamination that is the least harmful. Thiers[62] has discussed the problem of the contamination of biological samples in trace element analysis and has suggested means of controlling it; his comments are applicable to rock and mineral samples as well. Volborth[63] grinds a duplicate split of the sample in both ceramic and hardened steel pulverizers and analyzes both splits in order to determine the contamination from each pulverizer and to obtain the true composition.

3.4.1. ERRORS IN SAMPLE PREPARATION

Rock samples are usually submitted in the form of chips (about $1 \times 1 \times \frac{1}{2}$ in.), or as hand specimens (about $3 \times 3 \times 2$ in.), ranging in weight from about 50 g to 20 lb or more. It is necessary to obtain from this an aliquot portion of 20–50 g that will represent the larger sample, as the latter is supposed to represent the larger mass.

There may be special circumstances which permit the selection of one or two chips as being representative, but it is generally necessary to treat the whole sample. Grout[19] obtained good agreement between the SiO_2 content of both single specimens and composite samples of banded gneisses except for one sample in which, by error, the whole of the composite sample was not crushed and quartered; the SiO_2 content of the hand specimen was 71% but only 66% SiO_2 was obtained for the poorly sampled composite.

Laffitte[29] has emphasized the importance of grain size (see Sec. 3.2.1) in determining the size of aliquot that should be taken and in governing the magnitude of the relative error to be expected in the determination of constituents. With care, the error introduced by the sample preparation can be made negligible; for a 100-g sample having an average grain size of 4 mm, the relative error introduced into the determination of a major constituent could amount to 8%, but if the grain size is reduced to 1 mm, the relative error is reduced to about 1%. The premature quartering of samples should thus be avoided if the precision of the sampling is to be the same as that of the chemical analysis, particularly for those elements present in the less abundant minerals.

Grillot[18] studied the effectiveness of various types of crushing and grinding apparatus in reducing samples of both soft and hard rock to −200 mesh.* He found that there is an unavoidable loss of fine dust during this reduction, particularly during the mechanical grinding; a loss of 5 g was found for 200 g of both hard and soft rock after about 15 min of grinding. A 100 g sample of hard rock powder having an average diameter of about 0.3 mm, after being twice passed through the grinder, yielded about 10 g of flour having a diameter of less than 0.08 mm (200-mesh).

The process of coning and quartering is probably the most popular method of reducing the sample to a small representative portion. It is tedious, however, and unless done carefully can lead to gross errors from segregation, as the heavier fragments tend to roll to the base of the pile and do not always distribute themselves equally about the base.

3.4.2. TYPES OF EQUIPMENT

There are several types of equipment that may be used for the various stages of the sample preparation[32,38,58,65] and the choice is

---

* The convention used throughout to indicate particle size is as follows: particles smaller than a given lineal inch mesh size are designated as "minus," those larger as "plus"; e.g., material which is described as being −150 to +200 mesh in size will pass through a screen having 150 meshes to the lineal inch but not through one having 200 meshes.

dependent upon the use that is to be made of the sample. It is impossible to avoid the introduction of some foreign material during the sample preparation and the best alternative is to choose the type of contamination that will be the least objectionable. Samples must first be reduced in a jaw crusher to a particle size suitable for further reduction in a grinder of one type or another and, because of the hardness of the sample material, abrasion of the crusher or grinder will occur. The ideal arrangement would be to crush the sample to the desired particle size between blocks of similar material, but this is not practical. It is also possible to heat the sample in an electric furnace to about 600°C and then to quench it in distilled water; the sample will disintegrate readily to a grain size suitable for grinding in an agate mortar.[1,41] This method is not only undesirable because of the possible effect of the heating and quenching upon the constituents of the sample but it may also be dangerous; explosive decomposition of the material may occur when it is withdrawn from the furnace, even with slow heating.[44]

The reduction of the sample to manageable size can be done with a rock trimmer or hydraulic rock crusher, or by a simple bucking board and muller. The latter may be made of steel or ceramic material, and Myers and Barnett[41] have shown that contamination is introduced by the bucking board, as well as by automatic crushers and grinders. Bloom and Barnett[5] ground both quartzite and massive quartz to less than 100-mesh on bucking boards made of both high alumina ceramic and steel, and also in an agate mortar. The steel bucking board added objectionable amounts of Fe, Mn, Cr, Cu, Ni and V to the samples; only 0.003% Mg was added to the quartzite and quartz by the ceramic bucking board and muller, and 0.0001% Ti to the quartz alone. About 15 min were required to reduce 30 g of quarter-inch material to −100 mesh.

Further crushing of the material may be done in large, hardened-steel mortars of the Plattner (or "diamond") type in which the diameter of the pestle is only slightly less than that of a collar which in turn fits snugly into a depression in the base of the mortar. Sandell[50] studied the contamination of quartz and microcline by crushing them in a Plattner mortar and found that the amount of iron introduced from the mortar was less if the collar was not used; the crushing was also more efficiently done, but a cardboard screen must be used to prevent loss of material. Because Cu, Cr, and Ni were also introduced, the steel used for the mortar should be low in these constituents. Storm and Holland[60] found no significant contamination by Ni when quartz was crushed in a steel mortar to a size between 120 and 200 mesh. The Ellis mortar[24] is larger and heavier, and the crushing may be done entirely by the pestle without the use of a hammer; the pestle is also much smaller than the collar, and abrasion is minimized.

Jaw crushers are widely used to reduce the large fragments to pea size or smaller. Grillot[18] studied the efficiency of a jaw crusher, a cylinder grinder and an automatic agate mortar for the grinding of both hard and soft rocks, and his results are given in Table 3.2.

TABLE 3.2

Efficiency of Crushing and Grinding
(After Grillot,[18] pp. 9–10)

Equipment used	Nature of rock	First pass	Second pass	Third pass
Jaw crusher	soft	50% <1 mm		
	hard		62% <1 mm	
Cylinder grinder	soft	42% >0.3 mm	5.5% >0.3 mm	<1% >0.3 mm
	hard	30% >0.3 mm	2.5% >0.3 mm	1% >0.3 mm
Automatic agate mortar	soft (20 g, 0.3–0.08 mm)	6 g >0.14 mm 5 g 0.14–0.08 mm 9 g <0.08 mm	5 g >0.14 mm 3 g 0.14–0.08 mm 12 g <0.08 mm	
	hard (20 g, 0.3–0.08 mm)	2 g >0.14 mm 3 g 0.14–0.08 mm 15 g <0.08 mm	2 g >0.14 mm 3 g 0.14–0.08 mm 15 g <0.08 mm	

Jaw crushers are a major source of contamination because burring of the steel plates will occur with the consequent introduction of slivers of steel (tramp iron) into the sample. Face plates should be of hardened steel and should be changed when they become roughened; economy in operation at this point can be wasteful of effort at a later stage. An hydraulic rock crusher, in which the jaws used for breaking specimens are replaced by two steel blocks, will produce material of about 1 mm in diameter.[58]

Roller crushers are also employed at this stage. The equipment used at the Department of Geology, University of Manchester, has been described by Smales and Wager,[58] and was designed to reduce small pieces of rock to a powder with a minimum of contamination. It consists of two sets of rollers, one set of which reduces the coarse material to about 80 mesh while the other continues the reduction to −200 mesh.

There are several devices that may be employed for the reduction of the sample to a particle size between 80 and 100 mesh. Cylindrical grinders or pulverizers use opposing plates, one of which is stationary while the other revolves at a variable distance from it. The sample is fed from a funnel into the gap between the plates, and the desired reduction is achieved by narrowing the gap. Table 3.2 shows the results that may be obtained in this way. Steel plates will introduce considerable metal contamination, and ceramic plates are much to be preferred.[3] A cylinder mill that will handle as little as 0.5 g of sample, and is easily cleaned, has been described.[51] The cone grinder, hammermill[49] and rotary beater are other forms of pulverizers that can be used; it was found at the Geological Survey of Canada that excessive loss of fine material occurred with the rotary beater, and the metal screens used with this device are a major source of contamination. All of these machines require relatively large samples.

Further reduction of the sample to 150 mesh or finer can be achieved by various methods. The Schwingmühle (swing-mill) of Siebtecknik, Mühlheim, Germany, is used by Danielsson and Sundkvist[9] for the preparation of samples for the tape machine technique of spectrographic analysis, where uniformity of particle size is particularly important. This equipment may be obtained with a single large mill, or with as many as six smaller mills. Each mill consists of a container, a ring and/or solid grinders and a cover; these should be made of hardened steel or, preferably, the container should be lined with tungsten carbide and the ring and grinders also made of this alloy. The Bleuler Mill (rotary-, or swing-mill, Willy Bleuler Company, Switzerland) and the Shatterbox (Spex Industries, Michigan, U.S.A.) are similar machines and are both used at the Geological Survey of Canada. A ball mill, which is a mechanically rotated porcelain container holding about one-third of its volume of flint pebbles or porcelain balls, will produce a very fine powder and also thoroughly mix the sample. It is, however, somewhat difficult to clean and requires a large sample because of the amount that is retained on the surface of the balls and on the walls of the container. Related to the ball mill is the "paint-shaker,"[32] which uses ceramic balls to do the grinding; the sample is placed in a ceramic cylinder with one or more balls, six of these containers are clamped in a device used to mix the contents of paint tins and is vigorously shaken. Ballard, Oshry and Shrenk[2] have investigated the grinding and mixing efficiency of stainless steel ball mills for the preparation of samples for X-ray diffraction and spectrographic analysis.

Mortars are commonly used to grind the sample to the desired final particle size.   There are many types—agate, mullite, tungsten carbide, glass and porcelain (these latter two are now rarely used).   They may be used manually (some laboratories prefer this, believing it to be faster than automatic grinding[54]) or be operated mechanically.   Grinding under water is occasionally done[52]; the fines are decanted periodically into a reservoir and when all of the sample has been satisfactorily reduced, the water is removed by decantation and evaporation.   This method also reveals any "tramp iron" present as a surface scum. Allen (Ref. 24, p. 813) found that a weight loss of 0.291 g by a mortar and pestle occurred when 200 g of quartz sand were reduced to pass through a 150-mesh screen; the abrasion of the mortar and pestle is greater in automatic grinding than in manual grinding.   Hempel[24] found that steel and glass mortars were abraded less than agate but in rock analysis the contamination from steel and glass is undesirable. The use of a boron carbide mortar has been discussed by Boulton and Eardley (Supp. Ref.).

There are other types of equipment which can be used, some of which are designed for specialized problems, such as the mica pulverizer described by Neuman.[42]   A Waring Blendor is used to pulverize mica in the writer's laboratories.   A special container which permits the crushing of iron meteorites at liquid air temperature has been described.[4,12]

Sieving of the ground material is a step that must be given careful consideration.   Frequent sieving during the grinding removes the finer fractions and speeds up the grinding process, and also insures that the bulk of the powder will be of the desired size.   Care must be taken, however, to avoid contamination by the material of the sieve; if trace elements are to be determined, brass sieves should be avoided, and stainless steel sieves used instead.   The type of sieve commonly used for work in which contamination must be minimized consists of silk or nylon bolting cloth stretched over a glass or plastic holder.   The importance of the thorough mixing of material that has been sieved is shown by Lundell and Hoffman[34]; the composition of the material can change significantly with sieve size.

Mixing can be done manually by rolling the powder from one side to the other on a rubber cloth or sheet of glazed paper; the material must tumble over itself and must not just slide along the surface of the sheet. Mechanical mixers vary from simple to elaborate devices.   O'Neil[45] places a taped sample jar in sawdust in a pebble mill container and rotates the latter for 24 hours.   A spiral mixer than can be rotated on a lathe (a smaller model is turned by a stirrer motor) is used at the

Geological Survey; the sample bottles (which should be no more than half-filled) are held at a 45° angle and the rotation of the mixer imparts both a rotary and vertical movement to the contents. This device has been expanded to a large wooden drum capable of mixing 144 samples at once. Another type employs a wooden cylinder which rotates freely on two rubber rollers turned by means of a small motor; the cylinder has an inner stainless steel container fitted with baffles which cause the powder to be thoroughly tumbled as the cylinder rotates. There are also more elaborate mixers or blenders which can be used for quantities of material from several pounds to less than one pound. Whatever method is used, however, the mixing must be thoroughly done.

There remains only the final step of reducing the bulk sample to a size that will meet the needs of the analysis. A sample of 40–60 g is usually a very ample quantity, leaving sufficient material after the analysis for storage against future needs; this quantity is not always available, however, particularly for mineral samples. Large samples (one to several pounds) can be reduced to 100 grams by coning and quartering; again, there is the danger that unequal segregation of heavier material will occur during the rolling and coning of the material. A riffle, or Jones splitter, is a device having an even number of narrow sloping chutes, with alternate chutes discharging in opposite directions; these vary in size from those able to handle several pounds, to micro-splitters for small mineral samples which are vibrated electromagnetically. The sample is carefully poured in a stream across the top of the riffle and alternate portions are collected in containers placed below each set of chutes. This requires that a uniform stream of material be poured on the surface, and there is also considerable loss by dusting. A rotary splitter consists of a rotating table carrying six pie-shaped containers; the sample is poured into a conical receiver and falls slowly through a chute terminating just above the sample containers which are rotating beneath it.

3.4.3. SAMPLE PREPARATION PROCEDURES

There are many factors which govern the choice of procedure and often the final choice must be a compromise. Some contamination is inevitable and that procedure which will keep it to a minimum should be selected. The degree of magnitude of the contamination will be directly related to the final particle size, because the longer grinding period required to produce a finer particle size will naturally result in a greater amount of wear on the grinding material. The mesh size should be fine enough to insure that the sample will be as nearly homogeneous as possible; broadly speaking, the sample should pass through a 100-mesh screen for chemical analysis, whereas for other types of analysis, such as spectrographic or X-ray fluorescence, 200–300 mesh material is desirable because of the small sample aliquot that is used.

The size of the sample will often dictate the method to be used. A small sample is better prepared by hand; too much loss occurs when larger, automatic equipment is used. Very small samples can be ground directly in automatic pulverizers such as the Mixer-Mill or Wig-L-Bug used in the preparation of samples for optical emission spectrographic analysis.

The effect of grinding upon such constituents as ferrous iron and water must also be considered. Mauzelius[37] showed that excessive grinding oxidized a major part of the ferrous iron and recommended the use of the coarsest powder permissible. This was confirmed by Hillebrand,[23] who believed that the oxidation was due more to the local heat created by the friction of the pestle than to the increased surface that is exposed to air; he also demonstrated that there is a considerable increase in the water content with increasing fineness of the powder. Hillebrand et al.[24] discussed the problem in some detail, including the possibility of doing the grinding in an organic medium such as alcohol or carbon tetrachloride. Some analysts prefer to do the ferrous iron determination on a separate, coarser portion of the sample but this is objectionable because of the differences in moisture content, state of oxidation and homogeneity that may exist between the portions. If two portions are used, the moisture must be determined on both in order to correct for its effect on the ferrous iron, as well as on the other constituents determined on the finer portion. Hillebrand et al.[24] prefer to grind the whole sample in a non-oxidizing medium rather than to use a separate portion of coarser material.

Each laboratory will have its own preferred procedure which is dictated by both the type of analysis to be done and by the equipment available. The following procedure is used routinely at the Geological Survey of Canada.

Break the sample into pieces (approximately 1 in. cubes) in a rock trimmer or with a hardened steel hammer on a hardened steel plate. Crush these pieces in a jaw crusher* to about ⅛ in. or less, starting with

* If a jaw crusher is not available, crush the sample in an Ellis-type mortar, taking care not to grind the material and not to strike the collar with the pestle; the sample can be reduced directly to 1 mm or less if the Ellis mortar is used, and should be removed to a sheet of glazed paper, or to a rubber mat before new pieces are added to the mortar.

the jaws set wide apart to minimize loss of particles, and reducing the gap width on successive passes; three or four passes should be sufficient, and the steel plates of the jaws should be changed when they begin to show wear.

Cone and quarter the material obtained from the jaw crusher or steel mortar as follows: roll the material from one corner of the paper or mat to the opposite corner by raising one corner and causing the material to tumble over upon itself; repeat the process by raising each corner in succession until thorough mixing of the sample has been achieved; gather the material as a flat cone in the center of the sheet (flatten the cone, if necessary, with a spatula: if a steel spatula is used it should be demagnetized to prevent the removal of magnetic grains); divide it into four quarters and remove opposite quarters; repeat the mixing and quartering and remove opposite quarters, each time alternating the quarters that are removed. Continue until a sample of twice the desired size is reached (generally 20–40 g); quarter this and reserve one set of opposite quarters for further grinding; the other set may be retained as a reserve sample.*

At the Geological Survey the final sample (10–20 g) is transferred to a "paint-shaker" type of pulverizer and reduced to a particle size between 100 and 200 mesh by shaking in a ceramic mill for 15–20 min. If the grinding is to be done instead by hand, transfer the sample to a 100 mesh sieve (silk bolting cloth is preferable), and sieve it gently over a piece of glazed paper. Grind the oversize portion in an agate or mullite mortar; again sieve, and repeat these operations until all of the sample has passed through the screen. At no time must a particular sieve fraction be taken for the analysis; differences in the susceptibility to grinding of various minerals make it almost certain that compositional differences will exist between sieve fractions.[1]

Transfer the final powder to a numbered plastic vial, seal with a tight friction cap, and store it until needed. Before use, mix the sample for 20–30 min in an automatic mixer, pour the contents of the mixer vial on a piece of glazed paper, quarter the pile and, with a spatula (demagnetized steel, bone or plastic), return the quarters successively to the original sample vial; do *not* pour the sample into the vial, as this causes a segregation of particles according to density. This mixing should be repeated before use whenever the sample has been stored for any length of time, in order to eliminate the segregation that may occur during storage.

* It is a useful practice to keep a small supply (10–20 g) of coarse material as a reserve against future contingencies, particularly if the sample is special or unusual. The residue of the sample on which the analysis was made should *always* be retained.

3.5. Preparation of mineral samples for analysis

3.5.1. ERRORS IN SAMPLE PREPARATION

Unlike rock samples, the preparation of mineral samples does not usually involve the analyst in other than the final grinding of them. The problems involved in the preparation of a sufficiently pure sample (it is doubtful whether any mineral concentrate is really pure) are chiefly mineralogical and as such are better dealt with by a mineralogist; the analyst should have some knowledge of these problems, however, and a brief discussion of them is included here.

Some mineral samples can be checked for impurities by microscopic examination and an allowance made for the estimated content of these impurities in the analysis; hand picking of the crushed material with the help of a binocular microscope will often produce a reasonably pure concentration with a minimum of labor. Most minerals, however, exist in close company with other minerals and their separation is not an easy task.

A mineral may be intergrown with another mineral, often of the same grain size, and the exclusion of these "joined" grains from the sample could be a potential source of error. In order to separate minerals, by one means or another, it is necessary to crush the sample to a particle size that is smaller than the smallest mineral grain to be separated; this must not, however, be carried too far or a product of such fineness will be produced that surface attraction will make a clean separation virtually impossible. Mackenzie and Milne[35] have shown the considerable change in mica that is caused by excessive grinding. Another source of error is the occurrence of very small mineral grains as inclusions in larger grains, such as rutile needles or zircon grains in micas, and chlorite in feldspar; again, it is almost impossible to obtain a pure concentrate.

As a general rule, the bulk sample must be crushed to a sufficiently fine particle size, commonly -100 to $+200$ mesh, before beginning the concentration of the desired mineral fraction. Concentration is achieved usually by one or both of two methods, the one using the magnetic susceptibility of the mineral as a means of effecting a separation, the other its specific gravity.

Magnetic separation will separate a highly magnetic mineral, such as magnetite, from which is moderately magnetic (e.g., chromite), from those which are only weakly susceptible (e.g., olivine), and from those which are not attracted to a magnet, such as quartz. Highly magnetic minerals may trap less magnetic ones, and precautions must be taken to minimize this.

The differing specific gravities of minerals are utilized as a means of separation by suspending the mineral concentrate in a heavy liquid (bromoform, tetrabromoethane, methylene iodide) of known specific gravity which permits certain minerals to sink while others remain in suspension. Again, clusters of heavy minerals may trap grains of lighter minerals. Care must be exercised in the choice of heavy liquid used, e.g. Clerici solution (thallous formate–thallous malonate) should not be used if Tl is to be determined in the mineral concentrate. When heavy liquids are used, it is necessary to insure that no trace of them remains before proceeding with the analysis; if the loosely stoppered vial is placed in an oven at 105°C for a minute or so, the presence of organic compounds can be readily detected.

When the mineral to be studied is present in small amount (1% or less) it is necessary to start with a large sample in order to allow for the losses, as much as 40%, that will occur during the concentration procedure. Once the mineral concentrate has been obtained, a further purification is necessary. The use of acid or basic solutions to clean up the mineral grains prior to a microscopic examination is a hazardous step; the mineral may be attacked to an unknown degree and the by-products may cause a very misleading statement of the analysis. Hand-picking under a binocular microscope is laborious but it is often the only safe way to achieve a mineral concentrate of the desired purity.

A brief but thorough coverage of mineral separation and its attendant errors is given by Wager and Brown [(Ref. 58, Chapter II) and by Milner,[38] and in less detail by Ahrens and Taylor.[1]

3.5.2. TYPES OF EQUIPMENT

The types of equipment such as jaw crushers, pulverizers and automatic mortars needed to reduce the rock to the desired particle size, have been discussed in Sec. 3.4.2.

The Frantz isodynamic separator is the most popular of the devices used to separate minerals on the basis of their magnetic susceptibility; the powder moves down a vibrating chute parallel to the pole pieces of an electromagnet which separates the more magnetic and less magnetic particles into two streams which fall into separate collector vessels at the foot of the chute. By varying both the direct current to the electromagnet and the inclination of the chute, the forces applied to the particles may be controlled. Prior to treatment with the Frantz separator, the Carpco direct-roll magnetic separator can be used to make a preliminary coarse separation, thus reducing very considerably the bulk of the material which must be treated in the Frantz.

The concentration of heavy minerals may be done with a Wilfley concentration table. The bulk sample is fed by gravity to an inclined table bearing a number of riffles; the heavy minerals are caught by the riffles while the lighter ones float over them. A similar but more refined separation is achieved with the superpanner. In this device a thin film of water flows down a sloping flat surface; the surface layer of the film moves more rapidly than its bottom layer and thus minerals are separated along the sloping surface according to their specific gravity. By varying the slope of the surface, separations can be made among minerals whose specific gravities vary only slightly.

Simple devices can also be used to make separations, such as a small hand magnet enclosed in a piece of paper and lightly drawn over the surface of a powder, or centrifuge tubes and heavy liquids of the type described by Nickel,[43] or Marshall and Jeffries.[36] For handpicking grains under the microscope, Murthy devised a suction apparatus in which an intravenous needle connected to a soft plastic bottle sucks the mineral grains into a collecting tube when the pressure on the bottle is released.[40] A similar arrangement, but with constant suction, utilizes a thin plastic hose fitted with a similar intravenous needle and connected to a plastic bottle which in turn is connected to a simple water aspirator.

The physical properties and specific gravity ranges of various liquids used for laboratory sink-float separations of rocks or minerals are covered in some detail by Browning[6]; the equipment necessary for carrying out these separations, and the difficulties and necessary safety precautions, are briefly described.

3.5.3. SAMPLE PREPARATION PROCEDURES

The procedure to be followed will again be dictated by the equipment available and the quantity of material that must be processed in order to yield the desired concentrate. Fairbairn[14] has described a procedure, and has listed the equipment needed for the concentration of heavy accessories from large rock samples of the order of 50 lb in which use is made of a Carpco separator for the preliminary separation of such major constituents as biotite and hornblende, followed by removal of the nonmagnetic fraction by continuous heavy liquid separation. The heavy mineral concentrate thus obtained is further purified by magnetic and heavy liquid treatment. The coverage given by Smales and Wager,[58] Milner[38] and Browning[6] has already been mentioned.

3.6. Preparation of special samples for analysis

The foregoing material has been concerned chiefly with the routine treatment of rock and mineral samples. There are many other types of samples which require modification of these routine procedures; no attempt will be made to cover all of the possibilities but a few of the more obvious procedures will be mentioned.

The handling of wet material is facilitated by first drying the material, but some samples such as soils and clays have a high water content at room temperature which changes readily in response to atmospheric changes. It is usually necessary to dry these materials at 140°C to prevent caking during the grinding; the sample should then be spread out on a piece of glazed paper on the balance table, covered lightly to protect it from dust, and allowed to come to equilibrium with its surroundings before it is placed in a sample vial. This applies to other samples also having a high water content; the recommended procedure for weighing out sample aliquots of hydroscopic material is given in Sec. 9.1. Marshall and Jeffries[36] discuss ways of cleaning soils prior to mechanical analysis; a number of special centrifuge tubes for heavy liquid separation are also described. Further details are given by Searle and Grimshaw.[53]

Soils should be passed through a coarse screen to remove roots, leaves and other material; lumps are crushed, but care should be taken not to crush rock fragments at this stage. The weight of the material that remained in the screen, as well as that which passed through it, should be recorded. After thorough mixing, portions of the sample are removed and ground to pass through a screen having openings 0.5 mm in diameter (pH, soluble P, exchangeable cations), a 100-mesh screen (total major constituents) and a 200-mesh screen (trace elements).

Methods for the sampling of water, and the handling of the samples prior to analysis, are described in detail by Rainwater and Thatcher.[48]

The problem of obtaining a truly representative sample of a meteorite is a considerable one, and differs in magnitude from the iron to the stony types. Iron meteorites can be sampled by either drilling or milling and the purity of the material can be checked by means of a binocular microscope. Some analysts prefer to sample the metal and inclusions separately; a bulk composition is derived by analyzing the fractions separately and combining them on the basis of an areal analysis of a large slice cut from the surface.   Others prefer to use thin slabs free from visible inclusions which are totally dissolved without further treatment.

In stony meteorites the analyst is faced with the presence of three or four phases—silicate, metal, sulfide and phosphide; drilling may be used to isolate small samples for study, but it is not helpful in obtaining a sample for bulk analysis.   The sampling problem is related directly to the nature of the sample in that the amount of sample that must be taken is governed by its homogeneity.   Magnetic separation is, next to handpicking, the simplest way to effect a separation, but the use of a magnet results in two fractions, a magnetic (metal) one with some silicate grains either joined or trapped, and a nonmagnetic (silicate) fraction with some metal.   Separate analyses of these fractions must each be corrected for the presence of some of the other fraction, often on the basis of certain assumptions that are not strictly valid.   Various reagents such as mercuric chloride have been used to remove the metal fraction, but their use always raises the question of their effect on the other phases. A technique that promises to be very useful is one developed at the British Museum (Natural History); dry chlorine is passed over the heated sample, resulting in removal of the metal as volatile chlorides and leaving the silicate phase in an untouched form.[39]   Onishi and Sandell[46] dissolved the metal and sulfide phases of chondrites in aqua regia to separate the silicate phase, but some of the latter is decomposed in the process; troilite can be dissolved in hydrochloric acid saturated with bromine.   Berry and Rudowski[4,12] have designed a special container for crushing chondritic meteorites at liquid air temperature; embrittlement of the metal phase (Ni $<20\%$) makes it possible to crush the whole sample.

References

1. Ahrens, L. H., and Taylor, S. R.   *Spectrochemical Analysis*, 2nd ed., 1961. Reading, Mass.: Addison-Wesley Publishing Co. pp. 41–46.
2. Ballard, J. W., Oshry, H. I., and Schrenk, H. H.   Sampling, mixing and grinding techniques in the preparation of samples for quantitative analysis by X-ray diffraction and spectrographic methods.   *J. Opt. Soc. Am.* **33**, 667–675 (1943).
3. Barnett, P. R., Huleatt, W. P., Rader, L. F., and Myers, A. T.   Spectrographic determination of contamination of rock samples after grinding with alumina ceramic.   *Am. J. Sci.* **253**, 121–124 (1955).
4. Berry, H., and Rudowski, R.   The preparation of chondritic meteorites for chemical analysis.   *Geochim. Cosmochim. Acta* **29**, 1367–1369 (1965).
5. Bloom, H., and Barnett, P. R.   A new ceramic buckboard and muller.   *Anal. Chem.* **27**, 1037–1038 (1955).
6. Browning, J. S.   Heavy liquids and procedures for laboratory separation of minerals.   *U.S. Bur. Mines Inf. Circ.* **8007**, 1961, 14 pp.

7. Burks, H. G., and Harpum, J. R. Sampling for granite investigations in Tanganyika: the Imagi experiment. *Records Geol. Sur. Tanganyika* **10**, 64–68 (1963).

8. Cochran, W. G., Mosteller, F., and Tukey, J. W. Principles of sampling. *J. Am. Statistical Assoc.* **54**, 13–35 (1954).

9. Danielsson, A., and Sundkvist, G. The tape machine. II. Applications using different kinds of isoformation. *Spectrochim. Acta* **13**, 126–133 (1959).

10. Davis, G. R. Results of comparative sampling methods at Kilembe Mine, Uganda, during the exploration phase. *Trans. Inst. Min. Metall. (London)* **72**, (*Bull. Inst. Min. Metall. (London)* **673**, Dec 1962), 145–164 (1962–1963); *idem.*, **674**, 255–267 (1963).

11. Dixon, W. J., and Massey, F. J., Jr. *Introduction to Statistical Analysis*, 1951. New York: McGraw-Hill.

12. Easton, A. J., and Lovering, J. F. The analysis of chondritic meteorites. *Geochim. Cosmochim. Acta* **27**, 753–767 (1963).

13. Exley, C. S. Quantitative area modal analysis of granitic complexes: a further contribution. *Bull. Geol. Soc. Am.* **74**, 649–654 (1963).

14. Fairbairn, H. W. Concentration of heavy accessories from large rock samples. *Am. Mineralogist* **40**, 458–468 (1955).

15. Fairbairn, H. W., Ahrens, L. H., and Gorfinkle, L. G. Minor element content of Ontario diabase. *Geochim. Cosmochim. Acta* **3**, 34–46 (1953).

16. Flinn, D. An application of statistical analysis to petrochemical data. *Geochim. Cosmochim. Acta* **17**, 161–175 (1959).

17. Galle, O. K. Comparison of chemical analyses based upon two sampling procedures and two sample preparation methods. *Trans. Kansas Acad. Sci.* **67**, 100–110 (1964).

18. Grillot, H. Analyse chimique des roches et des eaux. *Bur. Recherches Géol., Géophys. Minières, Publ.* **20**, 1957, 40 pp.

19. Grout, F. F. Rock sampling for chemical analysis. *Am. J. Sci.* **24**, 394–404 (1932).

20. Groves, A. W. *Silicate Analysis*, 2nd ed., 1951. London: George Allen and Unwin. p. 17.

21. Hawkes, H. E., and Webb, J. S. *Geochemistry in Mineral Exploration*, 1962. New York: Harper and Row, Publishers. Chapter 3.

22. Hill, W. E., Jr. Methods of chemical analysis for carbonate and silicate rocks. *State Geol. Surv. Kansas Bull.* **152**, Pt. I, 1961, 30 pp.

23. Hillebrand, W. F. The influence of fine grinding on the water and ferrous iron content of minerals and rocks. *J. Am. Chem. Soc.* **30**, 1120–1131 (1908).

24. Hillebrand, W. F., Lundell, G. E. F., Bright, H. A., and Hoffman, J. I. *Applied Inorganic Analysis*, 2nd ed., 1953. New York: John Wiley and Sons, pp. 813–814.

25. Huleatt, W. P. Automatic sample preparation saves time and money for the U.S.G.S. *Eng. and Min. J.* **151**, 62–67 (1950).

26. Joseph, M. E. Statistical methods in sampling. *Proc. Australasian Inst. Mining and Met.* **202**, 81–101 (1962).

27. Kolthoff, I. M., and Elving, P. J. (Eds.). *Treatise on Analytical Chemistry*, Pt. I, Vol. 1, 1959. New York: Interscience. Chapter 4, "Principles and Methods of Sampling" (W. W. Walton and J. I. Hoffman).

28. Kolthoff, I. M., and Sandell, E. B. *Textbook of Quantitative Inorganic Analysis,* 3rd ed., 1952. New York: Macmillan Co. pp. 700–701.
29. Laffitte, P. Étude de la précision des analyses de roches. *Bull. Soc. Géol. France* **3**, 6th Ser., 723–745 (1953).
30. Lamar, J. E., and Thomson, K. B. Sampling limestone and dolomite deposits for trace and minor elements. *Illinois State Geol. Surv. Circ.* **221**, 1956, 18 pp.
31. Larsen, E. S. The accuracy of chemical analyses of amphiboles and other silicates. *Am. J. Sci.* **35**, 94–103 (1938).
32. Lavergne, P. J. Preparation of geological materials for chemical and spectrographic analysis. *Geol. Surv. Canada Paper* **65–18**, 1965, 23 pp.
33. Lovering, J. F., Nichiporuk, W., Chodos, A., and Brown, H. The distribution of gallium, germanium, cobalt, copper and chromium in iron and stony-iron meteorites in relation to nickel content and structure. *Geochim. Cosmochim. Acta* **11**, 263–278 (1957).
34. Lundell, G. E. F., and Hoffman, J. I. *Outlines of Methods of Chemical Analysis,* 1938. New York: John Wiley and Sons, pp. 21–23.
35. Mackenzie, R. C., and Milner, A. A. The effect of grinding on micas. I. Muscovite. *Mineralog. Mag.* **30**, 178–185 (1953).
36. Marshall, C. E., and Jeffries, C. D. Correlation of soil types and parent materials. *Soil Sci. Soc. Am. Proc.* **9**, 397–405 (1945).
37. Mauzelius, R. The determination of ferrous iron in rock analysis. *Sveriges Geol. Undersökn. Arsbok* **1**, 3–11 (1907).
38. Milner, H. B. *Sedimentary Petrography,* 4th ed. rev., Vol. 1, 1962. London: George Allen and Unwin, 54–75, 101–104.
39. Moss, A. A., Hey, M. H., and Bothwell, D. I. Methods for the chemical analysis of meteorites. I. Siderites. *Mineralog. Mag.* **32**, 902–916 (1961).
40. Murthy, M. V. N. An apparatus for handpicking mineral grains. *Am. Mineralogist* **42**, 694–696 (1957).
41. Myers, A. T., and Barnett, P. R. Contamination of rock samples during grinding as determined spectrographically. *Am. J. Sci.* **251**, 814–830 (1953).
42. Neumann, H. A pulverizer for micas and micaceous minerals. *Norsk. Geol. Tidsskr.* **36**, 52–54 (1956).
43. Nickel, E. H. A new centrifuge tube for mineral separation. *Am. Mineralogist* **40**, 697–699 (1955).
44. Oertel, A. C. Spectrographic analysis of mineral powders. *Internal Rept. Division of Soils,* C.S.I.R.O., Australia, 1961.
45. O'Neil, R. L. Analytical procedures applicable to fine-grained sedimentary rocks. *J. Sed. Pet.* **29**, 267–280 (1959).
46. Onishi, H., and Sandell, E. B. Gallium in chondrites. *Geochim. Cosmochim. Acta* **9**, 78–82 (1956).
47. Pitcher, W. S., and Sinha, R. C. The petrochemistry of the Ardara Aureole. *Quart. J. Geol. Soc. London* **113**, 393–408 (1957).
48. Rainwater, F. H., and Thatcher, L. L. Methods for collection and analysis of water samples. *U.S. Geol. Surv. Water Supply Paper* **1454**, 1961, 301 pp.
49. Ross, W. H., and Hardesty, J. O. Grinding of fertilizer samples for analysis. *J. Assoc. Offic. Agr. Chemists* **25**, 238–246 (1942).
50. Sandell, E. B. Contamination of silicate samples crushed in steel mortars. *Anal. Chem.* **19**, 652–653 (1947).

51. Schlesinger, M. D., Nazaruk, S., and Reggel, L. Laboratory cylinder-mill for small samples. *J. Chem. Ed.* **40**, 546 (1963).
52. Schoeller, W. R., and Powell, A. R. *Analysis of Minerals and Ores of the Rarer Elements*, 3rd ed., 1955. New York: Hafner Publishing Co. pp. 1–2.
53. Searle, A. B., and Grimshaw, R. W. *The Chemistry and Physics of Clays and Other Ceramic Materials*, 3rd ed., 1959. London: Ernest Benn.
54. Sergeant, A. G. The chemical analysis of silicate rocks. *Intern. Manual Geol. Surv. Great Britain*, (1962).
55. Shaw, D. M. Manipulation errors in geochemistry. *Trans. Roy. Soc. Can.* **55, 3rd. Ser.**, Sect. IV, 41–55 (1961).
56. Shaw, D. M., and Bankier, J. D. Statistical methods applied to geochemistry. *Geochim. Cosmochim. Acta* **5**, 111–123 (1954).
57. Sichel, H. S. New methods in the statistical evaluation of mine sampling data. *Trans. Inst. Min. Metall. (London)* **61**, 1951–52 (*Bull. Inst. Min. Metall. (London)* **544**, Mar., 1952), 261–288.
58. Smales, A. A., and Wager, L. R. (Eds.). *Methods in Geochemistry*, 1960. New York: Interscience Publishers, pp. 4–32.
59. Smith, C. H. Notes on the Muskox intrusion, Coppermine River area, District of Mackenzie. *Geol. Surv. Can. Paper* **61–25**, 1962, 16 pp.
60. Storm, T. W., and Holland, H. D. The distribution of nickel in the Lambertville diabase. *Geochim. Cosmochim. Acta* **11**, 335–347 (1957).
61. Strouts, C. R. N., Wilson, H. N., and Parry-Jones, R. T. *Chemical Analysis— the Working Tools*, rev. ed., 1962. Oxford: Clarendon Press. pp. 46–77.
62. Thiers, R. E. Contamination in trace element analysis and its control. *Methods of Biochemical Analysis, Vol. 5*, pp. 273–335 (D. Glick, Ed.), 1957. New York: Interscience.
63. Volborth, A. Dual grinding and X-ray analysis of all major oxides in rocks to obtain the true composition. *Appl. Spectr.* **19**, 1–7 (1965).
64. Washington, H. S. *Manual of the Chemical Analysis of Rocks*, 2nd ed., 1910. New York: John Wiley and Sons. pp. 43–58.
65. Willard, H. H., and Diehl, H. *Advanced Quantitative Analysis*, 1943. New York: D. Van Nostrand Co. pp. 16–23.
66. Wilson, A. D. The sampling of silicate rock powders for chemical analysis. *The Analyst* **89**, 18–30 (1964).
67. Wilson, C. L., and Wilson, D. W. (Eds.). *Comprehensive Analytical Chemistry*, Vol. IA, 1959. New York: Elsevier Publishing Co. Chapter II, "Sampling" (R. C. Tomlinson).
68. Youden, W. J. *Statistical Methods for Chemists*, 1951. New York: John Wiley and Sons.

Supplementary References

Baird, A. K., McIntyre, D. B., and Welday, E. E. Geochemical and structural studies in batholithic rocks of Southern California. Part II. Sampling of the Rattlesnake Mountain pluton for chemical composition, variability and trend analysis. *Bull. Geol. Soc. Am.* **78**, 191–222 (1967). This contains a very detailed discussion of hierarchical sampling and a study of variance.

Baird, A. K., McIntyre, D. B,. Welday, E. E. and Morton, D. M. A test of chemical variability and field sampling methods, Lakeview Mountain tonalite, Lakeview

Mountains, Southern California batholith. *Short Contributions to California Geol., Spec. Rept.* **92**, *California Div. Mines and Geol.*, 11–19 (1967). Diamond drill and hammer specimens, collected from 30 of 150 grid points spaced 2000 ft apart, are analyzed by X-ray fluorescence spectroscopy and the data treated statistically.

Boulton, J. F., and Eardley, R. P. The preparation of analysis samples of hard materials with a boron carbide mortar. *The Analyst* **92**, 271–272 (1967). The degree of contamination introduced by grinding fused alumina, sillimanite, silicon carbide and magnesite in boron carbide appears to be an order of magnitude lower than when a 95% alumina mortar is used.

Kleeman, A. W. Sampling error in the chemical analysis of rocks. *J. Geol. Soc. Australia* **14**, 43–47 (1967). It is recommended that for samples analyzed in duplicate, the rock should be crushed to pass a 120-mesh sieve, and for portions <0.5 g, the powder should be crushed more finely. Rock powders for use as reference standards should be crushed to −230 mesh.

Miesch, A. T. Theory of error in geochemical data. *U.S. Geol. Survey Prof. Paper* **574-A**, A1–17 (1967). "The adoption of rigid experimental design techniques in field geochemistry does not necessarily require radical departure from methods and concepts now being used but instead requires mainly a greater awareness of things we already know." The discussion is intended to help in interpreting a few basic statistical principles in terms of the particular situations in field geochemistry.

Zussman, J., Ed. *Physical Methods in Determinative Mineralogy*, 1967. London: Academic Press; Chapter 1, "Mineral Separations" (L. D. Mueller), pp. 1–30. Techniques of crushing, sieving, heavy liquid, electrostatic and magnetic separations, dielectric and high tension methods, froth flotation, electrochemical elutriation and hand-picking are described, with numerous references.

CHAPTER 4

THE WORKING ENVIRONMENT

The classical and rapid schemes of rock and mineral analysis, as outlined in Part III, do not require a specialized laboratory for their operation and they can be carried out in any inorganic analytical laboratory that is provided with the standard services. Nor is the use of relatively complex instrumentation mandatory for the final measurement, although the availability of such equipment has made it possible to materially reduce the analytical time involved while at the same time maintaining or improving the accuracy of the methods. The improved quality of the glassware and reagents in recent years has eliminated many precautions formerly necessary to maintain the overall "blank" at an acceptable level.

The following brief discussion is concerned with the working environment of the rock and mineral analyst; there are already available several detailed discussions of a more general nature in recent texts,[15,17] to which reference should be made. The evolution of the rock analysis laboratories of the Geological Survey of Canada is illustrated in Figs. 4.1 and 4.2.

4.1. Facilities

4.1.1. THE LABORATORY

The amount of space provided for this work will naturally depend upon the nature and volume of the latter. Large laboratories are more suited to routine analytical work, particularly when the samples are analyzed in batches, but for the more complex work of mineral analysis a smaller laboratory is preferable; traffic flow is minimized and space is provided for the specialized apparatus that is often needed.

At the Geological Survey of Canada the chemical laboratories are located on the seventh floor of an eight story building. This has the advantage of minimizing the exposure of the laboratories to airborne dust and other contaminants and, because the fan-room is located directly above the laboratories, insures a good exhaust in the fume hoods. Structural vibration, as it affects delicate instrumentation such as an analytical balance, has not been a serious problem.

Fig. 4.1. The chemical laboratory of the Mineralogy Division, Geological Survey of
Canada, in 1926. Photograph by Photographic Division, Geological Survey.

A convenient module is one that is about 20 ft wide × 30 ft deep,
which will provide adequate space for 2 or 3 analysts. At the Survey
the more specialized work is done in single modules and half-modules;
double modules are used for work of a routine nature on large numbers of
samples. The balance room is placed between two double module
laboratories, thus affording a convenient access for the largest number
of people.

The laboratories should be well-lighted, preferably by northern
exposure, and should have adequate bench and fume hood space. A
fume hood should not be less than 6 ft wide; a large part of the work is
done in the hood and there must be room for a steam or water bath and
a hot plate, as well as the usual services supplied in fume hoods. Provision should be made for at least one special fume hood designed to
permit the safe use of perchloric acid;[6] this should include facilities for
flushing the duct and walls of the fume hood with water, the use of
explosion-proof electrical outlets, and the exclusion of gas outlets.
The bench space should include some tables set at desk height to serve

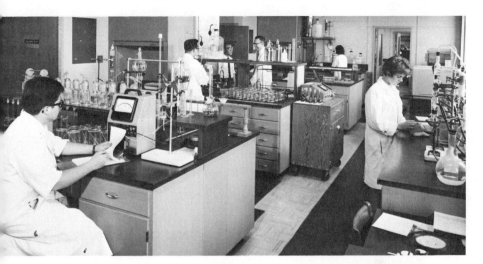

Fig. 4.2.   Part of the Rapid Methods Laboratory, Analytical Chemistry Section, Geological Survey of Canada, in 1965.   The door at the rear to the right leads to the Balance Room, which in turn opens into the Classical Analysis Laboratory. Note the spectrophotometer mounted on a movable bench.   Photograph by Photographic Section, Geological Survey.

as writing desks, and also to accommodate some pieces of equipment which are more conveniently operated at this height.   In planning the layout of the laboratory it is wise to leave some free wall space for the storage of such necessary items as gas cylinders, filing cabinets and laboratory carts.

There is much to be said for the use of built-in equipment.   By suspending such items as gas burners and filter stands from the back of the laboratory bench, the surface of the bench is more easily cleaned. The utilization of the floor of the fume hood for built-in steam or water baths, hot plates and furnace wells makes for efficient use of the available space.   Threaded holes in the back wall of the fume hood, into which support rods can be screwed, provide a useful means for keeping the fume hood floor free of support stands and bulky structures.

Some analysts prefer to have a separate instrument room where various instruments are kept free of the corroding effect of the normal laboratory atmosphere.   This is not necessary nor always desirable; at the Survey it has been found useful, both from the standpoint of economy of bench space and flexibility of operation, to mount certain pieces of equipment in general use, such as a spectrophotometer and muffle furnace, on movable benches which can be easily positioned for the convenience of the analyst.

It is advantageous to have one large sink in each double module for the cleaning of large pieces of equipment. Smaller sinks are suitable for the cleaning of glassware. Cup sinks are not very useful but they do provide for the use of water aspirators, and they supply cold water to condensers and similar items.

Cupboards of floor length size and preferably fitted with sliding glass doors should be provided for the storage of reagents and solutions. Wall cupboards, again with sliding glass doors, are also useful but should not be mounted over a workbench where they are difficult to reach.

A titration bench is a worthwhile addition to a large laboratory. This is a canopied area which is painted white, and has fluorescent or other lights mounted in the canopy; it may also have a frosted glass area in the floor of the bench with lights mounted underneath it to aid in the identification of the end point in some titrations. This assures uniform lighting conditions for all titrations.

It is important to have a proper balance table that will insulate the analytical balance from harmful vibration. Unwanted vibrations may be damped out by alternating layers of felt, or rubber, and lead or steel, and it is preferable to place the balance on a platform that is suspended from an outside wall; a table may then be built around but not touching this platform. At the Geological Survey there are two types of balance tables in use on the fifth, sixth and seventh floors. In the one type, the balance platform is fastened to a metal plate, the latter in turn fastened to the top of a large steel cylinder containing about 150 lbs of steel slugs. The cylinder hangs freely inside a wooden pedestal which rests upon hard rubber supports; the weight of the cylinder pulls the balance platform against hard rubber supports on the top of the pedestal. The whole is surrounded, but not touched by a separate work table. The second type is a simple but effective terrazo table that is cast in one piece and weighs about 400 lbs, designed by L. E. Moore, Division of Applied Chemistry, National Research Council; the table is strengthened internally by brass rods and the dimensions are such as to give the maximum damping effect, aided when necessary by isolation pads placed under each of the two side slabs.

4.1.2. SERVICES

The electrical outlets most frequently used are those for 110 V and/or 220 V; special needs may be supplied by special wiring as required, and it is advantageous to make provision for bringing in these special lines when the services are installed. Line fluctuations are a common occurrence and for smaller instruments such as spectrophotometers it will be necessary to provide voltage stabilizers; larger instruments are generally equipped with some means of voltage stabilization. Electric hot plates, ovens, and muffle furnaces are usually provided with some type of temperature control but for accurate work, such as in the sodium peroxide sinter procedure, it is wise to measure critical temperatures with a reliable pyrometer.

Electrically operated furnaces, ovens and hot plates have many advantages but the common laboratory gas burner remains one of the best means for providing a wide range of temperature control. The gas may be obtained from the municipal supply or from a separate source such as a bank of propane tanks; the latter has the advantage of being less subject to fluctuations in gas pressure. This is particularly important to the operation of a flame photometer, in which fluctuations in the gas pressure can be a serious source of errors. A steam line will find many uses, such as heating steam baths and operating water stills.

A supply of compressed air is required for many operations, from providing the air necessary for the air–propane mixture used in a flame photometer to the blowing out of dust from sample and grinding vials. A glass-wool filter should be inserted in the line to remove the droplets of oil and particles of dust that accompany the current of air. Vacuum outlets are also very useful for many operations, although water aspirators attached to the faucet of a cup sink will usually suffice.

A plentiful supply of hot and cold water is, of course, a necessity; the faucets of the cup sinks and fume hoods only need to supply cold water. The type of faucet that permits the mixing of hot and cold water is preferable and for those operations where close control of the temperature is necessary, as in photographic darkrooms, special mixing, heating and refrigeration devices must be provided. When water is circulated as a cooling medium, provision should be made to filter out the suspended or entrained solid matter that is commonly carried by the water as it is forced through the pipes at high pressure.

A central supply of distilled water has both advantages and disadvantages and is usually not provided unless the scale of laboratory operation is large enough to warrant it. While it does eliminate the need for individual water stills which require separate operation and cleaning, it does not always permit the degree of quality control that is desirable. At the Geological Survey a portion of the condensate from each water-still is passed through a mixed anionic-cationic resin column (which has a very long life) to provide both distilled, and distilled and deionized, water. The latter is used for the analytical work except when the possible presence of organic material is undesirable; the distilled water is used for the rinsing of glassware, and in water baths to minimize the formation of boiler scale. A periodic check should be made on the quality of the water used.

4.2. Equipment

Because the equipment used in rock and mineral analysis is that common to most inorganic analysis, it is not necessary to discuss it in detail here; particularly good coverage of materials, reagents and apparatus is given by Wyatt,[17] Beamish and McBryde[17] and by Strouts et al.[15] It will be useful, however, to discuss briefly a few items about which some information is necessary as a guide to proper usage.

4.2.1. GLASSWARE

Borosilicate or chemical resistant glass, of which Pyrex is the most common type, is the most suitable material for beakers, flasks, funnels and other similar items; it has a high resistance to chemical attack, and to mechanical and thermal shock. Alkaline solutions will attack it readily, particularly when boiled in it for any length of time, and solutions containing free hydrofluoric acid should not, of course, be used in it unless the introduction of varying amounts of silica and other constituents is of no consequence. Other solutions, including water, will also attack the glass in time and it is a good practice to retire glassware to less important usage when signs of wear become visible. The attack on the glass by acids other than hydrofluoric is generally negligible and solutions which are to be left standing in glass for any length of time should be acidified. Care must be taken not to defeat the use of chemical resistant glass by the unwitting use of less resistant glass in the form of stirring rods and cover glasses. *Vycor*, a silica glass containing approximately 96% SiO_2, and *fused silica* or *quartz* glass, either transparent or translucent, are useful when contamination by alkalies must be avoided. Both of these glasses can be used at temperatures much above the working limit for Pyrex, and the fused silica glass has a very high resistance to thermal shock; they have the disadvantage of high cost.

The same care and attention that are given to beakers and flasks should be given also to burettes, pipettes, dispensers and the other articles of glassware in common use. These should also be made of chemical resistant glass and should be treated with the respect that they deserve. Burettes and separatory funnels fitted with Teflon stopcocks are superior to those with glass stopcocks. Burettes in burette stands should be fitted with caps (small polyethylene thimbles) to keep out dust and grease when they are not in use; filling them with water results in attack on the glass over a period of time. The use of rubber policemen as guards for burette tips, particularly for those of the semimicro variety, will prevent chipping which may render the burette unfit for use. Care should be taken with the tips of pipettes; they are easily chipped and the rate of flow may change considerably as a result.

The method used to clean glassware is always subject to personal preference. The removal of the film of grease that forms on glass walls is necessary for reasons both esthetic and practical; drainage is seriously impaired if the surface is not clean (Ref. 10, p. 502). A mixture of concentrated sulfuric and chromic acids will remove a film of grease but will also leave an adsorbed film of chromium ion which is removed only with difficulty. A hot solution of sodium hydroxide and potassium permanganate will remove organic material resistant to the sulfuric–chromic acid mixture. Warm dilute nitric acid, hydrochloric acid or aqua regia will remove stains, but the glass surface will likely become sensitized and will retain adsorbed ionic films. Detergents are also used. Unless kept in sealed cupboards, glassware will accumulate a film of dust and grease on standing; the writer prefers to wash and dry glassware after use and to clean it immediately before putting it into use again.

4.2.2. PORCELAIN

Porcelain ware does not find extensive use in rock and mineral analysis, but casseroles are used for the dehydration of silica, and precipitates are occasionally ignited in porcelain crucibles. Porcelain is more resistant than glass to the action of basic solutions, provided that the inner glazed surface is in good condition, but it is not as resistant to acids. It can, of course, be used at much higher temperatures than glass, but care must be taken to avoid any interaction between the glazed surface and the crucible contents. Platinum crucibles are sometimes heated to a high temperature (above 1200°C) while resting in a porcelain crucible, and spot melting of the glaze can severely alter the weight of the platinum crucible. A common chemical reaction is that between the porcelain and the alkali salts in incompletely washed precipitates, producing a fused lump that is completely useless. When the inner glaze of the porcelain ware becomes dulled through chemical attack and physical abrasion, it should be discarded or reserved for less important use.

4.2.3. PLASTICS

The last 15 years have seen a great increase in the use of plastic ware in rock and mineral analysis.

Polyethylene (polythene) is now used to fabricate a wide variety of chemically resistant, common laboratory items that make the handling of solutions containing hydrofluoric acid much easier than was formerly possible. Ordinary polyethylene should not be heated much above 50–60°C, but an irradiated form is now available which will withstand heating to 100°C without deformation. Polyethylene wash bottles offer convenience of manipulation and close control of delivery; the fine jet is an easy means of filling semimicro burettes, for example, without the formation of trapped air bubbles. Polyethylene cover-glasses, which can be easily made from the base of empty hydrofluoric acid bottles, give useful service at water bath temperatures. Concentrated nitric and sulphuric acids discolor polyethylene and many organic solvents attack it. It also has a tendency to adsorb ions and to release them at a later date, and standard solutions that are stored in polyethylene for other than a short period should be restandardized before use. For other than standard solutions, and those materials which attack it, polyethylene storage containers have many advantages.

Teflon (polytetrafluoroethylene), also called Fluon and Gaflon, is the newest of the plastic materials to find general use and is an adequate substitute for platinum in some ways. The material is white, opaque and very strong, and can be machined; it can be used at temperatures up to 200°C and is chemically inert to most laboratory reagents. Because it is costly (about one-third the price of platinum) it is used chiefly for crucibles, evaporating dishes, covers and rods. A Teflon still for the purification of hydrofluoric acid has been described.

Perspex, or lucite, are trade names for polymethyl methacrylate, a transparent, hard, colorless plastic that is very useful for the fabrication of apparatus, and as safety shields and special hoods and boxes. It is easily machined and the edges can be bonded together by various solvents.

4.2.4. PLATINUM

At one time the most expensive consideration in the equipping of a laboratory for rock and mineral analysis was for platinum ware such as crucibles, dishes, tongs and forceps. Today it is possible to get along with less, thanks to the advent of polyethylene and Teflon ware. There are still certain operations that are either necessarily or better done in platinum and it offers many advantages over other materials, notably its ability to withstand high temperatures and ease of cleaning. Because of the extensive use of platinum ware in rock and mineral analysis its properties, uses and maintenance will be considered in detail; see also the discussion by Foner (Supp. Ref.).

The analyst is offered a choice between platinum and platinum alloys, and among platinum substitutes as well. These varieties of ware are discussed in the references cited previously, and also by Kolthoff and Sandell.[10] Suffice it to say here that pure platinum (actually, small amounts of other platinum metals are present to insure a sufficiently rigid product) has advantages over the ware containing one to several percent of rhodium or iridium; increased rigidity can be obtained by the use of reinforcing rims on crucibles and dishes* and the "scrap" value of platinum is high. Of the platinum substitutes in use, the best known is an alloy of palladium–gold (palau); while less expensive than platinum, it is more easily attacked by various substances than is pure platinum, and is less reclaimable. It has been said for it that fusion cakes are more readily removed from palau crucibles than from those of platinum, but opinions vary on this matter.

Platinum, because of its softness, is easily distorted and damaged and requires more care than is usually given to it. It is easily scratched; glass rods should be used with caution in removing fusion cakes from crucibles and a rubber policeman or a plastic rod are better tools for such an operation. Any deformation caused by dropping or squeezing (especially by tongs; the sharp-nosed variety can produce flutings in the rim of a crucible or dish that are difficult to remove) should as soon as possible be corrected by reshaping; all platinum ware should be provided with a hardwood shaper of appropriate size and shape. If deformations are not corrected they will become centers for the development of cracks. Platinum ware should always be kept clean and bright on both the inside and outside (the cleaning process will be described later) as a protection for the considerable investment of money involved, and in order to insure its being in proper condition for analytical use. Recrystallization resulting from prolonged heating at a high temperature will give the

---

* The use of welded reinforcement rims, now generally discontinued, should be avoided; dirt is trapped between the two layers and is difficult to remove. Similarly, tongs with solid platinum feet are preferable to those with platinum shoes which also tend to entrap dirt and will soon become worn.

ware a greyish, speckled look and, unless removed by burnishing, will lead to the development of pinholes and cracks; this is particularly so on the outside of the crucible bottom which receives the full force of the heat, and which is too often ignored in the cleaning operation.

Prolonged heating of platinum at temperatures above 1200°C should be avoided because of the volatilization of platinum that will occur; when such heating is necessary, a correction for the loss in weight must be made. Oxidizing conditions should prevail at all times. At no time should the reducing portion of a gas flame be in contact with the platinum; acetylene is particularly bad in this respect and on one occasion in our laboratory a hole was rapidly made in the bottom of a 30 ml crucible which was unknowingly in contact with the sharp blue cone of an acetylene flame. Platinum ware should preferably be handled only with platinum-tipped tongs when hot (pure nickel will also serve) and should rest on supports of fused silica, clay or platinum, never on iron or other metal supports. It is appropriate to comment here on one type of deformation that develops because of the unequal expansion of platinum during heating; expansion occurs more readily in the upper part of the crucible than in the lower part which is confined in the triangular support, leading to the development of a "cuspidor" shape that is most undesirable. Platinum basins and dishes should rest on porcelain or plastic-covered rings during the evaporation of solutions and not directly on copper or brass rings. Hot platinum ware should be cooled on a nonmetallic surface, such as a polished stone slab.

Although it is relatively chemically inert to most reagents and chemical compounds, platinum is attacked to some degree by a number of substances, and seriously by a few of them. Some precautions that should be taken during some common operations are as follows:

(1) *Fusion:* sodium and potassium carbonates, borax and boric acid may be fused at high temperatures with little attack on the container, but lithium carbonate or other lithium compounds cannot be so fused because they are adsorbed by the platinum;[15] alkali fluorides and bifluorides do not react, but pyrosulphate (and bisulphate) fusions will dissolve a small amount of platinum if the fusion is heated above about 700°C, or prolonged unduly. Nitrates, nitrites and peroxides should not be fused alone but a small amount mixed with sodium or potassium carbonates will do no harm. Alkali hydroxides, sulfides (tellurides and selenides) and cyanides must be avoided.

(2) *Ignition:* the first rule is that an unknown substance should not be heated in platinum; the unsuspected presence of elements such as lead, antimony or arsenic, especially if there are reducing substances

present, can result in the ruination of a crucible or dish. The oxides, carbonates, sulfates and oxalates of those elements which are not easily reduced, such as Si, Al, Ca, Mg, Mn, Ti, Zr, Th and rare earths, may be safely ignited in platinum at reasonable temperatures (1200°C); if suitable precautions are taken to maintain an oxidizing atmosphere, the compounds of elements such as Fe, Ni and Co may also be ignited safely. BaO and Li_2O attack platinum and should not be heated in it, nor should their compounds be heated in the presence of reducing substances. Filter papers should be burned off at a low temperature before the crucible is heated strongly; carbon, at high temperatures, will form platinum carbides and may cause embrittlement of the platinum. The paper should not be allowed to ignite; this not only creates undesirable drafts which may cause material to be lost from the crucible but also establishes a reducing atmosphere that may cause damage to the crucible, either from carbon monoxide or by reducing some substance, such as phosphate, arsenate or antimonate, which will subsequently attack the platinum. The ignition of magnesium ammonium phosphate is particularly prone to trap carbon, as will be discussed later. Wyatt[17] points out that even silicates may be a source of trouble if heated at about 1000°C in the presence of sulfur and carbon, because brittle platinum silicide is formed. Volatile halides must also be avoided.

(3) *Acid decomposition:* at normal temperatures, platinum is affected only by *aqua regia* and thus hydrochloric and nitric acids, or other mixtures which liberate nascent chlorine, should not be used in platinum. If potassium nitrate is used as an oxidant in an alkali carbonate fusion, followed by dissolution of the cake in hydrochloric acid, the evaporation must not be done in platinum, or several milligrams of the basin will be added to the analysis; it is thus preferable to use sodium peroxide as the oxidant in the fusion so that platinum may then be used if it is desirable. At water bath temperature the only precaution necessary is to avoid the heating in platinum of those solutions which contain much ferric chloride (i.e., the HCl solution of the fusion cake from highly ferruginous material) because attack will occur. At high temperatures phosphoric and sulfuric acids will attack platinum, but only slightly, after prolonged heating; there is no appreciable attack during the relatively brief heating of sulfuric acid to fumes of SO_3 in the volatilization of silica with HF and H_2SO_4, and in the recovery of silica from the pyrosulfate fusion of the R_2O_3, mixed oxides.

The cleaning of platinum ware is accomplished by a combination of acid treatment, fusion, ignition and burnishing. When dents and other deformations have been removed, the piece should be examined for stains. Digestion in hot concentrated HCl will remove most iron stains, although two or more such treatments, interspersed with ignitions (5–10 min) of the piece at red heat under oxidizing conditions, may be necessary. If the result is not entirely satisfactory, a brief fusion with a small amount of alkali bisulfate or pyrosulfate at a dull red heat is generally all that will be necessary and the piece, after removal of the fusion cake, is ready for burnishing. Clean sand, not too coarse (which scratches the surface unnecessarily) nor too fine (which is not sufficiently effective) is used; it should be free from dirt and relatively free of other minerals, and the grains should be well-rounded rather than angular. A well-sorted, clean sea sand is most useful. A small amount of sand is placed in the piece, moistened, and rubbed vigorously around the sides and over the bottom, holding the ware in such a way as to minimize deformation by squeezing; after 1 or 2 min of scouring, the sand is scraped into the palm of the hand and the piece is carefully rubbed over the sand, with particular attention being paid to the bottom. The lid is also scoured, again using the sand in the palm of the hand. Rubber policemen are sometimes used but should be avoided; the rubber is worn rapidly and exposure of the glass rod can lead to serious scratching of the platinum. The index finger is the most flexible tool for this work. When the burnishing is finished, the piece should be washed free from sand and then washed with soap and water to remove the film of "platinum black" that is present. Finally it is reshaped and ignited at red heat for about 5 min, cooled and examined for the absence of stains and recrystallization. New platinum ware should be ignited and examined for iron stains, for iron is often present as a surface contamination; it is a useful precaution to digest new ware in warm, concentrated hydrochloric acid before its initial use.

Platinum-tipped tongs are cleaned by rubbing the platinum feet briskly against sand held in the palm of the hand; the metal legs are cleaned with emery paper. The tongs should be cleaned when they become discolored and always after they have been used to hold a crucible or dish during a pyrosulfate fusion; during the latter operation there is always a transfer of a trace of the fusion mixture to the tongs and this could be a serious source of contamination in future operations.

4.2.5. OTHER MATERIALS

The only other materials that have significant application in rock and mineral analysis are silver, nickel and iron, in the form of crucibles for fusions. The melting point of silver is too low (960°C) to permit its use at high temperatures but it is suitable for fusions with sodium and potassium hydroxides, having the advantage over nickel that there is only slight attack on the crucible; any silver dissolved is easily precipitated and removed when the fusion cake is dissolved in dilute hydrochloric acid.   There is a tendency for the contents to creep and if the fusions are done in a muffle furnace, they must be handled with care in removal from the furnace to avoid loss of material.   Shell[13] prefers to use silver crucibles for the fusion of silicates on which a precise determination of iron is to be made, because crucibles of Pt or Pt–Rh retain an indefinite amount of iron which can only be recovered by repeated ignition of the crucible under oxidizing conditions and leaching in hot acid.

Nickel crucibles have a relatively long life when used for alkali hydroxide fusions and the amount of nickel that is introduced into the solution is not usually objectionable.   They are also superior to iron crucibles for fusions with sodium peroxide.   Iron crucibles have the advantage of low cost and, where the introduction of iron is of no consequence, are widely used for fusions with sodium peroxide.

4.2.6. HEATING EQUIPMENT

There are several different types of heating equipment required during the course of an analysis.   The sample must first be dried, usually at 105–110°C, in a drying oven equipped with a temperature control; the oven should be capable of giving temperatures up to about 250°C, in order to handle certain hydrous minerals which yield their water (water of constitution as well as moisture) only at higher temperatures. Evaporations are usually carried out on a steam, or water bath and little attention need be given to them during the course of the evaporation;* the bath should be capable of 24 hour operation without danger of becoming dry, so that such evaporations may be made overnight.   Baths with either 4 or 6 holes can be accommodated in a 6 ft wide fume hood; little use is made now of single-hole water baths.   A convenient and effective substitute for steam or water baths, when solutions are to be evaporated in dishes, is provided by tall form beakers of 400 ml capacity resting on a hot plate; the beakers are half-filled with water, a few carborundum boiling chips are added and the dishes are supported directly on the rim of the beaker, or on rings.   Infrared heating lamps,

---

* We have found that replacement of the brass or copper top of the bath by a sheet of bakelite will minimize the formation of copper salts.   Spraying of the copper or brass surfaces with clear plastic is also useful but the film does not last verv long and the surface tends to be sticky when hot.

either with internal reflectors or inserted in reflectors, are a rapid and efficient means of evaporation; the combination of steam bath, water bath, sand bath or, less preferably, hot plate and the heat lamp will considerably hasten the evaporation. Surface heating is less subject to bumping than is the conventional bottom heating and more use should be made of it. It is necessary that the lamp and reflector be wiped free of accumulated deposits before use.

Hot plates, which usually can be operated up to about 250°C if electrically heated, and to higher temperatures if gas-heated, find a multitude of uses. Their surface is not usually completely smooth, however, and thus uneven heating of vessels will occur, leading to the danger of loss of material by bumping. Because they tend to heat unevenly, the analyst must get to know the vagaries of a particular hot plate and be able to select areas most suited to his needs. Squares of thin asbestos paper help to produce a finer gradation of temperature but may cause harm to the hot plate because of the development of hot spots beneath the insulating paper.

Sand baths, heated either by a gas burner or, preferably, by a hot plate, offer some advantages over the use of a hot plate. The temperature is uniformly distributed throughout the sand, and the loose nature of the sand enables crucibles and dishes to be nested securely, at the same time subjecting a larger area to the heat. In the Geological Survey laboratories it has been found that enamel photographic trays, approximately 18 in. × 10 in. × 2 in. and two-thirds filled with coarse sand, are very convenient receptacles for up to twelve 50 ml platinum dishes; the tray is placed on a hot plate, and is easily removed when the hot plate is needed for another purpose. Care must be taken to remove any adhering sand from the bottom of the crucibles or dishes so heated.

The most useful and versatile source of heat in the laboratory is the gas burner and it is preferable to a hot plate when temperature must be carefully controlled. The Tirrill burner, with its gas regulatory needle valve, is very useful for low temperature work (boiling of solutions, rapid evaporations, initial charring of papers and some ignitions) while the meker, or Fisher, type of burner is more suitable for higher temperatures. The meker burner presents little hazard for the heating of platinum ware but, as pointed out previously, at no time must the inner blue cone of the Tirrill burner come into contact with the platinum. Earthenware chimneys, supported by steel tripods, serve both as a support for the triangles used to hold crucibles, and as a shield against drafts. Metal evaporators,[10] supported by iron rings, are efficient air baths for evaporating the contents of crucibles and small dishes but they offer no particular advantage over a sand bath.

Heating at high temperature is best done in an electrically heated muffle furnace which will give temperatures of 1200°C or more, sufficient to complete the heating of hydrous alumina to constant weight. Such furnaces, preferably large enough to take at least four 30 ml crucibles, give reasonably uniform heating conditions and freedom from reducible substances such as one encounters in the gas flame. It must be borne in mind, however, that the temperature indicated by the pyrometer scale is that existing at the tip of the thermocouple and there may be a difference of as much as 100–200°C between the temperatures at the front and back of the furnace. There must also be access for air in order that oxidizing conditions will prevail. The heating elements must *not* be exposed; even when covered they contribute a fine metallic dust to the inside of the heating chamber, and crucibles should not be placed on the floor of the chamber but rather on a removable refractory plate or tray which is cleaned before use. The charring of filter paper prior to the ignition of a precipitate is conveniently done by starting with a cold muffle furnace and allowing it to heat slowly; when more than one crucible is placed in the muffle, however, the crucible covers should be three-quarters in place to prevent both the contamination of the contents of one crucible by those of another, and the loss of material because of the drafts created by the combustion.

Blast lamps find some use in rock and mineral analysis, but are unnecessary if a muffle furnace capable of giving temperatures up to 1200°C or better is available. A glass blower's oxygen-gas hand torch is a very convenient and easily controlled source of high temperatures, such as is needed to complete the final heating of the glass bulb in the Penfield method for the determination of total water.

Some mention should be made of the types of tongs that will be found useful. For handling hot platinum ware it is essential that the tongs be equipped with platinum feet or shoes (pure nickel is an acceptable substitute but the platinum is better); by inserting suitable pieces of stainless steel between the lower and upper parts of the tongs they can be increased to 18 in. or more, a convenient length for use with a muffle furnace. Sharp-nosed tongs are very useful for gripping dishes and crucibles but if a lid is present it must first be removed; platinum claws, capable of grasping both 25 and 30 ml crucibles and fitted to the lengthened tongs previously described, will facilitate the introduction and removal of crucibles during heating without the need for displacement of the lids. Care is necessary when the sharp-nosed tongs are used; if they are squeezed too tightly the effect is rather to open the jaws and crucibles or dishes may be dropped.

4.2.7. Filter paper

There are various brands of filter paper on the market, each available in several grades which are usually comparable in thickness, porosity and wet strength. Of these the Whatman, and Schleicher and Schuell (S&S), papers are probably the best known; because they are the ones with which the writer is most familiar they will be used to illustrate the size and grade of paper when needed. Detailed discussions of these, and other papers, are given in recent textbooks[10,15,17] to which reference should be made.

Qualitative papers have no application in rock and mineral analysis. The quantitative papers used must be of the "ashless" variety (ash content generally less than 0.1 mg per 9 cm circle), papers that have been washed with both hydrochloric and hydrofluoric acids; hardened quantitative papers, which have a greater wet strength and increased rapidity of filtration, find only occasional usage. For coarse precipitates, such as the separated silica, the Whatman No. 41 (S&S 589 Black Ribbon) is most useful. Medium-grained precipitates such as calcium oxalate are best filtered through Whatman No. 40 (S&S 589 White Ribbon) while the fine-grained precipitates such as barium sulfate require the use of Whatman No. 42 (S&S 589 Blue Ribbon), or even of Whatman No. 44 (S&S 589 Red Ribbon).

The diameter of the filter paper for a filtration should be governed by the volume of the precipitate to be retained by it, since the precipitate should not occupy more than about one-half of the volume of the filter paper cone. If a paper larger than necessary is used, it requires additional washing to free the entrained soluble matter but there is nothing more frustrating than a filter paper which turns out to be too small for the job! Paper circles are available in diameters ranging from 5.5 to 15 cm; the 7, 9 and 11 cm circles will be found most useful, with the 9 cm diameter paper being the most readily adaptable.

The use of filter paper pulp is recommended by some analysts as an aid in promoting more rapid and efficient filtration. In the double precipitation of the R_2O_3 group elements its addition, as the macerated paper used in the first precipitation, is unavoidable and it does insure a more easily filtered precipitate. The indiscriminate use of paper pulp is not justified, however, because the improved filtering and ignition qualities that it imparts to the precipitate are offset by certain disadvantages. It contributes to the final weight of the ignited precipitate and, if a correction must be made for it, the weight of the paper pulp used must be known; it also presents an additional surface for the trapping of foreign ions and necessitates additional washing, which may be undesirable if the precipitate is more than slightly soluble in the wash solution. The presence of paper pulp, intimately mixed into the precipitate, may also foster the development of the undesirable reducing conditions mentioned previously, because of the difficulty of insuring that all of the organic material is charred at a low temperature. In hot acid solutions a degradation of the paper can occur, and some elements may, as a result, be reduced to lower valence numbers; the subsequent precipitation of these elements may be interfered with, or rendered incomplete. There are other ways of improving the filtering qualities of precipitates, and the use of paper pulp should be a last resort.

4.2.8. MICROCHEMICAL EQUIPMENT

The use of true microchemical methods in the analysis of minerals has received relatively little attention (see Alimarin and Frid, Supp. Ref.), and yet this is a fertile field for the application of these methods because of the scarcity of sample material that is usually available and the relatively uncomplicated nature of minerals. The mere scaling down of "macro" methods to handle centigram quantities is not microanalysis and it is to be hoped that more attention will be given to the use of these methods. For descriptions of apparatus and equipment, most of which is of simple construction, reference should be made to recent monographs by Belcher and Wilson,[3] and Von Nieuwenburg and Van Ligten;[16] the monograph by Alimarin and Petrikova[2] on ultramicroanalysis demonstrates what can be done when this type of analysis is carried to its ultimate conclusion.

4.2.9. SPECIALIZED ITEMS

There are many minor pieces of specialized equipment which serve a humble but very useful purpose and which are unknown outside of the laboratory in which they were developed or utilized. Because they may prove useful to other analysts, a few such items are described here.

Long forceps, 8–10 in. in length and made of nickel-plated steel, are a convenient means of handling weighing dishes and crucibles without contamination from the fingers, and they are not as heavy nor as clumsy as crucible tongs. Smaller forceps, 6 in. in length and fitted with platinum tips, are used to handle filter papers for transfer to crucibles. If the semicircular jaws of light wire tongs (a variety which opens when squeezed) are bent at right angles to the legs, and the wire jaws covered with thin rubber tubing, these will be found very useful for removing crucibles from desiccators and balance pans; regrettably, such simple tongs are also very difficult to obtain.

Wooden sample trays, approximately 12 in. × 8 in. × 1 in., with holes of the appropriate size drilled to a depth of about ¾ in., will keep sample vials in order and facilitate their movement in the balance room, particularly for batch analysis.   Temporary gummed labels on the trays will help to identify batches of samples at a glance.   Trays for the convenient and safe transportation of crucibles, such as the large nickel or platinum crucibles used in the preparation of solutions A and B in the "rapid" methods of analysis, may be made from two pieces of transite, approximately 14 in. × 7 in., separated by 1 in. wooden strips at the center and at each end; 12 holes, 1¾ in. in diameter, can be drilled into the upper transite layer.

In the Pratt method for the determination of ferrous iron the sample is decomposed in a platinum crucible with a mixture of sulfuric and hydrofluoric acids kept at a gentle boil for 5–10 min.   This requires a very small flame to avoid undesirable bumping or too rapid boiling and, since the decomposition must be done in a fume hood, the flame must be protected from drafts.   A conical chimney fastened to the burner will give some protection but at the Geological Survey a simple device is used which not only allows close control of the flame and protects it from drafts, but also provides a safe support for the crucible, a necessary requirement in view of the hazardous nature of its contents.   The burner housing and crucible support stand, shown in Figs. 4.3 and 4.4, consists of a small but heavy plywood box, 8 in. × 7 in. × 7 in., having one side missing; it is lined with transite and covered with a sheet of transite in which is cut a 3½ in. hole.   A silica or vitreosil triangle, large enough to support a 45 ml tall form platinum crucible, is fastened securely over this hole.   A micro gas burner, with attached chimney, is fastened to the base of the box beneath the central hole and is connected to a gas source by means of a needle valve inserted through a hole in one side of the box; a Tirrill burner, used for the initial heating of the crucible, can be connected by a Y joint to the same gas source.

A special support stand for use in the determination of total water by the Penfield method (Fig. 4.5) was designed by Mr. Serge Courville.[4] The details of the support stand are shown in Fig. 4.6.   It consists of a plywood base and tray holder which support a plastic overflow dish and metal cooling tray; the glass tube for the water determination rests, at a slight downward inclination, in slots cut into the sides and is supported just above the surface of the water–ice mixture in the cooling tray.   A transite shield is fastened to one side to serve as a heat reflector.

Fig. 4.3.   The apparatus used for the determination of ferrous iron (modified Pratt method) by potentiometric titration.   The decomposition of the sample is done in a fume hood, using the special burner housing and crucible support stand shown at the right.   Photograph by Photographic Section, Geological Survey.

4.3. Instrumentation

A survey of modern analytical instrumentation is outside the scope of this work and reference should be made to recent textbooks, both general[7,14,15] and specific,[1,5,11,12] dealing with this subject.   There may be some value, however, in describing briefly the types of instruments that are helpful, although not absolutely necessary, in the analysis of rocks and minerals.   It is still possible, given the necessary time, to do first class analytical work with little more in the way of instrumentation than a photoelectric colorimeter but when the rapidity with which the analysis can be made becomes important, then the use of more sophisticated instrumentation is indicated.   In general, the capability of an instrument to perform a particular function will exceed that of the hand or eye, and the analyst is enabled thereby to make measurements that otherwise would be impossible without lengthy separations and purifications.   Provided that such instruments are recognized as being no more than laboratory tools able to do only the jobs that they are

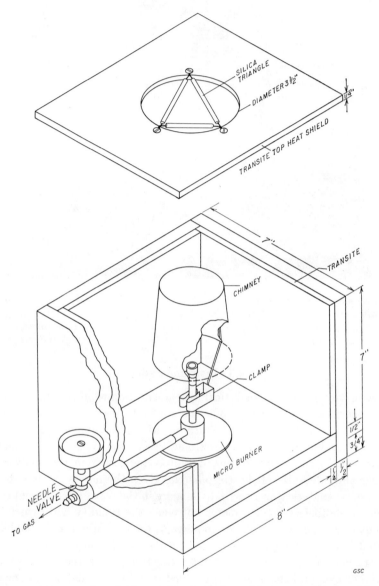

GSC

Fig. 4.4.   Diagram of special burner housing and crucible support stand used in the decomposition of the sample for the ferrous iron determination.

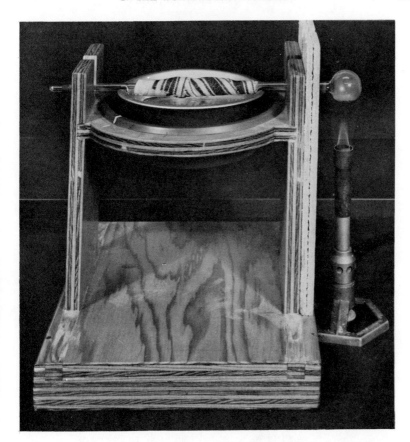

Fig. 4.5. Apparatus for the determination of total water by the Penfield method, showing the special support stand. Reproduced with permission of *The Canadian Mineralogist.*

set to do, then the modern rock and mineral analyst has available to him a host of laboratory helpers.

First and foremost among the needs of the laboratory are analytical balances, for both rough and precision weighing. There are many models available and the analyst must make his choice according to his needs, and must equate speed of weighing against such important balance characteristics as sensitivity and reproducibility (see Kolthoff and Elving, Supp. Ref.). There is much to be said for the types with built-in weights, particularly if the balances are to be used by those with little training and experience. At the Geological Survey the so-called "single pan" balance has been found reliable, durable and easy to

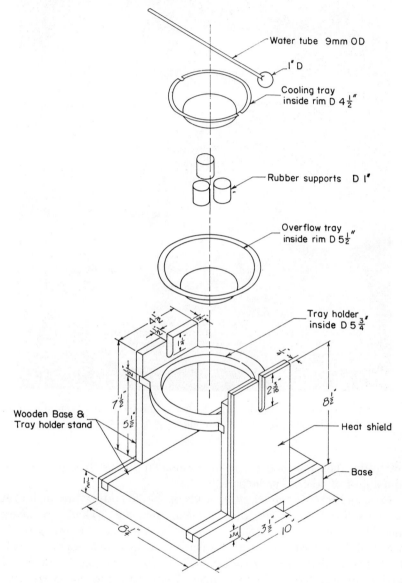

Fig. 4.6.   Diagrammatic view of special support stand shown in Fig. 4.5.   Reproduced with permission of *The Canadian Mineralogist*.

operate, even by those with little or no previous experience. Balances with a loading capacity of 200 g and a sensitivity of about 0.2 mg per scale division are used in the chemical laboratories, while those with a loading capacity of only 100 g and a sensitivity of 0.05 mg per scale division are used for the emission spectrographic work. A standard two-pan chainomatic balance, again with a capacity of about 200 g, is used when an object has to be tared. A torsion balance having a capacity of about 250 mg is a convenient type for weighing the small samples used, for example, in the X-ray fluorescence analytical procedure. A microbalance is a necessity for semimicro and microchemical work with samples weighing 100 mg or less.

For absorptiometric measurements there is need for a spectrophotometer which will enable measurements to be made routinely in the ultraviolet and visible regions, that is, in the range from about 200–1000 mμ. The available instruments range from the simple to the complex in operation and capability; there is an advantage in having a less expensive instrument for routine absorptiometric measurement and one capable of greater resolution and precision for special analysis and for method development work. A recording attachment which will provide a visual spectral record is a very useful addition to this instrument.

The Pyrex, or fused silica, cells used for absorbance measurements require careful handling in order to avoid scratching the optical surfaces and they should, of course, be kept free from surface films and fingerprints. Unless matched cells are available, however, it is unlikely that the absorbance of distilled water will be the same for each cell and it is necessary for accurate work to determine a correction for each cell, relative to one cell in which the absorbance is arbitrarily considered to be zero. Because wear is inevitable, the cells should be numbered and the relative cell correction checked each time that the cells are used. When not in use the cells should be kept immersed in a beaker of water containing a mild detergent. Before use they should be rinsed in distilled water, the outside surfaces patted dry with absorbent tissue, and the optical surfaces polished with lens paper. Once the cells are placed in the cell holder they should not be removed until the series of absorbance measurements is finished; a pipette, connected to a water aspirator pump, can be used to remove the solutions and water rinses from the cells.

Flame emission spectroscopy, which is particularly applicable to the determination of the alkalis and some of the alkaline earths, is one instrumental method which has found a place in rock and mineral

analysis. Flame attachments may be obtained for the spectrophotometers mentioned previously and, when used in conjunction with a pen
recorder, it is possible to determine traces of one alkali in the presence of
large amounts of the other alkalies. There is a definite advantage,
particularly if daily use is made of the instrument, in having a separate
flame photometer to obviate the switching back and forth between flame
attachment and lamp housing; there are a number of such instruments
available. Atomic absorption spectroscopy, a relatively recent development, is free from many of the problems that afflict flame emission
spectroscopy and has found much use as a rapid, highly sensitive and
selective technique (see Robinson, Supp. Ref.). It has had a considerable application in soil analysis but only recently has there been
significant use of it in the analysis of rocks and minerals (see Althaus,
Supp. Ref.).

A pH meter finds limited application in rock and mineral analysis for
the measurement of pH but can usefully serve as a potentiometer in
potentiometric titrations, such as in the determinations of iron and
manganese. It is only necessary to replace the glass electrode with a
bright platinum wire electrode, easily fabricated in the laboratory.
Again, there are several types available; it is preferable to have one
which has both pH and mV scales.

Automatic titrators, either for electrometric titrations alone or for
both colorimetric and electrometric titrations, find little use except in
the determination of calcium and magnesium by titration with EDTA.

Other instruments which have been used to a varying degree in the
determination of some elements are the fluorimeter and polarograph.
The fluorimeter is used for the determination of trace quantities of
uranium and beryllium; the preparation of the fused beads for final
measurement is empirical in nature and requires a special burner and
quenching apparatus. The polarograph has also been used chiefly for
trace element analysis but methods for the determination of major
constituents such as titanium and aluminium have been developed.

Optical emission spectrographic and X-ray fluorescence spectroscopic
analysis are techniques which require the use of large and expensive
instruments. They play a very important role in the qualitative and
semiquantitative determination of the major constituents, and optical
emission spectroscopy has for many years been one of the foremost
techniques for the determination of trace elements. The quantitative
determination of major constituents by emission spectroscopy (alone or
combined with chemical methods) has been mentioned (see Sec. 1.1) and
more will be said about the similar application of X-ray fluorescence

spectroscopy in a later section. Suffice it to say that these and similar techniques have become an integral part of any laboratory which must meet the increasingly detailed demands of those who require rock and mineral analyses. The optical spectrometer, be it one using a photographic process with attendant darkroom facilities and densitometers or a direct-reading instrument with its bank of electronic photocells and complex equipment for the automatic presentation of the photoelectric data in the form of dial or chart readings, will generally require an air-conditioned laboratory and much auxiliary equipment for successful operation; the X-ray fluorescence spectrometer has similar requirements. The need to invest in complex instrumentation of this type will be governed by the volume and nature of the work to be done.

4.4. Reagents

Because the analysis of a rock or mineral is a lengthy process, made up of a series of operations which are in part mutually dependent and which involve the use of many reagents, there is an unavoidable accumulation of impurities in certain steps of the analysis. Carrying a reagent blank determination through the analysis at the same time as the sample does not entirely solve the problem of determining the magnitude of these impurities. On individual determinations, such as the flame photometric determination of the alkalies or the absorptiometric determination of phosphorus, reagent blanks can be reliably employed, but in the analysis of the main portion there are serious objections to the validity of any blank obtained in this way. Fortunately, reagents of high quality are now available and this, combined with a few precautions on the part of the analyst, has eliminated reagents blanks as a significant problem.

There are two grades of reagents that are of particular importance. The first, the primary standard reagents, are of invaluable service in many ways and are supplied with a complete statement of the analysis of each bulk lot; the analyst generally can accept these statements at face value. The other type is the "analyzed reagent"; these are supplied with either an actual analysis of the bulk lot, or with a specification giving the maximum limits of impurities that may be expected. Other grades of reagent will likely require preliminary purification, depending on the use to which they are to be put, although there are numerous occasions when the use of a technical grade chemical is entirely acceptable.

Of the common reagents needed in rock and mineral analysis, only one really requires special preparation. Aqueous ammonia (ammonium hydroxide), as commercially supplied in glass bottles, has a high reagent blank and its use should be restricted to those occasions when its impurities will either be of no consequence or will be eliminated by subsequent steps in the analysis. Aqueous ammonia, in any concentration up to 15 M, is easily and readily prepared by bubbling ammonia gas through distilled and deionized water in a polyethylene bottle immersed in an ice water bath. Saturation of the solution, indicated by a steady stream of bubbles rising to the surface, occurs when the concentration is about 15 M.

Apart from chemical impurities, one must guard against solid impurities such as wood, dust and paper fibers introduced during the manufacturing processes, and solutions prepared from solid reagents should be filtered before use. At the same time, care should be taken in the use of spatulas and scoops to remove reagents from bottles, for hygroscopic reagents will attack glass and metal, and hard, caked reagents will abrade them.

Rock and mineral analysis requires the use of wash solutions such as ammonium nitrate and ammonium oxalate. These should be prepared fresh when needed and any unused portion discarded. The addition of certain reagents, such as ammonium chloride, is best made in the form of a solution but the solution should be freshly prepared and filtered. Further precautions having to do with the preparation and use of reagent solutions will be discussed in appropriate later sections.

A special warning should be given about the use of organic reagents, because their purity is usually much inferior to that of the inorganic reagents commonly used, and is often surprisingly low. It is well to check the purity of a new supply of reagent before using it (and preferably *before* the previous satisfactory supply is exhausted); because organic reagents are subject to deterioration on standing, periodic checks should be made.

For a fuller consideration of the problem of reagents reference should be made to the recent texts listed in Sec. 4.2.

References

1. Ahrens, L. H., and Taylor, S. R. *Spectrochemical Analysis*, 2nd ed., 1961. Reading, Massachusetts: Addison-Wesley Publishing Co., 454 pp.
2. Alimarin, I. P., and Petrikova, M. N. *Anorganische Ultramikroanalyse*, 1963; (translation of Neorganicheskii ul'tramikroanaliz, Moscow, 1960). New York: The Macmillan Co.
3. Belcher, R., and Wilson, C. L. *Inorganic Microanalysis*, 2nd ed., 1957. Toronto: Longmans, Green and Co., 153 pp.

4. Courville, S. Apparatus for total water determination by the Penfield method. *Canadian Mineral.* **7**, 326–329 (1962).
5. Dean, J. A. *Flame Photometry*, 1960. New York: McGraw-Hill Book Co., 353 pp.
6. Dieter, W. E., Cohen, L., and Kundick, M. E. A stainless steel fume hood for safety in use of perchloric acid. *U.S. Bur. Mines Rept. Invest.* **6473**, 1964, 12 pp.
7. Ewing, G. W. *Instrumental Methods of Chemical Analysis*, 2nd ed., 1960. New York: McGraw-Hill Book Co., 454 pp.
8. Hillebrand, W. F., Lundell, G. E. F., Bright, H. A., and Hoffman, J. I. *Applied Inorganic Analysis*, 2nd ed., 1953. New York: John Wiley & Sons. pp. 17–51.
9. Lundell, G. E. F., and Hoffman, J. I. Atomic weight of gallium. *J. Res. Natl. Bur. Std.* **15**, 409–420 (1935).
10. Kolthoff, I. M., and Sandell, E. B. *Textbook of Quantitative Inorganic Analysis*, 3rd ed., 1952. New York: The Macmillan Co. pp. 183–201, 502.
11. Liebhafsky, H. A., Pfeiffer, H. G., Winslow, E. H., and Zemany, P. D. *X-ray Absorption and Emission in Analytical Chemistry*, 1960. New York: John Wiley & Sons, 357 pp.
12. Milner, G. W. C. *The Principles and Applications of Polarography and Other Electroanalytical Processes*, 1957. New York: Longmans, Green and Co. 729 pp.
13. Shell, H. R. Determination of total iron in silicates and other nonmetallic materials. *Anal. Chem.* **22**, 326–328 (1950).
14. Stock, J. T. *Amperometric Titrations*, 1965. New York: Interscience Publishers. Chapters 6–10.
15. Strouts, C. R. N., Wilson, H. N., and Parry-Jones, R. T. *Chemical Analysis; the Working Tools*, Vol. I, (1962). Oxford: Clarendon Press. pp. 78–109.
16. Van Nieuwenburg, C. J., and Van Ligten, J. W. L. *Quantitative Chemical Micro-analysis*, 1963. New York: Elsevier Publishing Co. 181 pp.
17. Wilson, C. L., and Wilson, D. W. (Eds.). *Comprehensive Analytical Chemistry*, Vol. 1A, 1959. New York: Elsevier Publishing Co. "Materials," G. H. Wyatt, pp. 13–33; "Techniques, Reagents and Apparatus," F. E. Beamish and W. A. E. McBryde, pp. 430–545.

Supplementary References

Alimarin, I. P., and Frid, B. I. *Quantitative Mikrochemische Analyse der Mineralien und erze*, 1965. Dresden and Leipzig: Verlag Theodor Steinkopff, 372 pp.

Althaus, E. Die Atom-Absorptions-Spektraphotometrie—ein neues Hilfsmittel zur Mineralanalyse. *Neues Jahrb. Mineral. Monatsh.* 259–280 (1966).

Foner, H. A. The use and care of platinum laboratory apparatus. *Laboratory Practice*, 944–948 (1965).

Kolthoff, I. M., and Elving, P. J. (Ed.). *Treatise on Analytical Chemistry*, Part 1, Vol. 7, 1967. New York: Interscience Publishers; "Measurement of Mass" L. B. Macurdy, pp. 4247–4277.

Robinson, J. W. *Atomic absorption spectroscopy*, 1966. New York: Marcel Dekker. 204 pp.

CHAPTER 5

DECOMPOSITION OF THE SAMPLE

5.1. General

The first choice that must usually be made by the analyst is that of the best method for the decomposition of the sample, for few of these will be water-soluble. It is a selection that must be carefully made, because it sets the course of the subsequent analytical work and can save many unnecessary steps. The factors that influence the decision will be considered here and the various types of decomposition media discussed in some detail. Because there are almost as many of the latter reported in the literature as there are chemical compounds, the emphasis is placed upon the more common ones.

No general rule regarding the most advantageous choice can be given, because the latter is governed by both the nature of the sample and the nature of the analytical work which is to follow. Obviously, the method must effectively decompose the sample; in dealing with a rock sample, however, it must be borne in mind that a rock is a mixture of minerals and, depending upon the ultimate goal of the analysis, *effective* decomposition does not necessarily imply *complete* decomposition, although the careful analyst will be happier when the two are synonymous. In the determination of ferrous iron, for example, the sample is heated near the boiling point with a mixture of hydrofluoric and sulfuric acids for a fixed period of time; in most cases there remains a small residue, composed chiefly of quartz, which can be safely ignored. Continuation of the heating until the residue dissolved would not only serve no useful purpose but would, in this instance, increase the opportunity for air-oxidation of the ferrous iron. Alternatively, one may ignore a small residue if this residue will be further attacked in the course of the analysis; a small residue left after the initial fusion with Na_2CO_3 and collected with the dehydrated SiO_2 will in almost all cases be decomposed during the treatment of the impure SiO_2 with hydrofluoric acid and no appreciable error will be introduced.

The introduction of large quantities of material, which may be troublesome in subsequent stages of the analysis, should be avoided. In this respect the use of acids has a definite advantage over the use of a flux because the excess acid can usually be more readily removed. Too, fluxes are not so readily obtained in as pure form as are acids, and this is a matter of some concern when trace element analysis is contemplated.[2] In the same vein, the decomposition medium should not seriously attack the fusion vessel, with the introduction of potentially troublesome material into the analysis and accompanied, in some cases, by the retention of other material in the vessel's walls. Acids again have an advantage over fluxes in that the attack on the vessel is usually less because of the lower temperature that is employed. The advantages and disadvantages of various types of crucibles have been discussed in an earlier section (Sec. 4.2) and need not be repeated here.

The method of decomposition used should not result in the loss of constituents by volatilization when the determination of these constituents is to follow. Careful studies have been made of this and the results suggest that the analyst must be on his guard against unsuspected losses. Chapman et al.[10] state that, as a result of their study of the volatility of elements from mixed hydrofluoric and perchloric acid solutions, prediction of the volatility of many elements from mixed acid solutions is a virtual impossibility. Under their conditions Si, B, Ge, As, Sb, Cr, Se, Os, Ru, Re, and Mn could be volatilized. Hoffman and Lundell[28] added HCl or HBr to boiling (200–220°C) perchloric or sulfuric acid solutions of various metals; quantitative distillation from $HClO_4$ of Sb, As, Cr, Ge, Os, Re, Ru, and Sn, and from H_2SO_4 of Ge, As, Sb, Se, Sn and Re, was obtained, with a partial loss from both acid solutions of Bi, B, Zn, Mo, Te, and Tl. Gases formed by reaction are easily driven off, such as carbon dioxide, hydrogen sulfide and phosphine (if phosphides are present). Boric acid will be lost from a boiling aqueous solution and boiling concentrated sulfuric or perchloric acid will volatilize phosphoric acid.

The methods of decomposition can be divided into three groups: (1) decomposition by acids, both oxidizing (HNO_3, $HClO_4$, hot concentrated H_2SO_4) and non-oxidizing (HCl, HF, H_3PO_4, HBr, dilute H_2SO_4 and dilute $HClO_4$); (2) decomposition by fluxes, either by fusion with acid or alkaline compounds, or by sintering; (3) decomposition by other means, such as in bombs, in sealed tubes and by chlorination. The acid fluxes include the bisulfates, pyrosulfates and acid fluorides; the losses that occur during fusions with these fluxes result from the same general conditions that cause losses from acid solutions. The hydroxides, peroxides, carbonates and borates are the most common basic fluxes and there is less danger of loss by volatilization during fusions. At fusion temperatures, the alkaline peroxides are very powerful oxidizers. Oxidation occurs during fusion with alkali carbonates chiefly by bringing the elements into a more intimate contact with air than is possible by simple ignition; bisulfates have a more restricted oxidizing power. An additional advantage of the use of a flux is that the color of the solidified cake is often indicative of the presence of certain constituents. The elements are usually raised to their highest valence state; the most highly colored melts are those formed by the transition group elements.[52]

The specialized methods are employed usually when other methods fail, but they also have certain advantages, such as a short period of decomposition or the production of a simple, relatively salt-free solution, that commend them to more general use and outweigh the disadvantage of the specialized equipment that is required.

Many rocks and minerals are readily acid-soluble, such as carbonates, phosphates, many sulfides and sulfates, chlorides and borates; oxides are more readily dissolved by the non-oxidizing acids. Silicates which contain a high proportion of bases such as calcium are decomposed by strong acids such as HCl, but not so those of an acidic nature with much Al or Fe^{3+} present; complete decomposition is claimed, however, with mixtures of HCl, HNO_3, H_2SO_4 and $HClO_4$.[56] Hydrofluoric acid will, of course, effectively decompose most aluminosilicates. There are some minerals, such as kyanite, beryl, zircon and tourmaline, which are attacked only slightly by the common acids and fluxes at normal operating temperatures and pressures. For these, recourse is had to the more specialized methods listed previously. It is also possible, by preliminary treatment, to make a sample more amenable to decomposition by ordinary means; Hoffman et al.[26] found that kaolin is quickly decomposed by dilute HCl if it is roasted at 700°C for 15 min, whereas previously it was only slowly attacked by the acid.

Decomposition is generally expedited by using a very finely ground sample (-200 mesh) but preparation of this powder may not be entirely advantageous, as discussed in Sec. 3.4.3. Antweiler[3] has attempted to avoid the inaccuracies introduced during the crushing, grinding and sieving of rock samples by using rock fragments. He employs three methods—decomposition with HF, fusion with a mixture of Na_2CO_3 and H_2BO_3 and fusion with Na_2O_2; the first method is generally preferred.

In the brief discussion of the various types of decomposition media that follows, an effort has been made to give some guidance in the choice of a suitable acid, flux or both. Detailed discussions of the problems of decomposition are readily available[24,34,37] and these, and the other references given in this chapter, should be consulted for additional information.

5.2. Decomposition with acids

5.2.1. WITH HYDROFLUORIC ACID

The use of hydrofluoric acid is a time-honored procedure in silicate analysis and there are few silicates that are not more or less decomposed by it (see Langmyhr and Sveen, Supp. Ref.). Of these few, kyanite, axinite, beryl, zircon and some tourmalines and garnets are the most common, but under special conditions these can also be decomposed.[32] It will also dissolve many difficultly soluble oxides.[33] J. Lawrence Smith used it to decompose such earth acid minerals as samarskite, thus at the same time effecting an almost complete separation of the insoluble rare earth fluorides from the soluble earth acid and alkaline earth fluorides.[69] Minerals containing elements which form insoluble fluorides, such as magnesium and lead, may dissolve only slowly because of the shielding effect of the precipitated fluoride. Decomposition of rocks and minerals at room temperature by hydrofluoric acid in the presence of an oxidizing agent is the basis of a method for the determination of ferrous iron[73]; maximum decomposition is usually reached in less than 24 hr.

The acid used in the laboratory is approximately 48% HF (w/w), with a specific gravity and molarity of approximately 1.19 and 29.0, respectively. Caution is necessary in its use at all times; it is very poisonous and contact between the cold concentrated acid and the skin will produce irritating white spots, while the hot acid produces yellow, ulcerous sores that require immediate, special medical treatment. Vigorous flushing of the affected area with cold water is recommended as a preliminary treatment. Gloves, either rubber or plastic, should be worn when using this acid.

Langmyhr and Graff[35] have shown that no detectable loss of Si occurs when SiO_2 is decomposed by HF, even at 100°C; only a small amount is lost initially when the solution of fluorosilicic acid is evaporated to a smaller volume (0.2 mg SiO_2 is lost from 46 mg SiO_2 present when a 1:3 HF solution is evaporated from 25 to 2 ml at 100°C). This has been made the basis of an analytical scheme in which the sample is dissolved in HF alone and weight aliquots are taken for the subsequent determination of various constituents (see Sec. 1.1, Ref. 11). When solutions are taken to dryness, or to fumes of a higher-boiling acid, the loss of Si is quantitative. Other elements such as B, As, Ge and Sb form volatile fluorides also.[10] Clear solutions are seldom obtained when minerals are decomposed with HF; the residues were first thought to be undecomposed material but are now known to be precipitates formed by secondary reactions (Langmyhr and Kringstad, Supp. Ref.).

The acid should never be added to a dry powder, such as dehydrated and ignited SiO_2, because of the danger of loss due to the rapidity of the reaction. The sample should be moistened with water, the acid added cautiously and the covered container allowed to stand without additional heating until the initial reaction has taken place. This reaction may be accompanied by the generation of much heat, and highly siliceous material such as chert or glass sand will dissolve without the need for additional heating (5 g of chert (-100 mesh) will dissolve in as many seconds). It may be necessary to slow down the reaction by the addition of cold water.

Hydrofluoric acid is seldom used alone because of the need to expel the last traces of fluorine from the solution or residue, and both sulfuric and perchloric acids are used for this purpose. Perchloric acid has the advantage over sulfuric acid that almost all of the perchlorates are easily soluble; the formation of insoluble sulfates can be very troublesome in subsequent analytical work, as well as hindering the expulsion of residual fluorine. Sulfuric acid is, however, more effective in removing the latter. Failure to remove fluorine completely will prevent, for example, the subsequent quantitative precipitation of hydrous Al_2O_3; at least two, and preferably three, evaporations to dense fumes of either sulfuric or perchloric acid are necessary and the addition of a small amount of pure silica is recommended (Ref. 24, p. 850). It is sometimes convenient to add boric acid to the solution to render the residual fluorine harmless (as slightly dissociated fluoroboric acid); the complexing is not totally effective[21] and aluminum chloride has been found more reliable.

Mixtures of hydrofluoric acid with hydrochloric or nitric acids also find use. At the Geological Survey of Canada an HF–HCl mixture (containing a few drops of $HClO_4$) is used to decompose samples prior to the flame photometric determination of the alkalies; excess fluorine is expelled by heating the residue to fumes of $HClO_4$, and it is only rarely that filtration is necessary following this decomposition. Yaschenko and Varsharskaya[74] use an HF-saturated oxalic acid solution to decompose mica for subsequent determination of potassium as the perchlorate.

5.2.2. WITH SULFURIC ACID

Relatively little use is made of sulfuric acid alone to decompose rocks and minerals, chiefly because of the formation of insoluble sulfates which are often very difficult to redissolve. In the concentrated form it is oxidizing in nature and thus most oxides are not decomposed by it. Phosphates, fluorides and sulfides will dissolve in it, subject to the limitation mentioned previously; it finds preferred use, however, in the decomposition of rare earth minerals such as monazite (Schoeller and Powell,[58] p. 108).

As a mixture with other acids such as HF it finds many uses. Its high boiling point (about 340°C for the 96% acid) enables it to expel most other volatile compounds, including fluorine. Its dehydrating effect on silica is a useful step when removal of dissolved silica is desired. Ignition of the sulfates and subsequent leaching with hot water will yield a solution containing little more than the alkalies and some magnesium.[1]

The concentrated acid used in the laboratory is approximately 96% (w/w) with a specific gravity of 1.84 and a molarity of 18.0. The normal care required in its use is too well-known to require repetition here.

5.2.3 WITH HYDROCHLORIC ACID

Hydrochloric acid, as the approximately 37% (w/w), $12M$ solution having a specific gravity of 1.19, is probably the most commonly used halide acid. It dissolves most sulfides (pyrite is a notable exception), phosphates and carbonates, is the best solvent for oxides, and the resultant chlorides are, for the most part, water-soluble. Volatile chlorides are formed by Ge, Hg, Sb, As and Sn.

As mentioned in Sec. 5.1, hydrochloric acid will decompose silicates containing a high proportion of strong or moderately strong bases such as calcium, i.e., low-silica materials, to form chlorides with the cations present and liberating silica as a hydrous suspension. The effectiveness of this acid can be increased materially by its use at a temperature higher than that of its boiling point (a 31% solution has a constant boiling point of 110°C); this is accomplished in sealed tubes and few minerals resist decomposition in this manner (see Sec. 5.5.1).

5.2.4 WITH NITRIC ACID

The decomposition of sulfides, with attendant oxidation of the sulfide to sulfur or sulfate, is the most common application of strongly oxidizing nitric acid, with or without the addition of bromine, in rock and mineral analysis. Dissolution may be accompanied by separation, as in the formation of tungstic oxide or metastannic acid; antimony and molybdenum also form insoluble products. Nitric acid is generally ineffective with oxides but some thorium and uranium minerals, such as thorianite and uraninite, are dissolved with the simultaneous oxidation of U^{4+} to U^{6+}. Because of its low boiling point (86°C), excess nitric is easily expelled.

Nitric acid is more often used in conjunction with other acids. It is always used with unknown samples, or those containing organic material,

to oxidize any easily oxidizable material prior to the addition of perchloric acid. With hydrochloric acid (HNO_3 : HCl = 1 : 3) it forms
aqua regia which has, because of the presence of free, or nascent, chlorine,
a greater oxidizing power than nitric acid alone. Mercuric sulfide, for
example, is dissolved by aqua regia but not by nitric acid alone. A mixture of HF and aqua regia, or of HF and HNO_3, is effective in opening up
many minerals; Brooks[8] uses the former mixture to decompose large (40
g) samples of silicate rocks for trace element determinations.

The approximately 70% (w/w) acid, as used in the laboratory, has a
specific gravity of 1.42 and is about $16M$. The use of gloves is also
recommended in handling this acid; it can produce some unpleasant and
long-lasting yellow burns when in contact with the skin.

5.2.5. WITH PERCHLORIC ACID

The almost universal solubility of perchlorates (with the notable and
useful exception of K, Rb and Cs perchlorates) favors the use of perchloric acid for decompositions and it has replaced sulfuric acid in popu-
larity, even though the latter is often more efficient. It is one of the
strongest acids known and, while non-oxidizing in a diluted hot or cold
state, is a powerful oxidant when hot and concentrated. Mixtures of
hot concentrated sulfuric and perchloric acids have an even greater oxidizing power, equivalent to anhydrous $HClO_4$. Because of the oxidizing
power of perchloric acid, care must be exercised in its use[40,65] if serious
accidents are to be avoided. Contact of the boiling, concentrated acid
or hot vapor with either organic matter or easily oxidized inorganic
matter may lead to explosive reactions, and such reactions have been
known to occur in the cold also; the use of nitric acid (Sec. 5.2.4) as a
preliminary oxidizer before the temperature is raised above 100°C is
recommended.

Many silicate minerals can be decomposed with hot concentrated perchloric acid with subsequent separation and dehydration of the silica.
A mixture of HF and $HClO_4$ is very effective; when the sample solution
is evaporated to dryness, the excess fluorine expelled and the residue
ignited at about 550°C, thermal decomposition of the perchlorates occurs, resulting in a mixture of chlorides (alkalies), basic chlorides (alkaline earths) and oxides; hot water leaching of this residue, as described
previously for HF–H_2SO_4, will effect a useful separation.[38] Willard et
al.[72] used a distillation procedure with perchloric acid to expel residual
fluorine and at the same time to dehydrate the silica, prior to the determination of potassium as the perchlorate. Marvin and Woolaver[42] leached
the ignited perchlorates with hot water, followed by removal of the alka-

line earths with oxalate and oxine, in order to obtain the alkalies alone as chlorides; the use of beakers as weighing media for the alkali chlorides has been criticized.[23] Edge and Ahrens[14] use an HCl leach of the residue from the HF–HClO$_4$ decomposition of silicate rocks for a combined cation exchange—spectrochemical method for the determination of rare earths. The possible loss of some elements by volatilization when HF–HClO$_4$ solutions are heated must be borne in mind.[10]

A mixture of perchloric acid and phosphoric acid is a powerful solvent for some oxides, notably chromite; the reason probably lies in the conversion of the orthophosphoric acid to metaphosphoric acid.

Perchloric acid is available in both the stable 60% and 72% concentrations (w/w). The 72% acid, with a specific gravity of 1.67 and approximately 12M, boils at about 200°C, and is in general use; it can be obtained in a state of very high purity.

5.2.6. WITH OTHER ACIDS

Phosphoric acid, as the 85% (w/w) laboratory reagent (specific gravity 1.70, approximately 15M) is frequently used to decompose oxides and even chromite will be eventually dissolved. The formation of pyro- and metaphosphoric complexes at high temperature aids in dissolution. Ingamells[29] uses a mixture of phosphoric acid and sodium pyrophosphate to dissolve minerals such as biotite, garnet, chromite, olivine and bauxite, as well as Mn and Fe ores. A mixture of phosphoric and sulfuric acids is used to dissolve fused Al$_2$O$_3$[44] and chromium ores.[7]

Acetic acid is used by Epshtein and Ginberg[16] to dissolve carbonate and phosphate minerals prior to the determination of niobium. Hydrobromic acid is a good solvent for oxides and, as a mixture with HF, is used by Cuttitta[12] to prepare a solution of Mn ores from which thallium is determined by extraction as bromide; hydriodic acid is used to decompose pyrite, chalcopyrite, stibnite, and arsenic sulfides with the formation of H$_2$S, which is then determined.[48]

5.3. Decomposition by fusion

5.3.1. WITH SODIUM CARBONATE

Anhydrous Na$_2$CO$_3$ is the most commonly used of the alkaline fluxes. It will decompose silicates, oxides, sulfates, phosphates, fluorides and carbonates, to mention only the very familiar sample types, either by the formation of a definite compound (silicate, vanadate, aluminate, chromate) or by rendering the material more amenable to attack by acids. Fusions, which should be made in vessels of platinum or platinum alloy, usually require a temperature of 1000–1200°C for a flux to sample ratio of about 5:1; it is well to swirl the molten mass at some stage of the fusion process to insure complete exposure of the sample to the flux.   Usatenko et al.[68] by-passed the need for platinum ware by fusing the sample and flux, contained in a filter paper "package" resting on a channel-shaped nickel plate in a muffle furnace at 1000°C, apparently with excellent results.

The degree of oxidation that occurs during fusion with alkaline fluxes can be materially increased by the addition of a small amount of an oxidant to the flux (it follows that during these fusions care must be taken to exclude, or at least minimize, the entry into the crucible of reducing gases if a flame is used, or to maintain an oxidizing atmosphere if an electric muffle is used).   Malhotra[39] decomposed chromite with Na_2CO_3 (flux to sample ratio of 10:1) in a platinum crucible with the aid of a current of oxygen impinging on the surface of the melt from a bent platinum tube (a platinum Rose crucible would serve as well).   A current of air introduced through a hole in the crucible lid is equally effective; Fukusawa et al.[19] used this procedure to fuse carbonaceous material with Na_2CO_3.

A mixture of Na_2CO_3 and KNO_3 (1:2 to 1:4) is often used as an oxidizing alkaline flux, usually for the decomposition of sulfides, arsenides and similar reducing substances; the proportion of the KNO_3 should not be great enough to damage the platinum crucible, nor to cause too rapid a reaction with such readily oxidizable substances as arsenides.   One must not, however, treat the cooled mass with hydrochloric acid in platinum because of serious attack by the aqua regia formed. Na_2O_2 is an acceptable substitute for KNO_3 and creates no problems during dissolution of the fusion cake; it is the author's practice to mix a small amount of Na_2O_2 (approximately 0.1 g) with the Na_2CO_3 in all fusions as a precautionary measure.   Fukusawa et al. (op. cit.) use a 1:1 mixture of Na_2O_2 and Na_2CO_3 for the fusion of ilmenite. $KClO_3$ may also be used in place of either KNO_3 or Na_2O_2.

A mixture of Na_2CO_3 and $Na_2B_4O_7$ (1:1) has been used to decompose cassiterite[57] for the determination of indium; May and Grimaldi[43] used a 3:1 mixture to fuse ores and rocks prior to the determination of beryllium.

5.3.2.   WITH POTASSIUM CARBONATE

Because K_2CO_3 is more hygroscopic than Na_2CO_3 (it must be dehydrated immediately before use) it is seldom used except when the fusion products of some elements are more soluble than the corresponding sodium salts (e.g., niobium). According to Hoffman and Lundell,[27] however, the cake from a K_2CO_3 fusion of a sample containing much SiO_2 and elements which form insoluble carbonates (e.g., Ca, Pb, Mg) disintegrates more readily in hot water than does one from a fusion with Na_2CO_3.

An intimate mixture of anhydrous sodium and potassium carbonates in the ratio of their molecular weights is called "fusion mixture" and has a lower melting point than either of the single salts. This is an advantage for the fusion of samples in which chlorine or fluorine is to be determined, but is little used otherwise because of the general need to operate at temperatures much above the melting point of the mixture in order to be sure of complete decomposition of the sample. A disadvantage in the use of K_2CO_3 is the addition to the solution of a large amount of potassium salts, which tend to contaminate precipitates more readily than do those of sodium.

Polezhaev[53] used a 1:1 mixture of $KHCO_3$ and KCl (flux to sample ratio of 50:1) to decompose silicates prior to the determination of free silica; the fusion was done in a steel crucible.

5.3.3. WITH SODIUM AND POTASSIUM HYDROXIDES

Sodium and potassium hydroxides are very strong alkaline fluxes but they are used only under special circumstances. Their relative unpopularity derives partly from the fact that fusions cannot be made in platinum and recourse must be had to gold, silver, nickel (below 500°C) or, possibly, zirconium crucibles, and partly from the fact that they are not as easily obtainable in the pure state as are some of the other fluxes. There is also the disadvantage (for KOH) of the addition to the solution of a large amount of potassium salts (see Sec. 5.3.2).

These compounds tend to froth and spit when they are heated, because of their excessively hygroscopic nature. It is the usual practice to melt the flux first, allow it to cool, add the sample and then continue with the fusion. In the preparation of the sample solution used for the determination of SiO_2 and Al_2O_3 in the Shapiro and Brannock analytical scheme, 5 ml of 30% NaOH solution (approximately 1.5 g) are added to a nickel crucible and evaporated to dryness in a sand bath; the sample is then added and the fusion made. Another approach is to place separately the sample and solid NaOH on either side of the bottom of the nickel crucible; the crucible is placed over a gas burner at an angle sufficient to keep the sample and NaOH separated until the latter has melted, after which the crucible is set upright and the fusion continued.

5.3.4. WITH SODIUM PEROXIDE

Sodium peroxide, a powerful flux and oxidant, is not used much in rock analysis, chiefly because it is difficult to find a vessel in which the fusion can be made and still not seriously contaminate the analysis. It is also, because of its very hygroscopic nature, somewhat difficult to handle, but obtaining reagent grade material is no longer a problem. Fusions are made at 600–700°C and platinum is seriously attacked at this temperature, although it may be used if the crucible is first lined with a thin layer of fused sodium carbonate and the fusion is made rapidly at as low a temperature as is permissible. Silver, gold and zirconium crucibles resist attack by the fused peroxide. Usually the fusion is made in crucibles of nickel or iron, with accompanying contamination of the fusion by the constituents of the crucible. It is thus favored for analyses in which the contaminating substances are not to be determined. Muehlberg[47] used a mixture of sodium peroxide and sugar carbon for an explosive fusion that is very rapid and which attacks the nickel crucible only slightly; the very large amount of flux (15 g) required for a 0.5 g sample is a disadvantage and there is usually left a small unattacked residue which must then be fused with Na_2CO_3. Marvin and Schumb[41] studied the procedure in detail and found that the most effective mixture was Na_2O_2:C = 15:1, for a flux to sample ratio of 15:1. They also found that the residue remaining after dissolution of the fusion cake could be attributed to the spattering of sample on the lid of the crucible at the high temperature (1450°C) reached during the ignition; covering the mixture with a layer of Na_2O_2 reduced the amount of spattering. No loss in weight of the Ni crucible occurred and they successfully applied the method to a number of refractory minerals such as cassiterite, corundum, zircon and ilmenite. Theobald found that a mixture of NaOH and Na_2O_2 was more effective in the decomposition of pyrite and chromite[67] than either Na_2O_2 alone or with Na_2CO_3.

5.3.5. WITH BORAX AND BORIC OXIDE

Borax (sodium tetraborate) and boric oxide are an alkaline and acid flux, respectively, that are relatively little used at present, in spite of their having definite advantages to offer. Their neglect is probably a result of the inherent suspicion of the rock analyst towards the addition of boron to the analysis, even though its removal is a relatively simple process.

A disadvantage of borax, compared with boric oxide, is that it adds a quantity of sodium salts to the analysis, but Hillebrand et al.[24] believe it to be a better flux than boric acid if alkalies are not to be determined on the sample. Jeffery[33] has found that refractory oxides, particularly the earth acid and rare earth minerals, yield completely to borax fusion and it has been noted as satisfactory for the decomposition of zircon and chromite. He fuses a 0.5 g sample with 5 g of fused borax (intumescence of the material makes such a preliminary fusion mandatory), usually for a period of about 2 hr; dissolution of the fusion cake in HF will separate elements such as Fe, Ti, Nb, Ta and B, and evaporation of the filtrate with H_2SO_4 will remove fluorine and boron (as BF_3), leaving a pyrosulfate melt. The melt is a very viscous one and a platinum rod must be kept in the crucible to permit occasional stirring of the mass to insure complete contact between sample and flux. Lithium tetraborate is used extensively in the fusion of samples prior to pelletizing them for analysis by optical emission and x-ray fluorescence spectroscopy and has been proposed as a general flux prior to chemical analysis; a sodium fluoride–boric acid mixture is said to be more effective (Biskupsky, Supp. Ref.). The use of lithium metaborate ($LiBO_2$) has already been discussed (see Sec. 1.1 Refs. 9 and 20).

Boric oxide is a high temperature flux that is effective with those minerals that succumb also to borax fusion. It is prepared by slowly fusing boric acid in a platinum crucible and chilling the crucible rapidly in order to shatter the cooled cake. Fusions with boric oxide are done at 1000–1100°C and again the crucible must be cooled quickly in order to loosen the cake. As in the case of borax, the boron may be removed by volatilization as methyl borate.

Hillebrand et al.[24] consider the necessity for complete removal of the boron a serious liability in the use of these fluxes; they also warn that the volatilization of sodium borate is a serious possibility at the high temperature needed for fusion and when alkalies are to be determined, this possible loss must be considered.

Mention should be made here of the disagreement over the use of a boric acid fusion for the determination of SiO_2 in fluorine-bearing samples. In contrast to previous claims, Hoffman and Lundell[27] found that losses of silica occur when the sample is either fused with boric oxide, or with Na_2CO_3 and followed by evaporation of the H_2SO_4 or $HClO_4$ solution in the presence of added boric acid. Further, the silica is contaminated with boron and while the latter is not completely expelled by treatment with methyl chloride, it is expelled as BF_3 during the HF treatment of the impure silica, giving high results. A similar unsatisfactory result was obtained when H_3AsO_3 was used as a flux.

5.3.6. WITH POTASSIUM PYROSULFATE

Under this heading, consideration will be given also to the use of the bisulfate and to sodium pyrosulfate

Initial heating of the bisulfate will result in the loss of a molecule of water with the formation of the pyrosulfate. Because it is this latter compound that is the effective fluxing agent, and because the bisulfates are prone to frothing and spitting during heating, the bisulfate is seldom used now (but see Harden and Tooms, Supp. Ref.), unless it is for the preparation of the pyrosulfate because a commercial supply of the latter flux is not available.

Potassium pyrosulfate is a more effective flux than is sodium pyrosulfate because the latter tends to lose SO_3 more readily and thus reduces its effectiveness (once the pyrosulfate has been converted to the normal sulfate it is without effect on undecomposed sample). Again, the sodium compounds tend to be more soluble than those formed by potassium but this advantage is offset by the tendency of the melt to crust over more readily, thus making it difficult to see when decomposition of the sample is complete. Potassium pyrosulfate is much more widely used, and finds particular use for the fusion of the ignited oxides of the R_2O_3 group. The pyrosulfates are particularly effective with oxides and less so with silicates (quartz and opal are insoluble in sodium pyrosulfate and fusion with this flux is used to separate them from the silicates in a rock[4]).

Potassium pyrosulfate does not attack porcelain or silica vessels and fusions can be made in these when the introduction of platinum into the analysis must be avoided (this will occur even at a relatively low temperature). Platinum is significantly attacked when the fusion is made at a high temperature (medium to bright red heat) and this flux is effectively used to remove stains from crucibles; 2–3 mg of platinum may be dissolved during a prolonged fusion. Schoeller and Powell[58] obtained complete fusion of the finely divided ignited R_2O_3 oxides in 5–10 min; in the author's laboratories the bulk of the ignited oxides are transferred from the platinum crucible to one of Vycor in which the main fusion is to be done, and a brief fusion (1–2 min) suffices to fuse the small amount of residue clinging to the platinum crucible. Prolonged fusion will result in the formation of the normal sulfate and the crusting over of the fused mass; the flux can be regenerated by the addition of a drop or two of H_2SO_4 (1:1) to the cooled mass, followed by gentle heating to a molten state. Hillebrand et al.[24] point out that this is often necessary toward the end of a prolonged fusion in order to convert certain sulfates, which may have been formed during the conversion of the pyrosulfate to normal sulfate, to a more soluble form.

It is often useful, following the decomposition of a sample with hydro-fluoric acid, to add sulfuric acid and expel the fluorine by heating to fumes, and finally to fuse the residue with potassium pyrosulfate.[63]

5.3.7. WITH ALKALINE FLUORIDES

These (KF, KHF_2, NH_4F) are low-temperature fluxes and are used chiefly for the opening up of refractory silicates and oxides, such as some niobium, tantalum, beryllium and zirconium minerals, through the formation of soluble fluoride complexes. Although the temperature is lower, the time of fusion is usually shorter than that required for other fusions, and of course, platinum vessels must be used. As mentioned in Sec. 5.1, the losses that occur here and during fusion with other acid fluxes result from the same conditions that cause losses from acid solu-tions, i. e., the formation of volatile fluorides. Such fusions should be carried out in a hood because of the loss of HF that occurs during the heating.

Zircon can be decomposed by fusion with KHF_2 alone at 800–900°C[36] but in our laboratories it has been found useful to begin the attack with a mixture of KF and HF; the crucible is gently heated until initial frothing has ceased, then heated at 800–900°C for about 90 sec to insure complete decomposition of the zircon. Athavale et al.[5] used HF and KHF_2 for the decomposition of low-grade uranium ores.

NH_4F has the advantage of easy removal of the excess flux by volatili-zation after decomposition is complete.[59]   It has been used to decompose beryl and tourmaline[62] at 300–400°C, combined with subsequent heating with oxalic acid.

5.3.8. WITH OTHER FLUXES

There are numerous other fluxes, generally mixtures of two or more compounds, to be found in the literature and nothing is gained by at-tempting to catalog all of them. Mention is made of only a few to give some idea of the variety that has been used for special purposes.

A mixture of ZnO and Na_2CO_3 is used frequently[22,60] for the decompo-sition of silicates prior to the separation and determination of fluorine. Hillebrand et al. (Ref. 24, p. 836) mention the use of lead oxide and carbonate, and basic bismuth nitrate, citing as advantages their ready removal after the fusion and the opportunity of determining the alkalies in the fused portion. Crystalline pyrophosphoric acid, heated at 250°C, is used to separated free and combined silica.[9] Ammonium halides and nitrate were used by Isakov[31] to decompose minerals and ores; he found that a mixture of NH_4Cl and NH_4NO_3 (solid aqua regia) was most effective.

5.4. Decomposition by sintering

5.4.1. WITH SODIUM PEROXIDE

Fusion with Na_2O_2, and the disadvantages thereof, has been discussed in Sec. 5.3.4. Rafter[55] has proposed a sintering procedure that will, at a comparatively low temperature, easily and completely decompose those minerals resistant to decomposition by the usual methods, without the slightest attack on a platinum crucible, and very slight attack on iron or nickel vessels. He found it particularly useful in the decomposition of earth-acid and rare-earth minerals; Ta_2O_5 and tantalum minerals have been decomposed completely and the cake dissolved in cold water.

Control of the temperature and time of the sintering is very important. The finely ground sample (0.2–1 g) is sintered for 7 min with 1.2–3.0 g Na_2O_2 in a platinum crucible at 480 ± 10°C.

5.4.2. WITH SODIUM CARBONATE

In speaking of modified classical techniques in silicate analysis, R. C. Chirnside states that "it is quite extraordinary that the most significant advance in silicate analysis which has taken place as long ago as 1930 has been ignored for years by most analysts in this field" (Ref. 11, p. 11 T). He was referring to the work of Finn and Klekotka,[18] who found that many aluminosilicates were readily decomposed by sintering them with a small amount of Na_2CO_3 (0.6 g sodium carbonate for 0.5 g sample, sintered at 875°C for 2 hr). The reduction in the amount of sodium salts thus introduced into the analysis is an obvious advantage. Hoffman[25] continued this investigation and applied it to various rocks and minerals, but used a 2:1 ratio of flux to sample and obtained a fused cake rather than a sinter. The mixture, 0.5 g 60-mesh sample and 1.0 g Na_2CO_3, is heated in a 75 ml platinum dish for 15 min at 1200°C in an electric muffle furnace. Subsequent steps in the separation of silica are carried out in the same dish. The procedure has much to recommend it to the analyst.

Ruby and sapphire powders have been sintered in a platinum crucible with a flux consisting of anhydrous Na_2CO_3 and anhydrous $Na_2B_4O_7$, prior to the determination of iron (about 0.002 %). Only 50 mg of flux and 20 mg of sample were used and the mixture was heated in an electric muffle furnace at 1050°C for 10 min. The lower temperature and shorter heating time required minimized the exchange of iron between the melt and the crucible (Chirnside et al. Supp. Ref.).

A mixture of ZnO and Na_2CO_3 was used as a sintering medium by Moriya[45] for the separation and determination of boron in silicates (10 min at 900°C); Peck and Tomasi[51] sintered silicates with a mixture of $ZnO + Na_2CO_3 + MgCO_3$ at 800°C for 30 min for the subsequent determination of chlorine. Other mixtures that are proposed are $Na_2CO_3 + MgO + KNO_3$ for the determination of ferric iron in pyritic residues,[6] $Na_2CO_3 +$ oxalic acid as a general mixture for silicate decomposition[15] and $Na_2CO_3 +$ potassium oxalate for the decomposition of iron ores.[54]

5.4.3. WITH CALCIUM CARBONATE AND AMMONIUM CHLORIDE

This mixture, usually consisting of 0.5 g NH_4Cl and 4 g $CaCO_3$ for a 0.5 g sample, is that commonly used in the J. Lawrence Smith procedure for the determination of the alkali metals in silicates. This procedure, in which the silicates are decomposed by sintering them with the calcium oxide and calcium chloride formed from the thermal decomposition of the calcium carbonate and its reaction with the ammonium chloride, replaced for some years the Berzelius procedure which began with decomposition of the sample by a mixture of hydrofluoric and sulfuric acids. In recent years, there has been a return to this latter method, which lends itself readily to the flame photometric determination of the alkali metals.

The complete separation of the alkali metals as chlorides is not achieved. A small amount of magnesium will likely be present in the final residue and under some conditions a small amount of calcium may be present also. Conversely, not all the lithium will be present, although Shneider[61] found that this sintering procedure, rather than the Berzelius decomposition, gave accurate results for Li_2O in spodumene.

There have been many changes suggested for the sintering mixture, ranging from the elimination of the NH_4Cl to the use of $BaCO_3$ instead of $CaCO_3$. Stevens[66] suggested the substitution of $BaCl_2 \cdot 2H_2O$ for the NH_4Cl, in order to remove sulfates both present and formed during the ignition and prevent them from accompanying the final chloride residue. Its use also eliminates the need for slow initial heating required to expel the NH_3 formed during the sintering reaction. Esikov et al.[17] fused those samples having more than 0.5% Rb with $CaCl_2$ alone in a graphite crucible for the flame photometric determination of Rb and Sr.

Simon and Grimaldi[64] used a mixture of $CaCl_2 \cdot H_2O + CaO + MgO$ for decomposing mineralized rocks and molybdenites prior to the determination of rhenium. A mixture of CaO and KNO_3 has been used to decompose various sulfides in which germanium was to be determined.[13]

5.5. Decomposition by other methods

5.5.1. By sealed tubes

The acid attack on difficultly soluble minerals at high pressures and temperatures in sealed glass tubes is a method that has found relatively little use in rock and mineral analysis, although the solution and oxidation of organic compounds is often done in this way. The lack of attention to this method which, in addition to its ability to solubilize many difficultly soluble materials, yields a solution free of added salts and having only minimum contamination, is likely owing to the specialized equipment seemingly necessary and the somewhat hazardous aspect of the procedure. In point of fact, the equipment needed is not unduly specialized and the hazards can be confined to loss of sample only.

Originally the sample was placed in a platinum tube which in turn was placed inside a glass tube and the latter sealed, both tubes being filled with dilute HCl; the tubes were heated inside a steel tube which contained a volatile organic substance to equalize the initial pressure developed by the heated acid. Hillebrand et al. (Ref. 24, p. 850) have pointed out the difficulties of this procedure, the chief of which is the incomplete decomposition of the sample; a rather large amount (several milligrams) of platinum is also dissolved.

The only related work of recent years is that of Wichers, Schlecht, and Gordon who applied the method first to the attack of refractory platiniferous material[70] and then to the decomposition of refractory oxides, ceramics and minerals.[71] The special techniques and apparatus involved in the use of the sealed glass tube are discussed in detail[20] and the whole has been summarized in a recent work (Ref. 34, p. 1038). Briefly, a 0.1 g sample is placed in a borosilicate tube, 4 mm i.d. and 2 mm wall thickness, 20 cm in length and containing 1 ml HCl, and heated at 250–300°C for periods of 16–48 hr, without external protection; larger samples are heated in tubes having an internal diameter of about 15 mm and enclosed in a steel tube containing CO_2 to provide an external pressure similar to the internal one generated in the glass tube. Among the minerals successfully decomposed in this way were cassiterite, muscovite, amphibole, tourmaline, cordierite, sillimanite, spinel, bauxite and chromite; talc, diaspore, allanite, garnet and sphene were almost completely attacked, while partial attack was obtained for beryl, topaz and dumortierite. Of the plagioclase feldspars, only anorthite was completely decomposed; bytownite and andesine were partially attacked, albite and oligoclase not at all. Two acid rocks (andesite and rhyolite) were tried and found to be only partially attacked, even after 2–3 days of heating at 300°C.

5.5.2. BY BOMBS

The use of bombs is more widespread than the use of sealed glass tubes, particularly that of the Parr peroxide bomb.

Parr[49] first described the extension of his calorimeter bomb to the decomposition of samples (0.25–0.50 g) for analytical purposes in 1908; he obtained excellent decomposition of galena, sphalerite and pyrite, and also of shale, fireclay and titanium ores. Since then, many new applications have been described and bombs have been developed in sizes from 2.5 ml to 42 ml, thus accomodating micro, semimicro and macro samples. The method involves the ignition, by an electric hot wire or flame-induced hot spot, of a mixture of sample, Na_2O_2 and an accelerator (commonly $KClO_4$) in a nickel cup enclosed in a steel bomb; the reaction, which usually requires less than one minute, results in the formation of a molten mass which quickly solidifies to a cake containing the sodium salts of most of the elements in the sample, plus excess sodium oxide and hydroxide. For a description of the apparatus and operating procedures, the appropriate instruction manual should be consulted.[50]

Ingles[30] used an electric ignition bomb for the decomposition of refractory uranium ores, such as euxenite, by the Meuhlberg method[47] in order to reduce the loss of sample during the explosive reaction and to minimize the danger inherent in this method. He found this to be a useful general method for decomposing ores. He also found that a ratio of Na_2O_2 to sugar carbon of 15:1 resulted in a higher temperature of fusion with more effective attack on the ore; even with the higher temperature no evidence of distortion or attack of the bombs was found.

A very promising method is that devised by Ito,[32] who used a mixture of 1:1 H_2SO_4 and 48% HF in a steel bomb fitted with a Teflon liner to decompose a number of refractory minerals for the determination of ferrous iron and the alkalies. He used a finely powdered 0.5 g sample, and heated the bomb at 240°C for 3–4 hr. Tourmaline, kornerupine, staurolite, garnet, chrysoberyl, axinite, magnetite, ilmenite, chromite, tantalite, columbite, baddeleyite, rutile and corundum are totally decomposed; pyrite is only partly decomposed, as is zircon, but complete decomposition of the latter is achieved after 10 hr of heating with HF alone at 240°C. Contamination from the steel bomb is negligible; the Teflon liner is attacked slowly. Riley (see Sec. 9.8.2) used a Teflon bomb, heated at 150°C, to decompose refractory minerals for the preparation of solution B. Langmyhr and Graff (Sec. 1.1 Ref. 11) and May and

Rowe (Supp. Ref.) have described Teflon- and platinum-lined bombs for the decomposition of samples prior to the determination of SiO_2 in the resulting solution.

5.5.3. By chlorination

It is appropriate to conclude this brief summary of methods for the decomposition of samples with reference to an elegant method that is deserving of more attention than it has received. This is the use of chlorine, passed over a sample which is heated to 250–300°C, to separate the constituents by selective volatilization of their chlorides. Some early use was made of this technique for the decomposition of iron meteorites. Recent work[46] has shown that, provided the chlorine used is dry and free from HCl, no attack of feldspar, olivine, pyroxene, apatite, magnetite or chromite occurs, and thus it is particularly useful in the separation of the metal and sulfide phases from the silicate phase of siderites. The sample, which need not always be pulverized, is placed in a porcelain boat inside a special Pyrex reaction tube; the chlorine is passed rapidly over it and the reaction tube is heated until the reaction begins ($FeCl_3$ sublimes), the temperature finally being raised to 350°C and maintained for at least 1 hr. After cooling, the residue in the boat is leached with water and the residue recovered. The water leach of the residue will contain Ni, Co, Cr and Cu (and traces of Fe); Fe, S, P and Ge (Ga?) distill quantitatively. A mixture of sulfur dichloride and chlorine has been used to decompose tantalum, niobium, titanium and rare earth multiple-oxide minerals (Butler and Hall, Supp. Ref.).

References

1. Abbey, S., and Maxwell, J. A. Determination of potassium in mica: a flame-photometric study. *Chem. Can.* **12**, 37–41, (1960).
2. Ahrens, L. H., Edge, R. A., and Brooks, R. R. Investigations on the development of a scheme of silicate analysis based principally on spectrographic and ion exchange techniques. *Anal. Chim. Acta* **28**, 551–573 (1963).
3. Antweiler, J. C. Methods for decomposing samples of silicate rock fragments, *U.S. Geol. Surv. Profess. Papers* **424-B**, 322–324 (1961).
4. Asta'vfev, V. P. Determination of quartz and opal in rocks. *Akad. Nauk SSSR* **1958**, 51–53; *Ref. Zh. Khim.* **1959** (10), Abstr. 34, p. 630.
5. Athavale, V. T., Patkar, A. J., and Rao, B. L. Modified rapid procedure for the determination of uranium in low-grade ores. *J. Sci. Ind. Res. (India)* **B21(5)**, 231–233 (1962).
6. Balyuk, S. T. Rapid method for the determination of ferric oxide in pyritic residues. *Ogneupory* **1960** (12), 576–577; *Ref. Zh. Khim.* **1961**(11), Abstr. 11D120.

7. Brežný, B. Determination of FeO in chromium ore and in chromium magnesite. *Hutnicke Listy* **15**, (7), 552–554 (1960).

8. Brooks, R. R. Dissolution of silicate rocks. *Nature* **185**, 837–838 (1960).

9. Bulycheva, A. I., and Mil'nikova, P. A. Determination of free silica in the presence of silicates with pyrophosphoric acid. *Akad. Nauk. SSSR* **1958**, 23–32; *Ref. Zh. Khim.* **1959** (11), Abstr. 38, p. 328.

10. Chapman, F. W., Marvin, G. G., and Tyree, J. Y. Volatilization of elements from perchloric and hydrofluoric acid solutions. *Anal. Chem.* **21**, 700 (1949).

11. Chirnside, R. C. Silicate analysis: a review. *J. Soc. Glass Technol.* **18**, No. 210, 5T–29T, (1959).

12. Cuttitta, F. Dithizone mixed-color method for determining small amounts of thallium in manganese ores. *U.S. Geol. Surv. Profess. Papers* **424**–C, 384–385 (1961).

13. Dekhtrikyan, S. A. Determination of germanium in molybdenites and other sulphides. *Dokl. Akad. Nauk Arm. SSSR* **28**, (5), 213–216 (1959); *Ref. Zh. Khim.* **1960** (8), Abstr. 30, p. 469.

14. Edge, R. A., and Ahrens, L. H. The determination of Sc, Y, Nd, Ce and La in silicate rocks by a combined cation exchange–spectrochemical method. *Anal. Chim. Acta* **26**, 355–362 (1962).

15. Efros, S. M., and Bilik, O. Ya. Checking the method of sintering for decomposition of silicates. *Sb. Stud. Rabot Leningr. Tekhnol. Inst. Lensoveta, Leningr.* **1956**, 13–17; through *Chem. Abstr.*, **54**, 8434ᵃ (1960).

16. Epshtein, R. Ya, and Ginberg, G. P. Spectrophotometric determination of niobium in carbonate minerals. *Tr. Nauch. Issled. Inst. Geol. Arktiki* **119**, 84–90; (1961) *Ref. Zh. Khim.* **1961** (19), Abstr. 19D61.

17. Esikov, A. D., Beschastnova, G. S., and Yakovlev, G. N. Flame photometric determination of rubidium and strontium. *Byull. Komiss. Opredel. Absolyut. Vozrasta Geol. Formats. Akad. Nauk SSSR* **1962** (5), 76–81; *Ref. Zh. Khim.* **1963**, 19GDE, (3), Abstr. 3G47; through *Anal. Abstr.* **10**, 3088 (1963).

18. Finn, A. N., and Klekotka, J. F. On a modified method for decomposing aluminous silicates for chemical analysis. *J. Res. Natl. Bur. Std.* **4**, 809–813 (1930).

19. Fukasawa, T., Takabayashi, Y., and Hirano, S. Photometric determination of vanadium in various materials by the sodium diphenylaminesulphonate method. *Japan Analyst* **8**, 292–298 (1959).

20. Gordon, C. L., Schlecht, W. G., and Wichers, E. Use of sealed tubes for the preparation of samples for analysis, or for small-scale refining: pressures of acids heated above 100°C. *J. Res. Natl. Bur. Std.* **33**, 457–470 (1944).

21. Graff, P. R., and Langmyhr, F. J. Studies in the spectrophotometric determination of silica in materials decomposed by hydrofluoric acid. II. Spectrophotometric determination of fluosilicic acid in hydrofluoric acid. *Anal. Chim. Acta* **21**, 429–431 (1959).

22. Grimaldi, F. S., Ingram, B., and Cuttitta, F. Determination of small and large amounts of fluorine in rocks. *Anal. Chem.* **27**, 919–921 (1955).

23. Hall, W. T. Thermal decomposition of perchlorates and determination of sodium and potassium in silicates. *Ind. Eng. Chem. Anal. Ed.* **17**, 816 (1945).

24. Hillebrand, W. F., Lundell, G. E. F., Bright, H. A., and Hoffman, J. I. *Applied Inorganic Analysis*, 2nd ed. (1953). New York; John Wiley and Sons, pp. 835–851.

25. Hoffman, J. I. Decomposition of rocks and ceramic material with a small amount of sodium carbonate. *J. Res. Natl. Bur. Std.* **25**, 379–383 (1940).
26. Hoffman, J. I., Leslie, R. T., Caul, H. J., Clark, L. J., and Hoffman, J. D. Development of a hydrochloric acid process for the production of alumina from clay. *J. Res. Natl. Bur. Std.*, **37**, 409–428 (1946).
27. Hoffman, J. I., and Lundell, G. E. F. Determination of fluorine and of silica in glasses and enamels containing fluorine. *J. Res. Natl. Bur. Std.* **3**, 581–595 (1929).
28. Hoffman, J. I., and Lundell, G. E. F. Volatilization of metallic compounds from solutions in perchloric or sulfuric acid. *J. Res. Natl. Bur. Std.* **22**, 465–470 (1939).
29. Ingamells, C. O. Titrimetric determination of manganese following nitric acid oxidation in the presence of pyrophosphate. *Talanta* **2**, 171–175 (1959).
30. Ingles, J. C. The application of the electric ignition bomb to the sugar carbon–sodium peroxide method for dissloving refractory ores. *Topical Rept.* TR–131/55, 1955, 12 pp: Radioactivity Division, Mines Branch, (Canada).
31. Isakov, P. M. Decomposition of minerals and ores by ammonium salts. *Vestn. Leningr. Univ., Ser. Biol. Geogr. i. Geol.* **1955**, 117–123, through *Chem. Abstr.* **49**, 15605[i] (1955).
32. Ito, J. A new method of decomposition for refractory minerals and its application to the determination of ferrous iron and alkalies. *Bull. Chem. Soc. Japan* **35**, 225–228 (1962).
33. Jeffery, P. G. Decomposition of oxide minerals by fusion with borax. *Analyst* **82**, 66–67, (1957).
34. Kolthoff, I. M., and Elving, P. J. (Eds.). *Treatise on Analytical Chemistry*, Part I, vol. 2, 1961. New York: Interscience, pp. 1027–1050.
35. Langmyhr, F. J., and Graff, P. R. Studies in the spectrophotometric determination of silicon in materials decomposed by hydrofluoric acid. I. Loss of silicon by decomposition with hydrofluoric acid. *Anal. Chim. Acta* **21**, 334–339 (1959).
36. Luk'yanov, V. F., Savvin, S. B., and Nikol'skaya, I. V. Photometric determination of thorium in zircon by means of the new reagent Arsenazo III. *Zavodsk. Lab.* **25**, 1155–1157 (1959).
37. Lundell, G. E. F., and Hoffman, J. I. *Outlines of Chemical Analysis*, 1938. New York: John Wiley & Sons, pp. 24–29.
38. Lundell, G. E. F., and Knowles, H. B. The analysis of soda–lime glass. *J. Am. Ceram. Soc.* **10**, 829–849 (1927).
39. Malhotra, P. D. A modified method for the decomposition of chromite. *Analyst* **79**, 785–786 (1954).
40. Manufacturing Chemists' Association, Inc., General Safety Committee. *Guide for Safety in the Chemical Laboratory*, 1954. Princeton: D. Van Nostrand, pp. 117–119.
41. Marvin, G. G., and Schumb, W. C. The Na_2O_2–carbon fuison for the decomposition of refractories. *J. Am. Chem. Soc.* **52**, 574–580 (1930).
42. Marvin, G. G., and Woolaver, L. B. Determination of sodium and potassium in silicates. *Ind. Eng. Chem. Anal. Ed.* **17**, 554–556 (1945).
43. May, I., and Grimaldi, F. S. Determination of beryllium in ores and rocks by a dilution—fluorimetric method with morin. *Anal. Chem.* **33**, 1251–1253 (1961).

44. Mendliva, N. G., Novaselova, A. A., and Rychkov, R. S. Dissolution of fused Al₂O₃ and the determination of impurities. *Zavodsk. Lab.* **25,** 1293–1294 (1959).

45. Moriya, Y. Separation of boron in silicate by sintering with zinc oxide and sodium carbonate. *Japan Analyst* **8,** 667–671 (1959).

46. Moss, A. A., Hey, M. H., and Bothwell, D. I. Methods for the chemical analysis of meteorites. I. Siderites. *Mineral. Mag.* **32,** 802–816 (1961). But see Moss et al. Supp. Ref.

47. Muehlberg, W. F. An explosive method for peroxide fusions. *Ind. Eng. Chem.* **17,** 690–691 (1925).

48. Murthy, A. R. V., and Sharada, K. Determination of sulfide sulfur in minerals. *Analyst* **85,** 299–300 (1960).

49. Parr, S. W. Sodium peroxide in certain quantitative processes. *J. Am. Chem. Soc.* **30,** 764–770 (1908).

50. Parr Instrument Company. Peroxide Bomb Apparatus and Methods. *Parr Manual* **121,** 47 pp. (1950).

51. Peck, L. C., and Tomasi, E. J. Determination of chlorine in silicate rocks. *Anal. Chem.,* **31,** 2024–2026 (1959).

52. Pittwell, L. R. Color of fusion melts of the elements. *Chemist-Analyst* **51,** 94, 96 (1962).

53. Polezhaev, N. G. New method for determining free silica in the presence of silicates. *Akad. Nauk SSSR* **1958,** 33–43 ; *Ref. Zh. Khim.* **1959** 10, Abstr. 34620.

54. Quadrat, O. Solution of iron ores sparingly soluble in acids. *Chem. Anal. Warsaw* **4,** 405–409 (1959); through *Chem. Abstr.* **53,** 16803ᵃ (1959).

55. Rafter, T. A. Sodium peroxide decomposition of minerals in platinum vessels. *Analyst* **75,** 485–492 (1950).

56. Rossól, S. A new method for dissolving aluminosilicates. *Chem. Anal. Warsaw* **3,** 865–869 (1958); through *Chem. Abstr.* **53,** 19671ᵃ (1959).

57. Rozbianskaya, A. A. Determination of indium in cassiterite. *Tr. Inst. Mineral., Geokhim., i. Kristallochim. Redk. Elementov Akad. Nauk SSSR* 1961 (6), 138–141; *Ref. Zh. Khim.* **1962** (15), Abstr. 15D 54.

58. Schoeller, W. R., and Powell, A. R. *The Analysis of Minerals and Ores of the Rarer Elements,* 3rd ed., revised, 1955. New York: Hafner, pp. 2–4.

59. Shead, A. C., and Smith, G. F. The decomposition of refractory silicates by fused NH₄F and its application to the determination of silica in glass sands. *J. Am. Chem. Soc.* **53,** 483–486 (1931).

60. Shell, H. R., and Craig, R. L. Determination of silica and fluorine in fluorsilicates. *Anal. Chem.* **26,** 996–1001 (1954).

61. Shneider, L. A. Determination of lithium in spodumene ores. *Obogashchenie Rud* **3,** 31–42 (1958); *Ref. Zh. Khim.* **1959** (9), Abstr. 30, p. 971.

62. Shukolyukov, Yu A., and Matveeva, I. I. Determination of small amounts of potassium by an isotope-dilution method. *Zh. Anal. Khim.,* **16,** 544–548 (1961).

63. Sill, C. W., and Willis, C. P. Beryllium content of meteorites. *Geochim. Cosmochim. Acta* **26,** 1209–1214 (1962).

64. Simon, F. O., and Grimaldi, F. S. Spectrophotometric catalytic determination of small amounts of rhenium in mineralized rocks and molybdenites. *Anal. Chem.* **34,** 1361–1364 (1962).

65. Smith, G. F. The dualistic and versatile reaction properties of perchloric acid. *Analyst* **80,** 16–29 (1955).

66. Stevens, R. E.   Extraction of alkalies in rocks.   *Ind. Eng. Chem. Anal. Ed.* **12**, 413–415 (1940).
67. Theobald, L. S.   Attack by fusion of pyrites and chromite.   *Analyst* **67**, 287–288 (1942).
68. Usatenko, Yu. I., and Bulakhova, P. A.   Rapid method for the decomposition of silicates.   *Zavodsk. Lab.* **16**, 745–746 (1950); through *Chem. Abstr.* **49** 12197° (1955).
69. Wells, R. C.   Note on the J. Lawrence Smith method for the analysis of samarskite.   *J. Am. Chem. Soc.* **50**, 1017–1022 (1928).
70. Wichers, E., Schlecht, W. G., and Gordon, C. L.   Attack of refractory platiniferous materials by acid mixtures at elevated temperatures.   *J. Res. Natl. Bur. Std.*, **33**, 363–381 (1944).
71. Wichers, E., Schlecht, W. G., and Gordon, C. L.   Preparing refractory oxides, silicates and ceramic materials for analysis by heating with acids in sealed tubes at elevated temperatures.   *J. Res. Natl. Bur. Std.* **33**, 451–456 (1944).
72. Willard, H. H., Liggett, L. M., and Diehl, H.   Direct determination of potassium in a silicate rock.   *Ind. Eng. Chem. Anal. Ed.* **14**, 234 (1942).
73. Wilson, A. D.   VI. A new method for the determination of ferrous iron in rocks and minerals.   *Bull. Geol. Survey Gr. Brit.* **9**, 56–58 (1955).
74. Yashchenko, M. L., and Varsharskaya, E. S.   Determination of potassium in mica by the perchlorate method.   *Zavodsk. Lab.* **26**, 275–276 (1960).

Supplementary References

Biskupsky, V. S.   Fast and complete decomposition of rocks, refractory silicates and minerals.   *Anal. Chim. Acta* **33**, 333–334 (1965).   Samples are fused with a 2:3 mixture of boric acid and lithium fluoride in a platinum crucible at 800–850°C, and the melt dissolved in dilute hydrochloric or nitric acids.

Biskupsky, V. S.   Photometric determination of gallium via a sodium fluoride–boric acid fusion.   *Chemist–Analyst* **56**, 49–51 (1967).   A 0.25 g sample of mineral is fused in a platinum crucible with 0.5 g boric acid and 1.5 g NaF.

Butler, J. R., and Hall, R. A.   The separation of total rare earths and thorium from some multiple-oxide minerals.   *Analyst* **85**, 149–150 (1960).   Chlorine is bubbled through a flask of sulfur dichloride at 40–50°C and passed over the sample contained in a silica boat in a silica tube which is gradually heated to 500°C and maintained at this temperature for 1 hr.   Ti, Nb, Ta and Fe volatilize, and uranium divides itself between sublimate and residue.

Chirnside, R. C., Cluley, H. J., Powell, R. J., and Proffitt, P. M. C.   The determination of small amounts of iron and chromium in sapphire and ruby for maser applications.   *Analyst* **88**, 851–863 (1963).   For the determination of Cr the sample is fused with a 2:1 mixture of anhydrous Na_2CO_3 and anhydrous $Na_2B_4O_7$ in a platinum crucible, with flux-sample ratio of 50:1 to 10:1; for Fe, in order to minimize exchange of Fe between crucible and melt, the sample is sintered with above flux but with a flux-sample ratio of only 2.5:1.

Govindaraju, K.   Ion-exchange dissolution method for silicate analysis.   *Anal. Chem.*, **40**, 24–26 (1968).   The sample is fused with K_2CO_3 and H_3BO_3, the cooled melt finely ground and agitated for 30 min in an ion-exchange column with water and a strongly acidic cation exchanger.   Both the effluent and eluate ($2M$ HCl) may be analyzed by emission spectrography or by atomic absorption spectroscopy; B, Si and traces of Mo, Sn and Zr are in the effluent. The resin can also be dried for use with the tape machine attachment of a direct-reading spectrograph.

Harden, G., and Tooms, J. S. Efficiency of the potassium bisulphate fusion in geochemical analyses. *Bull. Inst. Mining Met. No.* **697** 129–141, (1964). Fusion of soil samples with bisulfate liberates all of the Cu, Ni, Co and Zn from feldspars and micas; almost all of the Cu but only a limited proportion of the Ni, Co and Zn are liberated from pyroxenes and amphiboles.

Langmyhr, F. J. The removal of hydrofluoric acid by evaporation in the presence of sulfuric or perchloric acids. *Anal. Chim. Acta*, **39**, 516–518 (1967). $HClO_4$— recommended temperature of evaporation is 180°C; amount of residual fluorine is considerably reduced by double evaporation. H_2SO_4—recommended temperature of evaporation is 250°C; a single evaporation is as effective as a double evaporation with $HClO_4$, a double evaporation is slightly better.

Langmyhr, F. J., and Kringstad, K. An investigation of the composition of the precipitates formed by the decomposition of silicate rocks in 38–40% hydrofluoric acid. *Anal. Chim. Acta* **35**, 131–135 (1966). CaF_2, $MgAlF_5 \cdot xH_2O$, $NaAlF_4 \cdot xH_2O$ and, probably, $Fe(II)(Al, Fe(III))F_5 \cdot xH_2O$ were formed when both pure chemicals and rocks (granite-diabase) were decomposed with hydrofluoric acid. Magnesium is not precipitated as MgF_2, and the possibility of the formation of insoluble Fe(II) compounds (especially for basic samples) must be considered during the determination of Fe(II).

Langmyhr, F. J., and Sveen, S. Decomposability in hydrofluoric acid of the main and some minor and trace minerals of silicate rocks. *Anal. Chim. Acta* **32**, 1–7 1965). Decomposability of 28 minerals was investigated and 21 were successfully decomposed at water-bath temperature. The remainder, except for topaz, were decomposed in a Teflon-lined bomb. Hydrofluoric acid alone is more effective than when mixed with other acids.

May, Irving, and Rowe, J. J. Solution of rocks and refractory minerals by acids at high temperatures and pressures. *Anal. Chim. Acta* **33**, 648–654 (1965). Samples (20–100 mg) are completely decomposed in a modified Morey bomb containing a removable nichrome-cased 3.5 ml platinum crucible, in HF, HF + H_2SO_4 or HCl, at temperatures up to 425°C and pressures up to 6000 psi.

Moss, A. A., Hey, M. H., Elliott, C. J., and Easton, A. J. Methods for the chemical analysis of meteorites. II. The major and some minor constituents of chondrites. *Mineral. Mag.* **36**, 101–119 (1967).

Nebesar, B., and Norman, D. M. A rotating oven for general sealed-vessel sample decomposition technique. *Anal. Chem.*, **40**, 663–664 (1968). The oven, which holds eight reaction tubes enclosed in steel pipes, is mounted vertically and is rotated slowly about its short axis. The thorough mixing of the charge assists sample decomposition at 250°C.

PART II

Methods of Analysis—Discussion

CHAPTER 6

SILICATES

The term "silicate" refers to a very large group of minerals character-ized by the tetrahedral coordination of Si(IV) by oxygen in different frameworks resulting from the various ways in which these SiO_4 tetra-hedra are linked together. It also refers to the rocks formed by these minerals, ranging from highly siliceous granite and rhyolite to ultrabasic peridotite and dunite. Within this lithological range there is a wide variation in the concentrations of the 25 or so elements commonly en-countered by the rock and mineral analyst.

In the following pages these constituents are discussed in the order in which they are usually determined in the course of an analysis.

6.1. Moisture (H_2O^-)*

Moisture or, as it is variously designated, hygroscopic water, minus water, non-essential water, H_2O^- or $H_2O^{-110°C}$, should be part of any rock or mineral analysis that lays claim to completeness. It is part of the total hydrogen content of the sample, as discussed in detail by Hillebrand et al.[2], who class it as non-essential water (excluding the pos-sible presence of H_2O as water of crystallization). It is water held by surface forces such as adsorption and capillarity, and its magnitude is related more to physical properties and sample treatment than to com-position. It is sometimes considered to be an indicator of the freshness of the sample material. Manheim[3] has discussed the problems en-countered with the determination of moisture in sediments; Drysdale et al.[1] have done the same for natural glasses.

The division into "minus" and "plus" water is an arbitrary one and the determination of moisture provides the petrographer with a means for reducing analyses to a common moisture-free basis.† It is essential, however, that the moisture content of the sample remain constant during the analysis unless all portions needed for the analysis are weighed at the one time. The fineness of the powder, which governs the amount of

* References for this section will be found on pp. 116, 117.

† Because of the differing atmospheric conditions that exist from place to place, Stevens and Niles[4] recommend that all analyses should be reported on a moisture-free basis.

surface exposed, has a great effect on the initial amount of adsorbed moisture, but later variations in the moisture content are caused chiefly by changes in humidity. It is thus preferable that the various portions of the sample needed in the analysis be weighed as nearly together as possible.

Grinding of the sample will cause changes in the moisture content; the latter will tend to increase because of an increase in exposed surface, but the heat and pressure developed during the grinding will have an opposing effect. The sample should be ground to that fineness which will enable the expulsion of the hygroscopic water in reasonable time, but excessive grinding should be avoided.

Heating at 100–110°C will drive off most of this non-essential water, although on occasion temperatures much above this are necessary. This is an indirect method of determination, because the loss in weight is assumed to be due to moisture alone. The actual temperature of heating is not important since the amount given up at either temperature differs little, but it is important that the temperature used should be stated. There is nothing to be gained from drying the sample before bottling; a change occurs each time that the bottle is opened to moisture-laden air, and if the sample contains hydrous minerals, serious errors may result. It is preferable to allow the sample to reach equilibrium with the atmosphere and then to weigh all portions as nearly together as possible.

The determination of moisture may be done either as a separate step, or as the initial step in a detailed analytical scheme, and the type of container used for the determination will be governed by the subsequent treatment of the sample. Crucibles of platinum, porcelain and nickel have been used; the advantages of platinum over these others have been stated and, in addition, the sample is then ready for fusion. Porcelain or nickel crucibles, or a glass weighing bottle, will do if the sample is not to be used for further work. The use of a watch glass is not recommended because of the ease with which losses may occur.

In some minerals the moisture content may be the major part of the total water present and failure to determine the moisture, present chiefly as adsorbed water, will give an entirely misleading significance to the values for H_2O^+. Variations in the H_2O^+ content of samples analyzed cooperatively by various laboratories may be due as much to incomplete drying of the sample as to the determination of the total water content.

References

1. Drysdale, D. J., Lacy, E. D., and Tarney, J. "Water minus" (H_2O^-) in natural glasses. *Analyst*, **88**, 131–133 (1963).
2. Hillebrand, W. F., Lundell, G. E. F., Bright, H. A., and Hoffman, J. I. *Applied Inorganic Analysis*, 2nd ed., (1953). New York, John Wiley & Sons, pp. 814–822.

3. Manheim, F. Water determination in sediments and sedimentary rocks. *Acta Univ. Stockholm*, **6** (1960), 127–143.
4. Stevens, R. E., and others. Second report on a cooperative investigation of the composition of two silicate rocks. *U.S. Geol. Surv. Bull.*, **1113**, 10 (1960).

Supplementary References

Redman, H. N. An improved weighing bottle for use in weighing of hygroscopic materials. *Analyst*, **92**, 584–586 (1967). A platinum crucible is placed in the weighing bottle; a small hole drilled through the neck and lid of the bottle allows equilibration of the contents without removing the lid.

6.2. Silica*

The determination of silica is usually, but not necessarily, the first in a series of analytical steps which provide for the determination of the major constituents of the sample. The classical gravimetric procedure is still widely used, and various modifications have been proposed in recent years. The colorimetric determination of silica, especially when present as a major constituent, has received much study and a combination of gravimetric and colorimetric methods is preferred by some analysts.[23] The use of the optical spectrograph for the determination of minor and trace amounts of silica has long been established, and recent developments in X-ray fluorescence spectroscopy have made possible the determination of major amounts of silica as well; an isotope dilution method has been proposed.[13] The detailed discussion by Shell,[24] which also includes a detailed description of analytical methods, is the most recent general coverage available; Andersson[2] has studied a number of important factors affecting the gravimetric determination of silica. Washington,[42] Hillebrand *et al.*,[19] Kolthoff and Sandell,[25] and Lundell and Hoffman,[26] the latter in a very useful tabular form, summarize much of the earlier work with emphasis on gravimetric methods; Wilson and Wilson[43] discuss gravimetric and titrimetric procedures. Boltz,[7] Mullin and Riley,[33] Ringbom *et al.*,[36] and Morrison and Wilson[32] provide information about the colorimetric methods. In this section only gravimetric, titrimetric and colorimetric methods will be considered in detail.

6.2.1. GRAVIMETRIC METHODS

(a) General

The separation and gravimetric determination of silica (the first step in a very old scheme of analysis) is not necessarily done just to determine this chief constituent; it also permits the subsequent determination of other major constituents with which the silica would interfere seriously if not first removed. The solution of the sample, which was either de-

* References for this section will be found on pp. 129–131.

composed directly by acid or rendered into an acid-decomposable form by fusion with a flux, is evaporated; the solution becomes supersaturated with respect to silica, present as silicic acid, $Si(OH)_4$, and the latter coagulates through transformation into high molecular weight aggregates. On further evaporation, this hydrated silica loses H_2O and, on evaporation to complete dryness, is rendered "insoluble" in a form which displays only a slight tendency to revert to a colloidal form when treated with dilute acid. A single dehydration will usually separate out 97–99% of the silica; a second dehydration will leave less than 1 mg in solution, and this is coprecipitated with the constituents of the ammonium hydroxide group, from which it may be recovered. Because of the adsorptive nature of the silica hydrogel that forms during the dehydration, significant adsorption of many other elements occurs[35] and these cannot be removed by prolonged washing; the material that is obtained after ignition of the separated silica is thus contaminated with small amounts of these and other elements, and a further step is necessary to give a true statement of the silica content of the sample. The residue is dissolved in hydrofluoric acid and a little sulfuric acid and the silica volatilized by evaporating the solution to dryness. The percentage of silica is obtained by difference.

This is a simplified version of the procedure and the behavior of the other constituents in the sample during this separation must be considered. The acid that is most commonly used for the dissolution of the sample and subsequent evaporation is hydrochloric; Table 6.1, reproduced from *Outlines of Methods of Chemical Analysis* by Lundell and Hoffman (Ref. 26, Table 28), summarizes the behavior of the acid group and other elements during evaporation with hydrochloric acid.

During the evaporation with hydrochloric acid, *germanium* and *trivalent arsenic* (which escaped oxidation to the quinquevalent state during decomposition of the sample) are entirely lost; other elements, such as *antimony, tin, mercury, selenium,* and *tellurium* will be lost in part. It is only rarely that any of these elements are present in quantities of analytical interest and their possible loss can usually be ignored.

Of more importance are those elements which interfere through the formation of insoluble compounds with other elements and which separate with the silica only to cause difficulties in the subsequent ignitions. Of these, *sulfur* is the most common and may form *barium* sulfate or *lead* sulfate if these elements are present in sufficient amount; these will interfere in the volatilization of the silica and ignition of the residue. *Calcium* and *strontium* may also form insoluble sulfates but rarely will this be encountered. *Phosphorus* may form insoluble phosphates with *titanium* and *zirconium*, or even with *thorium*, and there is a possibility

TABLE 6.1

Separation of the Acid Group by Evaporation with Hydrochloric Acid

N																	He
Li	Be											B	C	N	O	F	Ne
Na	Mg											Al	Si	P	S	Cl	A
K	Ca	Sc	Ti	V	Cr	Mn	Fe	Co	Ni	Cu	Zn	Ga	Ge	As	Se	Br	Kr
Rb	Sr	Y	Zr	Cb	Mo	—	Ru	Rh	Pd	Ag	Cd	In	Sn	Sb	Te	I	Xe
Cs	Ba	La	Hf	Ta	W	Re	Os	Ir	Pt	Au	Hg	Tl	Pb	Bi	Po	—	Rn
—	Ra	Ac	Th	Pa	U												

* Also elements 58–71.

Heavy solid blocks inclose elements that may be precipitated practically completely.

Heavy broken blocks inclose elements that interfere seriously and must be removed before the Acid Group is separated.

Light solid blocks inclose elements that may cause difficulties. Of these, carbon is usually removed at the start of the analysis.

Light broken blocks inclose elements that are definitely lost in preparing the solution for analysis or during the evaporation.

that the phosphorus will then be lost during the fuming with sulfuric acid that follows the hydrofluoric acid treatment of the ignited, impure silica. *Silver,* and to a lesser extent *lead* and *thallium,* may form insoluble chlorides; they will probably be weighed as chlorides in the first ignition but as some other weighing form after the evaporation with $HF-H_2SO_4$ and second ignition, and thus may introduce an error into the result for SiO_2.

Titanium and *zirconium* will also separate with the silica if hydrolysis occurs but, unless present in unusually large amounts, will not interfere in the subsequent treatment of the silica beyond adding to the weight of non-volatile residue. Similar behavior is found for *niobium, tantalum* and *tungsten,* but tungstic oxide will volatilize at 850°C; if the ignitions are done at this lower temperature, this can result in slightly high values for silica because of incomplete ignition. It has been recommended in

Part I that a semiquantitative, or even qualitative, spectrographic analysis should precede the chemical analysis; the presence of any of these elements in amounts likely to cause difficulties will thus be indicated and suitable precautions can be taken to eliminate them, or to guard against losses resulting from their behavior during the subsequent steps. For example, lead should be removed first of all by dehydration of the solution of the sample with H_2SO_4, followed by extraction of the $PbSO_4$ with ammonium acetate.

The presence of unattacked sample material in the separated silica is not a likely occurrence if the decomposition procedure has been carried out correctly. If present, however, the weight of the material will change during the first and second ignitions because of attack by the hydrofluoric and sulfuric acids used in the volatilization of the silica, and an error will be introduced into the results.[15]

Another source of error lies in the possible reactions between the silica and non-silica portions of the residue during the first ignition. Sulfates [e.g., barium (if the ignition temperature is greater than 1000°C), lead and the alkaline earths] are prone to this behavior and they will be weighed first as a silicate and secondly as an oxide or sulfate.

The chief source of difficulty that is usually encountered lies in the presence of boron and/or fluorine in significant amounts in the sample. The presence of boron will be shown by spectrographic analysis but there is no simple, reliable qualitative test for fluorine and its presence or absence should be determined by mineralogical examination. *Boron* will in part accompany the silica and be volatilized with it (as BF_3), thus giving high results for silica; it must be removed at the beginning of the analysis by treatment with methyl alcohol. *Fluorine* interferes by forming volatile SiF_4 in strong acid solution, which is lost when the temperature is raised above 100°C and thus causes low results for silica; the fluorine must be prevented from interfering before the solution of the fusion cake is acidified and evaporated. There are divergent views about the quantity of fluorine which can be tolerated without significant loss of silica (fluorine will cause the loss of about three-quarters of its own weight of SiO_2), but generally no special precautions need be taken if the fluorine content is less than 2%.

Other acids may be used instead of HCl but they all have certain disadvantages. *Sulfuric acid* encourages the separation of insoluble sulfates and thus lead, barium, and possibly, calcium and strontium will separate out with the silica, as mentioned in the discussion of the use of hydrochloric acid. Other elements, such as chromium, iron, nickel and aluminum, form anhydrous sulfates that are only slowly soluble and thus may be filtered off with the silica to increase the involatile residue to a point where it becomes difficult to handle. Similar difficulties may be had with tin, antimony and germanium, which tend to hydrolyze when the sulfuric acid solution is diluted prior to filtration. Phosphorus, in addition to the formation of insoluble phosphates as mentioned previously, may be lost if the fuming period is prolonged; arsenic may also be lost in this way. Other elements, especially boron and fluorine, will behave as previously described. Sulfuric acid is not much used at this stage in silicate analysis, unless the presence of certain elements so directs; when used, fuming to the point of formation of a paste should be avoided, and when the fuming is completed an equal amount of water should be added to the still hot mixture and the whole boiled before further dilution is attempted.

Nitric acid finds little use, unless the need to avoid the precipitation of an element such as silver is paramount, because it results in the formation of difficultly soluble salts.

Perchloric acid offers certain advantages because of the greater solubility of its salts (potasium is usually the only element present in sufficient concentration that forms an insoluble salt) and it is used when it is necessary to avoid the formation of insoluble sulfates or chlorides. However, tungsten, niobium and tantalum will accompany the silica, as will tin and antimony under certain conditions; molybdenum, germanium, vanadium, bismuth, and manganese reportedly may precipitate in part, and any insoluble sulfates formed will decompose only slowly in fuming perchloric acid. Boron and fluorine will behave as described previously. According to Kolthoff and Sandell,[25] any platinum dissolved from the crucible will be completely precipitated as well. A single evaporation with $HClO_4$ is not as effective as a single evaporation with HCl, and some analysts prefer to do the first evaporation with HCl and the second with perchloric acid. Care must of course be taken in its use, and nitric acid should be added at the beginning of the evaporation before a temperature of 100°C is reached; thorough washing of the filter paper is also necessary to remove any trace of perchloric acid and perchlorates before the paper is ignited.

Carron and Cuttitta[8] have used a gravimetric method which employs the direct volatilization of SiO_2 and its determination by difference. Tektites and similar glasses are almost completely free of volatile constituents such as CO_2 and H_2O, and it is only necessary to make a correction for the oxidation of FeO to Fe_2O_3 after the final ignition to constant weight; the temperature of ignition is critical.

Bennett and Reed (Supp. Ref.) have recently described a rapid method in which a new coagulating agent is used to separate the bulk of the silica without dehydration, and which permits the subsequent colorimetric determination of residual silica.

(b)   In the presence of fluorine

When a solution containing silica and fluorine is made strongly acid, a strong complex, SiF_6^{2-}, is formed.   If the temperature of the solution is raised above 100°C, the SiF_6^{2-} complex breaks down to form SiF_4 which is volatilized.   If either Si or F are being determined, low results will thus be obtained.   It is necessary either to prevent the escape of the SiF_4 or to effect a separation of the Si and F before the formation of the SiF_6^{2-}.

It has been suggested that the presence of boron, which forms BF_3, will prevent the loss of silica in the presence of acid fluoride; the boron may be derived from boric oxide or sodium tetraborate which is used as a flux to decompose the sample, or it may be present as boric acid in the acid mixture in which the alkaline fusion cake is dissolved.   Although the loss of silica is minimized by the presence of boron, it is not prevented, and there is, furthermore, the need to expel the excess boron which, if not removed, will accompany the silica during the separation of the latter (see Sec. (c) for further discussion of this point).

Berzelius achieved a separation of silica from fluorine by precipitating the silica in a variety of ways, in alkaline solution.   Some insoluble silicates are formed during the leaching of the alkaline fusion cake and are removed by filtration; some soluble silica is precipitated with $(NH_4)_2CO_3$ and the rest with ammoniacal zinc oxide from a neutral solution.   The residues are combined, the filter papers destroyed and the silica separated by the usual double dehydration with hydrochloric acid.   The silica determination must be done separately from the main portion analysis, of course, and is a lengthy, tedious, and difficult procedure and one that does not always yield satisfactory results.   Because there are less arduous methods now available, the Berzelius procedure will be omitted in Part III; Hillebrand et al. (Ref. 19, pp. 939, 943) give the details of the method.

Hoffman and Lundell[20] omitted the $(NH_4)_2CO_3$ precipitation and separated the silica by the use of the Zn(II) ion.   The residues were combined and a double dehydration used to separate the silica.   A third recovery of the silica was made by adding Al(III) to the filtrate, precipitating it with aqueous ammonia to scavenge the SiO_2, then dissolving it in H_2SO_4 and recovering the SiO_2 by dehydration.

Shell and Craig[40] found that only about 1% of the SiO_2 remained in the first filtrate when the sample was fused with a mixture of ZnO and Na_2CO_3 and the soluble SiO_2 precipitated by dissolving the fusion cake in ammoniacal ZnO solution.   Only one filtration is made and the residual SiO_2 is determined spectrophotometrically in the filtrate.   The ZnO was also found to give a better fusion than did Na_2CO_3 alone; no significant loss of fluorine occurred when the fusion was done at 1050°C or below.   The results obtained for SiO_2 (50–72% range) were always about 0.20% low; the reason for this is not known, although it is thought that it may be due to the presence of about 0.15% fluorine, which could not be washed out of the silica residue.   Shell and Craig recommend that a sample having a known SiO_2 content be analyzed at the same time to provide a basis for correction.

Shell[39] observed that it was possible to make a good recovery of SiO_2 from some samples containing much fluorine, without having to remove the latter, because of the presence of a large amount of Al(III); in a study of lepidolite micas, which contain much Al(III), Shell (Ref. 24, p. 145) found that a fluoride content up to 2.3% did not interfere. Briefly, the method is as follows: the sample, 0.500 or 1.000 g, is fused with 5 g of anhydrous Na_2CO_3, and the fusion cake dissolved in 200 ml HCl (1:3) containing 500 mg Al (approximately 4.5 g $AlCl_3 \cdot 6H_2O$) in a platinum dish.   The solution is evaporated to dryness in a water bath with additional surface heating by an infrared lamp to insure complete dehydration of the residue.   Acidification with HCl, filtration and washing are done as usual, and a second dehydration is made; filter paper pulp is added to help collect the small precipitate obtained in the second dehydration.   The combined papers are ignited and weighed as usual, and the impure SiO_2 treated with HF and H_2SO_4.   As a result of further investigation it is suggested that, under the proper conditions, SiO_2 may be determined in the presence of even larger amounts of fluorine by the addition of Al(III).   Czech et al.[12] recommend fusion with Na_2CO_3, $Na_2B_4O_7$ and Al_2O_3 for the determination of a small amount of silica in a sample containing much phosphate and fluorine.

(c)   In the presence of boron

Boron interferes seriously in the determination of silica because it will separate and be volatilized with it, thus causing high results to be obtained.   It is not usually present in a significant amount in most rocks unless the sample happens to be tourmaline-rich, of which boron is an essential constituent; other boron-bearing silicates, such as axinite and datolite, occur chiefly in basic rocks but are only rarely encountered.

The boron is more likely to come from a boric oxide or sodium tetraborate flux used to decompose the sample.

Hazel[17] used a $Na_2CO_3-Na_2B_4O_7$ fusion mixture to decompose refractory samples and claimed that, with proper washing, no boron contamination of the silica occurred. He also found no interference from boron in the precipitation of the R_2O_3 group. Hillebrand *et al.* (Ref. 19, p. 750) disagree with this and state that the presence of boron will cause high results to be obtained for both SiO_2 and Al_2O_3. Shell (Ref. 24, p. 152) thinks that the interference of boron may be disregarded in routine work but must be considered when highly accurate values are being sought.

Boron is most easily removed by volatilization from solution as methyl borate at water bath temperature. The procedure for this has been outlined in Sec. 9.2.3.

A new method for the gravimetric determination of silicon in the presence of boron involves the precipitation of the silica with gelatin from an acid solution which has been dehydrated by boiling with glycerol.[34] The recommended range is from 2 to 500 mg Si, and the accuracy and reproducibility of the procedure are said to be as good as or better than those of the conventional gravimetric methods, with the added advantage that boron need not first be separated. It has been used for the determination of silicon in iron and steel samples, and also for slags and ores. The original paper should be consulted for details of the method.

6.2.2. TITRIMETRIC METHODS

Wilson[44] proposed a titrimetric procedure for the rapid determination of silicon in which the silicon was precipitated as quinoline silico-12-molybdate, filtered, redissolved in a known volume of $1N$ NaOH, and the excess NaOH back-titrated with HCl. Armand and Berthoux[3] studied the conditions of precipitation in detail and found that the precipitate, dried to constant weight at 150°C, corresponded exactly to the formula $SiO_2 \cdot 12\,MoO_3 \cdot 4\,(C_9H_7N) \cdot 2H_2O$. Bennett,[4] Bennett and Hawley (Ref. 5, p. 52, 155), and Bennett, Hawley and Eardley (Ref. 6, p. 2) report that the Wilson method proved unsatisfactory because of occasional unaccountable errors, and they in turn studied the application of the procedure of Armand and Berthoux[3] to the rapid determination of silica in silicate materials. Shell (Ref. 24, p. 181) describes their procedure, including the titrimetric modification, in detail

The sample is fused with NaOH (to put the silica in the monomeric form) and dissolved in HCl; Bennett *et al.*[6] experienced some low results for high-silica materials and recommend that the weight of silica taken should not exceed 35 mg. The volume of the suspension formed by leaching the NaOH melt in water should be kept at about 175 ml before adding it to the HCl; local precipitation of SiO_2 may occur if the acid concentration is too high. Once the silica solution has been made acid, there should be no delay in the addition of ammonium molybdate and quinoline, but the precipitate formed may stand overnight without serious error. It is not necessary to redistill the quinoline before using it for routine work. The method is one that requires careful adherence to detail for the accurate formation of the complex and its precipitation. Phosphate and vanadate will cause high results to be obtained because they will behave as silicate; iron $> 5\%$ Fe_2O_3 tends to cause slightly low results. Fluoride and borate do not interfere.

Coulometric titration of the excess 8-quinolinol used to precipitate the quinoline molybdosilicate has been proposed for the rapid, indirect determination of silica.[41]

Miller and Chalmers[30] investigated the use of substituted quinolines and analogous compounds in place of quinoline and selected 2,4-dimethyl-quinoline as precipitant in their micro-determination of SiO_2 in silicate rocks (2–5 mg sample). They prefer to weigh the precipitate after heating it at 500–560°C, as nonhygroscopic silicomolybdic anhydride.

A titrimetric method for the determination of silica in glass, sand, and silicate materials, including refractories, has been described by Maxwell and Budd.[28] A 0.1 g sample is dissolved in H_2SO_4 and HF under pressure in a stoppered polyethylene reaction tube at 65°C, and the silica is precipitated as potassium fluorosilicate, filtered, dissolved in hot water and titrated with $0.1N$ NaOH solution, using bromthymol blue as indicator.

Interference from precipitated calcium fluoride is prevented by the addition of ammonium chloride, and any precipitated potassium boro-fluoride (boric acid is added to complex excess HF) is redissolved by the alcohol–water–KCl solution used as wash liquid. Aluminum will interfere by also being precipitated by a fluoride. An important feature of the method is its applicability to the analysis of glasses containing fluorine.

The method lends itself to routine operation and the agreement between duplicates, for the range 64–99.7% SiO_2, is within 0.3%. Duplicate determinations can be made in less than 3 hr.

McLaughlin and Biskupski[29] fuse a 0.1 g sample with Na_2O_2, dissolve the fusion in HCl + HNO_3 and precipitate the K_2SiF_6 with NaF and KCl. Interference from Al and Ti is prevented by the addition of calcium chloride before the precipitation. Phenolphthalein is used as the indicator in the titration with NaOH. The method has been applied successfully to ceramic materials, rocks and minerals. Morris has applied it to silicomanganese and ferromanganese slag.[31]

6.2.3. COLORIMETRIC METHODS

When silica is put into solution as a result of alkaline attack, it is present in an ionic monomeric form (non-colloidal) that is amenable to complex-formation and determination. In acid solution this silicic acid will react with ammonium molybdate to form a yellow silico-12-molybdate complex that may exist in two forms, an *alpha* form that is stable at lower acidities and another, the *beta* form, that is produced at higher acidities but which irreversibily changes to the alpha form in a matter of hours. The alpha form is the one usually employed, often as its reduction product, molybdenum blue. There is a considerable literature extant on the colorimetric determination of silica, much of which results from an imperfect understanding of the system and the effect of other ions on it. For a concise summary of the subject the reader is referred to the discussions by Shell (Ref. 24) and Chalmers.[9] Morrison and Wilson[32] have studied the two forms of molybdosilicic acid in considerable detail: Mullin and Riley[33] investigated a number of possible compounds for use in reducing the yellow complex to the more intense blue form. Details of methods using both the yellow and blue complexes are given by Boltz[7]; Hedin,[18] Corey and Jackson,[11] and Shapiro and Brannock[38] used the colorimetric determination of silica as part of their scheme for the rapid analysis of rocks. Only a brief discussion will be given here and the works cited should be consulted for further details.

The yellow complex of molybdosilicic acid, also called silico-12-molybdate, is formed usually at pH 1.3–1.6, but under special conditions a pH of 3.0–3.7 has been used.[36] The maximum absorption of the complex is found at 352 mμ, but it is usually more convenient to measure it at 400 mμ; at 329 mμ the α- and β-form have the same extinction coefficients and the measurement is thus independent of the form of the complex.[14] The silica may be separated by dehydration and then redissolved by alkali fusion, in order to remove extraneous interfering ions, or the determination may be carried out in the presence of these ions, provided that precautions are taken to prevent or minimize their interference. Chalmers and Sinclair[10] form the complex in the presence of sufficient acetone to stabilize the β-form.

The reduced form of the silico-12-molybdate complex, called heteropoly blue or molybdenum blue, is used when greater sensitivity is desired (see, however, Ringbom and Österholm[37]). Silico-12-molybdate is a strong oxidizing agent and as such is readily reduced to the molybdenum blue form by weak reducing agents, among which $SnCl_2$, Na_2SO_3, NH_2OH, metol (p-methylaminophenol sulfate), ascorbic acid and 1-amino-2-naphthol-4-sulfonic acid are the most popular. The latter is generally favored for use, but Mullin and Riley[33] selected metol as the best reductant. The maximum absorption occurs at 800-820 mμ, with 815 mμ being the wavelength commonly used, but some analysts prefer to use 615 mμ instead. The two forms of molybdosilicic acid, alpha and beta, are both reduced to blue compounds that differ with the reducing agent used; Morrison and Wilson[32] investigated their relative amounts and stabilities and found that the final product is affected by such factors as acid concentration, molybdate concentration, neutral salt concentration and the reagent used.

The silico-12-molybdate is but one of a group of heteromolybdates in which the silicon may be replaced by P, As, Ge, Se, Ni, Co, W, Cu, Ti, or Zr; the molybdenum may also be replaced by W, U, V, Ta, or niobium. Of these other elements only phosphorus and arsenic, and perhaps vanadium and germanium, will interfere seriously in the determination of silicon. Interference may also occur in other ways, such as through the adsorption of the blue reduction form by precipitates of such metals as Bi, Pd, and Sb. Volatilization from boiling HCl or from HBr–$HClO_4$ and HBr–H_2SO_4 mixtures will remove As, Ge, Sb, Sn, and Cr(III); the silica may also become partly dehydrated and must be separated and fused with Na_2CO_3 before continuing with its colorimetric determination. The interference from As and P, the latter of which is the most likely source of interference in silicates, may be eliminated by the addition of organic acids (citric, oxalic or tartaric) to the solution *after* the yellow complex has been formed; phosphate may also be removed by prior precipitation with, for example, magnesia mixture. The silico-12-molybdate complex can be reduced at a higher pH than that required for the reduction of the phosphomolybdate and arsenomolybdate complexes.

Other means that may be used to remove interferences include extraction (e.g., by chloroform–cupferrate which removes such elements as Sb, Bi, Fe, Mo, Nb, Ta, Sn, Ti, W, V, and Cu but does not extract silicon), ion exchange, and mercury cathode electrolysis; dehydration of the silica, as a means of separating it, has already been mentioned. Heating of the separated silica in a stream of dry hydrogen chloride will volatilize any Sb, Nb, Ta, Sn, W, Bi, Ge, Mo, and V that separated with it.

Organic compounds, and reducing compounds such as sulfides, should not be present during the formation of the yellow complex because of their reducing effect on the molybdate. Oxidizing agents obviously are not tolerated either, but Shell (Ref. 24, p. 167) states that successful use has been made of $KMnO_4$ as an oxidizing agent for ferrous iron and sulfides.

Fluorine, in the range of concentration usually encountered in silicates, presents no problem, but for larger amounts of fluorine it is necessary to complex it with boron (e.g., by fusing the sample with Na_2CO_3 and $Na_2B_4O_7$) or it will interfere seriously in the formation of the silico-12-molybdate. If Al(III) is present in excess of that needed to form AlF_6^{3-}, the boron may not be needed (Shell, Ref. 24, p. 170). Greenfield[16] has reported that, for a 0.1 g sample of a mineral in 2.2–3N HCl and using $SnCl_2$ as reducing agent, no interference is found for fluorine and phosphorus present to the extent of 6 and 15% (w/w), respectively.

One very important source of error, and one easily overlooked, is in the condition of the silica at the time of complexing it. Silica polymerizes readily and the colloidal form is not reactive; an alkali fusion at some stage is necessary to ensure that the silica is in the proper form.

The molybdenum blue reaction is the final step in the determination of silica by two very different methods. Alon et al.[1] use a microdiffusion technique to separate silica from the other constituents in solution; $HF + H_2SO_4$ is added to an aliquot (about 0.5 ml, containing up to 70 µg SiO_2) of the solution of a silicate in the reaction compartment of a polypropylene microdiffusion cell, and the SiF_4 formed diffuses to an inner compartment containing ethylene glycol as an absorbing agent. Hozdic[21] increased the amount of H_2SO_4 used and obtained better recoveries of SiO_2 from talc-like materials; he later[22] modified the method and apparatus by transferring the SiF_4 from the reaction vessel to an absorbing solution (ammonium molybdate) by means of a current of dry air. The silicate samples were fused with Na_2CO_3, leached and aliquots containing 5–50 µg SiO_2 used for the determination, which can be completed in 4 hr, compared to the 6 hr or more needed for the diffusion step in the Alon procedure. In the second method[27] a sample of 30–50 mg is decomposed with HF alone at temperatures up to 450°C and 6000 psi in a removable, nichrome-cased 3.5 ml platinum crucible in a modified Morey bomb; an aliquot of the solution, containing 100–200 µg SiO_2 and to which Al(III), as $AlCl_3$, is added to complex the excess hydrofluoric acid, is used for the determination.

References

1. Alon, A., Bernas, B., and Frenkel, C. The determination of silica by Conway's microdiffusion technique. *Anal. Chim. Acta* **31**, 279–284 (1964).
2. Andersson, L. H. Studies in the determination of silica. VII. Some experiments with the gravimetric silica determination. *Arkiv Kemi* **19**, 249–256 (1962).
3. Armand, M., and Berthoux, J. Dosage du silicium par précipitation du complexe silicomolybdique jaune par la quinoléine. *Anal. Chim. Acta* **8**, 510–525 (1953).
4. Bennett, H. A gravimetric method for the determination of silica. *J. Soc. Glass Technol.* **43**, No. 210, 59T–61T (1959).
5. Bennett, H., and Hawley, W. G. *Methods of silicate analysis*, 2nd ed., 1965. London: Academic Press, pp. 48–55.
6. Bennett, H., Hawley, W. G., and Eardley, R. P. Rapid analysis of some silicate materials. *Trans. Brit. Ceram. Soc.* **57**, 1–28 (1958).
7. Boltz, D. F. (editor). *Colorimetric determination of the nonmetals*, 1958. New York: Interscience. "Silicon," by G. V. Potter, pp. 47–74.
8. Carron, M. K., and Cuttitta, F. Determination of silica in tektites and similar glasses, by volatilization. *U.S. Geol. Surv., profess. paper* **450–B**, B78–79 (1962).
9. Chalmers, R. A. "Chemical Analysis of Silicates" in *The Chemistry of Cements*, Vol. 2, H. F. W. Taylor, Ed., 1964. London: Academic Press, pp. 171–189.
10. Chalmers, R. A., and Sinclair, A. G. Analytical applications of beta-heteropoly acids. Part II. The influence of complexing agents on selective formation. *Anal. Chim. Acta* **34**, 412–418 (1966).
11. Corey, R. B., and Jackson, M. L. Silicate analysis by a rapid semimicrochemical system. *Anal. Chem.* **25**, 624–628 (1953).
12. Czech, F. W., Hrycyshyn, T. P., and Fuchs, R. J. Determination of small amounts of SiO_2 in the presence of phosphate and fluoride. *Anal. Chem.* **36**, 2026–2027 (1964).
13. Filby, R. H., and Ball, T. K. The determination of silica in rocks and minerals by isotope dilution with silicon-31. *Analyst* **88**, 891–893 (1963).
14. Garrett, H. E., and Walker, A. J. The spectrophotometric determination of silicic acid in dilute solution. *Analyst* **89**, 642–650 (1964).
15. Goldich, S. S., Ingamells, C. O., and Thaemlitz, D. The chemical composition of Minnesota Lake marl—comparison of rapid and conventional chemical methods. *Econ. Geol.* **54**, 285–300 (1959).
16. Greenfield, S. Spectrophotometric determination of silica in presence of fluorine and phosphorus. *Analyst* **84**, 380–385 (1959).
17. Hazel, W. M. Silicate analysis. *Anal. Chem.* **24**, 196–197 (1952).
18. Hedin, R. Colorimetric methods for rapid analysis of silicate materials. *Proc. Swedish Cement Concrete Res., Inst. (Stockholm)*, **1947**, No. 8.
19. Hillebrand, W. F., Lundell, G. E. F., Bright, H. A., and Hoffman, J. I. *Applied Inorganic Analysis*, 2nd ed., 1953. New York: John Wiley & Sons, pp. 857–866.
20. Hoffman, J. I., and Lundell, G. E. F. Determination of fluorine and of silica in glasses and enamels containing fluorine. *J. Res. Natl. Bur. Std.* **3**, 581–595 (1929).

21. Hozdic, C. Determination of silicon in minerals by microdiffusion technique. *Anal. Chim. Acta* **33**, 567 (1965).

22. Hozdic, C. Rapid method for determination of microgram amounts of silicon by colorimetric procedures. *Anal. Chem.* **38**, 1626–1627 (1966).

23. Jeffery, P. G., and Wilson, A. D. A combined gravimetric and photometric procedure for determining silica in silicate rocks and minerals. *Analyst* **85**, 478–486 (1960).

24. Kolthoff, I. M., and Elving, P. J., (Eds.). *Treatise on Analytical Chemistry*, Part II, Vol. 2, 1962. "Silicon," by H. R. Shell, New York: Interscience, pp. 107–206.

25. Kolthoff, I. M., and Sandell, E. B. *Textbook of Quantitative Inorganic Analysis*, 3rd ed., 1952. New York: MacMillan Company, pp. 391–394, 702–703.

26. Lundell, G. E. F., and Hoffman, J. I. *Outlines of Methods of Chemical Analysis*, 1938. New York: John Wiley & Sons, Table 28, p. 43.

27. May, I., and Rowe, J. J. Solution of rocks and refractory minerals by acids at high temperatures and pressures. Determination of silica after decomposition with hydrofluoric acid. *Anal. Chim. Acta* **33**, 648–654 (1965).

28. Maxwell, J. M. R., and Budd, S. M. A volumetric method for the determination of silica in glass, sand, and silicate materials. *J. Soc. Glass Technol.* **40**, 509T–512T (1956).

29. McLaughlin, R. J. W., and Biskupski, V. S. Rapid determination of silica in ceramics. *Trans. Brit. Ceram. Soc.* **64**, 153–158 (1965); *idem.* The rapid determination of silica in rocks and minerals. *Anal. Chim. Acta* **32**, 165–169 (1965).

30. Miller, C. C., and Chalmers, R. A. Microanalysis of silicate rocks. Part II. The precipitation of silica as 2:4–dimethyl–quinoline silicomolybdate and its gravimetric determination as silicomolybdic anhydride. *Analyst* **78**, 24–32 (1953).

31. Morris, A. G. C. The volumetric determination of silica and its application to ferromanganese slag and silicomanganese analysis. *Analyst* **90**, 325–334 (1965).

32. Morrison, I. R., and Wilson, A. L. The absorptiometric determination of silicon in water, Pt. I. *Analyst* **88**, 88–99 (1963); Pt. II, 100–104; Pt. III, 446–455.

33. Mullin, J. B., and Riley, J. P. The colorimetric determination of silicate with special reference to sea and natural waters. *Anal. Chim. Acta* **12**, 162–176 (1955).

34. Pasztor, L. C. Gravimetric determination of silicon in the presence of boron. *Anal. Chem.* **33**, 1270–1272 (1961).

35. Rankama, K. On the examination of the residue from silica rock analysis. *Compt. Rend. Soc. Géol. Finlande* **14**, 1–33 (1939); *Bull. Comm. Géol. Finlande*, **126**, 3–35 (1941).

36. Ringbom, A., Ahlers, P. E., and Siitonen, S. The photometric determination of silicon as α-silicomolydic acid. *Anal. Chim. Acta* **20**, 78–83 (1959).

37. Ringbom, A., and Österholm, K. Differential method for precision colorimetric analysis. *Anal. Chem.* **25**, 1798–1803 (1953).

38. Shapiro, L., and Brannock, W. W. Rapid analysis of silicate, carbonate and phosphate rocks. *U.S. Geol. Surv. Bull.* **1144–A**, 1–56 (1962).

39. Shell, H. R. Determination of silica in fluosilicates without removal of fluorine. *Anal. Chem.* **27**, 2006–2007 (1955).

40. Shell, H. R., and Craig, R. L.   Determination of silica and fluoride in fluor-silicates. *Anal. Chem.* **26,** 996–1001 (1954).
41. Su, Yao-Sin, Campbell, D. E., and Williams, J. P.   Determination of silica in glasses, ceramics and refractories. *Anal. Chim. Acta* **32,** 559–567 (1965).
42. Washington, H. S.   *The Chemical Analysis of Rocks,* 2nd ed., (1910). New York: John Wiley & Sons, pp. 78–96.
43. Wilson, C. L., and Wilson, D. W., (Eds.).   *Comprehensive Analytical Chemistry,* Vol. 1C, *Classical Analysis,* 1962. "Silicon," by H. Thomas. New York: Elsevier, pp. 149–1962.
44. Wilson, H. N.   Volumetric determination of silicon. *Analyst* **74,** 243–248 (1949).

Supplementary References

Bennett, H., and Reed, R. A.   A coagulation method for determining silica (without dehydration) in silicate materials. *Analyst* **92,** 466–467 (1967).   The use of Polyox, a synthetic industrial coagulating agent, permits the colorimetric determination of residual silica, which is not possible when gelatin is used.

Van Loon, J. C., and Parissis, C.   Rapid scheme of silicate analysis based on the lithium metaborate fusion followed by atomic absorption spectrometry. I. Determination of silica. *Analytical Letters* **1,** 519–524 (1968).   The sample is fused with $LiBO_2$, dissolved in dilute HNO_3 and the silica determined by atomic absorption using a nitrous oxide flame.   Best results are obtained with 150–300 ppm SiO_2, and the accuracy and precision of the data given are generally better than 0.5%.

6.3. Precipitation with the sulfide ion*

Ordinarily, the next step in silicate rock or mineral analysis, following the separation of all but about 1 mg of silica, is the separation of those elements that are precipitated by aqueous ammonia at a pH just below 7, the R_2O_3 elements (R = Fe, Al, Cr, V, P, Ti and some others). Occasionally, however, it may be necessary to consider a precipitation in acid solution with hydrogen sulfide.

The acid filtrate (approximately $0.5N$ in HCl,) such as is obtained in Secs. 9.2.1, 9.2.2 or 9.2.3, will contain some platinum introduced from the platinum crucible and/or dish if these were used in the fusion and dehydration procedures.   The amount of platinum contributed by the crucible will usually be very small (<0.0005 g); the actual weight may be shown by the difference in the weight of the crucible before and after the fusion, provided that the crucible has not picked up iron or some other metal during the fusion.   Occasionally, the unsuspected presence of certain elements or compounds may result in abnormal attack on the crucible, and as much as 5 or 10 mg of platinum may pass into the solu-

* References for this section will be found on p. 135.

tion. Similarly, if a platinum dish was used for the dehydration with HCl and there was much ferric iron, manganate or vanadate present, several more milligrams of platinum may have been dissolved. The presence of appreciable platinum in the solution is a source of trouble in subsequent steps; if not removed prior to the addition of aqueous ammonia, the platinum will react with it to form a variety of complex ammines which, because of their stability, prevent the removal of the platinum by the usual precipitants. The platinum will then precipitate fractionally in successive steps, beginning with the precipitation of the R_2O_3 group. It is imperative that platinum in excess of that normally added to the analysis (which will largely escape precipitation) be removed before starting the precipitation with aqueous ammonia (Ref 2, pp. 341–343).

In most rock analyses (excepting those rocks which contain admixtures of ores), platinum will be the only element whose removal at this stage may be necessary. The qualitative or semi-quantitative spectrographic analysis that has been recommended as a very useful preliminary step will reveal the presence of others, such as copper, lead and zinc, and the analyst must decide whether or not the treatment with hydrogen sulfide is warranted (it is a somewhat messy business and is better avoided). In mineral analyses it is more likely that the step must be included, but again the preliminary spectrographic analysis will serve as a reliable guide.

It will be useful at this point to consider briefly the important features of precipitation with the sulfide ion. For a detailed discussion the reader is referred to Hillebrand et al. (Ref. 2, pp. 58–75) and to Lundell and Hoffman (Ref. 3, pp. 49–54, 59–63); Table 6.2 is reproduced from the latter text.

Precipitation is usually made in strong acid solution (HCl or H_2SO_4, although dilute HNO_3 and $HClO_4$ may be used) over the range 0.25–13N, although the lower normality is favored for most purposes. Under these conditions, the copper (Cu, Ag, Hg, Pb, Bi, Cd, Ru, Rh, Pd and Os) and arsenic (As, Au, Pt, Sn, Sb, Ir, Ge, Se, Te and Mo) groups will precipitate. Ga, In, Tl, V and W may also separate under certain conditions; other elements not usually separated here may precipitate through the formation of mixed sulfides, such as in the coprecipitation of Zn with copper. Of these elements, only Pt is likely to be of concern in rock analysis. In minerals and ores, of course, there may be several of these elements which should be removed, such as copper, lead (that portion not already removed during the separation of the silica), bismuth and molybdenum. Both Mo and Pt require special

TABLE 6.2

Members of the Hydrogen Sulfide Group

```
 •  •                                              •  •
 H
 •  •  •                          •  •  •  •  •  •  •
 Li  Be                                   N   O
 •  •  •                       •           •  •  •  •  •
 Na  Mg                        Al  Si   P   S   Cl
 •  •  •  •      •          •                         •
 K   Ca  Sc  Ti │ V │ Cr  Mn │ Fe  Co  Ni │ Cu │ Zn │ Ga  Ge  As  Se │ Br │
 •  •  •  •                                           •
 Rb  Sr  Y   Zr │Cb│ Mo  —  Ru  Rh  Pd  Ag  Cd │ In │ Sn  Sb  Te │ I │
 •  •  •  *  •        •  •  •                          •
 Cs  Ba  La  Hf │ Ta │ W │ Re  Os  Ir  Pt  Au  Hg │ Tl │ Pb  Bi  Po │ — │
 •  •  •  •
 —  Ra  Ac  Th  Pa  U
 •  •  •  •  •  •  •
```

* Also elements 58–71.

Heavy solid blocks inclose members of the Hydrogen Sulfide Group.

Heavy broken blocks inclose non-members of the group that may be precipitated in part under certain conditions.

Light solid blocks inclose elements that may still be present in small amount through incomplete precipitation of the Acid Group.

Light broken blocks inclose elements that are left in solution only under exceptional conditions.

Missing elements are those that have been removed in the General Procedure.

treatment to insure complete precipitation, i.e., a longer period of gassing and digestion with hydrogen sulfide. Except for Cd, In, Pb and possibly molybdenum, these elements will precipitate as well in $1N$ as in $0.2N$ acid solution.

The use of a higher pH (2–3), as in $0.01N$ acid solution, usually serves only to precipitate zinc. At a still higher pH of 5–6, achieved by use of organic acids and their salts (e.g., acetic acid and sodium or ammonium acetate), it is possible to separate Co and Ni from manganese; Tl and In are also completely precipitated, and partial precipitation of Fe may occur under certain conditions

Many elements are precipitable by sulfide ion under alkaline conditions, and the precipitation may be made in several ways. Table 6.3 (Ref. 3, Table 37) shows, however, that if it is preceded by both the separation of the copper and arsenic groups by hydrogen sulfide precipitation in strong acid solution and the removal of the R_2O_3 group by ammonia precipitation, then only Mn, Co, Ni and Zn will be of concern in rock analysis. Silicate minerals may, of course, contain much greater concentrations of elements such as vanadium and tungsten than are usually encountered in rocks, and these latter will also be precipitated here, provided that other members of the ammonium sulfide group are present.

TABLE 6.3

Members of the Ammonium Sulfide Group

H																	
Li												N	O				
Na	Mg											Si		S	Cl		
K	Ca			[V]	Mn	Co Ni	Zn							Br	
Rb	Sr													I	
Cs	Ba			[W]					Tl					—	
—	Ra														

Heavy solid blocks inclose members of the Ammonum Sulfide Group.

Heavy broken blocks inclose elements that may be carried down by members of the group.

Light blocks inclose elements that are left in solution only under exceptional conditions.

Missing elements are those that have been removed in the General Procedure.

It is possible, by the formation of complex anions of some elements, to effect separations within the foregoing groups. A well-known example of this is the separation of the arsenic group elements from most of those in the copper group, by the use of alkaline or polysulfide solutions. Similarly, the precipitation of such elements as aluminum and chromium can be prevented, during precipitation with ammonium sulfide, by the presence of tartaric acid.

The use of hydrogen sulfide is so widespread that little attention is paid to the availability of a substitute that will serve as well and at the same time be free of the obvious disadvantages of hydrogen sulfide. Of the many organic compounds that have been investigated, thioacetamide is the only one that has been found to be of general, practical use.[1,4]

Thioacetamide, CH_3CSNH_2, hydrolyzes readily in acid or alkaline solution to produce sulfide ion. The precipitates produced with it are more coarsely crystalline and dense and less contaminated with foreign ions than those formed by gassing the solution with hydrogen sulfide. The rate of hydrolysis, and thus the rate of precipitation, is controlled by regulation of the temperature (efficiency of hydrolysis increases rapidly with increasing temperature) and, because the amount of reagent needed to effect the precipitation can be calculated, the precipitation need not be done in a fume-hood. There is no need to use a gas delivery tube and this eliminates the troublesome loss of particles of the precipitate through adherence to the inner walls of the tube.

Procedures for qualitative analysis are basically the same as those for hydrogen sulfide, but close adherence must be made to specified acid and base concentrations, temperature and time of heating. The use of thioacetamide is particularly advantageous in microanalysis.

It has been pointed out[4] that one cannot assume that thioacetamide can be substituted generally for hydrogen sulfide without some modification of the procedure, and Flaschka[1] considers that it is of more use as a preliminary separation step than as a means of direct gravimetric determination

The reagent is soluble in water, alcohol and benzene, and a neutral aqueous solution is said to be stable for months, apart from the formation of a slight sediment which can be filtered off without impairing the usefulness of the reagent. For qualitative procedures the reagent is added, in the form of a 2–8% aqueous solution, to the unknown and the solution is warmed for several minutes at steam bath temperature.

References

1. Flaschka, H. Thioacetamide in analytical chemistry. *Chemist-Analyst* **44** (1), 2–7 (1955).
2. Hillebrand, W. F., Lundell, G. E. F., Bright, H. A., and Hoffman, J. I. *Applied Inorganic Analysis*, 2nd ed., 1953. New York: John Wiley & Sons, pp. 58–75, 341–343, 868.
3. Lundell, G. E. F., and Hoffman, J. I. *Outlines of Chemical Analysis*, 1938. New York: John Wiley & Sons, pp. 49–54, 59–63.
4. Wilson, C. L., and Wilson, D. W., (Eds.). *Comprehensive Analytical Chemistry*, Vol. 1A, 1959. New York: Elsevier, Chapter IV, p. 540.

6.4. Aluminum*

The determination of aluminum, directly or indirectly, is one of the most difficult steps in the analytical scheme, chiefly because so many other elements will interfere in the determination to a greater or lesser extent.

It is traditional for aluminum to be determined by difference. Aluminum and several other elements are precipitated together, for example, with aqueous ammonia, ignited and weighed as the R_2O_3 group of mixed oxides; the constituents other than aluminum are determined separately, some of them on other portions of the sample, and their percentages are subtracted from the total per cent R_2O_3 to give the percentage of alumina. It is obvious that any errors arising out of the precipitation itself, and from the determination of the other constituents of the mixed oxides, will be reflected in the final figure for alumina. It is not without good reason that the obtaining of an accurate value for Al_2O_3 is regarded as a test of analytical skill.

In recent years, there has been much effort devoted to devising methods for the direct determination of aluminum. These usually require, however, a preliminary separation of interfering elements, or are applicable only to samples that contain none or negligible amounts of these latter. Those methods which do permit the direct determination of aluminum without prior separations require rigid adherence to the specified conditions if erratic results are to be avoided; empirical corrections may have to be applied.

The analytical chemistry of aluminum has been recently discussed in detail by Farrah and Moss (see Kolthoff and Elving, Supp. Ref.); Berthelay (Supp. Ref.) investigated titrimetric, flame and optical emission spectroscopic methods for the determination of aluminum in rocks and minerals.

6.4.1. GRAVIMETRIC METHODS

(a) General

The precipitation of aluminum together with a number of other elements, by aqueous ammonia, has already been mentioned as the procedure most commonly used. The precipitation is done at a pH just under 7, in the presence of an excess of ammonium salt; one or more reprecipitations are usually necessary. The addition of an oxidizing agent such as bromine or ammonium persulfate will cause the simultaneous

* References for this section will be found on pp. 147–149.

precipitation of most of the manganese in the solution. Hummell and Sandell[9] obtained a good separation of as much as 40 mg of aluminum from an equal amount of iron by carrying out the precipitation with aqueous ammonia in the presence of thioglycolic acid; Ti and P are quantitatively precipitated (the latter if Al is present in excess) and thus the aluminum can be determined by difference. The subsequent precipitation of the aluminum as the 8-hydroxyquinolate can be successfully done in the presence of either Ti or P alone, but not when they are present together. Utilization of the thioglycolic acid filtrate for the determination of iron is not feasible.

Other procedures have found more limited use in the gravimetric separation and determination of aluminum.[6,11] At one time, the trivalent ions were precipitated as basic acetates from a buffered ammonium acetate-acetic acid solution, particularly if more than the usual amount of manganese or nickel was present, or if much iron was present together with cobalt and zinc. There are, however, a number of disadvantages inherent in the method, such as the partial precipitation of Zn and other divalent metals, and it is now seldom used. The use of benzoic acid-ammonium benzoate enables the quantitative precipitation of Fe, Al and Cr at a relatively low pH (3.8) with a minimum of interference from divalent metals; it is also employed by Milner and Woodhead[16] to separate aluminum for titration with EDTA, after prior separation of interfering elements with cupferron. Succinic acid, with ammonium chloride and urea in a boiling weakly acid solution, will precipitate aluminum homogeneously as a dense, basic succinate,[29] separating it from such elements as Mn, Ni, Co, Ca, Ba and Mg; phosphorus is partially precipitated also and must be determined in the ignited Al_2O_3. Some aluminum phosphate may also remain in solution, giving low results. Copper must be reduced to Cu(I) with hydroxylamine hydrochloride, and in the presence of much iron or zinc a double precipitation is necessary. The hydrolysis of urea, to form NH_3, can be used to separate the hydrous oxides as very dense precipitates; the urea hydrolyzes slowly in hot solution and the NH_3 that is formed removes the hydrogen ions present. Precipitation with 8-hydroxyquinoline (oxine) in a weakly acid solution will quantitatively separate aluminum and a number of other elements from the alkaline earths, magnesium and beryllium; aluminum is frequently precipitated in this form, after a prior removal of elements such as iron and titanium with cupferron, and either weighed as the hydroxyquinolate or oxide, or titrated bromometrically; when present in small amounts, the yellow aluminum hydroxyquinolate can be extracted with chloroform and determined colorimetrically.

Miller and Chalmers[15] precipitate aluminum finally as the 8-hydroxy-quinolate after the removal of Fe, Ti, V and Zr with cupferron.   The aluminum, together with beryllium, is extracted as the acetylacetonate from the filtrates from the cupferron separation and the aluminum then separated from beryllium as the hydroxyquinolate and weighed as such. They applied this method to the determination of aluminum in a 5 mg sample of a silicate rock and found that 0–40% Al_2O_3 could be determined with a reasonable degree of accuracy, the results in general being about 1% lower than those obtained by classical procedures.

Van Loon (Supp. Ref.) first separated aluminum from Fe by ion-exchange, then from other elements by a cupferron–chloroform extrac-tion, before determining it as the hydroxyquinolate; it is necessary to correct for coprecipitated manganese.

(b) Precipitation with aqueous ammonia

Lundell and Hoffman (Ref. 14, Table 35, p. 56) show in tabular form those elements that will be precipitated by aqueous ammonia (Table

TABLE 6.4

Members of the Ammonium Hydroxide Group

•	•															•	•
H																	
•											•	•	•	•	•		
Li	Be												N	O			
•														•	•		•
Na	Mg										Al	Si	P	S	Cl		
•	•									•					•		•
K	Ca	Sc	Ti	V	Cr	Mn	Fe	Co	Ni	Zn	Ga				Br		
Rb	Sr	Y	Zr	Cb						In			•	•	I		•
Cs	Ba	La*	Hf	Ta	W					Tl					—		
•	•		•	•		•	•	•	•	•	•	•	•	•			•
—	Ra	Ac	Th	Pa	U												
•	•																

* Also elements 58–71.

Heavy solid blocks inclose members of the Ammonium Hydroxide Group.

Heavy broken blocks inclose elements that may be caught in the precipitate (Si, P, V, and W more or less completely, and Co and Zn only partially).

Light solid blocks inclose elements that are still present in solution only under exceptional conditions.

Missing elements are those that have been removed in the General Procedure.

6.4). In most rocks, only Fe, Al, Ti, P and the residual SiO_2 remaining in solution will be of concern to the analyst, but in some samples there may also be appreciable amounts of Mn, Cr, Zr, Co, Zn and occasionally the rare earths that will have to be considered; minerals may have any number of these elements as major or minor constituents. Platinum dissolved from the crucible and/or dishes used for the separation of SiO_2, if not previously removed by precipitation with H_2S, will partly accompany the R_2O_3 group.[24] Information about the probable composition of the ammonium hydroxide group can be obtained from a preliminary qualitative or semiquantitative spectrographic analysis; if such information is not available some idea can be gotten from the color of the sodium carbonate fusion cake (Sec. 9.2.1), and from the size of the residue left after treatment of the ignited SiO_2 with HF and H_2SO_4. If the residue is small (2–3 mg) and dissolves readily in HCl after fusion with a small amount of Na_2CO_3, then all is well; if the residue is large and gives a cloudy solution in HCl, it is likely that significant amounts of such elements as Ta, Nb, Ti and Zr (the latter two elements either alone, or associated with phosphorus) are present.

There are two decisions* that must be made before starting the precipitation with aqueous ammonia. The first has to do with the residue left after the volatilization of the ignited SiO_2, the second concerns the treatment of the manganese present in the solution.

It is the writer's practice, when the nonvolatile residue is small, to ignite the R_2O_3 precipitate in the same crucible and thus obtain the total weight of the mixed oxides. This assumes that neither calcium nor magnesium were coprecipitated with the silica. If the residue is large, and if troublesome elements are known to be present, it is better to fuse it with about 1 g of anhydrous Na_2CO_3, dissolve it in dilute HCl (1:1) and add it to the filtrate from the SiO_2 determination (prior to the H_2S separation). The crucible is ignited and weighed as usual.

Much consideration is given in older textbooks to the precipitation of the ammonium hydroxide group with or without manganese, and

* Hillebrand *et al.* (Ref. 6, p. 867) suggest a third one, if the residual SiO_2 remaining in the solution is not to be recovered. Instead of making the standard separation of the R_2O_3 group, certain variations may be used. One can treat the filtrate from the H_2S separation with tartaric acid, make it ammoniacal and separate the Fe, Zn, Ni, Co and most Mn as sulfides; a cupferron separation of Nb, Ta, Zr, Ti and V can then be made on the acidified filtrate, or the latter may be made on one half of the oxidized filtrate from the silica or H_2S separations, with the usual precipitation with aqueous NH_3 being done on the other half; the cupferron separation will give a combined total for the Fe, Ti, Zr and V oxides, thus making possible a more accurate determination of Al by difference.

opinion is divided as to the best course to follow. Coprecipitation is achieved by the simultaneous addition of aqueous ammonia and an oxidizing agent such as ammonium persulfate or saturated bromine-water. The latter oxidant, the use of which was investigated by Holt and Harwood,[7] has the advantage over persulfate in that it adds nothing to the solution, but unfortunately not all of the manganese is precipitated. Persulfate does a better job[10] but much sulfate ion is added to the solution and may cause the precipitation of barium and strontium. Unless manganese is present in abnormally large amount, i. e., >1%, a double precipitation of the R_2O_3 group will give a final precipitate that is virtually free of it. If present in higher concentration, however, then the amount that accompanies the mixed oxides can be determined in an aliquot of the solution from the pyrosulfate fusion of the oxides. The manganese will then distribute itself among the remainder of the main portion constituents, with a very small amount (usually <0.05%) accompanying the calcium and the rest of it coprecipitating with the magnesium. Usually a small amount can also be recovered in the filtrate from the magnesium determination.[10] Some analysts prefer to accept this distribution and to determine it in all three places, or to determine that associated with the calcium and magnesium and, by comparison with the total manganese content found by analysis of a separate portion of the sample, to assume that the remainder has accompanied the mixed oxides. The writer prefers to precipitate the ammonium hydroxide group without manganese (correcting the R_2O_3 oxides if necessary for Mn_3O_4 as mentioned above) and then to remove the manganese as described in Sec. 9.9 before proceeding with the separation of the oxalate group. The precipitation of manganese with the ammonium hydroxide group will not be discussed, therefore, and the reader is referred to Groves (Ref. 5, pp. 56–60) for details of the use of bromine-water, and to Hillebrand et al. (Ref. 6, pp. 870–872) and Ingamells and Suhr (Supp. Ref., pp. 899–900) for the ammonium persulfate procedure.

When aqueous ammonia is added to the acid filtrate from the SiO_2 determination, Fe(III) begins to precipitate at about pH 2.5 and Al at about pH 3, with precipitation being complete at pH 6.6–6.7. The hydroxides of Co, Zn and Cu are also precipitated, but they redissolve in excess aqueous NH_3 to form complex ammino ions; if present in more than trace quantity, they may be difficult to redissolve and too great an excess, i.e., pH 9–10, may cause some of the precipitated hydrous alumina to redissolve also. This latter will then contaminate the calcium and magnesium precipitates at a later stage. The addition of the aqueous ammonia is continued to the change point of such indicators as methyl red (4.2–6.3) or bromcresol purple (5.2–6.8) and one or two drops are added in excess; the solution should smell faintly of NH_3, when the fumes have been blown away.

The precipitation is made in the presence of an ammonium salt, usually NH_4Cl, to prevent the precipitation of calcium and magnesium; the salt acts as a buffer to limit the alkalinity of the solution to pH <9. It also aids in coagulating the gelatinous precipitate. Similarly, the precipitate is washed with a hot solution of NH_4Cl or NH_4NO_3 to prevent peptization of the hydrous oxides; NH_4Cl is preferred if the filtrate from the R_2O_3 precipitation is to be acidified in a platinum basin, but if much iron is present it is better to use NH_4NO_3, because of the danger of the volatilization of iron chloride during the ignition of the precipitate.[26] The precipitation is done in a near boiling solution, and the filtration should be made with as little delay as possible, both to minimize absorption and occlusion of other ions by the gelatinous precipitate, and because the precipitate is less likely to peptize at high temperatures.

One reprecipitation must be done in order to remove the traces of calcium and magnesium that will usually be carried down with the precipitate; two reprecipitations should be made if the sample contains much magnesium (as in olivine, or an ultrabasic rock). It is a useful step, in any case, to make a third or clean-up precipitation on the combined filtrates after having destroyed the ammonium salts; the removal of this accumulation of salts also will make for a better subsequent precipitation of the calcium and magnesium.

Difficulty will often be experienced in the filtration, particularly the first one. The behavior of the precipitate and the ease of filtration are not always predictable, but in general the greater the ratio of iron to aluminum, the easier will be the filtration. Titanium, in more than the usual concentration, does not help the filtration, but the final behavior is dependent upon the relative proportions of all three constituents. It is important that the surface of the precipitate be kept wet at all times during the filtration; if the hydrous alumina is permitted to dry out, it will form a surface over which the wash solution will run but not penetrate. The second precipitation is usually much more satisfactory because of the presence of the macerated filter paper from the first filtration. The paper fibers also make for a more porous ignited residue and one in which the iron is more easily kept in the oxidized state. Groves[5] recommends the addition of a filter paper before the *first* precipitation. A correction can be applied to the R_2O_3 oxides for the weight of filter paper ash included in the ignited residue; in macro work this is usually negligible.

The ignition of the well-drained, or partly dried, precipitate must be carefully done, with final heating at 1200°C in order to render the Al_2O_3 non-hygroscopic; a free access of air must be provided during the ignition to prevent reduction of the ferric iron. The heating should not be prolonged unduly and should not exceed 1200°C. A correction for the volatilization of platinum at this high temperature should be applied when a high degree of accuracy is desired.

The procedure is subject to several sources of error unless all of the precautions mentioned in this discussion and in the procedure are observed. If the phosphorus content of the sample exceeds an equivalent amount of aluminum and iron, then the excess phosphate will precipitate the alkaline earths also; this may be avoided by the addition of a known amount of iron to compensate for the deficiency. The alkaline earths are precipitated as carbonates if the aqueous ammonia used in the precipitation contains $(NH_4)_2CO_3$; it is the practice at the Geological Survey to use aqueous NH_3 (approximately $15M$) prepared by saturating freshly boiled and cooled distilled water with ammonia gas. The commercial aqueous ammonia contains too much silica and other dissolved and suspended matter to permit its use at this stage of the analysis. The precipitate may contain not only coprecipitated ions such as Co, Zn and Cu, but also alkali salts if the washing was not properly done. The filter paper from the first precipitation must not be discarded but must either be macerated and combined with the precipitate, or ignited separately and then combined; it is very difficult to be sure that all of the hydrous alumina has been dissolved from the paper during the dissolution of the first precipitate. Attention has been drawn to the need for ignition of the oxides at 1200°C; it is also necessary that the crucibles used for the ignition should have a well-fitted cover, because the ignited alumina will take up moisture during the first few minutes of exposure to air, and it is necessary to keep the crucible tightly closed and to make the weighings as rapidly as possible.

The procedure for the recovery of residual SiO_2, by fusion of the ignited R_2O_3 oxides with potassium pyrosulfate and fuming of the solution with H_2SO_4, is given in Sec. 9.6.1. It is customary to make this recovery and to use the acid filtrate for the determination of titanium; the solution may then be used for the determination of total iron, if so desired. It is also possible, however, to utilize aliquots of this solution for the determination of a number of elements, particularly if the available sample is in short supply. Hillebrand et al. (Ref. 6, pp. 92–95) describe procedures for the determination of Zr, rare earths, Cr, U, V, Be and P as well.

6.4.2 TITRIMETRIC METHODS

Among the titrimetric procedures proposed for the direct determination of aluminum are those which give results of an accuracy equal to
or better than that obtainable by gravimetric methods, and those by
which results of lesser, although acceptable, accuracy may be obtained
in a much shorter time. Milner and Woodhead[16] (see Sec. 9.7.1) sintered
an aluminosilicate sample with Na_2O_2, separated Fe, Ti, Zr and Th with
cupferron–chloroform, precipitated the aluminum (in the presence of
thioglycolic acid) as the benzoate and then dissolved the precipitate in
a known excess of EDTA and back-titrated the excess with standard
iron solution, using sodium salicylate as indicator. Phosphate interferes if the ratio $Al_2O_3 : P_2O_5 < 4 : 1$. Mercy and Saunders (Supp.
Ref., p. 177), however, question the accuracy of results obtained by this
method. Bennett, Hawley, and Eardley[1] removed the SiO_2 as usual
after a Na_2CO_3 fusion of the sample and then extracted Fe and Ti from
an aliquot of the filtrate with cupferron-chloroform, as did Milner and
Woodhead; the aqueous solution, containing an excess of EDTA, was
buffered to a pH of about 4.4 with ammonium acetate and the excess
EDTA back-titrated with standard zinc solution, using an ethanolic
solution of dithizone as indicator, following the method developed by
Wänninen and Ringbom.[16,25] Manganese interferes stoichiometrically
but is usually in too low a concentration to be significant. For Al_2O_3
values from 8–53% the mean difference between duplicates was 0.065%
for gravimetric determination as the hydroxyquinolate and 0.05% for
the titrimetric procedure; the titrimetric values are, on the average,
0.03% lower than the gravimetric values. The procedure is described
fully in Sec. 9.7.2. Weibel[27] has described a similar procedure but decomposed the sample by treatment with HF and $HClO_4$; $BeSO_4$ is added
to complex the fluoride ion present. Pritchard (Supp. Ref.) simplified the
procedure by using a measured amount of 1,2-diaminocyclohexanetetra-
acetic acid (DCTA) to complex aluminum preferentially during the
precipitation and removal of other elements with sodium hydroxide;
the excess DCTA in the filtrate is titrated with standard zinc solution,
using Xylenol Orange as indicator. According to Mercy and Saunders
(Supp. Ref., p. 175) the method fails in the presence of much magnesium
(as in ultrabasic rocks); they prefer to remove most interfering elements
by cupferron–chloroform and to eliminate chromium by volatilization
as chloride before adding DCTA and titrating the excess with standard
lead solution.

A rapid control method for aluminum in opal glass, in which zinc is also an important constituent, is described by Cluley.[2]   A 0.1 g sample is decomposed with HF and $HClO_4$ and the aluminum, zinc and other metals present are complexed with an excess of EDTA, the excess being back-titrated with standard zinc solution using Xylenol Orange as the indicator; NaF is added to complex the aluminum and the released EDTA is again titrated with the standard zinc solution.   Duplicate determinations can be made in 1 1/2–2 hr; for 4% Al_2O_3, the standard deviation is 0.01%.   No attempt was made to apply this procedure to other than opal glass.   Dinnin and Kinser[3] also complex the aluminum in chromite with EDTA, following the separation of interfering elements by mercury cathode electrolysis and cupferron extraction, but back-titrate the excess EDTA with standard $FeCl_3$ solution, using Tiron (disodium-1,2-dihydroxybenzene-3,5-disulfonate) as indicator; phosphates must be absent.   Kristiansen[12] titrated aluminum directly with EDTA at pH 3, with 3-hydroxy-2-naphthoic acid serving as a fluorescent indicator; by lowering the pH to 2 and the temperature to near zero it is possible to titrate Fe(III) first and then, after heating the solution to 50°C and raising the pH to 3, to titrate the aluminum.   A double precipitation of aluminum as the hydroxyquinolate, after a cupferron-chloroform extraction of Fe and Ti, is employed by Ritchie[21] in his silicate analysis scheme; the precipitate is dissolved and titrated bromometrically.   Chablo (Supp. Ref.) has studied the applicability of several complexometric methods to the determination of aluminum.

6.4.3.   COLORIMETRIC METHODS

The development by J. P. Riley and his students of semimicro and micro methods of silicate analysis has resulted in many significant contributions to the armory of the rock and mineral analyst.   Particular attention has been paid to aluminum, and two procedures have been described in which final measurement of the aluminum is made colorimetrically, using the yellow complex formed by aluminum and 8-hydroxyquinoline in chloroform.   The first method[19,27] is applicable to ordinary silicate rocks and minerals; the sample is decomposed with HF and $HClO_4$, the solution buffered and elements complexed by the addition of hydroxylamine hydrochloride, sodium acetate, 2,2-dipyridyl and beryllium sulfate, and the aluminum extracted as the hydroxyquinolate in chloroform, made up to volume and the absorbance measured at 410 mμ.   When minerals containing appreciable concentrations of elements such as Cu, Co and Zr are analyzed, however, the first method fails because these metals also form colored complexes with 8-hydroxyquinoline; Riley and Williams[20] have made the method specific for aluminum by first extracting these interfering elements from solution with 8-hydroxyquinaldine in chloroform at pH 10 (if much titanium is present it is first extracted at pH 4 with the 8-hydroxyquinaldine) before extracting the aluminum at pH 4.5 with 8-hydroxyquinoline as described previously. Quinalizarin-sulfonic acid is added, if necessary, to complex zirconium, and uranium is removed as an anionic acetate complex by passage of the solution through an ion-exchange column. These methods are presented in detail in Sec. 9.8.2. Mercy and Saunders (Supp. Ref., p. 173) consider the Riley method to be the only acceptable photometric method for aluminum.

Shapiro and Brannock[23] have utilized the calcium–aluminum lake formed with Alizarin Red-S (sodium alizarin-3-sulfonate),[18] to determine aluminum in their rapid analysis scheme; iron interference is eliminated by the use of potassium ferricyanide and thioglycolic acid and an empirical correction is made for the titanium present which forms a colored complex absorbing at the same wave length, 475 mμ, as that of the aluminum. The determination is done on a small aliquot of their solution A, and a rigid adherence to fixed procedural detail is mandatory if satisfactory accuracy and reproducibility are to be obtained. This procedure is described in Sec. 9.8.1.

Aurintricarboxylic acid, or aluminon, is another lake-forming reagent that has found much use in the determination of aluminum.[22] Pa Ho Hsu[8] carried out an extensive investigation of the system in an effort to improve the precision and accuracy obtainable; he found that these latter are strongly influenced by the original pH of the sample and the presence of phosphate or silicate, and has stressed the importance of insuring at the beginning that all of the aluminum is in the ionic state. A preliminary acid and heat treatment has been found to give highly reproducible results. Wilson and Sergeant[28] used Pyrocatechol Violet to make a rapid evaluation of aluminum in milligram samples of minerals; the blue complex, in contrast to the yellow reagent color, does not have the high blank correction characteristic of other complex-forming reagents such as Alizarin Red-S and aluminon. Measurement is made in a solution buffered to 6.1–6.2 with acetic acid-ammonium acetate and Beer's law is followed for solutions containing up to 80 μg of Al_2O_3 in 100 ml, provided that sufficient excess of the reagent is present. Interference from Fe, Be and Cr can be eliminated by cupferron extraction, but the iron interference can be greatly diminished by the addition of hydroxylamine hydrochloride and o-phenanthroline; the use of the latter reagents, when the method is applied to ordinary rocks, leaves titanium as the only likely interference and an empirical correction can be applied.

No serious interference is caused by phosphate and fluoride ions when their concentration is less than that of the aluminum. When the only separation made is that of silica, the moderate accuracy attained is satisfactory for the rapid analysis of milligram samples (see also Mercy and Saunders, Supp. Ref., p. 175).

6.4.4. Other methods

Eshelman et al.[4] have developed a direct flame emission spectroscopic method that is sufficiently sensitive to determine aluminum in low concentrations in a variety of materials, including silicate minerals and glasses, using small samples. After decomposition of the sample with $HF-H_2SO_4$, the solution (1:9 in HCl) is extracted with cupferron in 4-methyl-2-pentanone to remove heavy metals, then buffered to pH 2.5–4.5 with ammonium acetate–aqueous ammonia and the aluminum selectively extracted with cupferron in 4-methyl-2-pentanone. The organic phase is aspirated directly into an oxyhydrogen flame. Fluoride and phosphate interfere seriously, presumably by preventing the extraction of the aluminum. Similarly, atomic absorption spectroscopy has not yet been successfully applied to the routine determination of aluminum in silicates; Amos and Thomas (Supp. Ref.) can detect <2 ppm Al in aqueous solution, using a flame mixture of nitrogen, oxygen and acetylene, but both iron and chloride ion interfere seriously and must be either removed or, in the case of iron, introduced into the standards.

A polarographic method for the routine determination of aluminum in rocks, over the range of 1–20% Al_2O_3, has been developed at the Geological Survey by Marilyn Levine.[13] The method is based upon the reduction of a sulfonated analog of Pontachrome Violet SW at the dropping mercury electrode; the low solubility of the latter dye limited its use to samples containing less than 7% Al_2O_3 and successful efforts were made to prepare a more soluble analog which turned out to be 1-(1-hydroxy-4-sulfonic acid-2-phenylazo)-2 naphthol-3,6-disulfonic acid. The sample is decomposed with HF and H_2SO_4, and then electrolyzed at a mercury cathode to remove interfering heavy metals. An aliquot of the electrolyzed sample is then evaporated to dryness with H_2SO_4 to expel any traces of fluoride ion present in solution, and the residue dissolved in perchloric acid. The pH is adjusted to 4.5–5.3, the dye solution added and the solution equilibrated at 54°C before cooling to room temperature prior to polarographing it. The method has been applied to a variety of rocks with a maximum error of 3%, which compares favorably with the maximum error found for the Alizarin Red-S colorimetric method.

Optical emission spectrography has been found useful for the quantitative determination of aluminum in amounts up to about 5%, particularly for samples containing large amounts of iron. The rapid determination of aluminum present as a major constituent can be done rapidly by X-ray fluorescence spectroscopy.

References

1. Bennett, H., Hawley, W. G., and Eardley, R. P.   The accurate volumetric determination of alumina in aluminosilicates.   *Trans. Brit. Ceram. Soc.* **61**, 201–206 (1962).

2. Cluley, H. J.   Rapid determination of Al_2O_3 in opal glass.   *Glass Technol.* **2**, 71–73 (1961).

3. Dinnin, J. I., and Kinser, C. A.   Indirect semiautomatic determination of alumina with EDTA.   *U.S. Geol. Surv. Profess. Paper* **424B**, 329–330 (1961).

4. Eshelman, H. C., Dean, J. A., Menis, O., and Rains, T. C.   Extraction and flame spectrophotometric determination of aluminum.   *Anal. Chem.* **31**, 183–187 (1959).

5. Groves, A. W.   *Silicate Analysis*, 2nd ed., 1951.   London: George Allen and Unwin, pp. 56–60.

6. Hillebrand, W. F., Lundell, G. E. F., Bright, H. A., and Hoffman, J. I.   *Applied Inorganic Analysis*, 2nd ed., 1953.   New York: John Wiley & Sons, pp. 75–96, 494–511, 866–879.

7. Holt, E. V., and Harwood, H. F.   The separation of manganese in rock analysis.   *Mineral. Mag.* **21**, 318–323 (1928).

8. Hsu, Pa Ho.   Effect of initial pH, phosphate and silicate on the determination of aluminum with aluminon.   *Soil Science* **96**, 230–238 (1963).

9. Hummel, R. A., and Sandell, E. B.   Separation of aluminum from iron with thioglycolic acid.   *Anal. Chim. Acta* **7**, 308–312 (1952).

10. Jeffery, P. G., and Wilson, A. D.   The precipitation of manganese in silicate rock analysis.   *Analyst* **84**, 663–665 (1959).

11. Kolthoff, I. M., and Sandell, E. B.   *Textbook of Quantitative Inorganic Analysis*, 3rd ed., 1952.   New York: Macmillan, pp. 77–80, 318–321, 703–704.

12. Kristiansen, H.   Direct titration of aluminum and stepwise titration of iron (III) and aluminum with EDTA and 3-hydroxy-2-napthoic acid as indicator.   *Anal. Chim. Acta* **25**, 513–515 (1961).

13. Levine, M.   The polarographic determination of aluminum in rocks.   *Geol. Surv. Canada Bull.*, **113**, 1964, 42 pp.

14. Lundell, G. E. F., and Hoffman, J. I.   *Outlines of Chemical Analysis*, 1938.   New York: John Wiley and Sons, Chapter IX, pp. 55–58.

15. Miller, C. C., and Chalmers, R. A.   Microanalysis of silicate rocks.   Part IV. The determination of alumina.   *Analyst* **78**, 686–694 (1953).

16. Milner, G. W. C., and Woodhead, J. L.   The determination of alumina in silicates (rocks and refractories).   *Anal. Chim. Acta* **12**, 127–137 (1955).

17. Nydahl, F.   The indirect complexometric titration of aluminum.   A study of the Wänninen-Ringbom method.   *Talanta* **4**, 141–146 (1960).

18. Parker, C. A., and Goddard, A. P.   The reaction of aluminum ions with alizarin-

3-sulphonate with particular reference to the effect of addition of calcium ions. *Anal. Chim. Acta* **4**, 517–535 (1950).

19. Riley, J. P. The rapid analysis of silicate rocks and minerals. *Anal. Chim. Acta* **19**, 413–428 (1958).

20. Riley, J. P., and Williams, H. P. The microanalysis of silicate and carbonate minerals. Part III. Determination of silica, phosphorus pentoxide and metallic oxides. *Mikrochim. Acta* **6**, 804–824 (1959); Part IV. Determination of aluminum in the presence of interfering elements. *Idem.* 825–830.

21. Ritchie, J. A. Methods of analysis of silicate rocks. *Dominion Laboratory Report* **D. L. 2049**, 1962, 32 pp.; Dept. Scientific and Industrial Research, New Zealand.

22. Sandell, E. B. *Colorimetric Determination of Traces of Metals*, 3rd ed., 1959. New York: Interscience, Chapter V, pp. 228–231.

23. Shapiro, L., and Brannock, W. W. Rapid analysis of silicate, carbonate and phosphate rocks. *U.S. Geol. Surv. Bull.* **1144–A**, A25–A27 (1962).

24. Smales, A. A., and Wager, L. R., (Eds.). *Methods in Geochemistry*, 1960. "Analysis by gravimetric and volumetric methods, flame photometry, colorimetry and related techniques," by E. A. Vincent. New York: Interscience, pp. 33–80.

25. Wänninen, E., and Ringbom, A. Complexometric titration of aluminum. *Anal. Chim. Acta* **12**, 308–318 (1955).

26. Washington, H. S. *The Chemical Analysis of Rocks*, 2nd ed., 1910. New York: John Wiley & Sons; pp. 97–110.

27. Weibel, M. Die aluminiumbestimmung in der chemischen silicatanalyse. *Z. Anal. Chem.* **184**, 322–327 (1961).

28. Wilson, A. D., and Sergeant, G. A. The colorimetric determination of aluminium in minerals by pyrocatechol violet. *Analyst* **88**, 109–112 (1963).

29. Wilson, C. L., and Wilson, D. W., (Eds.). *Comprehensive Analytical Chemistry*, Vol. 1A, 1959. "Precipitation from homogeneous solution," by Louis Gordon. New York: Elsevier, pp. 530–534.

Supplementary References

Amos, M. D., and Thomas, P. E. The determination of aluminum in aqueous solution by atomic absorption spectroscopy. *Anal. Chim. Acta* **32**, 139–147 (1965).

Berthelay, J. C. Dosage de l'aluminium dans les roches et les mineraux. *Ann. Fac. Sci. Univ. Clermont* **28**, 12–32 (1966).

Chablo, A. An evaluation of the complexometric methods of aluminum determination in silicates. *Chem. Anal. (Warsaw)* **9**, 501–508 (1964); in Polish.

Evans, W. H. Rapid complexometric determination of aluminum and total iron in silicate and other rock material. *Analyst* **92**, 685–689 (1967). Two separate aliquots of the sample solution, to which excess DCTA has been added, are back-titrated with 0.01M Cu solution (using o-dianisidine-N,N,N',N''-tetraacetic acid as indicator) under conditions which permit the determination of Al+Fe+Ti, and of Fe alone. Ti is determined colorimetrically on another aliquot and Al is obtained by difference.

Ingamells, C. O., and Suhr, N. H.   Chemical and spectrochemical analysis of standard silicate samples. *Geochim. Cosmochim. Acta* **27**, 897–910 (1963).

Kiss, E.   Chemical determination of some major constituents in rocks and minerals. *Anal. Chim. Acta* **39**, 223–234 (1967).   Various methods for the determination of aluminum were investigated and the complexometric titration with DCTA, following the extraction of Fe with MIBK and cupferron, of Ti with $CHCl_3$ and volatilization of Cr as chloride, is preferred.   A correction must be made for the Mn and Ni present.

Kolthoff, I. M., and Elving, P. J., (Eds.).   *Treatise on Analytical Chemistry*, Part II, Vol. 4, 1966. "Aluminum," G. H. Farrah and M. L. Moss.   New York: Interscience, pp. 367–439.

Mercy, E. L. P., and Saunders, M. J.   Precision and accuracy in the chemical determination of total Fe and Al in silicate rocks. *Earth and Planetary Science Letters* **1**, 169–182 (1966).

Pritchard, D. T.   Determination of aluminum in silicates with DCTA. *Anal. Chim. Acta* **32**, 184–186 (1965).

Van Loon, J. C.   Separation and accurate determination of iron and aluminum in a single sample of silicate rock. *Talanta* **13**, 1555–1560 (1966).

Van Loon, J. C.   Determination of aluminum in high silica materials. *Atomic Absorption Newsletter* **7**, 3–4 (1968).   The sample is decomposed with HF + H_2SO_4, taken up in HCl and buffered with 1% lanthanum prior to the determination of Al by atomic absorption spectroscopy.   Up to 2% Al was determined in silica brick, glass sand and silica refractory.

6.5. Calcium*

If the separations previously described have been made, the filtrate from the precipitation of the R_2O_3 group should now contain the alkaline earths (Ca, Sr, Ba), magnesium, manganese, a relatively large amount of sodium salts and small amounts of a number of elements which, for one reason or another, have escaped separation up until now.   It is possible, however, that some parts of the alkaline earth group may have already been removed.   If the SO_4^{2-} ion was present during the separation of SiO_2, then barium (and, rarely, strontium and calcium) may have accompanied the silica; the presence of fluoride ion during the precipitation with aqueous ammonia could have caused the precipitation of CaF_2, and that of carbonate ion the separation of varying amounts of all three of the alkaline earths.   If the PO_4^{3-} concentration exceeded that precipitable by the iron and aluminum present, then it is likely that some calcium phosphate was carried down with the R_2O_3 group.   Appropriate references have been made to these several possibilities, but if the methods previously described have been followed, all but a negligible amount of the alkaline earths will be present in the solution.   The

* References for this section will be found on pp. 160–163.

next step is the determination of this group, either directly after suitable precautions are taken to prevent the interference of other elements, or after first separating the alkaline earths from these elements and then separating them one from another. Calcium is usually preponderant over strontium and barium, and the two interfering elements which must be considered are manganese and magnesium. In the analytical scheme given here (Sec. 9.9) manganese is removed prior to the gravimetric separation of the alkaline earths.[20,30] A knowledge of the approximate concentrations of the alkaline earths and magnesium, such as is given by a preliminary semiquantitative spectrographic analysis, will be very helpful in choosing the most advantageous course to be followed next.

The analytical chemistry of calcium has recently been discussed in detail by Turekian and Bolter (see Kolthoff and Elving, Supp. Ref.). Of particular interest are the sections dealing with the determination of calcium by flame photometric and optical emission spectrographic methods. Reference should also be made to the discussion by Hillebrand *et al.*[17]

6.5.1. GRAVIMETRIC METHODS

The separation of calcium (and strontium), as calcium oxalate monohydrate ($CaC_2O_4 \cdot H_2O$), by neutralization with aqueous ammonia of an acid solution containing the calcium and ammonium oxalate, followed by the gravimetric determination of the calcium as carbonate or oxide, is the best-known and most widely used method. Precipitation is quantitative at pH 4 and the monohydrate, which is always obtained when precipitation is made from a hot solution, is stable under ordinary conditions; by adding ammonium oxalate to a hot, acid solution which is then neutralized with aqueous ammonia, a coarse precipitate which filters and washes readily is obtained, whereas that obtained from a cold, neutral or ammoniacal solution is finely divided and difficult to filter and wash. A second or third precipitation may be necessary to eliminate contaminating elements, particularly magnesium; in the presence of much magnesium it is preferable to so arrange conditions as to favor the complete removal of magnesium into the filtrate even though some calcium (and strontium) is thereby not precipitated and must be recovered later from the ignited magnesium precipitate. The ignited residue is then treated to separate the calcium from any strontium and barium that were coprecipitated with it. A fuller discussion of the oxalate procedure is given in Sec. 6.5.1a.

Calcium (and strontium) can also be precipitated as oxalate from a weakly acid solution containing organic acids such as oxalic, citric or acetic, thus obtaining a direct separation of it from such elements as Fe, Al, Ti and P in addition to magnesium. This method is useful for the determination of calcium in phosphate rock but finds little application in silicate analysis; manganese always interferes and magnesium may do so. Maynes (Supp. Ref.) found that 8-hydroxyquinoline-5-sulfonic acid is an effective masking agent in preventing the precipitation of Ti, Al and Fe from neutral and basic solutions; it also forms more stable complexes with manganese and magnesium than with calcium.

The alkaline earths can be separated as sulfates in a sulfuric acid–alcohol medium. It is not usually possible to make a direct separation of them as sulfate because of the presence of other elements which also form insoluble sulfates; too, alkali sulfates are relatively insoluble in alcohol, and sodium is probably present in abundance from the initial fusion of the sample. Thus the sulfate separation, although better in some respects than the oxalate separation, requires much additional manipulation in order to remove unwanted sulfates, and it is more commonly used to recover traces of the alkaline earths from the ignited magnesium precipitate.

Calcium is quantitatively precipitated as $Ca_3(PO_4)_2$ under the conditions favoring the precipitation of magnesium ammonium phosphate (Sec. 9.10.2). When the calcium concentration is very low ($<0.5\%$) and much magnesium is present, it is better to coprecipitate the calcium with the magnesium and to recover it as the sulfate from the ignited pyrophosphates.

(a) Precipitation as oxalate

The precipitation of the oxalate group is a deceptively simple step but it is possible, through insufficient appreciation of the likely sources of error, to obtain results that are far from being correct. Detailed investigations of the many factors which influence the form and purity of the final precipitate have been reported[18,22] and the optimum conditions for precipitation have already been suggested. The method to be discussed here is applicable when the CaO content of the sample exceeds 0.5% and calcium is preponderant over magnesium.

In the absence of manganese the possible coprecipitation, occlusion and post-precipitation of magnesium remain the chief sources of error. The coprecipitation of magnesium oxalate is minimized by introducing a large excess of ammonium oxalate into the solution whereby the magnesium oxalate tends to form a supersaturated solution. This supersaturation is broken by boiling and thus, particularly when much magnesium is present, boiling of the solution should be minimized or avoided. Kolthoff and Sandell (Ref. 22, p. 347) state that as much as 150 mg of magnesium may be present in 200–250 ml of solution without precipitation of magnesium oxalate taking place when the oxalate separation is made, as given in Sec. 9.10.1, but magnesium in excess of this amount may separate on standing, especially if the solution is kept warm. This post-precipitation of magnesium oxalate is to be avoided because even a triple precipitation of the calcium may not rid it of the accompanying magnesium. In most cases, however, this will not be a problem. The excess ammonium oxalate also serves to reduce the solubility of calcium oxalate in the presence of much magnesium; the latter forms a strong complex with oxalate ions and, unless an excess is present, may tie up oxalate needed to insure maximum precipitation of the calcium. A single precipitation of the calcium, in the presence of magnesium, will carry with it some magnesium as a result of occlusion; a double precipitation, which should always be made in accurate work, will eliminate the magnesium. A disadvantage of the use of a large excess of ammonium oxalate is that it will interfere in the subsequent precipitation of magnesium unless first removed; Holth[18] states that neither nitric acid nor aqua regia will decompose oxalate and recommends that the solution be evaporated to dryness and the residue heated with a free flame.

The double precipitation also serves to remove traces of alkali salts and silica, the latter derived from attack upon the glass by the ammoniacal solution, which are present in the first precipitation. It is unlikely, if the separation of the R_2O_3 group was done as described (Sec. 9.6), that any iron, aluminum or rare earths escaped precipitation, but if these elements are thought to be present in the first precipitate they can be recovered at this stage by destroying the oxalate and precipitating them with aqueous ammonia. Barium, if less than 5 mg are present, will be eliminated by the second precipitation, but the precipitation of strontium is almost quantitative, and it is weighed with the calcium. The barium and strontium which were not separated, together with most of the small amount of calcium still left in the solution, can be recovered as sulfates before the precipitation of magnesium; the unrecovered calcium will also be found with the magnesium and can be recovered from the pyrophosphate residue. Sulfate ion may give rise to the coprecipitation of calcium sulfate together with the oxalate and this will lead to high results if the calcium is weighed as the carbonate; ignition of the latter to the oxide will most likely decompose the sulfate also. Sulfide ion should be expelled from the acidified ammonium sulfide filtrate, if ammonium sulfide was used to separate manganese, prior to the precipitation of the calcium; any free sulfur formed and not removed will be collected by the oxalate precipitate and may be oxidized and fixed in the residue during the ignition. The ignited residue can, if necessary, be dissolved and the precipitation of the calcium repeated but it is less work to remove the hydrogen sulfide by boiling the acidified solution, oxidizing any remaining sulfide with bromine and filtering if so required.

Precipitation from homogeneous solution will yield a very coarse-grained precipitate of calcium oxalate;[14,21] either the hydrolysis of urea, or the hydrolysis of an ester of oxalic acid, such as dimethyl oxalate, can be used to give off the desired pH. Both of these procedures suffer from certain disadvantages and in most situations the standard method of precipitation will suffice.

There are several weighing forms of calcium that may be used but the oxide and carbonate are the ones most commonly employed. The properties, advantages and disadvantages of the monohydrate, anhyddrous oxalate, carbonate, oxide, sulfate and fluoride are discussed in detail by Kolthoff and Sandell (Ref. 22, pp. 342–346), who favor the use of the carbonate over the oxide because the former has a much higher molecular weight and is not appreciably hygroscopic. The calcium carbonate must, however, be collected in a Gooch or porcelain filter crucible rather than on filter paper, because the carbon of the paper will not be completely burned away at the ignition temperature used (475–525°C). If the oxide is used as the weighing form, the precipitate, after the paper has been burned off, is ignited at 1100–1200°C in a loosely covered crucible to facilitate the escape of carbon dioxide; the crucible must be kept tightly covered in a desiccator before weighing, because it quickly picks up water and slowly reacts with carbon dioxide, and the weighing must be made as rapidly as possible. Use of the carbonate as weighing form is recommended by others.[18,40] but Groves[15] considered that the need to have an accurate temperature control, if a residue of known composition is to be obtained, was too great a disadvantage to warrant its use. If adequate precautions are taken during the cooling and weighing of the crucible, ignition to the oxide is a convenient and reliable way of finishing the determination.

The ignited precipitate will contain, in addition to calcium, most of the strontium present in the sample and perhaps some of the barium. A correction may be applied to the value for CaO based upon the SrO content as determined spectrographically; alternatively, the ignited oxide can be dissolved and the strontium content determined by flame photometry. If these alternatives are not feasible, there remains a method of separation based upon the differential solubility of the dry nitrates of the alkaline earths.* Following the separation of barium as chromate, the dried nitrates are leached with either concentrated nitric acid (1.46–1.42 specific gravity) or with absolute alcohol and absolute ether; in either case the calcium dissolves preferentially and the insoluble residue of strontium nitrate can be filtered and weighed (see Sec. 9.33). Leliaert and Eeckhaut[25] studied the separation of calcium and strontium by various chemical procedures and concluded that the most reliable results are obtained with the nitric acid and alcohol–ether methods, although neither method gives a quantitative separation. They prefer to use an isotopic dilution method.

6.5.2. TITRIMETRIC METHODS

The $KMnO_4$ titration of the oxalate ions liberated when calcium oxalate is dissolved in dilute sulfuric acid, to the first permanent tinge of pink, is a well-established and reliable procedure capable of giving exact results when calcium alone is present. It is more rapid than a method involving the ignition of the precipitate to some particular weighing form and is particularly useful in the routine analysis of samples containing much calcium, such as limestones and dolomites. A double precipitation of the calcium is made following the prior removal of the R_2O_3 group, to give a precipitate relatively free of interfering ions. It is obvious that other elements present which form insoluble oxalates or are otherwise carried down with the calcium oxalate, such as manganese, strontium, barium and magnesium, will contribute to erroneously high results for calcium; a double precipitation will leave only strontium to accompany the calcium and a correction for this can be made. It is also important that the precipitated calcium compound have a $Ca^{2+} : C_2O_4^{2-}$ ratio of 1:1, such as is obtained when the precipitation is made from an acid solution as previously described; low results are found for precipitates produced in neutral or ammoniacal solutions, because of the possible formation of basic calcium oxalate or calcium hydroxide (Ref. 22, pp. 575–577).

It is preferable that the precipitate be collected on an asbestos or fritted-glass filter when a titrimetic finish is planned. Paper fibers, in acid medium, will reduce MnO_4^- and should be avoided; if it is necessary to use paper, the precipitate should be dissolved from the filter paper

* An anion exchange procedure for separating calcium and strontium has recently been described[12] but it processes only microgram quantities of the elements and does not appear to have much application to rock and mineral analysis.

with as little wear of the paper surface as is possible and the paper added to the titration flask only after the first permanent pink color has been obtained.

Much use is now made of the titration with ethylenediaminetetra-acetic acid (EDTA), usually as a 0.1–$0.01M$ solution of the disodium salt, using one of the numerous metal–indicators available to signal the end-point. Mention has already been made of the use of EDTA in the titration, directly or indirectly, of aluminum (Sec. 6.4.2) and the development of similar methods for the titration of calcium and of calcium plus magnesium in aliquots of a solution prepared by decomposing the sample with HF and H_2SO_4 (but see Ref. 19) has been an important part of the development of schemes for rapid rock analysis. Because of the non-specificity under normal conditions of EDTA towards metal ions, it is necessary to insure the absence of interfering elements by masking or separating them, and the many methods to be found in the literature are commonly variations on this theme. The use of EDTA as a titrant is now too complex a subject to permit detailed study here and the reader should consult more general references for further information and guidance.[1,34,37,44]

The usual procedure is to titrate the calcium alone, after removing the magnesium by precipitating it as the hydroxide at about a pH of 12, and then to titrate in a separate aliquot the calcium and magnesium at pH 10, thereby obtaining the magnesium by difference. The titrations are done in solutions from which the interfering ions are removed by precipitation or masked in some way to prevent their interference. While for the titration of magnesium (or calcium and magnesium) the metal indicator generally used is Eriochrome Black T (Ref. 1, p. 2), a variety of metal indicators have been and continue to be suggested for the titration of calcium.

Magnesium is usually precipitated with sodium hydroxide and the titration of the calcium is done in the presence of the precipitate. When much magnesium is present, this method is subject to two sources of error; (1) the precipitate occludes calcium ions and the much slower reaction of these occluded ions with EDTA results in an uncertain end point and usually low results and, (2) the precipitate also adsorbs some indicator which diminishes the sharpness of the end point. Baugh, Decker and Palmer[2] obtained poor results in the determination of calcium ($<0.5\%$) in magnesium oxide by the conventional EDTA procedure; better results were obtained when either the bulk of the magnesium was first removed by precipitation with aqueous ammonia and centrifugation, or the magnesium was precipitated slowly and as homogeneously as possible by adding a buffer solution (NaOH, KCN and NH₂OH·HCl). Sadek et al.[35] titrated calcium with EGTA (ethylene glycol-bis(β-aminoethyl ether)-N,N'-tetraacetic acid) which chelates with Ca selectively, using *zincon* (2-carboxy-2'-hydroxy-5'-sulfoformazylbenzene) as indicator (with added Zn-EGTA as indicator sensitizer). KCN and Eriochrome Black T were then added to the solution and the magnesium titrated directly with EDTA. Přibil and Veselý (Supp. Ref.) have used EGTA to determine calcium in limestone and dolomite, with calcein[8,43] as indicator; better accuracy is obtained if excess EGTA is added to mask calcium and the excess back-titrated with a standard calcium solution following the direct titration of magnesium with (1,2-cyclohexylenedinitrilo) tetraacetate (DCYTA), using methylthymol blue as indicator. By using a minimum amount of calcein in the titration of calcium, Goldich et al.[13] found that magnesium could be determined directly in the solution after the addition of ammonium. chloride and Eriochrome Black T.

The presence of metals such as iron, titanium, aluminum, copper, nickel, manganese and cerium will seriously affect the behavior of the metal indicators by oxidation or blocking. Oxidants may be reduced with ascorbic acid or hydroxylamine; manganese, iron and aluminum may be masked with triethanolamine, and copper and nickel with potassium cyanide. It is usually preferable to remove the interferences by precipitation and to mask any remaining traces of these elements with KCN and triethanolamine. The simplest procedure, although not the most desirable, is to precipitate the bulk of the interfering elements with aqueous ammonia; for most rocks this will leave only manganese still present as a source of interference and it can be masked, along with traces of the other heavy metals that have gone unprecipitated, by triethanolamine. It is likely that the R₂O₃ precipitate will adsorb some calcium and a double precipitation is usually necessary; this can be done rapidly by carrying out the precipitations in a centrifuge tube. Derderian[7] used zirconium, added as zirconyl oxychloride and precipitated at pH 5.5–6.5, to remove interfering phosphate in the titration of calcium and magnesium in plant material with EDTA; he found that the gelatinous hydrolysis product of the excess zirconium also removed heavy metals, probably by adsorption, and made the use of inhibitors unnecessary. No calcium or magnesium was lost when only a single precipitation was used. Kuhn[24] found that ammonium salts interfered in the titrations both by their buffering action and by making the precipitation of magnesium incomplete; he preferred to extract the cupferrates or acetylacetonates of iron and aluminum, and manganese diethyldithiocarbamate, with chloroform. Extraction of the 8-hydroxyquinolates of Fe and Al at pH 5.0 with chloroform was used by Cluley[6] and Riley[32,33]; the latter employed a continuous extractor, rather than a separatory funnel, to remove a large number of elements, and also used a preliminary precipitation with zirconium to remove phosphate, arsenate and selenate if present, the excess zirconium being removed during the subsequent extraction with 8-hydroxyquinoline. Riley precipitated manganese dioxide from the extracted solution by boiling with sodium chlorite.

The first metal indicator used for the complexometric titration of calcium was *murexide* (ammonium purpurate); the end point is difficult to see and it is customary now to screen it with another indicator for better end point detection. Naphthol Green B and murexide give a color change from pinkish brown to clear blue.[4] Chalmers[5] determined the end point spectrophotometrically at 610 mμ in a special cell designed for the microtitration of calcium in less than 5 ml of solution. *Calcein*,[8,43] or fluorescein complexone, shows a color change from yellow-green to pinkish-brown, best viewed in diffuse light. Below pH 12, both calcein and its calcium complex have a yellow-green fluorescence, but above pH 12 the indicator is brown, while the calcium complex remains fluorescent yellow-green; at the end point, the fluorescence and color of the calcium complex disappears and the end point is marked in ultraviolet light by the quenching of the fluorescence (see Diehl, Supp. Ref.). This latter is difficult to detect and it is better to use the color change of the indicator to locate the end point; an improved end point (green to purple) is obtained when a mixture of calcein, thymolphthalein and potassium chloride is used. *Calcon*[16] or Eriochrome Blue Black R, with a color change of pink to blue, has proved to be a good indicator for the titration of calcium, provided that some magnesium is present. Lott and Cheng[26] used it to detect the calcium end point in their stepwise titration procedure for calcium and magnesium, and later[27] found that the addition of polyvinyl alcohol reduced the adsorption of the indicator by the magnesium hydroxide precipitate; calcium is titrated directly with EDTA at pH 13 with Calcon as indicator, the pH is then adjusted to 10 and calcium plus magnesium is titrated with Eriochrome Black T as usual. Patton and Reeder[29] introduced *Calred*, which changes from wine-red to pure blue, as an indicator for calcium when titrated at pH 12–14 in the presence of magnesium. Belcher et al.[3] made a critical study of the complexometric titration of calcium in the presence of magnesium, with particular reference to the best choice of indicator; they omitted *Calred* from the comparison because they found it to be somewhat unstable in a strongly alkaline solution. *Murexide* was found to give a low recovery of calcium, *calcein* and *methylthymol blue* gave satisfactory results for pure calcium solutions only; *Calcon* was found to be the best indicator, particularly when increasing amounts of magnesium are present (e.g., for Mg:Ca = 1:10 to 1:5), and *Acid Alizarin Black SN* is recommended for the titration of calcium in pure solution and in solutions where the Mg:Ca ratio is less than 1:12. A new indicator that has been found useful at the Geological Survey is the sodium salt of thymolphthalexone.*

Visual titrations of calcium with EDTA are more easily done now, thanks to the new indicators that have been described. It still requires experience, however, to obtain consistently good end points and efforts have been made to automate the titration and thus eliminate operator bias as a source of error. Shapiro and Brannock[38,39] made an automatic photometric titration a part of their rapid analytical scheme; the change in absorbance during the EDTA titration of the calcium with murexide as indicator, and calcium plus magnesium using Eriochrome Black T, is plotted by a chart recorder, and the length of the titration compared with that required to titrate a known amount of calcium or magnesium is used to calculate the calcium or magnesium content of the sample. In silicate rocks, calcium and calcium plus magnesium are determined by the automatic titration method except for those samples in which the magnesium is less than 1%; in the latter, the magnesium is determined colorimetrically and the calcium is obtained by difference from the total of calcium plus magnesium. For carbonate rocks, a visual EDTA titration is preferred to the automatic one described previously; a solution of methyl red is added to screen the Eriochrome Black T, its yellow color emphasizing the complete disappearance of red at the end point. The calcium is obtained by difference from the determinations of calcium plus magnesium and of magnesium, the latter being determined either colorimetrically or by automatic titration following precipitation of the calcium as the oxalate. Malmstadt and Hadjioannou[28] used an automatic titrator designed to employ the automatic derivative technique for spectrophotometric end point detection to determine accurately both calcium and magnesium in limestone and dolomite. Calcon and Eriochrome Black T are used as the indicators for the titrations.

---

* This is 3,3'-(*bis-N,N*'-di(carboxymethyl)-(aminomethyl)-thymolsulfonphthalein ($C_{37}H_{44}N_2S$) and is conveniently available in tablet form (each containing 0.5 mg of the indicator) under the trade name, METAB.[11] The color change at pH 12 is blue to gray, and suitably sharp.

6.5.3. OTHER METHODS

Colorimetric methods for the determination of calcium are confined to the measurement of microgram quantities only and have found little use in rock and mineral analysis. The available methods were, until recently, of an indirect nature in which the anion associated with the calcium, rather than the calcium itself, is measured. For example, calcium phosphate is separated by precipitation, redissolved and the phosphate concentration determined by the molybdenum blue method[36]; calcium chloranilate, precipitated in a sodium chloranilate solution, is redissolved in EDTA disodium salt solution and the absorbance of the purple-red sodium chloranilate complex is measured.[10] Some direct methods are also available now, including the use of ammonium purpurate (murexide), o-cresolphthalein complexone,[36] glyoxal bis(2-hydroxyanil) (Williams and Wilson, Supp. Ref.) and Calcichrome, or cyclotris-7-(1-azo-8-hydroxy-naphthalene-3,6-disulfonic acid) (Herrero-Lancina and West, Supp. Ref.).

The determination of calcium by flame emission photometry is not an easy task and is one that does not appear to have gained much acceptance among rock and mineral analysts. The chief source of difficulty appears to lie in the anionic suppression of the calcium emission by sulfate, phosphate and aluminate. Stone and Thomas[41] have developed a rapid method for rocks having a low calcium content (up to 2.5% CaO) that utilizes an aliquot of solution B (Sec. 9.11.2) from which the R_2O_3 group is first removed by a double precipitation with aqueous ammonia, and which yields results that compare favorably with those obtained by EDTA titration. An empirical correction is made for the effect of sodium on the calcium emission; the magnesium interference varies with the calcium content[23] but in most acid rocks the effect is negligible. Others have suggested the use of releasing agents to eliminate the depressive effects of sulfate, aluminate and phosphate. Dinnin[9] found that Sr, La, Pr, Nd, Sm and Y are completely effective as such and using an oxygen–hydrogen flame, and with Sr as the releasing agent, determined calcium in a series of silicate rocks without the need for preliminary separations. Glycerol (10 % v/v) added to a $0.1M$ $HClO_4$ solution of the sample was found to act as a releasing agent in both oxygen–hydrogen and oxygen–acetylene flames by Rains et al.,[31] and they successfully applied the method to the determination of calcium in limestone, dolomite and phosphate rock, working with calcium concentrations of 0.1 to 10 $\mu g/ml$. Kramer[23] used a large excess of magnesium to suppress anionic interferences in solutions of phosphate, carbonate and silicate rocks; an oxygen–hydrogen flame was employed and the results obtained agreed to within $\pm$ 2% of those found by gravimetric methods in the 1–95% range. The lower limit of sensitivity of the method is 0.01% and it compares favorably with other methods in the 0–1% range. A radiation buffer is added to the sample and standard solutions to give final concentrations of 400 ppm Al, 250 ppm P, 900 ppm Mg, 50 ppm Fe, 80 ppm Na, 30 ppm K and 6 ml HNO_3 in each.

Abbey (Supp. Ref.) has described a simple flame emission photometric method for the determination of calcium, sodium and potassium in feldspars, using as little as 10 mg of sample; an excess of magnesium is added to minimize the depressant effect of aluminum on the calcium emission. Both flame emission photometric and optical emission spectrographic methods for similar determinations in feldspars have been given by Fraser and Downie (Supp. Ref.).

Atomic absorption spectroscopy is a relatively new analytical tool, and while much work has been done on the determination of calcium in soil and plant extracts, for example, little attention has been paid to its determination in rocks and minerals. Trent and Slavin[42] have investigated the determination of Na, K, Ca, Mg, Mn and Fe in a granite (G–1) and diabase (W–1); for 20 determinations on four separate aliquots (1:9 HCl) a standard deviation of 0.2% was found for CaO = 10.8%, giving a relative coefficient of variation of 2%, which includes errors due to sample preparation and the inhomogeneity of the samples. Lanthanum is used to suppress anionic interferences and it is necessary to match the density of the sample solution to that of the standards by adding an equal amount of HCl to the latter.

References

1. Barnard, A. J., Jr., Broad, W. C., and Flaschka, H. The EDTA titration: nature and methods of end point detection. *Chemist-Analyst* **45,** 86–93, 111–112 (1956); **46,** 18–28, 46–56, 76–84 (1957); *J. T. Baker Chemical Company Brochure,* November, **1957,** 30 pp.
2. Baugh, C. A., Decker, K. H., and Palmer, J. W. Determination of small amounts of calcium in magnesium oxide. *Anal. Chem.* **33,** 1804–1805 (1961)
3. Belcher, R., Close, R. A., and West, T. S. The complexometric titration of calcium in the presence of magnesium: a critical study. *Talanta* **1,** 235–244 (1958).
4. Bennett, H., Hawley, G. W., and Eardley, R. P. Rapid analysis of some silicate materials. *Trans. Brit. Ceram. Soc.* **57,** 1–28 (1958).
5. Chalmers, R. A. A spectrophotometric microtitration of calcium. *Analyst* **79,** 519–521 (1954).
6. Cluley, H. J. The rapid determination of lime and magnesium in soda–lime glasses. *Analyst* **79,** 567–573 (1954).

7. Derderian, M. Determination of calcium and magnesium in plant material with EDTA. *Anal. Chem.* **33**, 1796–1798 (1961).

8. Diehl, H., and Ellingboe, J. L. Indicator for titration of calcium in presence of magnesium using disodium dihydrogen ethylenediamine tetraacetate. *Anal. Chem.* **28**, 882–884 (1956).

9. Dinnin, J. I. Releasing effects in flame photometry. Determination of calcium. *Anal. Chem.* **32**, 1475–1480 (1960).

10. Fisher Scientific Company. A direct colorimetric reagent for calcium, zirconium, strontium and molybdenum. *Tech. Data* **TD–105**, 7 pp. 10/59.

11. Fisher Scientific Company. Metal Indicator tablets for acid-base and metal titrations. *Tech. Data* **TD–150**, 3 pp. 7/61.

12. Fritz, J. S., Waki, H., and Garralda, B. P. Anion exchange separation of calcium and strontium. *Anal. Chem.* **36**, 900–903 (1964).

13. Goldich, S. S., Ingamells, C. O., and Thaemlitz, D. The chemical composition of Minnesota lake marl—comparison of rapid and conventional chemical methods. *Econ. Geol.* **54**, 285–300 (1959).

14. Gordon, L., Salutsky, M. L., and Willard, H. H. *Precipitation from Homogeneous Solutions*, 1959. New York: John Wiley & Sons, pp. 51–59.

15. Groves, A. W. *Silicate Analysis*, 2nd ed., 1951. London: George Allen and Unwin, pp. 60–66.

16. Hildebrand, G. P., and Reilley, C. N. New indicator for complexometric titration of calcium in the presence of magnesium. *Anal. Chem.* **29**, 258–264 (1957).

17. Hillebrand, W. F., Lundell, G. E. F., Bright, H. A., and Hoffman, J. I. *Applied Inorganic Analysis*, 2nd ed. (rev.), 1953. New York: John Wiley & Sons, Chapter 40, pp. 611–631, 883.

18. Holth, T. Separation of calcium from magnesium by oxalate method. *Anal. Chem.* **21**, 1221–1226 (1949).

19. Hoops, K. G. The nature of the insoluble residues remaining after the HF–H$_2$SO$_4$ acid decomposition (solution B) of rocks. *Geochim. Cosmochim. Acta* **28**, 405–406 (1964).

20. Ingamells, C. O. Persulfate separation of Mn from R$_2$O$_3$ filtrate. Private communication, College of Mineral Industries, The Pennsylvania State University, University Park.

21. Ingols, R. S., and Murray, P. E. Urea hydrolysis for precipitating calcium oxalate. *Anal. Chem.* **21**, 525–527 (1949).

22. Kolthoff, I. M., and Sandell, E. B. *Textbook of Quantitative Inorganic Analysis*, 3rd ed., 1952. New York: MacMillan, pp. 337–351.

23. Kramer, H. The flame photometric determination of calcium in phosphate, carbonate and silicate rocks. *Anal. Chim. Acta* **17**, 521–525 (1957).

24. Kuhn, V. Progrès récentes dans les dosages complexométriques du calcium et du magnesium dans les laitiers. *Chim. Anal.* **40**, 340–344 (1958).

25. Leliaert, G., and Eeckhaut, J. Investigation of chemical separation methods for Ca and Sr. *Anal. Chim. Acta* **16**, 311–320 (1957).

26. Lott, P. F., and Cheng, K. L. Stepwise EDTA titration of calcium and magnesium with CI 202 and CI 203 as indicators. *Chemist-Analyst* **46**, 30–31 (1957).

27. Lott, P. F., and Cheng, K. L. Improved end point by addition of polyvinyl

162

ROCK AND MINERAL ANALYSIS

alcohol in the EDTA titration of calcium with Calcon as indicator. *Chemist-Analyst* **48**, 13 (1959).
28. Malmstadt, H. V., and Hadjioannou, T. P. Rapid and accurate automatic titration of calcium and magnesium in dolomites and limestones. *Anal. Chim. Acta* **19**, 563–569 (1958).
29. Patton, J., and Reeder, W. New indicator for titration of calcium with EDTA. *Anal. Chem.* **28**, 1026–1028 (1956).
30. Peck, L. C., and Smith, V. C. Removal of manganese from solutions prior to the determination of calcium and magnesium. *U.S. Geol. Surv. Profess. Papers* **424–D**, 401–402 (1961).
31. Rains, T. C., Zittel, H. E., and Ferguson, M. Elimination of anionic interferences in the flame spectrophotometric determination of calcium. *Talanta* **10**, 367–374 (1963).
32. Riley, J. P. The rapid analysis of silicate rocks and minerals. *Anal. Chim. Acta* **19**, 413–428 (1958).
33. Riley, J. P. The use of continuous extraction for the removal of interfering elements in the determination of calcium and magnesium. *Anal. Chim. Acta* **21**, 317–323 (1959).
34. Ringbom, A. *Complexation in Analytical Chemistry*, 1963. New York: Interscience, 395 pp.
35. Sadek, F. S., Schmid, R. W., and Reilley, C. N. Visual EGTA titration of calcium in presence of magnesium. *Talanta* **2**, 38–51 (1959).
36. Sandell, E. B. *Colorimetric Determination of Traces of Metals*, 3rd ed., 1959. New York: Interscience, pp. 366–380.
37. Schwarzenbach, G. *Complexometric Titrations*, 1957. London: Methuen, 132 pp.
38. Shapiro, L., and Brannock, W. W. Automatic photometric titrations of calcium and magnesium in carbonate rocks. *Anal. Chem.* **27**, 725–728 (1955).
39. Shapiro, L., and Brannock, W. W. Rapid analysis of silicate, carbonate and phosphate rocks. *U.S. Geol. Surv. Bull.* **1144–A**, 35–46 (1962).
40. Smales, A. A., and Wager, L. R., (Eds.). *Methods in Geochemistry*, 1960. New York: Interscience, p. 46.
41. Stone, M., and Thomas, J. E. Flame photometric determination of calcium in silicate rocks. *Analyst* **83**, 691–694 (1958).
42. Trent, D., and Slavin, W. Determination of the major metals in granitic and diabasic rocks by atomic absorption spectrophotometry. *Atomic Absorption Newsletter* **19**, 1–6 (1964); Perkin-Elmer Corporation, Norwalk, Connecticut.
43. Tucker, B. M. Calcein as an indicator for the titration of calcium with EDTA. *Analyst* **82**, 284–285 (1957).
44. Welcher, F. J. *The Analytical Uses of Ethylenediamine Tetraacetic Acid*, 1957. Princeton, N. J.: D. Van Nostrand, 366 pp.

Supplementary References

Abbey, S. Determination of potassium, sodium and calcium in feldspars. *Can. Mineral.* **8**, 347–353 (1965).
Diehl, H. *Calcein, Calmagite and o,o'-Dihydroxyazobenzene; Titrimetric, Colorimetric and Fluorometric Reagents for Calcium and Magnesium*, 1964. Columbus, Ohio: The G. Frederick Smith Chemical Company, pp. 21–29.

Fraser, W. E., and Downie, G. The spectrochemical determination of feldspars within the field microcline–albite–labradorite. *Mineral. Mag.* **33**, 790–798 (1964).

Herrero-Lancina, M., and West, T. S. Sensitive and selective spectrophotometric reaction for determination of trace amounts of calcium. *Anal. Chem.* **35**, 2131–2135 (1963).

Kiss, E. Chemical determination of some major constituents in rocks and minerals. *Anal. Chem. Acta* **39**, 223–234 (1967). Interfering elements (Fe, Ti, Al, Mn) are removed by chelation with acetylacetone and extraction with CCl_4, prior to the gravimetric or titrimetric (EDTA) determination of calcium and magnesium.

Kolthoff, I. M., and Elving, P. J., (Eds.). *Treatise on Analytical Chemistry*, Part II, Vol. 4, 1966. "Calcium," by K. K. Turekian and E. Bolter. New York: Interscience, pp. 107–152.

Maynes, A. D. The direct gravimetric determination of calcium in silicates. *Anal. Chim. Acta* **32**, 288–291 (1965).

Přibil, R., and Veselý, V. Determination of magnesium (and calcium) in calcareous materials and silicates. *Chemist-Analyst*, **55**, 82–84 (1966).

Williams, K. T., and Wilson, J. R. Colorimetric determination of ultramicro quantities of calcium, using glyoxal bis(2-hydroxyanil). *Anal. Chem.* **33**, 244–245 (1961).

6.6. Magnesium*

Frequent mention has been made of magnesium in the discussion of the determination of calcium (Sec. 6.5), usually with regard to its effect upon the latter, and reference should be made where appropriate to this material. A recent and thorough coverage of the analytical chemistry of magnesium is given by Wengert, Reigler and Carlson (Ref. 16, pp. 43–93); Chalmers[8] has summarized recent analytical methods for magnesium in silicates. Because many of the references cited in Sec. 6.5 deal as well with the determination of magnesium, appropriate citation is made of them in the following sections but they are not included in the References for Sec. 6.6.

6.6.1. GRAVIMETRIC METHODS

Magnesium is usually determined gravimetrically by precipitation with dibasic ammonium phosphate (($NH_4)_2HPO_4$) followed by ignition and weighing as the pyrophosphate ($Mg_2P_2O_7$). Less frequently it is precipitated as the 8- hydroxyquinolate and ignited to the oxide. Booth (Ref. 30, pp. 65–72) described the use of 2-hydroxy-1-naphthaldehyde as a reagent for the gravimetric determination of magnesium; the complex compound ($C_{22}H_{14}O_4Mg$) contains 6.63% Mg, is only slightly soluble

* References for this section will be found on pp. 172–174.

in water and ethyl alcohol and is particularly useful for the determination of semimicro amounts of magnesium. A preliminary separation of interfering elements is necessary. The precipitation at pH 3.5–4.0 of magnesium by EDTA, as a salt closely approximating to $MgC_{10}H_{14}O_8N_2 \cdot 6H_2O$, has also been proposed[6]; it can be dried at room temperature and weighed as the hydrate, or vacuum-dried at 100°C and weighed as the anhydrous salt. If Ca > Mg, a double precipitation is necessary.

(a) Precipitation with dibasic ammonium phosphate

If the separations of the various groups of elements have been made as previously described, it is likely that there will be present in the combined filtrates from the precipitation of calcium (see Sec. 9.10.1), in addition to magnesium, only alkali salts, perhaps a small amount of manganese and any calcium, strontium and barium that escaped precipitation in the ordinary scheme of analysis and was not separated as sulfate in ethanolic medium. This assumes, of course, that the ammonium salts accumulated during the analysis have been destroyed; it is preferable that the concentration of ammonium oxalate should not exceed 1 g/100 ml if a satisfactory precipitation of magnesium is to be obtained. Bodenheimer and Gaon[5] confirm the warning of Holth (Sec. 6.5, Ref. 18) that ammonium oxalate is not detroyed by boiling with nitric acid alone; it must either be volatilized from a dried residue or the solution fumed with HNO_3 and H_2SO_4. They found, however, that a satisfactory precipitation of as little as 100 μg of magnesium is obtained when the amount of dibasic ammonium phosphate added was equal to that of the ammonium oxalate already present. This latter is not easy to determine and removal of the excess ammonium oxalate is to be preferred. These authors showed, incidentally, that the precipitation of small amounts of magnesium in the presence of ammonium oxalate is progressively retarded as the magnesium content decreases.

Because it is virtually impossible to make an accurate, single precipitation of magnesium, a double precipitation is the general rule, with nearly ideal conditions being established for the second precipitation. Magnesium ammonium phosphate is relatively soluble as a precipitation medium, its solubility being determined by the concentration of the NH_4^+ and PO_4^{3+} ions present. Thus its solubility is considerably diminished by increasing the concentration of aqueous NH_3 to 5–10%. An excess of NH_4^+ salts, particularly of ammonium phosphate, will, however, result in the eventual precipitation of a number of magnesium compounds* instead of only one. Solubility losses are further reduced by keeping the volume of the solution small, by cooling the solution to 10°C or so before precipitation and by allowing the solution and precipitate to stand overnight in order to discourage the tendency of the precipitate to form a supersaturated solution. Thus it is preferable to carry out the first precipitation in such a way as to insure the minimum solubility loss, regardless of the composition of the precipitate, and then to dissolve it and reprecipitate the magnesium under carefully controlled conditions in which the amount of NH_4^+ salts in solution is kept to a minimum and only a slight excess of ammonium phosphate is present. The problems inherent in this precipitation have been studied in detail.[14,17]

Successful ignition of the precipitate to $Mg_2P_2O_7$ is a difficult and sometimes frustrating task, because of the tendency of carbon from the filter paper to become entrapped in the pyrophosphate and virtually "fireproofed." The charring of the paper and oxidation of the carbon must be done at a low temperature and must be done slowly; under no circumstances must the crucible be allowed to become even a dull red before all of the carbon has disappeared.† The use of a Gooch crucible is not recommended because of the attack on the asbestos by the strongly alkaline solution.

The final ignition of the crucible and contents should be done at about 1100°C. At temperatures much above this there is the possibility of reduction of phosphate with subsequent damage to the platinum crucible, or of loss of phosphorus; if temperatures much below 1100°C are used, the $Mg_2P_2O_7$ will only very slowly come to constant weight. The final residue is seldom white and more likely will have a greyish color.

Mention has been made of the possible errors introduced by the presence of compounds in the precipitate whose ignited form is not $Mg_2P_2O_7$. Another and particularly annoying source of error is in the unsuspected presence of coprecipitated compounds such as $Ca_3(PO_4)_2$, which, if present in sufficient quantity, may lower the melting point of the pyrophosphate sufficiently to enable it to fuse into a lump at temperatures

---

* Kolthoff and Sandell[17] list 11 different compounds that may be present; ignition of this mixture could lead to either positive or negative errors, depending upon the particular compound present in greatest concentration.

† Attempts to correct a carbon-bearing pyrophosphate residue by subsequent treatment with HNO_3 or oxygen have not been successful. If a correction for entrapped carbon is necessary, the writer prefers to dissolve the weighed residue in dilute H_2SO_4 (1:10) and to filter and weigh the carbon in a weighed, fine porosity, fritted glass crucible dried at 110°C.

below 1000°C. A very small amount of calcium, however, does not give this trouble, and indeed, it is preferable to coprecipitate such a small amount of calcium with the magnesium and then to recover it from the ignited pyrophosphates rather than to attempt to precipitate it as the oxalate in the presence of a large amount of magnesium (Sec. 9.10.2).

Any strontium not previously separated will, like calcium, be coprecipitated with magnesium and can be recovered from or determined in the ignited pyrophosphates. Barium behaves in a less predictable manner and must be removed before the precipitation (see Sec. 6.17).

The slow release of NH_3 by the reaction of NH_4Cl and methanolamine has been recommended for the homogeneous precipitation of magnesium ammonium phosphate.[10,16] In making the precipitation without prior separations, de Saint-Chamant and Vigier[10] complexed calcium with EDTA, and iron, aluminum and titanium with lactic acid.

(b)   Precipitation with 8-hydroxyquinoline

The precipitation of magnesium as the 8-hydroxyquinolate (pH 9.5–13) offers some advantages over the use of dibasic ammonium phosphate.[7] It is more rapid because one precipitation usually will suffice to separate the magnesium quantitatively and the lengthy digestion of the precipitate is eliminated; the precipitate filters easily and need not be ignited at a high temperature but can be dried and weighed as either the dihydrate, $Mg(C_9H_6NO)_2 \cdot 2H_2O$, or the anhydrous salt. The favorable gravimetric factor $(Mg(C_9H_6NO)_2 \rightarrow MgO = 0.1291)$ makes the method particularly useful for those samples, such as granites, in which the MgO content may be as low as 0.1%.

Because many metals form slightly soluble hydroxyquinolates over a range of pH conditions, it is better to have first removed all but the alkalies before precipitating the magnesium. The final solution from the successive separations described previously is thus very suitable for this purpose. A further advantage gained from the use of 8-hydroxyquinoline is that the excess reagent can be completely removed should it be desirable to make further use of the filtrate. Although small amounts of calcium, strontium and barium can be tolerated, a double precipitation of the magnesium should be made, as is the case also when large amounts of the alkali metals are present; it is preferable first to remove calcium as the oxalate* but care must be taken to avoid a large excess of oxalate which will make the precipitation of magnesium 8-hy-

---

* The presence of calcium as the oxalate does not interfere in the subsequent bromometric titration of the precipitated magnesium 8-hydroxyquinolate.

droxyquinolate less than quantitative. This excess oxalate can, of course, be destroyed before continuing with the precipitation of the magnesium. Under special conditions, magnesium can be precipitated in the presence of aluminum (Ref. 14 p. 635).

The precipitate, collected in a fritted glass filtering crucible, may be dried at 105°C and weighed as $Mg(C_9H_6NO)_2 \cdot 2H_2O$. This presupposes that only the desired compound is present and, because the reagent tends to be carried down with the precipitate (a large excess of reagent must be avoided) it may be preferable to dry the precipitate at 160°C, at which temperature the anhydrous salt, $Mg(C_9H_6NO)_2$, is formed and any excess 8-hydroxyquinoline is volatilized. The lower temperature is preferable because of the danger that some decomposition of the 8-hydroxyquinolate may occur at the higher one. If the precipitate is collected on a filter paper, it should be ignited to MgO only after very cautious charring of the paper.

The determination may be completed more rapidly if the separated magnesium 8-hydroxyquinolate is titrated bromometrically (see Sec. 9.13.1).

6.6.2. TITRIMETRIC METHODS

Magnesium can be determined by the bromometric titration of the precipitated 8-hydroxyquinolate. The latter is dissolved in HCl, a standard solution of $KBrO_3/KBr$ is added, followed by an excess of KI, and the iodine formed is titrated with standard sodium thiosulfate solution; the acid solution of the 8-hydroxyquinolate can also be potentiometrically titrated directly with the standard bromate solution.[30] Similarly, the precipitate obtained with naphthaldehyde[30] can be dissolved in an excess of acid and the excess titrated with a standard base solution.

Much use is now made of the titration of combined calcium and magnesium with EDTA, employing Eriochrome Black T as the metallochromic indicator, for the rapid determination of magnesium. The latter is arrived at by difference, following the titration of the calcium in a separate aliquot of the solution after separation, or removal, of the magnesium.* This titration has been previously discussed in some detail (see Sec. 6.5.2), to which reference should be made (Sec. 6.5, Refs. 4,7,13,24,28,32,33,38,39). The method is one that is capable of giving

---

* Goldich *et al.* (Sec. 6.5, Ref. 13) obtained an approximate value for magnesium in marls by first titrating the calcium present with EDTA, using a minimum amount of calcein as indicator; the pH is then adjusted and the magnesium titrated with EDTA, using Eriochrome Black T as indicator.

results that are comparable to those obtained gravimetrically and in much less time than is required for the latter; extreme care on the part of the analyst is necessary, however, especially if the amount of magnesium present is small.

Although Eriochrome Black T is largely favored as the indicator for this titration, and for the titration of magnesium alone,[16,25] it is not stable in solution and another o,o'-dihydroxyazo compound, Calmagite, has been proposed which can be used in place of Eriochrome Black T without change in procedure and the solution of which is stable indefinitely.[11] Other indicators have also been suggested. No screening is necessary when o-cresolphthalein complexone (phthaleincomplexone or metalphthalein) is used, as was found necessary for Eriochrome Black T (Naphthol Green B added); the color change is from mauve-pink to colorless.[3] Pouget[23] used phthaleincomplexone as indicator in the automatic photometric titration of calcium and magnesium, making the titration in an alcoholic solution which improved the sharpness of the end point. A methanolic solution was also used by Maier,[18] who titrated magnesium directly in the presence of large amounts of calcium; the indicator used was *magon* (1-azo-2-hydroxy-3-(2,4-dimethylcarboxanilido)-naphthalene-1'-(2-hydroxybenzene), which has a color change from yellow-red to blue-violet to blue.

The use of ethylene glycol-bis(β-aminoethyl ether)-N,N'-tetraacetic acid (EGTA) as a titrant for calcium was mentioned previously (Sec. 6.5.2, Ref. 35); it has also been used[13] to complex calcium, thus permitting the direct titration of magnesium with EDTA and Eriochrome Black T. Přibil and Veselý (Sec. 6.5, Supp. Ref.) also use EGTA to complex calcium but titrate the magnesium with (1,2-cyclohexylenedinitrilo) tetraacetate (DCYTA) using methylthymol blue as indicator.

Shapiro and Brannock[29] use an automatic photometric titration with EDTA and Eriochrome Black T of calcium plus magnesium as part of their rapid analysis scheme for silicate rocks. A "spike" aliquot of standard CaO + MgO solution is added to samples and standards to insure that a good end point will be obtained, and the titration is made without prior separation of R_2O_3 group elements. Hydroxylamine hydrochloride, potassium cyanide and triethanolamine are used to prevent interference with the end point of the titration. A small correction must be made for Mn(II) which is also titrated. A similar procedure for automatic photometric titration is described by Ritchie.[26]

6.6.3. COLORIMETRIC METHODS

Colorimetric methods for the determination of magnesium have received more attention than have similar methods for calcium, but because these methods have utilized the formation of lakes they have not been very popular. Sandell and others[16,27] have stressed the superiority of reactions which form soluble colored complexes rather than lakes, and suitable reagents for this purpose are now available.

The older methods employed the dyes Titan Yellow, Thiazole Yellow and Brilliant Yellow, and various dispersants, inhibitors, and color stabilizers have been proposed.[2,22,28] Shapiro[28] found that the EDTA titration of small amounts of magnesium (0.01–2%), particularly in the presence of much calcium, was not satisfactory and proposed a rapid photometric determination with Thiazole Yellow for low magnesium in silicates, carbonates and phosphates (either calcium or phosphate must first be removed); polyvinyl alcohol is used as the dispersant, and hydroxylamine and aluminum as color intensifiers. Meyrowitz[21] adapted Shapiro's method to the analysis of small samples of high-magnesium minerals, using Clayton Yellow. Detailed investigations of the nature and analytical usefulness of commercial Titan Yellow, with special reference to silicates and soil extracts, have been made recently (Hall et al., King and Pruden, Supp. Refs.).

The use of Eriochrome Black T in a photometric procedure for the determination of magnesium has been hailed as the first method to involve a true solution of a colored magnesium complex in which the color intensity is directly proportional to the magnesium concentration.[16] The solution containing the magnesium and dye is buffered to a pH of 9.5, and the intensity of the red complex is measured at 520 mμ. The color is stable for 1 hr and then decreases in intensity very slowly; the absorbance curve is linear for magnesium concentrations up to 100μg per 100 ml volume and the sensitivity of the method is reported as 1:50 million. However, calcium and some other metals will also form complexes, and thus the procedure is subject to a number of interferences.

Mann and Yoe[19,20] have described two reagents which form soluble reddish complexes with magnesium at pH 8.95 in the presence of ethanol. One of these is available commercially under the trivial name "magon" and the name "magon sulfonate" was proposed for the other by Sydney Abbey of the Geological Survey of Canada, who developed a photometric method for the determination of small amounts of magnesium (<1% MgO) in rocks, using this reagent.[1] This procedure will be discussed in some detail here.

The EDTA titration of magnesium (Sec. 6.6.2) is not very satisfactory for those samples containing little magnesium (<2%) especially if the calcium content is high. Satisfactory results are obtainable in this concentration, with reasonable rapidity, by a spectrophotometric procedure which utilizes a small aliquot of the solution B already prepared (see

Sec. 9.11.2) as part of the rapid analysis scheme.   After the addition of a small amount of calcium, the accumulated ammonium sulfate is removed by heating with HCl and HNO_3; the reagent, magon sulfonate, is added, followed by triethanolamine and a borax buffer solution, and the absorbance of the pink magnesium complex is measured at 510 mμ after 1 hr.

The removal of ammonium sulfate, formed when the aliquot of solution B is neutralized with aqueous ammonia to precipitate the R_2O_3 group elements (Sec. 9.11.2, Step 1), is necessary because it otherwise will prevent attainment of the proper pH of 8.95 on addition of the borax buffer, and also may produce a calcium sulfate turbidity in the ethanolic medium in which final measurement is made.

Calcium enhances the absorbance of the blank but not appreciably that of the magnesium complex; the effect is nearly constant over a range of 0.4–12 ppm of calcium.   The addition of calcium insures a constant calcium effect for all samples.

Traces of aluminum not removed by the precipitation with aqueous ammonia must be complexed with triethanolamine or a positive error will result.   Triethanolamine has a slight effect on the sensitivity and color stability, and a constant amount must be used with blanks, standards and samples.

The colored system that is obtained involves a mixture of the pink complex and the blue color of the excess reagent, the latter component decreasing in amount with increasing concentration of magnesium.   The sensitivity of the method can theoretically be enhanced by correcting for the absorbance due to this excess reagent, by measuring the absorbance of the mixed color system at the peak wavelength of the reagent, where the absorbance of the complex is negligible.   A linear relationship exists between the absorbances of the reagent at the two wavelengths.   When an attempt was made, however, to construct such a "correction curve" it was found to be not feasible; when the solution, after a one-hour wait, was placed in an absorption cell it underwent a gradual change that is thought to be related to the excess reagent, and this instability effect could not be eliminated.   Measurement of the absorbance due to excess reagent is thus not advisable because of this instability; the absorbance remains nearly constant for 5 min and should be measured immediately after filling each absorption cell.

Indirect methods for the colorimetric determination of magnesium have been described.[27]   Bodenheimer[4] made a single precipitation of magnesium with dibasic ammonium phosphate, dissolved the precipitate in dilute HNO_3 and measured the phosphorus present spectrophotometrically as the molybdovanado phosphoric acid complex. Calcon (Sec. 6.5.2) forms a stable chelate with magnesium and use is made of this to bleach the color of the less stable calcium complex.[24]

6.6.4. OTHER METHODS

Knutson[15] determined magnesium (2–10 pm) by flame photometry in plant material only, using a highly reducing, carburizing flame; phosphate and sulfate interfered negatively, but this interference was eliminated by the presence of excess Ca(II) in the reducing flame. Dean[9] mentions the severe repression that is also caused by silicate ions, and recommends the use of nonaqueous or aqueous-organic solvents for better emission intensity. Dinnin[12] showed that an 80% acetone solution partially releases the magnesium from the depressant effect of aluminum and phosphate, and when Ca or Sr are used as releasing agents in acetone medium, full release is obtained. Lanthanum will release the magnesium emission in the presence of a relatively high concentration of aluminum.

The determination of magnesium by optical emission spectrography is a useful method for trace amounts of magnesium; the determination of magnesium as a major constituent of rocks by X-ray fluorescence spectroscopy will be discussed in a later section.

Perhaps the most potentially useful new technique to be developed in the last decade is atomic absorption spectroscopy (West, Supp. Ref.; see also Sec. 4.3); it is only in the last five years that it has had much significance in rock and mineral analysis. Magnesium has received much attention and is now one of the elements most commonly determined by this technique. Only aluminum offers serious interference in the determination of magnesium through the formation of a compound (magnesium aluminate?) that is not decomposed completely in the flame. Rubeška and Moldan (Supp. Ref.) were able to determine accurately as little as 0.01% MgO by using a methanolic solution of 8-hydroxyquinoline and calcium as releasing agent; Nesbitt (Supp. Ref.) employed a hotter nitrous oxide-acetylene flame to overcome the interference of aluminum. At the Geological Survey of Canada, strontium (500 ppm) is used as releasing agent in a method developed by Sydney Abbey; the procedure utilizes an aliquot of a solution already prepared for the determination of sodium and phosphorus as part of an X-ray fluorescence–chemical rapid analysis scheme (Abbey, Supp. Ref.). In an aliquot containing 10 mg of sample, magnesium in the range 0–2% can be determined in the presence of as much as 20% of aluminum. The procedure is described in a later section.

References

1. Abbey, S., and Maxwell, J. A. Photometric determination of small amounts of magnesium in rocks. *Anal. Chim. Acta* **27**, 233–240 (1962).
2. Beater, B. E., and Maud, R. R. Rapid determination of readily soluble magnesium in soils. *Chemist-Analyst* **48**, 10–11 (1959).
3. Bennett, H., Hawley, G. W., and Eardley, R. P. Rapid analysis of some silicate materials. *Trans. Brit. Ceram. Soc.* **57**, 1–28 (1958).
4. Bodenheimer, W. Determination of magnesium oxide in silicates. *Bull. Res. Council Israel* **6C**, 73 (1957).
5. Bodenheimer, W., and Gaon, M. Colorimetric determination of magnesium in silicates. *Bull. Res. Council Israel* **7A**, 117–120 (1958).
6. Bricker, C. E., and Parker, G. H. Precipitation of magnesium with (ethylenedinitrilo)-tetraacetic acid. *Anal. Chem.* **29**, 1470–1474 (1957).
7. Chalmers, R. A., Ed., *Quantitative Chemical Analysis*, 11th ed., 1956. London: Oliver and Boyd, p. 361.
8. Chalmers, R. A., in "Chemical Analysis of Silicates," *The Chemistry of Cements*, Vol. 2, H. F. W. Taylor, Ed., 1964. London: Academic Press, pp. 171–189.
9. Dean, J. A. *Flame Photometry*, 1960. New York: McGraw-Hill, pp. 189–195.
10. De Saint-Chamant, H., and Vigier, R. A new method for the precipitation of magnesium ammonium phosphate. Application to the determination of the phosphate ion in phosphoric acid, alkali and alkaline earth phosphates and natural calcium phosphates. *Bull. Soc. Chim. France* **1954**, 180–188.
11. Diehl, H. *Calcein, Calmagite and o,o′-Dihydroxyazobenzene. Titrimetric, Colorimetric and Fluorometric Reagents for Calcium and Magnesium*, 1964. Columbus, Ohio: The G. Frederick Smith Chemical Company, pp. 55–61.
12. Dinnin, J. I. Use of releasing agents in the flame photometric determination of magnesium and barium. *U.S. Geol. Surv. Profess. Papers*, **424–D**, D391–D392(1961).
13. Fabregas, R., Badrinas, A., and Prieto, A. Volumetrische bestimmung von magnesium in gegenwart von calcium. *Talanta* **8**, 804–808 (1961).
14. Hillebrand, W. F., Lundell, G. E. F., Bright, H. A., and Hoffman, J. I. *Applied Inorganic Analysis*, 2nd ed., 1953. New York: John Wiley & Sons, pp. 632–645.
15. Knutson, K. E. Flame photometric determination of magnesium in plant material. *Analyst* **82**, 241–254 (1957).
16. Kolthoff, I. M., and Elving, P. J., (Eds.). *Treatise on Analytical Chemistry*, Part II, Vol. 3, 1961. "Magnesium," G. B. Wengert, P. F. Reigler and A. M. Carlson. New York: Interscience, pp. 43–93.
17. Kolthoff, I. M., and Sandell, E. B. *Textbook of Quantitative Inorganic Analysis*, 3rd ed., 1953. New York: Macmillan, pp. 352–363.
18. Maier, R. H. A new sensitive indicator for semimicro chelometric titration. *Nature* **183**, 461–462 (1959).
19. Mann, C. K., and Yoe, J. H. Spectrophotometric determination of magnesium with sodium 1-azo-2-hydroxy-3-(2,4-dimethylcarboxanilido)-naphthalene-1-(2-hydroxybenzene-5-sulphonate). *Anal. Chem.* **28**, 202–205 (1956).
20. Mann, C. K., and Yoe, J. H. Spectrophotometric determination of magnesium with 1-azo-2-hydroxy-3-(2,4-dimethylcarboxanilido)-naphthalene-1-(2-hydroxybenzene). *Anal. Chim. Acta* **16**, 155–161 (1957).
21. Meyrowitz, R. The direct spectrophotometric determination of high-level

magnesium in silicate minerals. A Clayton Yellow procedure. *Am. Mineralogist* **49**, 769–777 (1964).

22. Mitchell, T. A. The spectrophotometric determination of magnesium with thiazol yellow dyes. *Analyst* **79**, 280–285 (1954).
23. Pouget, R. Methode d'analyse chimique des roches silicatées. *Rept. C. E. A.* **2176**, 1962, 28 pp.; Centre d'études nucléaires de Fontenoy-aux-roses; Commissariat à l'energie atomique.
24. Reilley, C. N., and Hildebrand, G. P. Methods of indirect spectrophotometry. *Anal. Chem.*, **31**, 1763–1766 (1959).
25. Ringbom, A. *Complexation in Analytical Chemistry*, 1963. New York: Interscience, pp. 82–93.
26. Ritchie, J. A. Methods of analysis of silicate rocks. *Rept.* **DL 2049**, 21–22 (1962). Dept. Scientific and Industrial Research, Dominion Laboratory, New Zealand.
27. Sandell, E. B. *Colorimetric Determination of Traces of Metals*, 3rd ed., 1959. New York: Interscience, Chapter 25, "Magnesium."
28. Shapiro, L. Rapid photometric determination of low level magnesium in rocks. *Chemist-Analyst* **48**, 73–74 (1959).
29. Shapiro, L., and Brannock, W. W. Rapid analysis of silicate, carbonate and phosphate rocks. *U.S. Geol. Surv. Bull.* **1144–A**, 35–46 (1962).
30. Wilson, C. L., and Wilson, D. W., (Eds.). *Comprehensive Analytical Chemistry*, Vol. 1C (Classical analysis), 1962. "Magnesium," by E. Booth. Amsterdam: Elsevier, pp. 65–72.

Supplementary References

Abbey, Sydney. Analysis of rocks and minerals by atomic absorption spectroscopy: I. Determination of magnesium, lithium, zinc and iron. *Geol. Survey Canada Paper* **67–37**, 1967, 35 pp.

Evans, W. H. The determination of magnesium in silicate and carbonate rocks by the titan yellow spectrophotometric method. *Analyst* **93**, 306–310 (1968). The determination is made on an aliquot of the filtrate from the succinate precipitation of the R_2O_3 group. Polyvinyl alcohol is used as the stabilizing colloid, and sucrose is added to prevent interference from calcium. The method has been applied to rocks having a range of 0.2–20% Mg.

Hall, R. J., Gray, G. A., and Flynn, L. R. Observations on the use of titan yellow for the determination of magnesium with special reference to soil extracts. *Analyst* **91**, 102–111 (1966).

King, H. G. C., and Pruden, G. The component of commercial titan yellow most reactive towards magnesium; its isolation and use in determining magnesium in silicate minerals. *Analyst* **92**, 83–90 (1967).

Kiss, E. Chemical determination of some major constituents in rocks and minerals. *Anal. Chim. Acta* **39**, 223–234 (1967). See Sec. 6.5, Supp. Refs., for details of the method for the determination of Mg.

Nesbitt, R. W. The determination of magnesium in silicates by atomic absorption spectroscopy. *Anal. Chim. Acta* **35**, 413–420 (1966).

Přibil, R., and Veselý, V. Determination of magnesium (and calcium) in calcareous materials and silicates. *Chemist-Analyst* **55**, 82–84 (1966).

Rubeška, I., and Moldan, B.   Determination of magnesium in silicate and carbonate
    rocks by atomic absorption spectrophotometry.   *Acta Chim.* (*Hungary*) **44,**
    367–371 (1965).
West, T. S.   Atomic analysis in flames.   *Endeavor* **26,** 44–49, 1967.

6.7. Titanium*

6.7.1. GENERAL

The behavior of titanium during the analysis of the main portion has
been frequently noted, particularly in the sections dealing with silica
(Secs. 6.2.1a, 9.2.1) and with the precipitation of the R_2O_3 group by
aqueous ammonia (Sec. 6.4.1a).

In most aluminosilicates, the titanium content is low enough so that
no problems are encountered in the complete decomposition of the sample
by fusion, nor in the subsequent dissolution of the fusion cake in HCl
and evaporation of the solution.   When titanium minerals such as rutile
and ilmenite are present in more than the usual amount, however, they
will resist decomposition and will accompany the silica through the
stages of ignition and treatment with sulfuric and hydrofluoric acids.
A small amount of such undecomposed material will do no harm, but an
excessive amount can lead to a serious error in the final value for the
silica.   The titanium minerals are probably oxides and are weighed as
such after the ignition of the impure silica; the treatment with HF and
H_2SO_4 will likely convert the oxides to sulfates (the H_2SO_4 is added to
prevent loss of titanium through volatilization as fluoride) and this sulfate may not be decomposed on subsequent ignition of the residue remaining after the HF treatment, thus negatively affecting the silica
value.   Titanium sulfate, however, has a tendency to spatter during the
evaporation to dryness and will introduce a positive error into the silica
result as well as causing the value for TiO_2, if the latter is subsequently
determined in the solution of the ignited R_2O_3 precipitate, to be low.
This undesirable behavior on the part of titanium is accentuated by its
strong tendency to hydrolyze from even strongly acid solution; when
present in amounts exceeding about 0.5%, the hydrolysis product may
impart a somewhat milky look to the HCl solution of the fusion cake
which should serve as a warning to the analyst.   The solubility of the
hydrolysis product is dependent on the temperature at which the hydrous
oxide was precipitated; that formed by prolonged heating of the solution
can be redissolved only with much difficulty and will usually be filtered
off with the silica residue.   The finely divided precipitate will tend to

* References for this section will be found on pp. 179, 180.

pass in part through the paper, particularly during the washing of the silica with dilute HCl, and the addition of paper pulp will not insure a clear filtrate. This will do no harm, because it is preferable to retain as little as possible of the hydrous titanium oxide on the paper and that which escapes filtration will be collected with the other members of the R_2O_3 group in the subsequent precipitation with aqueous ammonia. If, however, a precipitation with hydrogen sulfide for the removal of platinum (Sec. 6.3) is to follow, then the filtrate must be clear if the subsequent determination of titanium in the R_2O_3 group residue is to represent the total titanium content of the sample. Under these circumstances it is probably more preferable to determine the titanium on a separate portion of the sample but that present in the R_2O_3 group will still have to be determined in order to obtain a correct value for aluminum by difference.

When both titanium and phosphorus are present in the sample, insoluble titanium phosphate will accompany the silica residue and will remain after the treatment with HF. This will usually be of little consequence but when both are present in amounts greater than usual it can result in an undesirably high nonvolatile residue.

When hydrofluoric acid is used in the decomposition of titanium minerals, the titanium and fluoride ions form a strong complex that will interfere in the colorimetric determination of titanium. The fluoride must be expelled completely; the solution should be evaporated to fumes of SO_3 (sulfuric acid is usually present), then cooled and diluted, all solid material dissolved, and the evaporation and fuming repeated. Failure to expel all of the fluoride will also affect any subsequent precipitation of the R_2O_3 group with aqueous ammonia; the precipitation of aluminum will be incomplete and magnesium may be coprecipitated if present in large amount.

The hydrolysis of titanium may be prevented by the use of strongly acid solutions (but prolonged heating is to be avoided) and by the use of a complexing agent such as hydrogen peroxide or citric acid. The effect of these preventive measure on the subsequent steps in the analysis must be considered and they will usually be employed only when the determination of titanium is being done on a separate portion of the sample.

A preliminary separation of titanium, prior to its final determination, is not usually needed. A separation from elements which will interfere in the gravimetric, titrimetric and colorimetric determination of titanium may, however, be achieved by such means as fusion of the sample with sodium carbonate and leaching of the cake in water, by precipitation of the titanium with sodium hydroxide or with organic precipitants, by solvent extraction and by selective complexation; these, and other means, are discussed in more detail in standard reference texts,[1,3,4,15,17] which should be consulted.

6.7.2. COLORIMETRIC METHODS

The measurement of the absorbance of the colored complex formed by the reaction of titanium with such reagents as hydrogen peroxide, tiron (disodium-1,2-dihydroxybenzene-3,5-disulfonate) or chromotropic acid (1,8-dihydroxynaphthalene-3,6-disulfonic acid), either spectrophotometrically or visually with Nessler tubes or Duboscq colorimeter, is particularly applicable to samples having a TiO_2 content of less than about 2%, which includes most rocks and all but a few minerals.

The determination of titanium is usually made on the dilute sulfuric acid solution of the R_2O_3 group, after fusion of the ignited oxides with potassium pyrosulfate. The solution will as a rule contain major amounts of aluminum and iron, together with minor amounts of the other constituents of the ammonia group, possibly some platinum from the crucible, and a very large amount of potassium salts. If the determination of the titanium is made following the titrimetric determination of the iron in the R_2O_3 group, there will be present also a relatively large amount of manganese or chromium from the permanganate or dichromate used as titrant.

The yellow anionic complex that titanium forms with H_2O_2 is thought to be $TiO_2(SO_4)_2{}^{2+}$ by some, or possibly $Ti(H_2O_2)^{4+}$ or a similar species.[14] A few elements also form colored complexes under similar conditions and a number of others interfere in various ways, but these can be eliminated or counteracted. The method is reliable and widely used, although it is not the most sensitive one for the determination of titanium. The effect of interfering elements and the precautions necessary to eliminate this effect have been discussed in detail by several authors,[1,3,4,6,14] and only a summary will be presented here.

The intensity of the colored complex is reduced by the presence of a large amount of alkali salts, by phosphate and by even traces of fluoride. The bleaching action of alkali salts is less in 10% sulfuric acid and an equal amount of the salts should be added to the standards also; the same applies to the phosphoric acid added to complex iron, but if much titanium is present, care must be taken to avoid the precipitation of titanium phosphate. Fluoride, even in trace amount, exerts a pronounced bleaching effect, and while the addition of beryllium will result in the preferential formation of a complex with the fluoride, it is better to remove the latter by repeated fuming of the sulfuric acid solution.

The intensity of the titanium peroxy complex will be enhanced by those elements which also form colored complexes with hydrogen peroxide, among which are iron, vanadium, molybdenum, niobium (in the presence of Fe) and uranium, as well as those elements such as chromium and nickel (and iron) which form colored solutions. The interference of the latter can be eliminated by using the unperoxidized solution as a blank (this is not readily done if the measurement of the color intensity is made visually with Nessler tubes or a Duboscq colorimeter) and applying a suitable correction to the subsequent measurements. Phosphoric acid will complex iron, but it must be added to both the sample and standard solutions.[13] By measuring the absorbance of the solution at both 460 and 330 mμ, the absorbance maxima for the vanadium and molybdenum complexes respectively, the contribution of the latter to the absorbance at 410 mμ can be determined; the presence of vanadium can also be detected by adding hydrofluoric acid to the solution after measurement to bleach the titanium peroxy complex and reveal the reddish brown coloration due to vanadium. The intensity of the colored solution increases with increasing temperature (the increase is due largely to the change in intensity of the ferric sulphate, rather than to the titanium complex[11]) and all measurements should be made at the same temperature.

Mercury cathode electrolysis[7,10] is a convenient method by which to remove many interfering ions without the addition of other reagents; large amounts of Fe, Cr, Ni and Mo (but not Nb and V) can be separated and it is a particularly useful tool when low concentrations of titanium must be determined in highly ferruginous materials. If phosphorus, vanadium, uranium or niobium are present in significant quantity, the titanium can be separated from them by precipitating it with sodium hydroxide (if Fe is not present it should be added), followed by mercury cathode electrolysis.

In most rocks the only interfering element of any significance is iron, and for $Fe_2O_3 < 10\%$, its effect is slight. A convenient way for determining the extent of the iron interference in routine work, if the iron content is known, is to prepare a correction curve for a constant amount of titanium in the presence of increasing amounts of iron up to 10% Fe_2O_3; for concentrations of Fe_2O_3 greater than 10%, it is better to remove the iron, or to counteract its effect by adding either phosphoric acid to the sample and standards, or to add an equivalent amount of ferric sulfate to the standard.

The use of disodium 1,2-dihydroxybenzene-3,5-disulfonate as a reagent for titanium was proposed by Yoe and Armstrong[18]; because the reagent is used for the determination of titanium and iron, it has been given the name "tiron." Titanium forms a stable, lemon-yellow complex with tiron; an intense blue complex is formed by ferric iron but this is bleached by sodium dithionite ($Na_2S_2O_4$) and titanium can be determined in its presence. Interference occurs from the colored complexes formed by Cr, Cu, Mo, V, Os and uranium; tungsten forms a colorless complex that absorbs strongly at the wavelength used for measurement. These elements are, for the most part, seldom encountered in significant concentration in most silicate rocks; aluminum and calcium consume reagent but their effect is overcome by the use of a suitable excess of it. Only fluoride ion interferes seriously among the anions.

Yoe and Armstrong buffered the final solution at a pH of 4.7 with a sodium acetate–acetic acid buffer and reported little change in the intensity of the colored complex over the pH range 4.3–9.6. Nichols[9] found an earlier "plateau" at pH 2.3–4.0 and preferred to measure the absorbance at a pH of 3.0 ± 0.2 without the use of a buffer; this lower pH has the advantage of insuring that all of the titanium will be in a reactive form (metatitanic acid is thought to be precipitated at pH>4), and at this pH, the ferric iron is readily reduced with either ascorbic acid or hydroxylamine hydrochloride. Sodium dithionite is not entirely stable, and at a pH of 4.7, decomposes slowly with the formation of sulfur. Nichols reported that the intensity of the colored complex increased with increasing solution temperature; the color is stable for at least 24 hr and Beer's law is obeyed up to at least 80 μg Ti per 100 ml.

Measurement of the absorbance was made at 410 mμ by Yoe and Armstrong; the maximum absorbance of the titanium complex is at 380 mμ, but because dithionite absorbs strongly at wavelengths <410 mμ, it was necessary to use a higher wavelength. A further disadvantage to the use of dithionite is that air oxidation of the reduced iron occurs readily, and vigorous stirring of the solution must be avoided.

Rigg and Wagenbauer[12] have further improved the method by the use of thioglycolic acid to reduce the ferric iron; because the reducing agent does not absorb at 380 mμ, the absorbance measurements can be made at the maximum for the complex. At this wavelength, and in a solution buffered at pH 3.8 (sodium acetate-acetic acid), Beer's law is strictly obeyed for concentrations of TiO_2 up to at least 200 μg in 50 ml of solution. The intensity of the colored complex was not found to be appreciably temperature-dependent, and to be stable for at least 24 hr; the color should be allowed to develop for at least 1 hr before the absorbance is measured.

Shapiro and Brannock[16] adapted the method of Yoe and Armstrong to their rapid method scheme for TiO_2 contents up to 3%; the reagent is added in solid form from a scoop (125 mg) and the absorbance measurement is made at 430 mμ.

Jeffery and Gregory (Supp. Ref.) have proposed the use of diantipyrylmethane for the photometric determination of titanium. While less sensitive than tiron and chromotropic acid, the reagent is less subject to interference from other metallic ions and to variations in the acid strength of the sample solution. Tin and uranium must be absent.

6.7.3. OTHER METHODS

For TiO_2 concentrations greater than 1%, the titrimetric method, in which the Ti(IV) is reduced to Ti(III) in a Jones reductor and the reduced form, or its equivalent, titrated with a standard oxidizing solution, is much used.

The gravimetric determination of titanium may be done in a number of ways, but all require either preliminary separation of interfering elements, or the use of complexing agents, or both.[4,15,17]

Polarographic methods have been proposed for the determination of titanium having a wide range of concentration in rocks and minerals[2,5,8] of very different compositions. The results are comparable in accuracy to those obtained colorimetrically but the method is not as sensitive as the colorimetric one, and the latter is to be preferred when the TiO_2 content is less than 0.05%. Electrolysis of the solution at a mercury cathode will remove most interfering elements[7,10]; the polarographic procedure is not affected by the presence of alkali salts or traces of fluoride which will seriously affect the formation of the colored complex.

Titanium, especially in trace amounts, is readily determined by optical emission spectrography. X-ray fluorescence spectroscopy is used at the Geological Survey of Canada to determine titanium, over the range 0.1–2% TiO_2 in a variety of geological materials with satisfactory accuracy; this will be discussed in more detail in a later section.

References

1. Codell, M. *Analytical Chemistry of Titanium Metals and Compounds*, 1959. New York: Interscience, 378 pp.
2. Graham, R. P., and Maxwell, J. A. Determination of titanium in rocks and minerals. Mercury cathode-polarographic method. *Anal. Chem.* **23**, 1123–1126 (1951).
3. Hillebrand, W. F., Lundell, G. E. F., Bright, H. A., and Hoffman, J. I. *Applied Inorganic Analysis*, 2nd ed., 1953. New York: John Wiley & Sons, pp. 576–587.

4. Kolthoff, I. M., and Elving, P. J., (Eds.). *Treatise on Analytical Chemistry*, Part II, Vol. 5, 1961. "Titanium," by E. R. Scheffer. New York: Interscience, pp. 1–60.
5. Kolthoff, I. M., and Lingane, J. J. *Polarography*, 2nd ed. (rev.), 1952. New York: Interscience, pp. 444–446.
6. Kolthoff, I. M., and Sandell, E. B. *Textbook of Quantitative Inorganic Analysis*, 3rd ed., 1952. New York: Macmillan, pp. 705–707.
7. Maxwell, J. A., and Graham, R. P. The mercury cathode and its applications. *Chem. Rev.* **46**, 471–498 (1950).
8. Milner, G. W. C. *The Principles and Applications of Polarography and Other Electroanalytical Processes*, 1957. London: Longmans, Green, pp. 465–466.
9. Nichols, P. N. R. The photometric determination of titanium with Tiron. *Analyst* **85**, 452–453 (1960).
10. Page, J. A., Maxwell, J. A., and Graham, R. P. Analytical applications of the mercury electrode. *Analyst* **87**, 245–272 (1962).
11. Peck, L. C. Systematic analysis of silicates. *U.S. Geol. Surv. Bull.*, **1170**, 36–37, 70–72 (1964).
12. Rigg, T., and Wagenbauer, H. A. Spectrophotometric determination of titanium in silicate rocks. *Anal. Chem.* **33**, 1347–1349 (1961).
13. Riley, J. P. The rapid analysis of silicate rocks and minerals. *Anal. Chim. Acta* **19**, 413–428 (1958).
14. Sandell, E. B. *Colorimetric Determination of Traces of Metals*, 3rd ed., 1959. New York: Interscience, pp. 868–882.
15. Schoeller, W. R., and Powell, A. R. *Analysis of Minerals and Ores of the Rarer Elements*, 3rd ed., 1955. New York: Haffner, pp. 113–133.
16. Shapiro, L., and Brannock, W. W. Rapid analysis of silicate, carbonate and phosphate rocks. *U.S. Geol. Surv. Bull.* **1144–A**, A29–A30 (1962).
17. Wilson, C. L., and Wilson, D. W., (Eds.). *Comprehensive Analytical Chemistry*, Vol. 1C (Classical analysis), 1962. "Titanium," by W. T. Elwell and J. Whitehead. Amsterdam: Elsevier, pp. 497–506.
18. Yoe, J. H., and Armstrong, A. R. Colorimetric determination of titanium with disodium-1, 2-dihydroxybenzene-3, 5-disulphonate. *Anal. Chem.* **19**, 100–102 (1947).

Supplementary References

Jeffery, P. G., and Gregory, G. R. E. C. Photometric determination of titanium in ores, rocks and minerals with diantipyrylmethane. *Analyst* **90**, 177–179 (1965).
Van Loon, J. C., and Parissis, C. The determination of titanium in silicates by atomic absorption spectrophotometry. *Analytical Letters* **1**, 249–255 (1968). An aliquot of the sample solution, following decomposition with HF + H_2SO_4 and removal of fluoride by evaporation with HCl, is added to $AlCl_3$ solution to give a final [Al] of 750–1000 ppm, and the atomic absorption measurements are made using a nitrous oxide–acetylene flame. This [Al] was found to stabilize the absorption and to mask the interference effects of other constituents.

6.8. Manganese*

The references made to the behavior of manganese in previous sections have been concerned with its possible precipitation with the R_2O_3 group (Sec 6.4.1a), its separation by ammonium sulfide or by ammonium persulfate (Secs. 9.9.1, 9.9.2), and the possible necessity of determining it in the precipitated R_2O_3 group (Sec. 9.6.1), calcium (Secs. 9.10.1, 9.10.2) and magnesium (9.12.1).

In most rocks the manganese concentration is usually very low (0.1–0.5%), seldom exceeding 1%. In some silicate minerals, such as spessartite and andradite garnet, the concentration may be several per cent; in the manganese silicates rhodonite ($MnSiO_3$), and braunite ($2Mn_2O_3 \cdot MnSiO_3$), it is a major constituent.

It is possible to determine the total manganese present in a sample by recovering it at various stages in the analysis, and when the amount of sample available for analysis is small, such a multiple determination becomes practical. If the manganese content is high (1% or more), a small amount will be found with the precipitated R_2O_3 group and a very small amount with the calcium; most of the manganese will be coprecipitated with the magnesium, and the remainder will be found in the filtrate from this precipitation. In the scheme of analysis previously described, the manganese is separated by either ammonium sulfide or ammonium persulfate; the precipitate may be discarded, but it is possible to redissolve the separated MnS or MnO_2 and to determine the manganese colorimetrically. The hydrolysis of the hydrous zirconium oxide used to collect the precipitated MnO_2 in the persulfate procedure will interfere in a colorimetric determination of the manganese but can be eliminated by centrifuging the solution. If the manganese concentration is high, it can be determined titrimetrically after destruction of the filter paper and oxidation of the manganese separated by either method. If sufficient sample is available, however, it may be less work to make the determination of total manganese on a separate sample.

The gravimetric determination of manganese, best done as manganese ammonium phosphate with final ignition to the pyrophosphate ($Mn_2P_2O_7$), is rarely used, having been supplanted by titrimetric and colorimetric methods. If used, it requires a preliminary separation of the manganese.

* References for this section will be found on pp. 186,187.

6.8.1. TITRIMETRIC METHODS

In small amounts ($<2\%$) the manganese can be determined easily, rapidly and reliably by a colorimetric procedure; if $>2\%$ the determination is best made titrimetrically. For the colorimetric determination we need consider no other colored complex than the permanganate cation (MnO_4^-). This latter is also used in the titrimetric procedures, but in recent years more attention has been paid to the use of the complex formed between trivalent manganese and pyrophosphate, which closely resembles the permanganate cation.

Divalent manganese is oxidized to Mn (VII) at room temperature by sodium bismuthate in nitric acid solution; the excess bismuthate is removed by filtration. A measured amount of ferrous sulfate is added and the excess is back-titrated with standard permanganate solution; with suitable precautions, as much as 500 mg Mn can be determined in this way. The method is subject to a number of serious interferences, some of which can be avoided by titrating the Mn(VII) directly with sodium arsenite. A detailed discussion of the method is given by Cooper and Winters (Ref. 2, pp. 491–495).

Less interference is encountered when the oxidation is done by ammonium persulfate in boiling nitric–phosphoric–sulfuric acid solution; a small amount of silver nitrate must be present to catalyze the oxidation. The latter also prevents the precipitation of MnO_2, up to a concentration of about 5 mg Mn per 100 ml; above this concentration its precipitation is prevented by phosphoric acid. Sandell[6] states that the oxidation of the manganese is incomplete if much titanium is present. The titration is made directly with sodium arsenite and the end point is a distinctive greenish-yellow ("apple-green") color; removal of the excess persulfate is not necessary because it is not reduced by arsenite. Because the reaction between Mn(VII) and arsenite is not stoichiometric, the arsenite must be standardized against standard samples. As much as 100 mg of Mn can be determined by this procedure but it is usually confined to the titration of 15–25 mg. For details of the procedure, see Ref. 2, pp. 488–491.

The oxidation can also be made with potassium periodate in hot nitric and phosphoric acid solution. Excess periodate must be removed by precipitation as mercuric periodate and filtration, and the procedure is subject to the interferences encountered in the bismuthate procedure.

The oxidation of Mn(II) to Mn(III) by Mn(VII) in neutral pyrophosphate solution, with consequent reduction of the Mn(VII) to Mn(III) also, has been made the basis of a very useful method for the

determination of manganese over a wide range of concentration.[3] This
method is applicable to the determination of manganese contents as
low as 2% if a 1 g sample is used, and correspondingly less if larger
samples are available. The titration is made with $0.02M$ KMnO$_4$ and
a 10-ml semimicro buret should be used for samples having Mn contents
of 2–10% (either the whole sample or an aliquot containing no more than
about 40 mg Mn, whichever is less, is titrated, requiring about 10 ml of
the $0.02M$ KMnO$_4$). For Mn >10% a 50-ml buret may be used for
aliquots containing 100–200 mg of manganese.

Because of the deep color of the pyrophosphate complex it is necessary
to determine the end point potentiometrically. A pH meter serves satisfactorily as a potentiometer, provided that it has a scale capable of
being read to ± 10 mV per scale division; for the titration cell a bright
platinum wire and a saturated calomel reference electrode are needed.
The initial potential of the cell is about +200 mV vs. SCE and it increases slowly, and quickly becomes constant, after each addition of
KMnO$_4$; the potential break at the end point is very large (at pH 6–7
for a pure Mn solution the potential at the end point is + 0.47 ± 0.02
V vs. SCE, and the maximal value of $\Delta E/\Delta V$ is 100–200 mV per 0.1
ml KMnO$_4$[3]). The titration requires about 15–20 min to complete.
Stock[9] describes the amperometric titration of Mn(II) with KMnO$_4$ in
a saturated sodium pyrophosphate medium (pH 6–8) with a Pt-Ag
electrode combination.

Reducing substances, such as oxides of nitrogen, interfere in that they
consume permanganate and cause high results to be obtained. When
nitric acid is used to dissolve the sample, the solution must be boiled
vigorously and a small amount (1 g per 250 ml of solution) of urea or
sulfamic acid is added prior to final dilution to volume. Magnesium
and aluminum, when present in large amounts, form precipitates that
may carry down some manganese and the method fails in this case.
Vanadium, if equal to or greater than the manganese content, will interfere but can be circumvented by doing the titration at pH of 3–3.5; this
is a situation that will rarely be encountered. Scribner,[7] in a study of
the possible interferences, has shown that low results are obtained for
manganese in the presence of Cu, Co, Fe or Ni if the initial acidity of the
sample solution is high; air oxidation of some of the manganese occurs
during the adjustment of the PH (6–7) and thus sample solutions should
contain a minimum amount of free acid at the start. If the pH obtained
on addition of the sample solution to the pyrophosphate solution is <6,
oxygen should be excluded as completely as possible from the solution
prior to the addition of sodium hydroxide. Chromium(III) will inter-

fere unless sufficient time (50–60 min) is allowed for it to form a stable complex with the pyrophosphate.

Ingamells[1] utilizes the oxidation of Mn(II) to Mn(III) by nitric or perchloric acids in hot concentrated phosphoric acid containing pyrophosphate (see also Knoeck and Diehl, Supp. Ref.), but titrates the Mn(III) with a standard Fe(II) solution, using diphenylamine sulfonate as indicator. Besides eliminating the need for an electrometric titration, it is possible to repeat the titration if there is doubt about the completeness of the reaction. Nitric acid is the preferred oxidant, and when the MnO content >5 mg, nitric acid must be present in order to insure that the oxidation is complete. The addition of nitric acid in the final stages of the procedure may be omitted when less than 5 mg MnO are present.

Sodium pyrophosphate is added to facilitate the oxidation of the Mn(II) when the MnO concentration is near the acceptable upper limit (approximately 2 mg per ml H_3PO_4); it may be omitted when the manganese content is low. When silicates are decomposed directly in the acid mixture, the $Na_4P_2O_7$ prevents the precipitation of silicic acid and thus aids in the dissolution of the sample.

Ingamells reports that chloride or perchlorate, in the presence of nitrate, will cause erroneous results; if hydrochloric acid is used in the decomposition of the sample, it should be removed by repeated evaporations with nitric acid before the phosphoric acid is added. In the presence of much chromium, sulfate should be absent or low results will be obtained, presumably because some manganese is carried down by the insoluble precipitate that forms. Vanadium is the only serious interference, being quantitatively oxidized to V(V) by nitric acid and reduced to V(IV) by Fe(II); a correction can be applied after it is determined separately.

Standardization of the ferrous ammonium sulfate solution is best achieved by carrying an aliquot of a standard manganese solution, prepared from pure manganese metal, through the whole procedure.

It is possible to decompose many silicate minerals by heating them directly with phosphoric acid and sodium pyrophosphate. Samples of biotite and garnet, ground to pass a 115-mesh screen, require 15 min and 1–2 hr of heating at about 200°C, respectively. It is a wise precaution to swirl the sample frequently during the initial stages of heating in order to prevent caking of the powder which prolongs the time required for the decomposition. A distinct advantage of the procedure is that if there is doubt as to whether or not the decomposition has been complete, which is not always easily ascertained, it is only necessary after the titration to evaporate the solution, to add more acid and to repeat the procedure.

6.8.2. COLORIMETRIC METHODS

The photometric determination of small amounts of manganese by oxidation of Mn(II) to Mn(VII) is one of the simplest determinations in silicate analysis, particularly when KIO_4 is used as the oxidant. Potassium periodate, either in solution or as the solid reagent, is added to a nitric acid solution of the sample to which also has been added either sulfuric or phosphoric acid; ferric periodate is insoluble in nitric acid but is soluble in either of the other two acids. Phosphoric acid is preferred because it also bleaches the color of the ferric iron. The solution is heated to near the boiling point (if the KIO_4 was added as a solid it is necessary to boil the solution in order to dissolve the reagent; by using a solution of the reagent it is sufficient to immerse the manganese solution in a boiling water bath) for about 1 hr, cooled and the absorbance measured at 525 or 545 mμ. The maximum absorbance of the permanganate is at 525 mμ but chromate also absorbs strongly at this wavelength and at 545 mμ the absorbance of the permanganate is only slightly less while that of the chromate is much less.

Sandell[6] states that the speed of oxidation is increased when the sulfuric acid concentration is 3.5N or greater (10 ml 36N H_2SO_4/100 ml solution), except when the manganese content is very low (a few micrograms per 10 ml of solution); in the latter case the sulfuric acid concentration should be <15 ml 36N H_2SO_4 per 100 ml of solution both for full color development and to prevent subsequent fading. For small amounts of manganese it is preferable to have sulfuric acid as the only acid present, and to add silver nitrate.

All reducing substances must be removed prior to oxidation of the manganese. Peck[4] ignites the sample briefly in a platinum dish in order to destroy any soluble organic matter that may be present; reducing substances will block the oxidation and may give entirely misleading results if the manganese content is low. Solutions in which no color develops should be allowed to stand for at least 24 hr before reporting the manganese as "not found."

The use of ammonium persulfate is preferred by Riley,[5] who reports that more consistent results are obtained for low manganese concentrations with persulfate than with periodate; as mentioned before, oxidation by KIO_4 is very slow when the Mn content is very low. Riley uses a 20 ml aliquot (50 mg) of his solution B (see Sec. 9.8.2) to which is added a mixed reagent (HNO_3, H_3PO_4, $AgNO_3$, $HgSO_4$) and ammonium persulfate. The mercuric sulfate is added to form a slightly dissociated complex with any chloride ion that may be present. If much titanium is present the oxidation of the manganese will be incomplete and periodate should be used instead of persulfate.[6]   Shapiro and Brannock[8] use a high concentration of $AgNO_3$ (50 mg per 50 ml) to obtain full color development at room temperature after one hour.   For MnO in the range 0–3%, a 25-ml aliquot (50 mg) of their solution B is used.

Beer's law holds for concentrations of manganese up to at least 150 mg/liter[6]; in a 1-cm cell the optimum concentration for measurement is approximately 1 mg Mn per 100 ml, and the preferred range is about 0.5–1.5 mg.

Manganese in concentrations up to 1% is readily and accurately determined by optical emission spectrography, and for amounts <0.01% this method is to be preferred.   It is particularly useful when the supply of sample is limited, and for checking such ignited residues as the R_2O_3 group, CaO and $Mg_2P_2O_7$ for the presence of manganese.   At the Geological Survey of Canada the routine determination of manganese, in the range 0.1–2% is done by X-ray fluorescence spectroscopy, as will be described in a later section.

References

1. Ingamells, C. O.   Titrimetric determination of manganese following nitric acid oxidation in the presence of pyrophosphate.   *Talanta* **2,** 171–175 (1959).
2. Kolthoff, I. M., and Elving, P. J., (Eds.).   *Treatise on Analytical Chemistry,* Part II, Vol. 7, 1961.   "Manganese," by M. D. Cooper and P. K. Winter.   New York: Interscience, pp. 425–502.
3. Lingane, J. J., and Karplus, R.   New method for determination of manganese.   *Ind. Eng. Chem. Anal. Ed.* **18,** 191–194 (1946).
4. Peck, L. C.   Systematic analysis of silicates.   *U.S. Geol. Surv. Bull.* **1170,** 45–48, 76–78 (1964).
5. Riley, J. P.   The rapid analysis of silicate rocks and minerals.   *Anal. Chim. Acta* **19,** 413–428 (1958).
6. Sandell, E. B.   *Colorimetric determination of traces of metals,* 3rd ed., 1959.   New York: Interscience, pp. 606–620.
7. Scribner, W. G.   Influence of metal ions and sample acidity on the determination of manganese.   A study of the Lingane-Karplus method.   *Anal. Chem.* **32,** 966–969 (1960).   Influence of chromium III on the determination of manganese.   *Idem.,* 970–972.
8. Shapiro, L., and Brannock, W. W.   Rapid analysis of silicate, carbonate and phosphate rocks.   *U.S. Geol. Surv., Bull.* **1144–A,** A30–A31 (1962).
9. Stock, J. T.   *Amperometric Titrations,* 1965.   New York: Interscience, pp. 538–540.

Supplementary References

Abdullah, M. I.   The automatic determination of manganese in silicate rocks and sediments.   *Anal. Chim. Acta* **40,** 526–530 (1968).   A Technicon Autoanalyzer is used for the determination of Mn as the orange-red complex formed by Mn (II) with formaldoxime in alkaline solution.

Johansen, O., and Steinnes, E.   Precision analyses of manganese in rocks by neutron activation analysis. *Anal. Chim. Acta* **40**, 201–205 (1968).

Knoeck, J., and Diehl, H.   Titrimetric, coulometric and spectrophotometric determination of manganese following perchloric acid oxidation in the presence of pyrophosphate. *Talanta* **14**, 1083–1095 (1967).   Mn (II) is oxidized to Mn(III) by a boiling mixture of perchloric and phosphoric acids, without the addition of sodium pyrophosphate.

6.9. Phosphorus*

6.9.1. GENERAL

The determination of phosphorus, usually present in the range 0.1–0.5% P_2O_5, is a necessary part of any silicate rock analysis; failure to determine it will cause a positive error in the value for Al_2O_3 if the latter is to be obtained by difference from the combined weight of the R_2O_3 group elements.   The determination is made on a separate portion of the sample, never on the solution prepared by the pyrosulfate fusion of the ignited R_2O_3 residue, nor on any solution that has been fumed strongly with sulfuric acid.   At the Geological Survey of Canada it is our practice to combine the determination of phosphorus with that of sodium in the rapid rock analysis scheme, and with that of manganese (Sec. 9.16.1) in the regular scheme of rock and mineral analysis; when the amount of sample is in short supply it is also determined in an aliquot of the filtrate from the separation of the silica.

In silicate rocks and minerals the phosphorus is usually present as orthophosphate (apatite is the most common mineral), and since this is the form required in the various methods available for its determination, it is only necessary to put the sample into solution.   The usual alternatives, acid decomposition and alkali fusion, will also serve to convert any pyrophosphate or metaphosphate present to the *ortho* form.   Phosphorus may be present, however, as monazite or xenotime and these complex rare earth minerals are not easily decomposed; a dark gritty residue of more than a few grains should not be ignored, even though it will probably consist for the most part of undecomposed quartz.   It is recommended (Sec. 9.16.1) that this residue, or any white, non-gritty residue (which may be titanium phosphate), should be recovered and fused with Na_2CO_3, leached with water, filtered and the filtrate acidified with nitric acid and combined with the main solution.

Phosphorus is partly volatilized when a solution containing it is fumed strongly with sulfuric acid; no significant loss occurs if the solution is heated to only the first fumes of SO_3, and solutions may be evaporated

* References for this chapter will be found on p. 193,194.

to dryness with other acids (HCl, HNO_3 and HF) without loss. Nitric acid is preferable but hydrofluoric acid is usually needed to decompose the silicate; the excess hydrofluoric acid should be removed by repeated evaporation of the solution to dryness with nitric acid, and the interference of fluoride prevented by the addition of boric acid to the solution.

In the conventional scheme of analysis that has been given, some of the phosphorus may be separated with the silica (Sec. 6.2.1a), especially if much titanium (or zirconium, thorium or tin) is present. If too much sulfuric acid is added and the volatilization of the silicon is done at a too high temperature, some phosphorus may also be volatilized, resulting in a negative error in the final value for Al_2O_3, and a positive error in that for silica. The nonvolatile residue is eventually combined with the main R_2O_3 precipitate (Sec. 6.4.1a) and the phosphorus passes into solution during the fusion with potassium pyrosulfate; phosphorus may be lost if the fusion is protracted, or if the temperature is too high, and for this reason it should not be determined in this solution.

If phosphorus is present in a relatively large amount, it may cause the precipitation of all or part of any of the alkaline earths present during the precipitation with ammonium hydroxide, thus introducing a very serious error into the whole main portion analysis.

For most silicate rocks and minerals the determination of phosphorus is most conveniently made by measurement of the intensity of the yellow complex formed by orthophosphate, vanadate and molybdate, or of the blue color formed by the selective reduction of the orthophosphate-molybdate complex. A preliminary separation of the phosphorus is sometimes made to eliminate the interference of silicate, arsenate and molybdate (which form similar heteropoly complexes) and of elements such as copper, nickel and chromium which form colored solutions, but usually it is only necessary to remove silicate by volatilization of the silicon with hydrofluoric acid. An occasional sample containing more than the usual amount of phosphorus (e.g., phosphatic sandstone) will be encountered and extreme dilution of the sample solution will be necessary to bring it into the proper concentration range for colorimetric measurement; in this instance the determination is better made gravimetrically as magnesium ammonium phosphate (with ignition to the pyrophosphate), either after a preliminary separation of the phosphorus, e.g., with ammonium molybdate, or by making the precipitation in the presence of citric acid. For small amounts of phosphorus it is possible to weigh the yellow ammonium phosphomolybdate directly, or the precipitate may be dissolved in a measured amount of standard alkali solution and the excess back-titrated with a standard acid solution.

Gravimetric and colorimetric methods will be covered in some detail in the following pages, but because they do not have much application in silicate analysis, the titrimetric methods will not be discussed. The analytical chemistry of phosphorus has been extensively reviewed[6]; other texts treat in detail the gravimetric and titrimetric[5,7,12] and colorimetric[3] procedures.

6.9.2. GRAVIMETRIC METHODS

The methods most often employed for the gravimetric determination of phosphorus involve its precipitation either as ammonium phosphomolybdate or magnesium ammonium phosphate, or a combination of these two precipitation forms.* Precipitation as yellow ammonium phosphomolybdate is usually employed for the separation of phosphorus from interfering elements, with subsequent dissolving of the precipitate in aqueous ammonia and reprecipitation of the phosphorus as magnesium ammonium phosphate using magnesia reagent and aqueous ammonia. It is possible, however, to employ the yellow precipitate or its ignition product as the final weighing form, just as it is possible, with suitable precautions, to precipitate the phosphorus directly as magnesium ammonium phosphate without preliminary separation.

The difficulties attending the use of ammonium phosphomolybdate as the final weighing form are connected with the uncertain composition of the final product, this latter being affected by the conditions of the precipitation. Much has been written about this[5–7] and the recent study of the conditions favoring the quantitative precipitation of ammonium 12-molybdophosphate,[1] using the radioisotope phosphorus-32, will eliminate much of the uncertainty. These authors show that quantitative precipitation is obtained after heating for 30 min at any temperature between 50 and 80°C, followed by digestion for 30 min at room temperature, with stirring at 15-min intervals; at 90°C, the precipitation of molybdic acid is noted. The precipitation is subject to interference from a number of ions, including fluoride, but the inhibiting effect of these may generally be circumvented by the use of a large excess of ammonium molybdate (twice the stoichiometric amount is recommended[1]). Vanadium(V) will be precipitated at elevated temperatures, but if reduced to V(IV) and if the precipitation is made at room temperature, it will not interfere; in general, its concentration in the sample will be

---

* Precipitation as quinoline phosphomolybdate is also advocated (Ref. 12, pp. 223–224) and appears to have certain advantages over the older methods. The procedure is not gravimetric, however, but involves a subsequent titration with NaOH.

negligible.   Heslop and Pearson (Supp. Ref.) studied the effect of arsenate and some transition-metal ions (Fe(III), Cr(III), Mn(II), Ni(II))
on the precipitation; under the conditions necessary for the quantitative
precipitation of phosphorus, arsenic is also precipitated by ammonium
molybdate.   Organic matter will retard the precipitation, and Peck[9]
prefers to ignite the sample before starting to decompose it, if the presence of organic substances is suspected.   Large amounts of silica must
be removed and this is readily accomplished by decomposing the sample
with a mixture of nitric and hydrofluoric acids; the fluorides are largely
expelled by evaporation with nitric acid alone.   In the procedure recommended by Peck[9] no more than 600 mg of fluorine or 40 mg of SiO_2 remain in the solution, and these concentrations can be tolerated.   The
precipitation is preferably made from a nitric acid solution (perchloric
acid is acceptable but hydrochloric and sulfuric acids interfere) with a
mixed reagent consisting of ammonium molybdate and nitric acid or
ammonium nitrate.   The presence of nitric acid is necessary for the
formation of the precipitate and ammonium nitrate (as a 5–15% solution) has been used to reduce the solubility of the precipitate and to
speed up the precipitation; the study by Archer et al.[1] shows, however,
that ammonium nitrate has no effect upon the efficiency of the precipitation.   The filtered precipitate must be washed with a dilute solution
of an electrolyte (ammonium nitrate, or potassium nitrate if the precipitate is to be dissolved and titrated) to prevent peptization; it is reported
that the use of a mixed ammonium molybdate–nitric acid reagent yields
a precipitate that will not peptize.[1]

   The composition of the yellow precipitate is ideally $(NH_4)_3PO_4 \cdot$
$12MoO_3 \cdot 2HNO_3 \cdot H_2O$ which, on drying at 110°C, loses nitric acid and
water to give $(NH_4)_3P(Mo_3O_{10})_4$, containing 3.78% P_2O_5.   This percentage may drop to as little as 3.73, depending upon the conditions of
the precipitation, and it is this uncertainty of composition that militates
against its use for the determination of large amounts of phosphate; for
the small amount present in most rocks, however, the error thus introduced would be negligible.   The precipitate may also be heated at 400–
500°C for 30 min to give a residue approximating $P_2O_5 \cdot 24MoO_3$.   If the
precipitate is to be dissolved and an alkalimetric titration used as the
finish, then it is important that the ratio of $PO_4 : MoO_3$ be maintained at
1:12; if the phosphorus is to be reprecipitated as the magnesium ammonium phosphate the exact composition of the yellow precipitate is
unimportant as long as the precipitation is quantitative.

   The problems inherent in the precipitation of magnesium ammonium
phosphate have been discussed in detail in the section dealing with the

gravimetric determination of magnesium (Sec. 6.6.1a) and similar problems may be encountered in the determination of phosphorus, even though the precipitation is made in the reverse manner. Many elements, if present, will interfere in the precipitation and it is often necessary to make a preliminary separation of the phosphorus with ammonium molybdate as previously discussed, or to complex these interfering elements so that they will not interfere. This latter is achieved by the addition of ammonium citrate (10–15%) to the solution and the procedure is as accurate as that involving preliminary separation, and is more rapid. For accurate work, regardless of the manner in which the precipitation is made, a second precipitation is mandatory to insure that the precipitate has the proper composition $(MgNH_4PO_4 \cdot 6H_2O)$; for most rocks, however, the phosphorus content is so low that the increased accuracy is not worth the time required for a double precipitation and a single precipitation will suffice. To minimize errors the precipitation should be made from an acid solution containing magnesia reagent, and also sufficient ammonium chloride to buffer the solution at approximately pH 10.5 when the solution is first slowly neutralized with aqueous ammonia and a few milliliters added in excess. Reference should be made to recent texts for a fuller account.[5-7]

6.9.3. Colorimetric methods

Two methods are in general use for the colorimetric determination of phosphorus. The one employs the yellow color of the mixed heteropoly acid formed when molybdate solution is added to an acidic solution (at least $0.7M$) containing orthophosphate and vanadate ions; the other utilizes the blue color which results from the selective reduction of molybdophosphoric acid, the intensity of the blue color being proportional to the concentration of the phosphate present in the latter. Use is also made of the yellow color of this molybdophosphoric acid complex[3,6] after its selective extraction, but it will not be considered here. Most of the rapid analysis schemes employ the colorimetric determination of phosphorus.

The yellow molybdovanadophosphoric acid complex is the basis of the method considered to be the most specific for phosphorus.[2,11] There are many conflicting data available to show the influence of various factors on the formation of the complex, such as the stability of the reagents and the use of separate reagents vs. one combined reagent, the choice of acid and acid concentration, changes in temperature of the solution and the wavelength of measurement; these have been summarized recently (Ref. 6, pp. 351–353). Phosphorus is not the only element which forms

a colored complex of this nature, however, and the methods in use eliminate the interference of such ions as silicate, arsenate, and germanate,
as well as of elements such as Fe, Ni and Cu which form colored compounds, by preliminary separation of the phosphorus, by choice of pH,
by removal of the interferences through selective volatilization or mercury cathode electrolysis and by choice of the wavelength of measurement.   Nitric acid is used most frequently, usually about 0.5N, but
HCl, HClO$_4$ or H$_2$SO$_4$ may also be used.   The molybdate and vanadate
reagents, about 5 and 0.25% solutions, respectively, may be added separately or as one solution, but the order of addition, if added separately,
must be acid, vanadate and molybdate; the final concentrations recommended for vanadate and molybdate are 0.002M and 0.04M, respectively (Ref. 6, p. 353).   These authors also recommend the addition
of Na$_2$SO$_4$ solution to minimize the effect of temperature changes on the
intensity of the colored complex.   The absorbance of the latter is measured at 400–460 mμ; at the higher wavelength the interference of iron-
(III), copper and nickel is minimized.   The interference of these ions
can also be compensated for by use of an aliquot of the sample solution,
without the addition of reagents, as the blank.   Fluoride ion interferes
in the color development and it is customary to bake the decomposed
sample, after evaporation to dryness with nitric acid, to expel as much
of the fluorine as possible; the baking should not be done at temperatures
greater than 250°C or some phosphorus will be lost by volatilization.
According to Boltz[3] the optimum concentration range is 5–40 ppm phosphorus, for a 1-cm cell.

The blue color formed by the reduction of the molybdophosphoric acid
is more intense than the yellow complex discussed previously and the
method is thus more sensitive, although there is at the same time a loss
of color stability and an increased need for a more rigorous control of
operating conditions.   Boltz (Ref. 3, p. 32) discusses the formation of
two blue complexes; that which forms at an acidity of about 1N and
has an absorbance maximum at 830 mμ, is called "heteropoly blue" to
distinguish it from "molybdenum blue," which forms at lower acidities
and which has an absorbance maximum at 650–700 mμ.   The choice of
reducing agent seriously affects the intensity and stability of the color
and the literature is replete with recommendations for particular reagents (Ref. 6, pp. 348–351); a similar discussion is given in the section
on the colorimetric determination of silica (Sec. 6.2.3).   A further modification involves the extraction of the molybdophosphoric acid into an
organic solvent such as isobutyl alcohol before reduction to heteropoly
blue; this is one means of eliminating the interference of many other ions.

Boltz[3] uses hydrazine sulfate as the reducing agent and adds it separately from the molybdate; the solution is then heated for 10 min at about 100°C. Riley[10] prefers ascorbic acid for the reduction and mixes it with the molybdate as a single reagent; the addition of potassium antimonyl tartrate to the reagent was later found to give increased sensitivity when applied to the determination of phosphorus in natural waters.[8] This method has been modified (van Schouwenburg and Walinga (Supp. Ref.)) for use with soils in which the interference from arsenic must be eliminated. The As(V) is reduced with a mixed reducing agent ($Na_2S_2O_3$ and $NaHSO_3$); the interference of Sn may be avoided but vanadium and tungsten still introduce slight errors. Chalmers[4] used the reduced form to determine phosphorus in silicate samples weighing only a few milligrams; ferrous ammonium sulfate is the reductant and the interference of vanadate and Fe(III) is eliminated by first passing the solution through a semimicro silver reductor. The silica is not removed but is prevented from interfering by having a final acid concentration of $0.76N$ H_2SO_4.

References

1. Archer, D. W., Heslop, R. B., and Kirby, R. The conditions for quantitative precipitation of phosphate as ammonium 12-molybdophosphate. *Anal. Chim. Acta* **30**, 450–459 (1964).
2. Baadsgaard, H., and Sandell, E. B. Photometric determination of phosphorus in silicate rocks. *Anal. Chim. Acta* **11**, 183–187 (1954).
3. Boltz, D. F., (Ed.). *Colorimetric Determination of Nonmetals*, 1958. "Phosphorus," by D. F. Boltz and C. H. Lueck. New York: Interscience, pp. 29–46.
4. Chalmers, R. A. Microanalysis of silicate rocks. Part III. The spectrophotometric determination of phosphoric oxide in the presence of silica. *Analyst* **78**, 32–36 (1953).
5. Hillebrand, W. F., Lundell, G. E. F., Bright, H. A., and Hoffman, J. I. *Applied Inorganic Analysis*, 2nd ed., 1953. New York: John Wiley & Sons, pp. 694–710, 895–897.
6. Kolthoff, I. M., and Elving, P. J., (Eds.). *Treatise on Analytical Chemistry*, Part II, Vol. 5, 1961. "Phosphorus," by William Rieman, III and John Beukenkamp. New York: Interscience, pp. 317–402.
7. Kolthoff, I. M., and Sandell, E. B. *Textbook of Quantitative Inorganic Analysis*, 3rd ed., 1952. New York: Macmillan, pp. 377–383.
8. Murphy, J., and Riley, J. P. A modified single solution method for the determination of phosphate in natural waters. *Anal. Chim. Acta* **27**, 31–36(1962).
9. Peck, L. C. Systematic analysis of silicates. *U.S. Geol. Surv. Bull.* **1170**, 45–48, 78–79 (1964).
10. Riley, J. P. The rapid analysis of silicate rocks and minerals. *Anal. Chim. Acta* **19**, 425–426 (1958).
11. Shapiro, L., and Brannock, W. W. Rapid analysis of silicate, carbonate and phosphate rocks. *U.S. Geol. Surv. Bull.* **1144–A,** A31 (1962).

194   ROCK AND MINERAL ANALYSIS

12. Wilson, C. L., and Wilson, D. W., (Eds.). *Comprehensive Analytical Chemistry*, Vol. 1C (Classical analysis), 1962. "Phosphorus," by S. Greenfield. Amsterdam: Elsevier, pp. 220–236.

Supplementary References

Brunfelt, A. O., and Steinnes, E. The determination of phosphorus in rocks by neutron activation. *Anal. Chim. Acta* **41**, 155–158 (1968).

Henderson, P. The determination of phosphorus in rocks and minerals by activation analysis. *Anal. Chim. Acta* **39**, 512–515 (1967).

Heslop, R. B., and Pearson, E. F. The effect of arsenate and of transition-metal ions, on the precipitation of phosphate as ammonium 12-molybdophosphate. *Anal. Chim. Acta* **33**, 522–531 (1965).

van Schouwenburg, J. C., and Walinga, I. The rapid determination of phosphorus in presence of arsenic, silicon and germanium. *Anal. Chim. Acta* **37**, 271–274 (1967).

6.10. The alkali metals*

The determination of sodium and potassium can seldom be omitted from a detailed silicate rock or mineral analysis, and usually these two elements comprise 1–10% or more of the sample. Lithium is usually present only in trace amounts in rocks but is more abundant in such minerals as feldspar, muscovite and beryl, and is a major constituent of lepidolite and spodumene. Rubidium is an abundant trace element that has a close geochemical association with potassium. Cesium is less abundant than rubidium and usually accompanies it; it does, however, occur as a major constituent of one silicate mineral, pollucite. Which of these elements, apart from sodium and potassium, should be determined will be governed by the nature of the sample and the purpose of the analysis; the determination of lithium, rubidium and cesium as part of a routine rock analysis is, however, seldom necessary. As has been mentioned repeatedly, a preliminary optical emission spectrographic analysis will usually indicate the presence of unusual amounts of these latter elements and the need for their determination, not only for the sake of completeness but also to correct for their effect upon the determinations of sodium and potassium.

The availability of detailed reviews of the analytical chemistry of the alkali metals[14,20,32] makes it unnecessary to attempt a detailed discussion here. The review by Kallman (Ref. 20, pp. 306–460), with 471 references, is particularly helpful and covers not only gravimetric and flame photometric methods, but spectrochemical, X-ray fluorescence, polarographic and radiochemical procedures as well. Other references describe

* References for this section will be found on pp. 200–202.

the direct application of atomic absorption spectroscopy,[3] optical emission spectrography[29] (see also Gurney and Erlank, Supp. Ref.) and neutron activation[4,5] to the determination of the alkali metals in rocks and minerals. The procedures to be discussed here are gravimetric and flame photometric only; the determination of potassium by X-ray fluorescence spectroscopy will be treated in a later section.

6.10.1. GRAVIMETRIC METHODS

The preliminary decomposition of the sample is usually done by one or another of two methods: by a combination of sintering and extraction; or by digestion with a mixture of acids, one of which is hydrofluoric acid. Fusion of the sample with H_3BO_3 + Li_2CO_3 and dissolution of the crushed and ground fusion cake in citric acid, prior to flame photometric determination of sodium and potassium, has been described by Govindaraju (Supp. Ref.).

The sintering–extraction procedure as originally proposed by J. Lawrence Smith[12,21,30,31] has changed but little in the intervening years (see Sec. 5.4.3). Although the full procedure is seldom followed now, the decomposition and extraction procedure is often used because it yields a chloride solution of the alkali metals that is relatively free from interfering elements and well suited to subsequent gravimetric analysis. Those elements which are likely to accompany the alkali metals in the chloride solution include Mg, Zn, Tl, B, F, S, Se, Te, Br and iodine.[22] In silicate rocks, only Mg, B, F, and S are likely to be of significance and all but a small part of the boron and fluorine present will be removed during the extraction and precipitation steps; additional precautions are taken to remove the small amounts of magnesium and sulfur (as SO_4^{2-}) that accompany the alkali metals. The procedure cannot be used if lithium is to be determined, for some of it is lost in the extraction and precipitation steps.

The use of hydrofluoric and sulfuric acids to decompose silicates was recommended by Berzelius and this mixture is widely used today (see Sec. 5.2); only a few silicates will resist these acids, among them zircon, beryl, topaz and tourmaline,[15] and the sintering–extraction procedure should be used when these minerals are known to be present. The rest of the involved Berzelius procedure for obtaining a pure solution of alkali chlorides is not used now, but the acid solution is particularly suitable for further treatment by one of several methods which lead to the determination of the alkali metals by flame photometry (Ref. 20, pp. 332–334). The weight of the mixed chlorides may also be determined, after acid decomposition of the sample, by placing the alkali metals on an ion-exchange resin and then eluting them with dilute HCl.[25] The solution of mixed chlorides is then ready for further separation and determination.

The direct gravimetric determination of *sodium* is readily done by precipitating it with zinc (or magnesium) uranyl acetate, a selective procedure that requires few preliminary separations[13,20,21]; lithium will accompany the sodium. A chloride solution such as that obtained from the J. Lawrence Smith decomposition procedure is ideal for the separation. The method is more suited to routine work because the high solubility of the triple acetate salt makes it difficult to obtain highly accurate results. Another approach is to convert the chlorides to perchlorates, from which sodium (and lithium if it present also) can be extracted with a mixture of normal butyl alcohol and ethyl acetate; after expulsion of the ethyl acetate, sodium chloride is precipitated by hydrogen chloride in butyl alcohol.[14] The most popular, indirect gravimetric method for the determination of sodium is that in which chloroplatinic acid is used to precipitate potassium (together with rubidium and cesium) from the weighed, mixed chlorides; the weight of the KCl-equivalent of the K_2PtCl_6 is subtracted from that of the mixed chlorides, and the sodium (and lithium) are obtained by difference. It is obvious that the accuracy of the latter determination depends very much upon the accuracy of the potassium determination, and on the absence of appreciable amounts of rubidium, cesium and lithium unless these are also to be determined. The two groups of salts can be weighed, dissolved in water and the platinum removed with formic acid, after which extraction procedures can be used to separate the individual alkali metals. A simpler procedure is to determine the lithium, rubidium and cesium in the two groups by flame photometry.

The separation of *potassium* (and rubidium and cesium if present) from sodium (and lithium, if present) with chloroplatinic acid suffers from two disadvantages; the reagent is expensive (and recovery of the platinum is a tedious procedure), and it is stated that the composition of the insoluble yellow precipitate may vary somewhat from the theoretical K_2PtCl_6, requiring the use of an empirical factor for $K_2O(0.3056)$ instead of the theoretical one (0.3067). Peck[24] reports that the theoretical factor may be used but that the values for smaller amounts of potassium will be slightly low, and those for large amounts, slightly high. The method involves the extraction of the soluble sodium salt with 80% ethyl alcohol, and the separation is not complete; accurate results are only obtained if the small amount of potassium that also dissolves is determined in the sodium chloroplatinate fraction. Preliminary separation of the alkali metals is necessary and the chloroplatinate method usually follows a J. Lawrence Smith sinter and extraction. Ammonium salts must be absent because ammonium chloroplatinate is also insoluble; the sulfate ion must be removed prior to the extraction with alcohol because sodium sulfate is insoluble in this solvent.

Conversion of the chlorides to perchlorates and extraction of the sodium and lithium perchlorates with n-butyl alcohol and ethyl acetate suffers from some of the disadvantages given for the chloroplatinic acid method. Ammonium and sulfate ions must be absent.

The quantitative precipitation of potassium from dilute acid (acetic or nitric) solution by cobaltinitrite may be done without preliminary separation, but because the composition of the precipitate varies with the conditions of precipitation, the method can be safely used only for small amounts of potassium (<0.1 g).

Sodium tetraphenylboron is a highly selective precipitant for potassium, and in dilute mineral acid solution no preliminary separation is necessary. Ammonium salts must be absent. Rubidium and cesium also form insoluble compounds. The method enjoys several advantages over those mentioned previously, and is applicable to both large and small amounts of potassium. The final determination of the potassium may be done either gravimetrically[6,20] or titrimetrically (Cluley, Supp. Ref.).

Lithium accompanies sodium in the separations mentioned previously and further separation is based upon the greater solubility of lithium chloride in organic solvents. The J. Lawrence Smith decomposition procedure cannot be used if lithium is to be determined, because of the loss of lithium during the extraction and precipitation steps, and acid decomposition of the sample is usually required. A novel method of separation, in which the sample is heated at 1200°C with a mixture of calcium carbonate and calcium chloride in a platinum tube, yields the lithium as chloride by distillation.[11]

Rubidium and *cesium* accompany potassium in the separations that have been described. Further separations can be achieved by the use of precipitants such as 9-phosphomolybdic and silicotungstic acids, but the use of flame photometric procedures is much to be preferred. Feldman and Rains[10] recently described the collection of cesium from an acid solution of a rock by adsorption on ammonium-12-molybdophosphate; the latter is dissolved in NaOH, the Cs extracted with sodium tetraphenylboron in 4-methyl-2-pentanone (hexone) and cyclohexane, and the organic extract is used for flame photometric determination of the cesium.

6.10.2. FLAME EMISSION AND ABSORPTION METHODS

Because of characteristically low excitation potentials, the alkali metals require only a relatively cool flame (a mixture of air and city gas or propane) for their flame photometric determination. Because few other elements are sufficiently excited at this temperature to offer significant interference, the flame photometric determination of the alkali metals is one of the simpler steps in rock and mineral analysis[2]; it is particularly applicable to "rapid" schemes of analysis.[26—28] The principles and problems of flame photometry have been discussed in detail[7,20] and only a few comments will be made here.

In most rocks, and many minerals, the possible effect of Li, Rb and Cs on the radiation intensities of *sodium* and *potassium* is negligible. A preliminary spectrographic analysis will indicate whether or not the former elements are present in significant amounts; it will also indicate other potentially troublesome elements whose presence in the acid solution of the sample might not be suspected. Sodium enhances the potassium radiation intensity significantly but potassium has only a slight enhancing effect on that of sodium; this interference can be compensated for by adding an equivalent amount of sodium, which is determined first, to the potassium standard solutions. Elements such as aluminum, iron, calcium and magnesium depress the radiation intensity of the alkali metals and steps must be taken either to remove these elements first, such as by the J. Lawrence Smith decomposition procedure mentioned previously, by thermal decomposition and leaching[1,23] or by ion exchange[25,27] (see also Olsen and Sobel, Supp. Ref.), or to compensate for their effect by the use of an internal standard, a radiation buffer, or by measurement of the background radiation.[8] It is also possible to utilize the greater depressing effect of one element on another to minimize the interference of the latter; Jackson and Smith[19] found that only calcium interfered significantly in the determination of sodium in coal ash and added aluminum to suppress this interference (see also Ref. 20, p. 383). Ammonium salts inhibit the radiation emission of the alkali metals and thus excess acid should be removed by means other than by neutralization with aqueous ammonia. Anion interference, particularly from free sulfuric acid, can be serious and the decomposition procedure must be adjusted to yield a constant amount of acid at its completion; both hydrochloric and nitric acids have much less effect. Too great an excess of acid is also undesirable from the standpoint of burner corrosion. The simple and rapid method for the determination of sodium, potassium and calcium of Abbey (Supp. Ref.) has already been mentioned (Sec. 6.5.3).

Horstman and others[9,16,17] have studied the determination by flame emission spectroscopy of *lithium* in silicate rocks in detail. At the Geological Survey of Canada a similar study using atomic absorption spectroscopy has been carried out by Sydney Abbey (Supp. Ref.). The sample is decomposed with a mixture of hydrofluoric, nitric and perchloric acids, fumed to dryness, and the dry perchlorates are then dissolved in dilute HCl (0.5N). The equivalent of 5% sodium or potassium in the sample has no effect upon the absorption signal for lithium but the concentration of free acid has a marked effect and must be carefully controlled. The effect of the total salt concentration on the absorption signal is not certain; results suggest that the latter is weakened with increasing total salt concentration but the depression is not directly proportional. It is recommended that "standard addition" tests be used to establish the magnitude of the matrix depressant effect for samples of unknown composition and a correction factor applied where feasible. The procedure, applicable to many types of rocks, is given in Sec. 12.3.2; the final solution should contain 5–125 ppm lithum.

With a few exceptions, the concentrations of *rubidium* and *cesium* encountered in most silicate rocks and minerals will be at the trace level. A method is given for their determination by flame photometry in the potassium chloroplatinate fraction derived from the J. Lawrence Smith decomposition and separation of the alkali metals (Sec. 9.20); Abbey's method (Sec. 9.21.3) gives the conditions for the measurement of rubidium radiation and a similar measurement can be made for cesium, at 852 mμ and a slit width of 0.15 mm. The combination of a large sample and a small final volume of solution will increase the sensitivity of the procedure. Mention has also been made of the precipitation of traces of cesium on a carrier, followed by organic extraction of the alkali metal and its flame photometric determination in the organic extract.[10]

Horstman extended the procedure used for the determination of lithium[9] to include rubidium and cesium as well.[16] Following the acid decomposition of the sample, precipitation of the ammonia group is effected by the addition of solid calcium carbonate, and calcium is later removed by precipitation as the sulfate in 50% ethyl alcohol. An accuracy of 5 ppm Li, 10 ppm Rb and 5 ppm Cs is claimed for the range from 10 to 200 ppm of the alkali metal.

The limit of detection for rubidium and cesium has been extended to about 2 ppm Rb_2O and 1 ppm Cs_2O by Ingamells.[18] Separate samples are used for sodium and potassium (the potassium content must be accurately known because it greatly enhances the rubidium and cesium emission in the flame), for rubidium and cesium, and for lithium. The "neutral leach" method of Abbey (Sec. 9.21.3) is used for the preparation of the sample solutions, excepting that for the lithium determination which is prepared by the method of Ellestad and Horstman (Sec. 9.21.4). The flame attachment described by these workers is modified to provide a highly stable flame, and the addition of an external galvanometer circuit to a Beckman Model DU spectrophotometer makes it possible to use differential flame spectrophotometry, resulting in improved precision of measurement. The determinations of potassium and sodium are made first, on the basis of which the potassium content of the rubidium–cesium sample solution is adjusted to 1000 ppm, after which the rubidium and cesium measurements are made. Large amounts of manganese will affect the final results and some interference is caused by magnesium. The method requires close control of the many variables that will affect the measurements, and the detailed outline and discussion given by Ingamells should be consulted.

References

1. Abbey, S., and Maxwell, J. A. Determination of potassium in micas. A flame photometric study. *Chem. Canada* **12**, 37–41 (1960).
2. Bennett, H., and Hawley, W. G. *Methods of Silicate Analysis*, 2nd ed., 1965. London: Academic Press, 81–84.
3. Billings, G. K., and Adams, J. A. S. The analysis of geological materials by atomic absorption spectrometry. *Atomic Absorption Newsletter* **23**, 1–7 (1964). Perkin-Elmer Corporation, Norwalk, Connecticut.
4. Butler, J. R., and Thompson, A. J. Different values for Cs in G–1. *Geochim. Cosmochim. Acta* **26**, 1349–1350 (1962).
5. Cabell, M. J., and Smales, A. A. The determination of rubidium and cesium in rocks, minerals and meteorites by neutron activation analysis. *Analyst* **82**, 390–406 (1957).
6. Cluley, H. J. The determination of potassium by precipitation as potassium tetraphenylboron, and its application to silicate analysis. *Analyst* **80**, 354–364 (1955). See also Supplementary References.
7. Dean, J. A. *Flame Photometry*, 1960. New York: McGraw-Hill, 354 pp.
8. Easton, A. J., and Lovering, J. F. Determination of small quantities of potassium and sodium in stony meteoritic material, rocks and minerals. *Anal. Chim. Acta* **30**, 543–548 (1964).
9. Ellestad, R. B., and Horstman, E. L. Flame photometric determination of lithium in silicate rocks. *Anal. Chem.* **27**, 1229–1231 (1955).
10. Feldman, C., and Rains, T. C. The collection and flame photometric determination of cesium. *Anal. Chem.* **36**, 405–409 (1964).
11. Fletcher, M. H. Determination of lithium in rocks by distillation. *Anal. Chem.* **21**, 173–175 (1949).
12. Grillot, H. Analyse chimique des roches et des eaux. *Bureau Recherches Géologiques, Géophysiques, Minières* **20**, 18 (1957).

13. Groves, A. W. *Silicate Analysis*, 2nd ed., 1951. London: George Allen and Unwin, 77–78, 176–181.
14. Hillebrand, W. F., Lundell, G. E. F., Bright, H. A., and Hoffman, J. I. *Applied Inorganic Analysis*, 2nd ed., 1953. New York: John Wiley & Sons, 646–670, 923–934.
15. Hoops, K. G. The nature of the insoluble residues remaining after the HF–H₂SO₄ acid decomposition (solution B) of rocks. *Geochim. Cosmochim. Acta* 28, 405–406 (1964).
16. Horstman, E. L. Flame photometric determination of lithium, rubidium and cesium in silicate rocks. *Anal. Chem.* 28, 1417–1418 (1956).
17. Howling, H. L., and Landolt, P. E. Determination of lithium in silicate minerals and leach solutions by flame photometry. *Anal. Chem.* 31, 1818–1819 (1959).
18. Ingamells, C. O. Determination of major and minor alkalies in silicates by differential flame spectrophotometry. *Talanta* 9, 781–793 (1962).
19. Jackson, P. J., and Smith, A. C. A rapid method for determining potassium and sodium in coal ash and related materials. *J. Appl. Chem.* 6, 547–559 (1956).
20. Kolthoff, I. M., and Elving, P. J., (Eds.). *Treatise on Analytical Chemistry*, Part II, Vol. 1, 1961. "The Alkali Metals," by S. Kallman. New York: Interscience, pp. 306–460.
21. Kolthoff, I. M., and Sandell, E. B. *Textbook of Quantitative Inorganic Analysis*, 3rd ed., 1952. New York: Macmillan, 395–403.
22. Lundell, G. E. F., and Hoffman, J. I. *Outlines of Methods of Chemical Analysis*, 1938. New York: John Wiley & Sons, p. 41.
23. Marvin, G. G., and Woolaver, L. B. Determination of sodium and potassium in silicates. An improved method. *Ind. Eng. Chem. Anal. Ed.* 17, 554–556 (1945).
24. Peck, L. C. Systematic analysis of silicates. *U.S. Geol. Surv. Bull.* 1170, 42–44, 63–76 (1964).
25. Reichen, L. E. Use of ion exchange resins in the analysis of rocks and minerals. Separation of sodium and potassium. *Anal. Chem.* 30, 1948–1950 (1958).
26. Rigg, T., and Wagenbauer, H. A. Analysis of silicate rocks. Part 1: Routine determination of major constituents. *Prelim. Rept., Res. Council Alberta* 64–1, 15–18 (1964).
27. Riley, J. P. The rapid analysis of silicate rocks and minerals. *Anal. Chim. Acta* 19, 424–425 (1958).
28. Shapiro, L., and Brannock, W. W. Rapid analysis of silicate, carbonate and phosphate rocks. *U.S. Geol. Surv. Bull.* 1144–A, 33–35 (1962).
29. Shaw, D. M., Wickremasinghe, O. C., and Weber, J. N. Spectrochemical determination of Li, Na, K and Rb in rocks and minerals using the Stallwood jet. *Anal. Chim. Acta* 22, 398–400 (1960).
30. Smith, J. Lawrence. Determination of the alkalies in silicates by ignition with carbonate of lime and sal-ammoniac. *Am. J. Sci.* (3rd Ser.) 1 (Whole No. 101), No. IV, 269–275 (1871).
31. Washington, H. S. *The Chemical Analysis of Rocks*, 2nd ed., 1910. New York: John Wiley & Sons, pp. 141–155.

32. Wilson, C. L., and Wilson, D. W., (Eds.). *Comprehensive Analytical Chemistry*, Vol. 1C (Classical analysis), 1692. "Alkali Metals," by K. Gardner. Amsterdam: Elsevier, pp. 9–55.

Supplementary References

Abbey, Sydney. Determination of potassium, sodium and calcium in feldspars. *Can. Mineral.* **8**, 347–353 (1965).

Abbey, Sydney. Analysis of rocks and minerals by atomic absorption spectroscopy. I. Determination of magnesium, lithium, zinc and iron. *Geol. Surv. Canada Paper* **67–37**, 1967, 35 pp.

Cluley, H. J. Determination of potassium oxide in glass. *Trans. Soc. Glass Technol.*, **43**, 62T–72T (1959).

Govindaraju, K. Rapid flame photometric determination of sodium and potassium in silicate rocks. *Appl. Spectr.* **20**, 302–304 (1966). This method is supplementary to the general scheme for the determination of the major constitutents of silicate rocks by optical emission spectroscopy discussed in Sec. 1.1.

Gurney, J. J., and Erlank, A. J. DC arc spectrographic technique for determination of trace amounts of Li, Rb and Cs in silicate rocks. *Anal. Chem.* **38**, 1836–1839 (1966).

Olsen, E. D., and Sobel, H. R. Selective retention of alkali metals on cation-exchange resins. *Talanta* **12**, 81–90 (1965). By precomplexing multivalent elements with EDTA, only the alkali metals are retained on the resin, from which they can be eluted and concentrated.

6.11. Iron*

Iron, in geological materials, exists in the ferrous, ferric and (rarely) metallic forms; it is very likely that it will be present as a mixture of two or more of these valence states. Its determination, particularly of the valence state(s) in which it occurs, is a matter of much importance; it is to be regretted that the methods available for the determination of these valence states still give cause for doubting the reliability of the results obtained. To the petrographer and mineralogist the ferric/ferrous ratio is an important parameter and the correct determination of this ratio is still a challenge to the analyst. It is to be supposed that in a mixture as complex as a rock or most minerals the chemical equilibrium that exists in the solid state will undergo adjustment when the sample is put into solution; the best that the analyst can do is to minimize those external influences that will change this equilibrium and seek to obtain values for the different valence states that are as near to the true values as possible.

The problems associated with the determination of the valence state(s) of iron begin with the preparation of the sample. The effect of grinding

* References for this section will be found on pp. 214–217.

on the state of oxidation of iron has long been under consideration[13,17] and this has already been discussed (see Sec. 3.4.3) in some detail. It is the usual practice at the Geological Survey of Canada to grind all samples to pass through a 100-mesh screen, without reserving a coarser portion for the determination of ferrous iron as is recommended by some analysts.

If the conventional scheme of analysis described previously has been followed, the total iron content of the sample will be found in the sulfuric acid solution of the ignited R_2O_3 precipitate (Sec. 9.6) which was fused with potassium pyrosulfate (Sec.9.6.1) and which was used for the determination of titanium (Sec. 9.15.1a). The iron can be conveniently determined titrimetrically in this solution (see Sec. 9.15.1 and second footnote), or in an aliquot of it, after the expulsion of the hydrogen peroxide by evaporation of the solution to small volume (Sec. 9.23.3), the precipitation of accumulated platinum with hydrogen sulfide and the expulsion of the latter, again by evaporation to small volume. One advantage of its determination at this stage is that complete decomposition of the sample will have been achieved; another advantage is that the value for this (usually) major constituent of the R_2O_3 group is obtained on the same portion of the sample used for the determination of the Al_2O_3 by difference, thus eliminating a possible inhomogeneity of the sample as a source of uncertainty in the value for Al_2O_3. Some iron may, however, be lost to the crucible during fusion of the sample with sodium carbonate, especially if the fusion is done over a burner (Shell[33] and Supp. Ref.). The determination may be made titrimetrically on a separate portion of the sample (for example, by combining it with the determination of ferrous iron), or it may be finished colorimetrically, polarographically, by X-ray fluorescence, atomic absorption or, particularly for low concentrations, by optical emission spectroscopy. The ferrous iron determination must be done on a separate portion of the sample and titrimetric or spectrophotometric methods are most frequently used. Ferric iron is seldom determined directly, but is usually obtained by difference from the separate determinations of the ferrous and total iron contents of the sample. Metallic iron is determined either colorimetrically or titrimetrically after selective solution of the metallic portion of the sample. These various approaches will be discussed in more detail in the following pages.

6.11.1. FERROUS IRON

The methods for the determination of ferrous iron can be divided into two types, those in which the released Fe(II) is determined in a later step and those in which the Fe(II) ions are oxidized upon release by an oxidizing agent present in excess during the decomposition of the sample, the excess then being determined. Other variations in method or technique stem from efforts to insure complete decomposition of the sample, and range from simple acid attack without further precautions to fusion or acid attack in sealed tubes or bombs. Previous work of this nature, including the sealed tube method of Mitscherlich, the fusion method of Rowledge, Cooke's acid decomposition procedure and Pratt's modification of it, have been described in detail by Hillebrand et al.,[13] Washington[39] and most recently by Schafer (Supp. Ref.).

Both Groves[10] and Juurinen[16] describe modifications of the fusion method which was developed to insure the complete breakdown of refractory silicates such as staurolite. Groves mixes the sample with sodium metafluoborate, $(NaF)_2B_2O_3$, in a platinum boat and fuses it in a silica tube at 950°C under an atmosphere of CO_2; the boat and the fused melt are dropped into a solution of boric and sulfuric acids, the melt dissolved by prolonged boiling under CO_2 and the Fe(II) titrated with $KMnO_4$. The long period of boiling needed to dissolve the glassy melt can be avoided by dissolving the coarsely crushed material in an excess of oxidant (Groves used the iodine monochloride method developed by Hey[12] (see p. 206)). Juurinen[16] substituted a "turbid" silica tube for one of Pyrex glass used by Rowledge and modified the dissolution procedure to exclude air; he found, however, that the small amount of oxygen present in the sealed tube during the fusion does not appreciably oxidize ferrous iron. Mikhailova et al.[21] fuse the sample with sodium fluoborate in a platinum crucible under an atmosphere of CO_2 and polarograph the solution of the melt, using sodium oxalate as supporting electrolyte; polarographic waves are obtained for both Fe(II) and Fe(III); a similar procedure was used by Bien and Goldberg (Supp. Ref.), who fused the sample with sodium metafluoborate under nitrogen, and dissolved the pulverized sample in sodium citrate–citric acid–potassium nitrate solution before polarographing it (+0.15 to −0.35 V vs. SCE).

The method of decomposition by acid attack under pressure in a sealed tube (Mitscherlich) is seldom used now but modifications of the procedure have recently been described. Riley and Williams[27] decomposed a 5 mg sample in a stoppered Teflon tube having a capacity of about 1 ml, with a mixture of HF and H_2SO_4 previously flushed free of oxygen with nitrogen. The acid mixture completely fills the stoppered tube which is heated to approximately 100°C in a boiling water bath; the determination of the ferrous iron is made colorimetrically using 2,2′-dipyridyl. The procedure is part of the microanalytical scheme proposed by Riley for the analysis of silicates. Complete decomposition of such refractory minerals as staurolite, tourmaline and sapphirine was achieved by Ito,[15] who heated 100–200 mg samples with HF and H_2SO_4 in a Teflon-lined steel bomb at 240°C for about 4 hr; the solution is washed into a boric acid solution and the determination finished titrimetrically with potassium permanganate. Contamination of the solution by the steel of the bomb is very slight (approximately 0.04 mg Fe_2O_3 after 4 hr), but the blank consumption of $KMnO_4$ was found to increase with time of heating and the latter should be limited to 4 hr at the most, if possible. No appreciable oxidation by the oxygen present in the air trapped inside the bomb was found; pyrite is partially decomposed by this procedure, and high results for ferrous iron are likely when it is present.

Decomposition of the sample by a boiling mixture of hydrofluoric and sulfuric acids at atmospheric pressure, with air excluded either by steam or by a stream of carbon dioxide, is probably the most widely used technique; the crucible and contents are then immersed beneath the surface of a sulfuric–boric acid solution, and the ferrous iron is titrated with a standard solution of $K_2Cr_2O_7$, either potentiometrically or visually using diphenylamine sulfonic acid (sodium or barium salt) as indicator. Some minor modifications of this procedure have been described, such as the addition of a "spike" of standard ferrous ammonium sulfate solution to the titration beaker in order to insure a satisfactory end point when the FeO content of the sample is small,[32] and the addition of a few milliliters of ferrous ammonium sulfate to the titration beaker, followed by titration as usual with $K_2Cr_2O_7$ to remove oxidizing agents in the H_2SO_4–H_3BO_3 solution prior to the addition of the crucible contents (this eliminates the need to boil and cool the water).[29] Pyrite and some refractory minerals (e.g., tourmaline, chromite) are not decomposed; it is sometimes possible for some of these difficultly soluble minerals to be dissolved by repeated digestion of the residue (with or without further grinding). Schafer (Supp. Ref.) decomposes the sample in H_2SO_4 and HF at 80°C in a special polyethylene vessel, purged before and during the decomposition with nitrogen, and titrates the Fe(II) potentiometrically in the decomposition vessel with $K_2Cr_2O_7$ and diphenylamine sulfonate. Fahey[8] dissolves magnetite and ilmenite in the presence of amphibole and pyroxene without decomposing the latter silicates, by heating them overnight at steam-bath temperature with 1:1 HCl in a tightly stoppered 50 ml erlenmeyer flask filled with carbon dioxide; some ilmenites require a second overnight acid treatment. Vincent and Phillips[35] used a semimicro version of the conventional procedure (15–20 mg samples in a 5 ml crucible) for iron–titanium oxide minerals. Langmyhr and Graff[9,19] attempted to use an aliquot of a solution prepared by dissolving the sample in hydrofluoric acid in a covered Teflon vessel, with the final measurement of the Fe(II) being made spectrophotometrically; good results were obtained for low values of Fe(II) but for high values a systematic negative error was found.

The determination of Fe(II) in the presence of Fe(III) by formation of a colored Fe(II) complex has been proposed. Walker and Sherman[38] decomposed a 0.1 g sample of soil with HF and H_2SO_4, added the mixture to a boric acid–sulfuric acid solution, filtered it and complexed the Fe(II) in a small aliquot of the solution with bathophenanthroline (4,7-diphenyl-1,10-phenanthroline); the red complex was extracted into nitrobenzene for spectrophotometric measurement. When the organic content of the soil exceeds 10%, high results are obtained for ferrous iron. Bathophenanthroline has also been used for the determination of Fe(II) in HCl-soluble iron oxides, but isoamyl acetate was employed for the extraction.[3] Shapiro[31] heated a 10-mg sample (200-mesh) of rock in a capped, 1-oz plastic bottle with HF–H_2SO_4 in the presence of about 20 mg of o-phenanthroline, and then after buffering the solution with sodium citrate, measured the intensity of the ferrous complex formed; because of the progressive fading of the color with time, the length of the heating period must be the same for samples and standards. An attempt was made[23] to extend the procedure to the determination of Fe(III) also but without success, chiefly because of failure to decompose all of the sample; it was found, however, that by dissolving the o-phenanthroline in ethyl alcohol (greater solubility) the reaction could be carried out at a low pH without need for a buffer. Many reagents have been proposed for the colorimetric determination of Fe(II); among the best known are thiocyanate, mercaptoacetic acid and 2,2-bipyridyl.

Mention has been made of the method in which the sample is fused with NaF and B_2O_3 in a sealed Pyrex tube. The difficulty experienced in dissolving this fused cake without oxidation of the Fe(II) led Hey[12] to carry out the dissolution in a mixture of hydrochloric acid and iodine monochloride; the I_2 formed during the oxidation of the Fe(II) was titrated with a solution of KIO_3. Nicholls[24] applied this technique to carbonaceous shales, following dissolution of the sample in a mixture of HF and H_2SO_4; the solution is first poured onto solid boric acid, then into a bottle containing HCl and ICl, after which the titration with KIO_3 solution is made as before. Unlike the Hey fusion method, however, the acid decomposition does not seriously attack any pyrite present. A semimicro procedure, in which the sample solution is added to $K_2Cr_2O_7$ solution and the excess $Cr_2O_7^{2-}$ is titrated with a standard solution of ferrous ammonium sulfate, has also been described.[20] Wilson[40] chose to eliminate the possibility of aerial oxidation of the ferrous iron by decomposing the sample in the presence of an oxidizing agent, in this case ammonium metavanadate (NH_4VO_3); the sample, in an HF–H_2SO_4–NH_4VO_3 mixture, is allowed to stand as long as is necessary for complete decomposition to take place, and is then washed into a beaker containing boric acid solution and the excess V (V) is titrated with a standard solution of ferrous ammonium sulfate, with barium diphenylamine sulfonate as indicator. Slowly soluble fluorides (probably Ca or Mg compounds) must be dissolved by stirring before the titration is made; Langmyhr and Graff[19] found that some precipitates formed in this way contained Fe(II). At the Geological Survey of Canada this procedure has been modified by the addition of an excess of ferrous ammonium sulfate to the sample solution, in order to be able to make the final titration with potassium dichromate. Wilson[41] scaled his procedure down to handle samples of 3–20 mg, omitting the boric acid in order to insure a sharp end point during the titration (made with a micrometer-syringe burette in a volume of 7 ml). He also used a colorimetric finish in which the V(IV) formed during the reaction at low pH is used to regenerate the Fe(II) when the pH is increased to 5; the Fe(II) reacts with 2,2'-bipyridyl which is added, together with beryllium sulfate to complex the free fluoride ions. Oxidizing and reducing substances, such as organic carbon, sulfides and the oxides of vanadium, will interfere. These procedures, with some modifications, have been extended to the determination of the oxidizing capacity of manganese compounds.[42] Reichen and Fahey[26] decompose the sample with HF and H_2SO_4 in the presence of $K_2Cr_2O_7$ and titrate the excess of the latter with standard Fe(II) solution; boric acid is omitted because it decreases the effective action of Fe(III) during decomposition of the sample and a correction, proportional to the amount of excess $Cr_2O_7{}^{2-}$ found, must be made for the $Cr_2O_7{}^{2-}$ destroyed by the hydrofluoric acid. Garnet was completely decomposed by this procedure but tourmaline and staurolite were unaffected. Van Loon (Supp. Ref.) uses KIO_3 as oxidant but boils the HF–H_2SO_4 mixture for about 15 min to speed up the decomposition and also to volatilize the iodine formed by oxidation of the Fe(II); KI is then added and the liberated I_2 is titrated with sodium thiosulfate. Chlorides must be absent. A different approach is described by Ungethüm (Supp. Ref.); the sample is decomposed in HF alone at room temperature in the presence of a measured amount of $AgClO_4$, and the excess Ag(I) is titrated potentiometrically with a standard solution of potassium bromide. Halogens will interfere;

up to 5% of pyrite is not significantly attacked (and the Fe(II) of the pyrite is thus not determined).

It has been suggested previously that the value found for Fe(II) may not be the true value, because of the presence of oxidizing or reducing substances which are released to react upon decomposition of the sample. Ingamells[14] suggests that it is better to report only a value which expresspresses the oxygen excess or deficiency in the sample (see also Wilson[42]) and describes a procedure in which the sample is dissolved in a mixture of phosphoric acid and sodium pyrophosphate in the presence of $Mn(VII)$ or $Cr(VI)$; the excess of the oxidant is then titrated with a standard Fe(II) solution. The method is limited to those samples that dissolve directly in the acid mixture, do not contain sulfur or organic carbon, and do not yield peroxides when dissolved. A similar procedure was applied to ferrites by Cheng,[1] who dissolved the sample in a $0.1N$ phosphatocerate solution at 280–300°C.

In addition to the possible errors associated with the sample preparation and decomposition, there are other sources of error that must be considered:

1. Some iron-bearing minerals are refractory and even prolonged boiling with H_2SO_4 and HF will not decompose them (tourmaline, staurolite, ilmenite, magnetite). If the FeO content is not determined this will, in addition to causing a low result for the total FeO in the sample, also cause the Fe_2O_3 value to be too high, if the total iron is determined on the solution of the R_2O_3 group as described in Sec. 9.23.3. The boiling should not be too prolonged, incidentally, because of the oxidizing nature of hot, concentrated sulfuric acid. Fine grinding of the initial sample will usually increase its vulnerability to acid attack, but it is better first to repeat the attack on the insoluble residue than to risk oxidation by excessive grinding; chromite will not usually succumb and Peck[25] has warned about the surprisingly resistant behavior of siderite.

2. The introduction of "tramp" iron from the crushing and grinding equipment must be avoided; not only will this iron be counted as ferrous iron, it may reduce some of the Fe(III) present as well.

3. Aerial oxidation has been generally considered the chief source of error in the determination of ferrous iron, particularly during the process of acid decomposition and, to a lesser extent, during the titration; if the determination is being done on the semimicro or micro scale, this error can be considerable.[27] Clemency and Hagner,[4] who determine total iron and ferric iron by coulometric generation of titanous ion and obtain the ferrous iron by difference, have shown experimentally, however, that immediately following the decomposition of the sample there is a marked

reduction of the ferric iron with a subsequent slow reoxidation; they suggest that this reduction, rather than aerial oxidation, may be responsible for erratic results and recommend that the period of decomposition be kept as short as possible. Various means of preventing aerial oxidation during decomposition have been suggested, usually involving the flushing of the decomposition vessel in some way with an inert gas, but if the acid mixture is heated before addition, is brought rapidly to boiling after addition and maintained at this temperature during a period of decomposition that is kept as short as possible, aerial oxidation of Fe(II) will be negligible and the use of an inert atmosphere is unnecessary. Similarly, it is customary to use water that has been boiled and cooled, or treated in some other way, to remove dissolved oxygen; Peck[25] has shown that such a precaution is unnecessary.

Hydrofluoric acid alone favors aerial oxidation of ferrous iron because of the formation of slightly ionized ferric fluoride with a consequent lowering of the oxidation potential of the Fe(III)–Fe(II) system.[18] Aerial oxidation proceeds much more slowly in HCl and H_2SO_4 solutions at room temperature. The harmful effect of fluoride is overcome by complexing it with boric acid, beryllium or aluminum[9] prior to the titration.

4. The presence of sulfide minerals in the sample makes an accurate determination of the ferrous iron an impossible task. If the sulfides are decomposable, whether by acid decomposition or by fusion, there is the strong possibility that the S(II) will reduce some of the Fe(III) present, thus giving a high value for Fe(II) and a correspondingly low one for Fe(III). The magnitude of the error depends upon the method of decomposition because, while most of the S(II) is lost harmlessly during the acid decomposition in a crucible, it will exert an effect proportionately much greater when the decomposition is carried out in a sealed tube at a high temperature and pressure; the oxidation of the S(II) to SO_3 has the capacity to reduce any Fe(III) present to give an apparent FeO equivalent to about 14 times the weight of S(II) present. Thus, even 0.01% of sulfur could cause a serious error if the decomposition is done in a closed system.

Pyrite is not usually decomposed except in a bomb or sealed tube, and its presence will usually mean a low value for the FeO and a correspondingly high one for the Fe_2O_3. The presence of Fe(III) is thought to increase the solubility of pyrite. If pyrite is the only sulfide mineral present in appreciable quantity a correction may be applied to the Fe(II) and Fe(III) values on the basis of the known sulfur content of the sample.

5. Carbon, if present as organic matter (graphite is without effect), will give high results for Fe(II) because it will tend to be oxidized during the titration, or will reduce an added oxidizing agent. The effect is greater when $KMnO_4$ is used as the titrant.

6. Trivalent vanadium will be oxidized during the titration with, or on addition of, an oxidant and thus will give a high value for Fe(II) but in most rocks and minerals it will be negligibly low. Pentavalent vanadium will oxidize ferrous iron.

In summary, incomplete decomposition will affect the results adversely, no matter what method is used, but the error caused by the presence of S(II) will be greater when decomposition is carried out in a closed system. Aerial oxidation, which can be kept to the minimum by careful and uncomplicated handling of the decomposition step, can be rendered negligible by carrying out the decomposition in the presence of an oxidant. The presence of other reducing substances, however, will still cause errors under these conditions, as well as in the direct titration of the ferrous iron by an oxidant. The substitution for the oxidant of a reagent that will form a colored complex with ferrous ions during the decomposition seems to offer the best opportunity for future investigation.

Both potassium permanganate and dichromate are commonly used as the oxidant. Permanganate, which does not require a separate end point indicator, is more affected by the presence of organic matter. A fading end point is also experienced in the presence of excess hydrofluoric acid, because of the formation of slightly ionizing manganic fluorride which encourages the oxidation by $KMnO_4$ of the Mn(II) formed during the titration; the addition of boric acid to the solution removes the fluoride ions from effective action. When a chloride solution is to be titrated with $KMnO_4$, Mn(II) (as $MnSO_4$) should be added to prevent the oxidation of chloride ion to chlorine.

6.11.2. FERRIC IRON

The ferric iron value is, as a rule, determined as the difference between the total iron, expressed as Fe_2O_3, and the Fe_2O_3 equivalent of the FeO value. It is thus a repository for errors inherent in either or both of the FeO and total iron determinations.

$$\% \text{ Total Fe (as } Fe_2O_3) - (\% \text{ FeO} \times 1.1113) = \% Fe_2O_3$$

The work of Clemency and Hagner,[4] who used the coulometric generation of titanous ions to determine the ferric iron directly, has been mentioned previously in the discussion of the possible errors affecting the ferrous iron determination. They note that there is an initial reduction of ferric iron following decomposition by, as seems most likely, some minor constituent of the sample, and recommend the use of a shorter period of decomposition, i.e., 5 min instead of the usual 10. A method has been described in which the sample is decomposed by HF and HCl in an inert atmosphere, and the Fe(III) titrated with EDTA solution, using xylenol orange as the indicator[36,37]; reference has been made in Sec. 6.11.1 to polarographic methods (Mikhailova *et al.*,[21] and Bien and Goldberg, Supp. Ref.). These approaches offer hope that the direct determination of Fe(III) will eventually become part of the analytical scheme for silicate analysis.

6.11.3. TOTAL IRON

There are many procedures available for the determination of the total iron content of silicates, in both major and trace amounts, but no attempt will be made to summarize them here. Reference should be made to recent reviews[17,43] and, for the spectrophotometric determination of trace amounts, to the authoritative text by Sandell.[30]

The determination of the total iron content may be made on a separate portion of the sample, on the solution of the sample portion used for the determination of ferrous iron, on the solution of the fused R_2O_3 group oxides, or on an aliquot of a general sample solution such as solution B. The methods used are, for the most part, either titrimetric or colorimetric, but the determination can also be made by controlled-potential coulometry,[4,22] atomic absorption spectroscopy (Trent and Slavin,[34] Abbey (Supp. Ref.)), emission spectrography and X-ray fluorescence spectroscopy.

The preliminary separation of the iron is seldom attempted now, but such a separation may be made by precipitating the iron with hydrogen sulfide, ammonium sulfide or sodium hydroxide, by fusion of the sample with sodium carbonate followed by water-extraction of the melt, or by ether extraction, to mention only some of the more common methods (Ref. 13, pp. 387–389). When a preliminary separation is not made, the potential errors listed for the determination of ferrous iron apply for the most part to that of the total iron as well, plus an additional source of error resulting from the use of reducing agents. It is most likely that the value obtained for the total iron content by titration will always be on the high side, assuming that complete decomposition of the sample has been achieved. This latter problem has already been discussed in some detail in previous sections. An acid decomposition method is most widely used, but except when the acid attack is done in a bomb or sealed tube, fusion of the sample is more certain to result in its complete decomposition and solution.   It is sometimes judicious to combine these two procedures, particularly when there are constituents present that may be harmful to platinum, by leaching the sample first with hydrochloric acid and then fusing the acid-insoluble residue with sodium carbonate or potassium pyrosulfate.

The reduction of the Fe(III) to Fe(II) can be done in several ways, but when the resulting Fe(II) is to be titrated with an oxidizing agent, the reducing agent must not itself be oxidizable.   The most common reductant for this purpose is stannous chloride, the excess of which is removed by the addition of mercuric chloride just prior to the titration of the Fe(II), or which can be oxidized first by a preliminary titration if a potentiometric end point is used.   Reduction of Fe(III) takes place rapidly in hot hydrochloric acid solution and only Pt and V will interfere, the latter in an erratic fashion; because the reduction of Pt(IV) to Pt(II) is not quantitative it must be removed (precipitation with H_2S).   Reduction with amalgamated zinc (Jones reductor) will give more accurate results than will reduction with $SnCl_2$, but many more elements, among them Ti, V, Cr and U, are also reduced.   The silver reductor is better (Cr and Ti are not reduced) and finds frequent use, particularly when the work is done on a semimicro scale[35]; Chalmers (Supp. Ref.) considers that this is the only method that can be wholeheartedly recommended for the determination of iron in any amount (see Miller and Chalmers, Supp. Ref.).   Mercy and Saunders (Supp. Ref.) investigated the use of both $SnCl_2$ and the silver reductor in their study of methods for the determination of total iron in silicates and consider titrimetric methods to be preferable to photometric ones.   Van Loon (1966) (Supp. Ref.) used an ion-exchange column to separate Fe from Al and V; the iron is eluted with $0.1M$ HCl, passed through a silver reductor, and titrated with $KMnO_4$.   Blanchet and Malaprade (Supp. Ref.) use $TiCl_3$ to reduce iron to Fe(II); the excess Ti(III) is titrated first with $K_2Cr_2O_7$ (arsenotungstic acid as indicator), which is then used to titrate the Fe(II), using barium diphenylamine sulfonate as indicator.

The use of potassium dichromate and permanganate has been discussed (Sec. 6.11.1); titration is also done with Ce(IV), using the o-phenanthroline ferrous sulphate complex as indicator.

The most deservedly popular reagent for the colorimetric determination of iron is o-phenanthroline; this reagent forms a strongly colored, reddish-orange complex with Fe(II) in weakly acidic, neutral or weakly alkaline medium (pH 2–9) that is very stable and follows Beer's law closely in its absorbance characteristics.   Tiron (disodium 1,2-dihydroxybenzene-3,5-disulphonate)[44,45] is also frequently used. Recent studies have used 4,7-diphenyl-1-10-phenanthroline (bathophenanthroline) for small amounts of iron because of its greater sensitivity[2,5,6]; the use of syn-phenyl 2-pyridyl ketoxime, which forms a sensitive ferrous iron complex in highly alkaline solutions, has also been described.[5] Mercy and Saunders (Supp. Ref.) investigated the use of thioglycollic and sulfosalicylic acids.

The determination of the total iron content of rocks by X-ray fluorescence spectroscopy is part of the routine rapid analysis scheme in use at the Geological Survey of Canada, and will be discussed in a later section. Recently, it has been demonstrated (Abbey, Supp. Ref.) that atomic absorption spectroscopy is a potentially very useful method for this determination, particularly when the amount of sample is limited; an apparent ionization interference, which resulted in a positive bias for iron values, is eliminated by the addition of 500 ppm Sr to standard and sample solutions. Because this same addition is made for the determination of magnesium, the method is conveniently used to determine magnesium and iron in one solution, and has been applied to determine these elements in biotite. The procedure is given in detail in a later section.

6.11.4. METALLIC IRON

It is unlikely that the rock and mineral analyst will need to determine the metallic iron content of a sample unless he is involved in the analysis of a meteorite. Not all meteorites contain a metal phase but when it is present it introduces further complications into an analysis already sufficiently complex and a reliable determination of the metal content becomes of first importance. It is a problem that has also been of much concern in the analysis of slags and ores.*

The oldest method is that in which the sample is digested with mercuric chloride, and sometimes with ammonium chloride as well, to dissolve the metallic phase; the mercurous chloride is removed and the Fe(II) is titrated as usual. Particle size plays an important part in the dissolution and the sample should be -200 mesh to insure complete extraction of the metal. Because of the presence of mercury salts, it is possible to determine only the iron in the metal phase. A similar pro-

---

* If it is only the removal of the metallic iron that is required, in order to permit the subsequent determination of Fe(II) and Fe(III) in the nonmetallic phase, this may be accomplished by shaking the sample with a nearly neutral solution of ferric chloride and allowing it to stand for several hours before filtering and washing the insoluble material.[37]

cedure involves the displacement of copper from cupric sulfate by metallic iron.

Both of these methods fail when iron sulfide and phosphide are present, and Riott[28] has introduced the use of cupric potassium chloride to overcome this drawback; a very dilute acetic acid solution of the reagent is used under an atmosphere of carbon dioxide, and no attack of the iron oxides and sulfides takes place. The precipitated copper reduces the excess $CuCl_2$ to $CuCl$, from which copper is removed by displacement with aluminum; the Fe(II) is titrated as before. Only 0.1 g of metallic iron can be safely handled, and iron carbides, if present, will cause the results to be high.

Habashy[11] made a comparative study of existing methods for the determination of metallic iron in the presence of iron oxides and found that the four older methods which were considered gave results equally as good as those obtained by his proposed procedure when the particle size was 250 mesh or finer; he found that his procedure alone was reliable when the particle size was coarser than 100 mesh. He uses a variation of the $CuSO_4$ method in which the sample is digested with $CuSO_4$ and mercury; the precipitated copper amalgamates with the mercury which is then separated, dissolved in nitric acid and the copper determined electrolytically.

A very different approach is proposed by Easton and Lovering[7] for the separation not only of the iron of the metallic phase but of the nickel and cobalt as well. The sample is digested with $HgCl_2$–NH_4Cl and the extract, made $10M$ in HCl, is placed on an anion exchange column from which the nickel is eluted first with $10M$ HCl, the cobalt with $6M$ HCl and the iron with $0.6M$ HCl; the mercuric salts are retained on the column. Final determination of the metals is made colorimetrically, o-phenanthroline being used for the iron.

References

1. Cheng, K. L. Determination of ferrous oxide in ferrites. *Anal. Chem.* **36,** 1666–1667 (1964).
2. Chirnside, R. C., Cluley, H. J., Powell, R. J., and Proffitt, P. M. C. The determination of small amounts of iron and chromium in sapphire and ruby for maser applications. *Analyst* **88,** 851–863 (1963).
3. Clark, L. J. Iron (II) determination in the presence of iron (III) using 4,7-diphenyl-1,10-phenanthroline. *Anal. Chem.* **34,** 348–352 (1962).
4. Clemency, C. V., and Hagner, A. F. Titrimetric determination of ferrous and ferric iron in silicate rocks and minerals. *Anal. Chem.* **33,** 888–892 (1961).
5. Cluley, H. J., and Newman, E. J. The determination of small amounts of iron. *Analyst* **88,** 3–17 (1963).

6. Cuttitta, F., and Warr, J. J. Use of bathophenanthroline for determining traces of iron in zircon. *U.S. Geol. Surv. Profess. Papers* **424**, C383–384 (1961).
7. Easton, A. J., and Lovering, J. F. The analysis of chondritic meteorites. *Geochim. Cosmochim. Acta* **27**, 753–767 (1963).
8. Fahey, J. J. Determination of ferrous iron in magnetite and ilmenite in the presence of amphiboles and pyroxenes. *U.S. Geol. Surv. Profess. Papers* **424**, C386–387 (1961).
9. Graff, P. R., and Langmyhr, F. J. Studies in the spectrophotometric determination of silicon in materials decomposed by hydrofluoric acid. II. Spectrophotometric determination of fluosilicic acid in hydrofluoric acid. *Anal. Chim. Acta* **21**, 429–431 (1959).
10. Groves, A. W. *Silicate Analysis*, 2nd ed., 1951. London: George Allen and Unwin, pp. 88–94, 181–186.
11. Habashy, M. G. Quantitative determination of metallic iron in the presence of iron oxides in treated ores and slags. *Anal. Chem.* **33**, 586–588 (1961).
12. Hey, M. H. The determination of ferrous iron in resistant silicates. *Mineral. Mag.* **26**, 116–118 (1941).
13. Hillebrand, W. F., Lundell, G. E. F., Bright, H. A., and Hoffman, J. I. *Applied Inorganic Analysis*, 2nd ed., 1953. New York: John Wiley & Sons, pp. 384–403, 907–923.
14. Ingamells, C. O. A new method for "ferrous iron" and "excess oxygen" in rocks, minerals and oxides. *Talanta* **4**, 268–273 (1960).
15. Ito, J. A new method of decomposition for refractory minerals and its application to the determination of ferrous iron and alkalies. *Bull. Chem. Soc. Japan* **35**, 225–228 (1962).
16. Juurinen, A. Composition and properties of staurolite. *Annales Acad. Sci. Fennicae, Ser. A, Pt. III* **47**, 19–22 (1956).
17. Kolthoff, I. M., and Elving, P. J., (Eds.). *Treatise on Analytical Chemistry*, Part II, Vol. 2, 1962. "Iron," by L. M. Melnick. New York: Interscience, pp. 247–310.
18. Kolthoff, I. M., and Sandell, E. B. *Textbook of Quantitative Inorganic Analysis*, 3rd ed., 1952. New York: Macmillan, pp. 711–712.
19. Langmyhr, F. J., and Graff, P. R. A contribution to the analytical chemistry of silicate rocks; a scheme of analysis for 11 main constituents based on decomposition by hydrofluoric acid. *Norges Geolog. Undersök.* **230**, 1965, 128 pp.
20. Meyrowitz, R. A semimicro procedure for the determination of ferrous iron in nonrefractory silicate minerals. *Am. Mineral.* **48**, 340–347 (1963).
21. Mikhailova, Z. M., Yarushkina, A. A., Mirskii, R. V., and Shil'dkrot, E. A. Determination of total ferrous and ferric iron in rocks. *Ref. Zh. Khim.* **19GDE**, 1964, Abstr. 4G122; through *British Abstracts* **12**, 114 (1965).
22. Milner, G. W. C., and Edwards, J. W. The determination of iron in metals and minerals by controlled potential coulometry. *Analyst* **87**, 125–133 (1962).
23. Moore, J. M., Jr., and Maxwell, J. A. Spectrophotometric method for determination of ferrous iron and total iron in silicates. Unpublished work, Geol. Survey Canada, 1961.
24. Nicholls, G. D. Techniques in sedimentary geochemistry. Determination of ferrous iron contents of carbonaceous shales. *J. Sed. Pet.* **30**, 603–612 (1960).
25. Peck, L. C. Systematic analysis of silicates. *U.S. Geol. Surv. Bull.* **1170**, 39–42, 72–73 (1964).

216   ROCK AND MINERAL ANALYSIS

26. Reichen, L. E., and Fahey, J. J. Improved method for the determination of FeO in rocks and minerals, including garnet. *U.S. Geol. Surv. Bull.* **1144–B**, 1–5 (1962).
27. Riley, J. P., and Williams, H. P. The microanalysis of silicate and carbonate minerals. I. Determination of ferrous iron. *Mikrochim. Acta* **4**, 516–524 (1959).
28. Riott, J. P. Determining metallic iron in iron oxides and slags. *Ind. Eng. Chem. Anal. Ed.* **13**, 546–549 (1941).
29. Ritchie, J. A. Methods of analysis of silicate rocks. *Dept. Sci. Ind. Res., Rept.* **D. L. 2049**, 19–20 (1962). New Zealand.
30. Sandell, E. B. *Colorimetric determination of traces of metals*, 3rd ed. (rev.), 1959. New York: Interscience, pp. 522–544.
31. Shapiro, L. A spectrophotometric method for the determination of FeO in rocks. *U.S. Geol. Surv. Profess. Papers* **400–B**, B 496–497 (1960).
32. Shapiro, L., and Brannock, W. W. Rapid analysis of silicate, carbonate and phosphate rocks. *U.S. Geol. Surv. Bull.* **1144–A**, A 48–49 (1962).
33. Shell, H. R. Determination of total iron in silicate and other nonmetallic materials. *Anal. Chem.* **22**, 326–328 (1950).
34. Trent, D. J., and Slavin, W. Determination of various metals in silicate samples by atomic absorption spectrophotometry. *Atomic Absorption Newsletter* **3**, 118–125 (1964).
35. Vincent, E. A., and Phillips, R. Iron–titanium oxide minerals in layered gabbros of the Skaergaard intrusion, East Greenland. Pt. I. Chemistry and ore microscopy. *Geochim. Cosmochim. Acta* **6**, 1–26 (1954).
36. Vorlicek, J., and Vydra, F. Direct complexometric determination of ferric iron in ores. *Hutnicke Listy*, **18**, 733–734 (1963); through *British Abstracts* **12**, 110 (1965).
37. Vydra, F., and Vorlicek, J. EDTA titration of iron (III) in the presence of iron (II) in ores and slags. *Chemist-Analyst* **53**, 103–105 (1964).
38. Walker, J. L., and Sherman, G. D. Determination of total ferrous iron in soils. *Soil Science* **93**, 325–328 (1962).
39. Washington, H. S. *The Chemical Analysis of Rocks*, 2nd ed., 1910. New York: John Wiley & Sons, 134–141.
40. Wilson, A. D. A new method for the determination of ferrous iron in rocks and minerals. *Bull. Geol. Surv. Gr. Brit.* **9**, 56–58 (1955).
41. Wilson, A. D. The micro-determination of ferrous iron in silicate minerals by a volumetric and a colorimetric method. *Analyst* **85**, 823–827 (1960).
42. Wilson, A. D. The titrimetric and spectrophotometric determination of the oxidizing capacity of manganese compounds. *Analyst* **89** 571–578 (1964).
43. Wilson, C. L., and Wilson, D. W., (Eds.). *Comprehensive Analytical Chemistry*, Vol. 1C (Classical Analysis), 1962. "Iron," by R. A. Chalmers. Amsterdam: Elsevier, pp. 635–655.
44. Yoe, J. H., and Armstrong, A. R. Colorimetric determination of titanium with Tiron. *Anal. Chem.* **19**, 100–102 (1947).
45. Yoe, J. H., and Jones, A. L. Colorimetric determination of iron with Tiron. *Ind. Eng. Chem. Anal. Ed.* **16**, 111–115 (1944).

Supplementary References

Abbey, S. Analysis of rocks and minerals by atomic absorption spectroscopy. I. Determination of magnesium, lithium, zinc and iron. *Geol. Surv. Canada Paper* **67-37,** 1967, 35 pp.

Bien, G. S., and Goldberg, E. D. Polarographic determination of ferrous and ferric iron in refractory minerals. *Anal. Chem.* **28,** 97–98 (1956).

Blanchet, M. L., and Malaprade, L. Méthode rapide de dosage des principaux éléments d'une roche silicatée. *Chim. Anal.* **49,** 11–27 (1967).

Chalmers, R. A., in "Chemical analysis of silicates," *The Chemistry of Cements,* Vol. 2, H. F. W. Taylor, Ed., 1964. London: Academic Press, pp. 178–179.

Kiss, E. Chemical determination of some major constituents in rocks and minerals. *Anal. Chim. Acta* **39,** 223–234 (1967). The method of Riley and Williams (see Ref. 27) for FeO was not found to be satisfactory for FeO $> 15\%$. Digestion of the sample in an All-Teflon apparatus under an atmosphere of N_2 is substituted and the titration of Fe(II) is made potentiometrically using a Pt–W electrode and recorder.

Mercy, E. L. P., and Saunders, M. J. Precision and accuracy in the chemical determination of total Fe and Al in silicate rocks. *Earth Planetary Sci. Letters* **1,** 169–182 (1966).

Miller, C. C., and Chalmers, R. A. A critical investigation of the use of the silver reductor in the micro-volumetric determination of iron, especially in silicate rocks. *Analyst* **77,** 2–7 (1952).

Pamnani, K. and Agnihotri, S. K. A modified method for determination of ferrous iron in rocks and minerals. *Laboratory Practice* **15,** 867 (1966). The method is similar to that of Wilson (see Ref. 40), except that $KMnO_4$ is used as the oxidant and the decomposition is carried out overnight at 80–90°C. Because some of the MnO_4- reacts with the HF present, a blank must be run with each batch.

Schafer, H. N. S. The determination of iron(II) oxide in silicate and refractory materials. Part I. A review. *Analyst* **91,** 755–762 (1966); Part II. A semi-micro titrimetric method for determining iron(II) oxide in silicate materials. *Idem.,* 763–770.

Shell, H. R. Possible loss of iron during sodium carbonate fusions of silicates and rocks. *Anal. Chem.* **26,** 591–593 (1954).

Ungethüm, H. Eine neue methode zur bestimmung von eisen(II) in gesteinen und mineralen, insbesondere auch in bitumenhaltigen proben. *Z. Angew. Geol.* **11,** 500–505 (1965).

Van Loon, J. C. Titrimetric determination of the iron(II) oxide content of silicates using potassium iodate. *Talanta* **12,** 599–603 (1965).

Van Loon, J. C. Separation and accurate determination of iron and aluminum in a single sample of silicate rock. *Talanta* **13,** 1555–1560 (1966).

6.12. Water*

The "water content" of a rock or mineral is generally taken to mean that amount of water which is expelled from the sample when it is heated

---

* References for this section will be found on pp. 223, 224.

at a temperature greater than 105–110°C and excludes the water (moisture, H_2O^-) which is driven off when the sample is heated at a temperature in or below this range (Sec. 6.1). If the sample is only air-dried, then the water obtained by the following methods is the *total water* content of the sample, assuming complete release and collection of the water. On subtraction of the H_2O^- (moisture) content of the sample, if known or subsequently determined, or if the sample used was first oven-dried (105–110°C), a value is obtained that is variously designated as H_2O^+, *water of constitution*, $H_2O > 110°C$ and *essential* or *bound water*. The comments made in Sec. 6.1 apply to this section as well and should be reviewed. Because of the many factors which affect the water content of a rock or mineral, it is unavoidable that the division of the total water content into temperature-controlled fractions must be an arbitrary one and this should be borne in mind when these values are considered. It has been shown that the loss of water is dependent more on the duration than upon the temperature of heating, at least for natural glasses[1,2]; Chalmers[26] believes that, in order to interpret the role of water released at any particular temperature, it is necessary to combine the information obtained from weight loss–time–temperature curves with that from differential thermal analysis curves. For general discussions on this subject, including the role of hydrogen in minerals, reference should be made to Hillebrand *et al.*, (Ref. 10, pp. 814–822), Groves (Ref. 5, pp. 269–274) and to the recent review by Mitchell.[12]

The methods for the determination of water in silicates are not numerous and of these only a very few, all of which are gravimetric, are in general use.[7,10,12] These methods have in common the use of heat to release and expel the water, which will be accompanied by any other volatile constituents similarly expelled. The water is collected, by condensation or absorption, and weighed after suitable precautions are taken to prevent the interference of other volatile constituents; the ideal arrangement would require the use of a vacuum extraction line by which the various volatile constituents could be separately condensed and purified, or otherwise specially treated,[4] but this would not lend itself to routine operation. Other techniques of measurement that have been proposed include the titration of the collected water with Karl Fischer and other reagents,[7,12,15,20] measurement of the dielectric constant,[7] and electrolysis[12]; in the latter the water vapor is carried by flow gas into a cell, adsorbed in an hygroscopic electrolyte and quantitatively electrolyzed, the magnitude of the electric current being proportional to the number of moles of water absorbed per unit time. A dc-arc spectrographic method has recently been described (Quesada and Dennen, Supp. Ref.).

The loss on ignition (usually at about 1000°C) of a sample is occasionally reported as a substitute for the determination of total water, but its use is fraught with danger and at best it is more useful as a relative than as an absolute value. Its significance, with examples of such, has been discussed in detail[7,10]; some analysts reject it as being completely misleading.[5,27] It is obvious, of course, that the value obtained by igniting the sample at 1000°C will represent the water content only in the absence of other similarly volatile constituents such as carbon dioxide, sulfur, fluorine, chlorine and, to some extent, the alkali metals; it also requires the absence of constituents that are readily oxidized or reduced during ignition, chief among which is ferrous iron. Riley[18] has shown that the amount of FeO remaining unoxidized after the ignition of silicates is very variable (0.16–3.0%) and a reliable correction for this oxidation cannot be applied; several examples are also given by Hartwig-Bendig.[7] The use of loss on ignition for the determination of water in sediments and sedimentary rocks has been investigated by Manheim,[14] who recommends 1000°C as the most suitable ignition temperature; he also mentions the possibility that the sample may become contaminated by material from the furnace walls during the heating.

All minerals do not give up their water readily; talc, topaz, staurolite, chondrodite and phlogopite are the worst offenders but titanite[21] and some varieties of epidote[25] are also troublesome. The chief problem lies in insuring that all of the water is expelled, which can be done by using very high temperatures capable of breaking down the most resistant structure, but which then requires a special furnace capable of attaining these elevated temperatures and a sample tube capable of withstanding the latter. Hartwig-Bendig[7] has discussed this problem in detail with abundant supporting data and, more recently, Peck[16] has presented a very practical evaluation of the factors affecting the determination. There are some minerals that will yield all of their water on simple ignition, but as a rule, it is better to employ a flux to break down the structure (thus making it possible to use a lower maximum temperature); the flux also serves to retain other volatile constituents such as fluorine and sulfur and provides a supply of oxygen to prevent any reduction of water to hydrogen during the ignition. Among the numerous fluxes that have been suggested,* there are three that find most use—sodium tungs-

* Anhydrous Na_2CO_3 has been used[7,10] with the advantage that the analysis can then be continued as usual on the fused sample.

tate, lead oxide (litharge) and lead chromate. Peck[16] tested these three fluxes with hornblende and found that, while no single flux was completely satisfactory, a mixture of two parts of PbO and one part of $PbCrO_4$ was sufficiently reactive to decompose hornblende and did not intumesce disadvantageously with samples containing carbon or carbon dioxide. The flux mixture is dried by heating it to 800°C and is then kept in a tightly sealed bottle. Sodium tungstate was found to be not sufficiently reactive, as well as being hygroscopic, and Peck does not recommend its use.

If a purification and absorption train is used in the determination of water, there will be little likelihood of interference from other volatile constituents not completely retained by the flux. When the water is to be weighed directly, as in the Penfield method, the effect of these other volatiles, particularly carbon dioxide, must be considered or serious errors may be introduced. Litharge, or the litharge–lead chromate mixture of Peck, will retain the sulfur, chlorine and fluorine present, unless these latter are present in unusually large amounts. The addition of freshly ignited calcium oxide to the flux–sample mixture will help to retain excessive amounts of fluorine. Carbon dioxide is not retained by the flux and, unless eliminated, may cause a two-fold error; (1) if present in large amount it may carry some water vapour out of the tube and, (2) if not removed from the tube before it is weighed, the CO_2 will be counted as H_2O and cause results to be high. The usual method of correction for carbon dioxide is to allow the tube to "drain," i.e., the tube is supported at a 45° angle with provision for retaining the water and a correction is applied to compensate for the water vapor lost during the "draining" period. Peck[16] prefers to omit the center bulb of the Penfield tube and to connect the tube horizontally to a special bulb containing water and air; after a time, the water-saturated carbon dioxide in the tube becomes so thoroughly diluted with water-saturated air that it no longer has any effect upon the results. The writer prefers to eliminate the carbon dioxide by means of a displacement procedure introduced by Dr. S. S. Goldich in the Rock Analysis Laboratory at the University of Minnesota. It is not possible to correct for the presence of organic carbon which may be oxidized to water and carbon dioxide during the fusion–ignition of the sample.

The use of an absorption train, usually in conjunction with a combustion tube furnace, has been described by several authors,[5,7,10,13,18] and will not be discussed in detail here. The sample, contained in a platinum, porcelain, alumina or silica boat and mixed with one of a variety of fluxes, is heated to about 1000°C in a silica or vitreosil tube while a current of dry air is drawn over it. The water is absorbed in a suitable agent such as calcium chloride, or anhydrone ($Mg(ClO_4)_2$). The flux will hold back most of the other volatile constituents but purification tubes can be inserted in the train ahead of the water absorption tube. It is recommended that the drying agent used to dry the air that is drawn through the system should be the same as that used to absorb the sample water, and that all weighings should preferably be made under the same atmospheric conditions. It is advantageous to keep the blank as small as possible. Jeffery and Wilson[11] devised a closed-circulation system in which the air is recycled by means of a small pump; not only is a very low blank obtained but it is only necessary to change the desiccant occasionally, instead of frequently as in the open system. Methods in which the macro procedure is adapted to the micro scale have also been described.[9,19]

The most practical method for routine use is the refreshingly simple one first suggested by G. J. Brush and developed later by S. L. Penfield.[17] This method, or a modification of it, has been described by many authors[3,7,10,13,16,27] and in careful hands it is capable of giving very satisfactory results. The sample, mixed with a flux, is heated in a bulb which is blown on the end of a length of borosilicate tubing; the expelled water is condensed and collected in the cooled, center portion of the tube, the fused bulb and sample are pinched off and the tube and collected water are weighed. The tube is then dried, reweighed and the loss in weight is that of the water expelled from the sample. One or more bulbs may be blown in the center area of the tube to act as retainers for the condensed water (Penfield describes a variety of such tubes), particularly if the tube is inclined at an angle (usually 45°) to allow any carbon dioxide present to escape by "draining." These additional bulbs increase the cost of the water tubes* and some analysts have eliminated the need for them.[3,16] The other pieces of equipment usually required include a long-stemmed funnel for the introduction of the sample–flux mixture, a capillary plug to minimize the flow of air in the tube during the ignition period, and a small cap with which to close the tube prior to the weighing of the tube and water, and which is removed during the weighing. If the tubes are to be allowed to "drain," a rack capable of supporting them at

---

* New tubes may be made by joining a length of borosilicate tubing to the used tube, after cutting off the sealed end, and blowing a new sample bulb. It is important that the walls of the sample bulb be uniform and of sufficient thickness to withstand a short period of intense heating without collapsing. At the Geological Survey it has been found preferable to purchase water tubes rather than to make them (see also Cruft et al., Supp. Ref., p. 590).

a 45° angle will also be needed.   Some more specialized items are described by Cruft et al. (Supp. Ref.) in their discussion of the determination of water in apatite.

The chief disadvantage of the Penfield method is that it is difficult to
attain an ignition temperature sufficiently high to expel all of the water
from the sample before the bulb collapses.   Another disadvantage is
that any carbon dioxide also present in the sample will interfere seriously
unless special precautions are taken to eliminate it.

Peck[16] has designed a furnace which will heat four samples concurrently
at 900°C and which requires little attention during the heating period
of about 50 min.   Penfield[17] also used a "sort of" furnace, made of fire-
brick and lined with charcoal, for minerals which would not yield their
water readily (no flux was used); the tube was wrapped in platinum foil
to support it.   The use of a propane–oxygen mixture in a glass-blower's
hand torch, properly regulated, will also yield a temperature of 900°C
before collapse of the bulb.*   The objection to the use of the Penfield
method on the ground that a sufficiently high temperature cannot be
obtained[5] is no longer valid.

Riley[18] believes that the Penfield method has a fundamental flaw, in
that not all the water that is liberated is condensed.   Peck[16] determined
the combined water in two amphiboles by his modification of the Penfield method and obtained results that were virtually identical with those
obtained by Friedman and Smith,[4] who heated the samples under vacuum in an induction furnace (1450°C), reduced the H_2O to H_2 over hot
U metal and measured the volume of the hydrogen.

The Penfield method has been adapted to the semimicro scale by
Guthrie and Miller,[6] and to microanalysis by Sandell.[22]   The latter
places a closely fitting helix of nichrome ribbon around the end of the
tube to prevent sagging when it is heated.

Harvey[5,8,10] proposed a method that combines features of both the
Penfield and combustion–absorption techniques in simplified form.   The
sample, in a platinum boat, is heated in a silica test tube supported by
an asbestos heat screen; a weighing bottle containing $CaCl_2$ is connected
to the test tube, and also a small $CaCl_2$ drying tube to remove moisture
from air drawn into the apparatus during the cooling period.   The
method requires the absence of significant amounts of volatile constituents

---

* A reasonable estimation of the maximum temperature obtainable with the hand-
torch was obtained by placing a thermocouple in a Penfield tube so that the end of it
was at the center of the sample bulb.   When the bulb was heated with the oxygen-
propane flame a maximum temperature of 1050°C was indicated by the thermocouple
as the glass collapsed around it, and 950°C thereafter.

other than water and carbon dioxide; the latter is removed by repeated vertical displacement. Wilson[28] has considerably improved the apparatus by substituting ground-glass joints sealed with Teflon sleeves for the rubber connections, and uses magnesium perchlorate in the absorption and guard tubes. No flux is used and the ignition is prolonged for one hour; no provision is made for removing carbon dioxide and other volatiles must be absent or some means of retaining them must be introduced into the apparatus.

Shapiro and Brannock[12,23,24] have included a somewhat similar method as a part of their rapid scheme of analysis. The sample is mixed with sodium tungstate and heated in a horizontal pyrex test tube, the upper part of which is cooled in a polyethylene jacket containing crushed ice and salt. The expelled water is absorbed on a preweighed roll of filter paper placed in the cold portion of the tube. The paper roll is weighed before and after the heating in a stoppered weighing tube. Cooling was originally done by means of a wet paper strip wrapped around the tube and an empirical correction was made to compensate for moisture that was not condensed; use of the ice bath makes this correction unnecessary. Any organic matter present will, as in all methods for the determination of water, cause high results to be obtained.

References

1. Butler, B. C. M. Metamorphism and metasomatism of rocks of the Moine series by a dolerite plug in Glenmore, Ardnamurchan. *Mineral. Mag.* **32,** 866–897 (1961).
2. Carmichael, I. S. E. The pyroxenes and olivines from some tertiary acid glasses. *J. Petrol.* **1,** 309–336 (1960).
3. Courville, S. Apparatus for total water determination by the Penfield method. *Can. Mineral.* **7,** 326–329 (1962).
4. Friedman, I., and Smith, R. L. The deuterium content of water in some volcanic glasses. *Geochim. Cosmochim. Acta* **15,** 218–228 (1958).
5. Groves, A. W. *Silicate Analysis,* 2nd ed., 1951. London: George Allen and Unwin, pp. 95–104, 269–274.
6. Guthrie, W. C. A., and Miller, C. C. The determination of rock constitutents by semimicro methods. *Mineral. Mag.* **23,** 405–415 (1933).
7. Hartwig-Bendig, M. Bestimmung des Gesamtwassers in der anorganischen Mineralanalyse. *Z. Angew. Mineral.* **2–3,** 195–223 (1939–1941).
8. Harvey, C. O. Simple method for determination of water in silicates. *Bull. Geol. Surv. Gr. Brit.* **1,** 8–12 (1939).
9. Hecht, F. Die mikroanalytische Bestimmung der Wassers in anorganischen Substanzen. *Mikrochim. Acta* **1,** 194–204 (1937).
10. Hillebrand, W. F., Lundell, G. E. F., Bright, H. A., and Hoffman, J. I. *Applied Inorganic Analysis,* 2nd ed., 1953. New York: John Wiley & Sons, pp. 814–835.

11. Jeffery, P. G., and Wilson, A. D.   Closed-circulation systems for determining water, carbon dioxide and total carbon in silicate rocks and minerals. *Analyst* **85**, 749–755 (1960).

12. Kolthoff, I. M., and Elving, P. J., (Eds.).   *Treatise on Analytical Chemistry*, Part II, Vol. 1, 1961. "Water," by John Mitchell, Jr. New York: Interscience, pp. 69–206.

13. Kolthoff, I. M., and Sandell, E. B.   *Textbook of Quantitative Inorganic Analysis*, 3rd ed., 1952. New York: Macmillan, pp. 298–300, 717–719.

14. Manheim, F.   Water determination in sediments and sedimentary rocks. *Acta Univ. Stockholm.* **6**, 127–143 (1960).

15. Mitchell, J., Jr.   Karl Fischer reagent titration. *Anal. Chem.* **23**, 1069–1075 (1951).

16. Peck, L. C.   Systematic analysis of silicates. *U.S. Geol. Surv. Bull.* **1170**, 17–21, 61, 83 (1964).

17. Penfield, S. L.   On some methods for the determination of water. *Am. J. Sci.*, 3rd Ser., **48**, 31–38 (1894).

18. Riley, J. P.   Simultaneous determination of water and carbon dioxide in rocks and minerals. *Analyst* **83**, 42–49 (1958).

19. Riley, J. P., and Williams, H. P.   The microanalysis of silicate and carbonate minerals. II. Determination of water and carbon dioxide. *Mikrochim. Acta* **1959**, 526–535.

20. Rulfs, C. L.   Volumetric microdetermination of water in minerals. *Mikrochemie* **33**, 338–343 (1948).

21. Sahama, Th. G.   The chemistry of the mineral titanite. *Bull. Comm. Géol. Finlande* **138**, 102–106 (1946).

22. Sandell, E. B.   Microdetermination of water by the Penfield method. *Mikrochim. Acta* **38**, 487–491 (1951).

23. Shapiro, L., and Brannock, W. W.   Rapid determination of water in silicate rocks. *Anal. Chem.* **27**, 560–562 (1955).

24. Shapiro, L., and Brannock, W. W.   Rapid analysis of silicate, carbonate and phosphate rocks. *U.S. Geol. Surv. Bull.* **1144-A**, 46–48 (1962).

25. Smethurst, A. F.   Anomalies in the analytical determination of water in epidote. *Mineralog. Mag.* **24**, 173–179 (1935).

26. Taylor, H. F. W., Ed.   *The Chemistry of Cements*, Vol. 2, 1964. "Chemical Analysis of Silicates," by R. A. Chalmers. London: Academic Press, pp. 171–189.

27. Washington, H. S.   *The Chemical Analysis of Rocks*, 2nd ed., 1910. New York: John Wiley & Sons, pp. 156–161.

28. Wilson, A. D.   Determination of total water in rocks by a simple diffusion method. *Analyst* **87**, 598–600 (1962).

Supplementary References

Cruft, E. F., Ingamells, C. O., and Muysson, J. Chemical analysis and the stoichiometry of apatite. *Geochim. Cosmochim. Acta* **29**, 581–597 (1965).

Quesada, A., and Dennen, W. H.   Spectrochemical determination of water in minerals and rocks. *Appl. Spectr.* **21**, 155–156 (1967).

The sample, mixed with a quartz buffer, is burned in a dc arc in a special electrode and the bandhead OH 3063.6 Å measured. No distinction can be made between OH and H_2O, and careful control of absorbed moisture is required.

6.13. Carbon*

Carbon is usually present in the form of independent, accessory carbonates such as calcite, dolomite and siderite but it may also be present as a carbonate in the structure of some silicates such as cancrinite, or in other minerals such as apatite. It may also occur as graphite, or even as carbide; in the form of diamond it is not likely to be encountered by the analyst.

The concentration of carbon dioxide, i. e., of carbonate carbon, is very low in most igneous rocks unless the latter have been subjected to much alteration (e.g., by weathering). As graphite or carbonaceous matter (a mixture of graphite and organic matter) carbon is also usually present only in small amount, although it may reach a concentration of several per cent in rocks such as graphite schists and slates. The geochemistry of carbon in igneous rocks has been discussed in detail by Hoefs.[12]

In non-carbonate sedimentary rocks, such as shale and sandstone, the carbon content is very variable but usually it is much higher than in igneous rocks. In the carbonate rocks it is, of course, a major constituent.

Carbon, as *carbonate*, seldom presents any problem in the determination of other constituents. Care must be taken to avoid mechanical loss when acid is added to a slurry containing carbonates, as in the determination of ferrous iron (Sec. 9.22.1). When present as *graphite* it may be more troublesome in a physical sense than in a chemical one, but as *carbonaceous matter*, a mixture of graphite and organic debris, or as *organic matter* alone, it will create several problems for the analyst. Fusion of such a sample with sodium carbonate in a platinum crucible may result in appreciable attack on the crucible (it should be noted that those samples that have organic matter present usually contain other elements, such as sulfur and arsenic, which are responsible for much of the damage done to the crucible) but this is eliminated, or reduced, by a preliminary roasting of the sample prior to fusion and by the addition of a small amount of an oxidant such as sodium peroxide to the flux (Sec. 6.2.1). Mention has also been made of its deleterious effect upon the accurate determination of ferrous iron (Sec. 6.11.1), and of water (Sec. 6.12).

In the following discussion of methods for the determination of carbon it is difficult to confine the discussion to silicates alone, since most methods are primarily intended for carbonate-rich materials. In general,

* References for this section will be found on pp. 232–234.

these methods are applicable to silicates simply by using an appropriately larger amount of sample.

6.13.1. "CARBONATE" CARBON

The comments made about the use of *loss on ignition* as a means of determining water (Sec. 6.12) apply equally as well to the determination of carbon dioxide by this method. It can be used as a rough check on the magnitude of the amount of carbonate that may be present in a sample, provided that other volatile constituents are either absent or have already been determined, but it is only under very special circumstances that much reliance may be placed upon the results. A method involving controlled loss on ignition is described by Galle and Runnels[7] for the determination of carbon dioxide in limestone and dolomite; the sample is ignited first at 550°C to remove all non-carbonate volatile constituents weighed and then ignited at 1000°C to decompose the carbonate minerals. Warne[26] investigated the method in some detail and found that ankerite could be treated similarly, but not magnesite and siderite, and possibly some other carbonates as well. Waugh and Hill[27] applied it to pyritic limestones of relatively high purity but a correction had to be applied for the oxidation of the sulfide to sulfate.

The most straightforward procedure is that in which the volume of carbon dioxide released by acid decomposition of the carbonates or other CO_2-bearing minerals is measured directly. Fahey[4] has described a simple and rapid procedure in which the sample is heated with HCl and NaCl in a Pyrex test tube and the volume of the carbon dioxide is measured by the displacement of mercury in a buret connected directly to the test tube; it has been found particularly suitable for rocks containing less than 10% carbon dioxide. Wolff[28] surrounds the reaction and measuring vessels with a thermostatically controlled water-jacket and keeps the temperature and pressure constant. In both of these methods it is imperative that no gases other than carbon dioxide be released, which is a serious disadvantage; an important advantage is the relatively large volume occupied by 1 mg of carbon dioxide (0.405 cm³, at STP[28]). A quantitative manometric method for the determination of calcite and dolomite in soils and limestones has also been described[23,24]; see also Iordanov, Supp. Ref.

Shapiro and Brannock[21,22] devised a simple and rapid modification of Fahey's method as part of their rapid silicate rock analysis scheme, which involves only the measurement of the volume of gas formed during treatment of the sample with hot acid. The sample is placed in a borosilicate test tube which is fitted with a side-arm tube at an angle of about

60°, and paraffin oil, $HgCl_2$ (both as a solid and as a saturated solution) and HCl are added; the tube is heated to about 120°C in an aluminum heating block* and the liberated carbon dioxide displaces the oil in the side-arm tube where its volume is measured and compared with that of a known weight of CO_2 liberated under similar conditions. The $HgCl_2$ is added to prevent the formation of hydrogen by reaction of the acid with any "tramp" iron present in the sample. The method is recommended for those samples which contain less than 6% carbon dioxide.

A further refinement of the volumetric technique is one in which the measured volume of liberated gas, which is mixed with a certain amount of CO_2–free air and water vapor, is passed through a solution which will preferentially absorb carbon dioxide; the volume of CO_2 is then found by difference when the gas volume is measured a second time. Goldich *et al.*[8] used this procedure to determine CO_2 in lake marls and obtained results that were comparable with those obtained by the standard gravimetric absorption train method. The volume of liberated CO_2, plus air and water vapor, is measured in a water-jacketed gas-measuring buret using acid sodium sulfate solution as the containing fluid; the CO_2 is absorbed in potassium hydroxide solution in an absorption pipette and the gas volume is measured again to give the volume of carbon dioxide by difference. Measurement of the temperature and pressure is done in order to apply a correction for the vapor pressure of the sodium sulfate solution. Shapiro and Brannock[22] have modified the apparatus of Goldich *et al.* somewhat and use this procedure for the rapid determination of CO_2 in those samples in which its concentration is probably greater than 6%; the need to correct for the vapor pressure of the sodium sulfate is eliminated by standardizing the method with a sample of known CO_2 content which is carried through the procedure concurrently with the samples.

The most popular method is that in which the carbon dioxide, liberated by acid treatment or ignition of the sample and purified by passage through absorbers which will remove volatile constituents other than CO_2, is absorbed in a suitable medium, the increase in weight of which is then determined. Air or nitrogen, freed of CO_2, is usually employed as a carrier gas. This procedure has been described in detail in standard texts.[10,11,16,17]

The size of sample used will depend upon the carbon dioxide content but for most silicates a sample of 2–3 g is preferable. When acid is used

* The heating was originally done with an electric heater, which was later replaced with an oil bath. The aluminum heating block is now preferred (L. Shapiro, personal communication).

to liberate the CO_2, it is usually hydrochloric that is selected because of the solubility of its compounds; its high volatility is a possible source of undesirably high blanks, however, and many analysts prefer to use phosphoric acid instead. A condenser attached to the decomposition flask will remove the bulk of the HCl–water vapor formed when the acid mixture is boiled to expel the liberated CO_2; if not so removed it will necessitate frequent changing of the desiccant ($CaCl_2$, P_2O_5, anhydrous $Mg(ClO_4)_2$) in the absorption train. A variety of substances are used to remove such unwanted constituents as HCl, H_2S and Cl_2, among which are anhydrous copper sulfate, chromic acid–phosphoric acid, silver arsenite and manganese dioxide.

The apparatus used for this procedure varies little from analyst to analyst and generally consists of a decomposition flask fitted with a small condenser and a separatory funnel for the addition of acid (an absorption tube containing a CO_2-absorbent, such as Ascarite or Mikhobite, removes CO_2 from the air drawn into the train through this funnel) or a combustion tube and furnace if thermal decomposition is to be used, followed by a series of absorption tubes containing absorbents to remove unwanted volatiles and vapors; one or, preferably, two weighed absorption tubes to absorb the evolved CO_2 and a final tube containing absorbents to protect the weighed absorption tubes from the accidental sucking in of air complete the apparatus. Gas bubbler tubes at the beginning and end of the absorption train serve to indicate the rate of passage of the flow gas, and also give warning of the presence of a leak in the system; water should not be used in these bubblers and a mixture of phosphoric and chromic acids will function doubly as a rate indicator and as a scavenger of hydrogen sulfide. Peck[18] describes a compact absorption train that features a combined reaction vessel and condenser.

Jeffery and Wilson[15] have modified the conventional absorption train into a closed circulation system (the apparatus, slightly modified, can be used for the determination of non-carbonate carbon as well); the air is recycled through the system by means of a small pump and significantly lower blank values were obtained by Jeffery and Wilson for this method (0.8 mg CO_2 per determination for HCl, <0.1 mg for H_3PO_4) than were obtained for the conventional method (1.5–2.0 mg CO_2 per determination). Riley[19] determines carbon dioxide and water simultaneously by heating the sample to $1100°–1200°C$ in a combustion tube through which a current of purified nitrogen is drawn, and collecting the liberated CO_2 and H_2O in suitable absorbents; an auxiliary tube containing copper wire which is heated to $700°–750°C$ is necessary to remove acidic oxides of nitrogen that form during the combustion of the sample.

The apparatus has been modified to permit the simultaneous determination of H_2O and CO_2 on as little as 10 mg of sample.[20]

The liberated CO_2 may be absorbed in a medium such as a solution of $Ba(OH)_2$ and the resulting precipitate of $BaCO_3$ either filtered off and weighed, or the excess $Ba(OH)_2$ determined by an acid-base titration. Hoefs[12] utilized a special absorption cell in which the CO_2 is absorbed in weakly alkaline $Ba(OH)_2$–$BaCl_2$ solution and determined by means of a potentiometric titration in the special cell; after the determination of carbonate carbon, the carbonaceous matter is oxidized with chromic acid and the determination finished in the same way. A non-aqueous titration has been proposed in which the liberated CO_2 is absorbed in formidimethylamide and then titrated with potassium methoxide in benzene-methanol using thymolphthalein as indicator.[9]

Jeffery and Kipping[13,14] have utilized gas chromatography for the determination of carbon dioxide and other volatiles in rocks and minerals. The sample (5 mg for carbonates, up to 1 g for other rocks) in a small (12 ml) reaction flask, is treated with orthophosphoric acid and the liberated CO_2 is carried by a stream of hydrogen through a chromatographic column packed with silica gel where a separation from other volatile constituents is achieved. Detection is by means of a thermal conductivity cell and the sensitivity for CO_2 is high, as little as 5 ppm in a 1 gm sample being detected. Calibration of the apparatus is done with pure $CaCO_3$ and must be done frequently. A dynamic sorption apparatus has been adapted by Thomas and Hieftje (Supp. Ref.) for the determination of CO_2; a thermal conductivity cell is used for the final measurement (see also Sec. 7.12).

6.13.2. CARBONACEOUS MATTER

The determination of non-carbonate carbon is similar in most respects to the determination of carbonate carbon but oxidation, as well as decomposition, is involved. Both of these operations must be completed, particularly if the determination is to be of the total carbon, i.e., both the carbonate and non-carbonate fractions. If a combustion method is used to decompose the sample, the oxidation of the carbonaceous material is achieved by carrying out the combustion (usually at 800–1000° C) in a stream of purified oxygen or air, or by mixing an oxidizing flux such as lead chromate, vanadium pentoxide, potassium chromate or dichromate with the sample (the flux also retains most if not all of other volatiles present), or by a mixture of both methods. Wet oxidation of the carbonaceous matter is achieved by heating the sample with an oxidant such as chromic acid; this is usually combined with a preliminary determination of the carbonate carbon by acid decomposition of the sample, the apparatus requiring little or no modification for the second step. Both combustion and wet oxidation methods require the prior removal of other volatile constituents before the final determination of the liberated CO_2 by any of the methods described in Sec. 9.26; combustion of the sample in air may result in the formation of oxides of nitrogen, in addition to those of sulfur if the latter is present in the sample, and these too must be removed.

Because of the usually small amount of carbon that is present in silicates, measurement of the volume of gas formed by oxidizing it to CO_2 is usually not feasible.

The determination of carbonaceous matter by difference from the determination of total and carbonate carbon is not very satisfactory when the sample contains much carbonate and little carbonaceous matter. A preliminary acid treatment of the sample with HCl, or HCl and HF, will eliminate the carbonate (and also sulfides)[3,5,11,25]; the residue is filtered onto an asbestos pad, dried at 105–110°C, and then the pad and residue are burned in a combustion tube in the usual way. Volatile or soluble components of the carbonaceous matter may be lost by this procedure; Frost reports no significant loss.[5]

The apparatus needed for the gravimetric determination of the CO_2 liberated by burning the sample is similar to that used for the determination of carbonate carbon, except for the substitution of one or more combustion furnaces for the acid decomposition flask and condenser, and the insertion of additional absorption tubes to remove volatiles not encountered in the acid-evolution procedure. Hillebrand et al.[11] have described an elaborate furnace and train in which purified air is drawn through the system and the sample is fused with a mixture of lead and potassium chromates; a heated, reduced copper spiral is used to reduce oxides of nitrogen, and granular lead chromate is packed into the exit end of the tube and heated to 300–400°C to remove sulfur and halogen compounds. A three-unit split electric furnace is very useful when different portions of the purification and absorption train must be maintained at different temperatures. Vovsi and Bal'yan[25] purify the oxygen used for the oxidation by passing it through granulated Cr_2O_3 heated to 600–650°C; nitrogen or sulfur oxides formed during combustion of the sample are removed by passing the gas through silica gel impregnated with chromic acid (any chromic acid carried over is caught in silica gel impregnated with a saturated solution of ferrous sulfate in 1:3 sulfuric acid).

The carbon dioxide may also be absorbed in a solution of barium hydroxide and the determination completed titrimetrically.[5,11]

The efficiency of the wet-oxidation procedure, using a phosphoric acid-chromic acid mixture, was investigated by Dixon[2] who found that a second oxidation flask was necessary to insure the complete oxidation of the gaseous carbon compounds which would otherwise escape absorption in the train. Mercuric oxide was found to enhance the activity of the acid mixture but was added only to the second flask because, at the final high temperature to which the first flask is heated, it shortens the active life of the oxidant. The acid mixture in the second flask also serves to trap unwanted volatile constituents. Groves[10] found that for the usually low carbon content of most rocks only one oxidation flask was necessary; he cautions against the use of rubber connections to the oxidation flask because of the high blank values that usually result. Jeffery and Wilson[15] modified Dixon's procedure by using a closed circulation system (see also Secs. 6.12 and 9.26) that gives a reagent blank of only 0.7 mg (C?) per determination compared to the 2–3 mg per determination found for Dixon's method. They also use a single oxidation flask* but after the initial oxidation appears to be complete, they recommend that the walls of the flask be rinsed down to recover graphite or other carbonaceous matter adhering to them, that more acid mixture be added and that the oxidation be repeated; low recoveries were obtained for samples containing a large amount of graphite when this precaution was not observed. They prefer to make the determination of carbonate carbon a separate step for such samples, rather than to combine the two determinations. Hoefs[12] applies the wet oxidation procedure only to samples containing >0.1% carbon, using a single flask and finishing the determination potentiometrically (see also Sec. 6.13.1).

Hillebrand et al.[11] used sulfuric and chromic acids as the oxidizing mixture but it is difficult to avoid the concurrent liberation of large amounts of unwanted SO_3. Phosphoric acid has an added advantage over sulfuric acid in that, unlike the latter, it does not exert an oxidizing effect at high temperatures and can be used for a preliminary determination of carbonate carbon without the danger of partial oxidation of some carbonaceous matter as well. The use of a mixture of $K_2Cr_2O_7$ in H_2SO_4, with silver sulfate, for the rapid wet oxidation of organic carbon at 150°C, has been discussed by Frost[6] as a means for the rapid and simple determination of carbon in sedimentary rocks, particularly in shale; the excess

* But a second flask is included in the train for the purpose of retaining the water distilling from the oxidation flask which contains the dilute H_3PO_4 used for the determination of carbonate carbon.

potassium dichromate is determined by titration with $0.5N$ ferrous sulfate, using orthophenanthroline as indicator. A correction must be aplied for oxidizable sulfur compounds that are also present in the sample; other reducing agents, such as ferrous iron, will also interfere but sulfur exerts the most effect.

Frost[5] has also investigated the applicability of three different methods for the determination of total and organic carbon in sedimentary rocks— (1) a gravimetric combustion method, (2) a gasometric method, with fusion in a Parr micro-bomb (the potentially high blank limits this to samples containing $>0.2\%$ organic carbon) and (3) a combustion method followed by absorption in barium hydroxide solution and titration of the excess OH^- with hydrochloric acid. The first is a control method and Frost prefers to do the routine determinations by (2) if the organic carbon content is greater than 0.5%; by (3) if it is less than this amount.

A two-stage optical emission spectrographic procedure for carbonate carbon and "equivalent graphite" carbon (graphite plus diverse organic material) has been described.[1] An aliquot of the sample, with added quartz, is burned in a copper electrode; another aliquot is treated with HCl to remove carbonate carbon, and the dried residue, again with added quartz, is burned as before. Measurement of the density of the cyanogen bands in the spectrum of the first burning gives the total carbon content; similar measurements on the second spectrum give the "equivalent graphite."

References

1. Dennen, W. H. Spectrographic determination of carbon in sedimentary rocks by using dc arc excitation. *Spectrochim. Acta* **9**, 89–97 (1957).
2. Dixon, B. E. The determination of carbon in rocks and minerals. *Analyst* **59**, 739–743 (1934).
3. Ellingboe, J. L., and Wilson, J. E. Direct method for the determination of organic carbon in sediments. *Anal. Chem.* **36**, 434–435 (1964).
4. Fahey, J. J. A volumetric method for the determination of carbon dioxide. *U.S. Geol. Surv. Bull.* **950**, 139–141 (1946).
5. Frost, I. C. Comparison of three methods for the determination of total and organic carbon in geochemical studies. *U.S. Geol. Surv. Profess. Papers* **400–B**, B480–483 (1960).
6. Frost, I. C. Evaluation of the use of dichromate oxidation to estimate the organic carbon content of rocks. *U.S. Geol. Surv. Profess. Papers* **424–C**, C376–377 (1961).
7. Galle, O. K., and Runnels, R. T. Determination of carbon dioxide in carbonate rocks by controlled loss on ignition. *J. Sed. Petrol.* **30**, 613–618 (1960).
8. Goldich, S. S., Ingamells, C. O., and Thaemlitz, D. The chemical composition of Minnesota lake marl—comparison of rapid and conventional chemical methods. *Econ. Geol.* **54**, 285–300 (1959).

9. Grant, J. A., Hunter, J. A., and Massie, W. H. S. The determination of carbon dioxide by non-aqueous titrimetry. *Analyst* **88**, 134–136 (1963).
10. Groves, A. W. *Silicate Analysis*, 2nd ed., 1951. London: George Allen and Unwin, pp. 109–117.
11. Hillebrand, W. F., Lundell, G. E. F., Bright, H. A., and Hoffman, J. I. *Applied Inorganic Analysis*, 2nd ed., 1953. New York: John Wiley & Sons, pp. 766–778, 934–935.
12. Hoefs, J. Ein Beitrag zur Geochimie des Kohlenstoffs in Magmatischen un Metamorphen Gesteinen. *Geochim. Cosmochim. Acta* **29**, 399–428 (1965).
13. Jeffery, P. G., and Kipping, P. J. The determination of constitutents of rocks and minerals by gas chromatography. Part I. The determination of carbon dioxide. *Analyst* **87**, 379–382 (1962).
14. Jeffery, P. G., and Kipping, P. J. *Gas Analysis by Gas Chromatography*, 1964. New York: Macmillan, pp. 107–115, 175–177.
15. Jeffery, P. G., and Wilson, A. D. Closed circulation systems for determining water, carbon dioxide and total carbon in silicate rocks and minerals. *Analyst* **85**, 749–755 (1960).
16. Kolthoff, I. M., and Sandell, E. B. *Textbook of Quantitative Inorganic Analysis*, 3rd ed., 1952. New York: Macmillan, pp. 372–375.
17. Lundell, G. E. F., and Hoffman, J. I. *Outlines of Methods of Chemical Analysis*, 1938. New York: John Wiley & Sons, pp. 180–181.
18. Peck, L. C. Systematic analysis of silicates. *U.S. Geol. Surv. Bull.* **1170**, 48–50, 79–80 (1964).
19. Riley, J. P. Simultaneous determination of water and carbon dioxide in rocks and minerals. *Analyst* **83**, 42–49 (1958).
20. Riley, J. P., and Williams, H. P. The microanalysis of silicate and carbonate minerals. II. Determination of water and carbon dioxide. *Mikrochim. Acta* **1959**, 525–535.
21. Shapiro, L., and Brannock, W. W. Rapid determination of carbon dioxide in silicate rocks. *Anal. Chem.* **27**, 1796–1797 (1955).
22. Shapiro, L., and Brannock, W. W. Rapid analysis of silicate, carbonate and phosphate rocks. *U.S. Geol. Surv. Bull.* **1144–A**, A14–15, A49–51 (1962).
23. Skinner, S. I. M., and Halstead, R. L. Rapid method for determination of carbonates in soils. *Can. J. Soil Sci.* **38**, 187–188 (1958).
24. Skinner, S. I. M., Halstead, R. L., and Brydon, J. E. Quantitative manometric determination of calcite and dolomite in soils and limestone. *Can. J. Soil Sci.* **39**, 197–204 (1959).
25. Vovsi, B. A., and Bal'yan, Kh. V. Rapid method for determining carbon in rocks. *Zavodsk. Lab.* **25**, 437–439 (1959).
26. Warne, S. St. J. Determination of carbon dioxide in carbonate rocks by controlled loss on ignition—additions and modifications. *J. Sed. Petrol.* **32**, 877–881 (1962).
27. Waugh, W. N., and Hill, Walter E., Jr. Determination of carbon dioxide and other volatiles in pyritic limestones by loss on ignition. *J. Sed. Petrol.* **30**, 144–147 (1960).
28. Wolff, G. Die bestimmung von Karbonat in Mineralen und Gesteinen. *Z. Angew. Geol.* **10**, 320–322 (1964).

Supplementary References

Iordanov, N. Manometric determination of carbonate in minerals and ores. *Talanta* **13**, 563–566 (1966). The small sample, 0.2–0.05 g, is decomposed with warm H_2SO_4 (20%) in a previously evacuated vessel, and the volume of CO_2 is measured manometrically. The method has been applied to samples having a range of 0.1–40% CO_2.

Thomas, J. Jr., and Hieftje, G. M. Rapid and precise determination of carbon dioxide from carbonate-containing samples using modified dynamic sorption apparatus. *Anal. Chem.* **38**, 500–503 (1966).

6.14. Sulfur*

6.14.1. GENERAL

The sulfur content of most silicate rocks is very low ($<0.1\%$) and the sulfur is usually in the form of a sulfide such as pyrite (FeS_2) or pyrrhotite (Fe_7S_8); the sulfur-bearing silicates, nosean ($Na_8(Al_6Si_6O_{24})$ SO_4) and häuyne ($(Na, Ca)_{4-8}(Al_6Si_6O_{24})(SO_4, S)_{1-2}$) are seldom encountered. The rocks may have been subjected to sulfide mineralization, of course, but the nature and source of the sample will usually give ample warning of this possibility. Stony meteorites may contain much sulfur present as troilite, FeS. Among the sulfates often encountered in small amounts are barite, celestite, gypsum and alunite ($K_2Al_6(OH)_{12}$ $(SO_4)_4$); barite is the most common of these minerals and a spectrographic analysis will again alert the analyst to the possibility of its presence. Sulfur may also occur in the elemental (native) form or in organic compounds but such occurrences are unusual.

The analytical problems raised by the presence of sulfur in its various guises are encountered chiefly at the beginning of the determination of the main portion constituents, and in the determination of ferrous iron. These have been discussed already in the appropriate sections dealing with these determinations. If the sulfur is present as a sulfide and is to be determined, the sample must not be ground too finely or some of the sulfide may be oxidized to SO_2 and be lost; the sulfide may also be oxidized to sulfate and this will be a source of error if the sulfide is to be determined by a method involving the evolution of hydrogen sulfide.

The form(s) in which the sulfur is present in the sample can be determined with some assurance by boiling about 1 g of the powdered sample with dilute HCl (1:4); the odor of hydrogen sulfide or the blackening of a piece of lead acetate paper will indicate the presence of an *acid-soluble sulfide* such as pyrrhotite. If on the addition of a few drops of 5%

---

* References for this section will be found on pp. 242, 243.

barium chloride solution to the filtrate from this acid digestion a white precipitate is obtained, it is likely that *acid-soluble sulfates* are present in the sample. If the remaining residue is again boiled with acid, this time dilute nitric acid (1:4) containing some bromine, and the filtrate is again tested with barium chloride, a white precipitate indicates the presence of *acid-insoluble sulfides* such as pyrite, or possibly *elemental sulfur*. Fusion of the residue with sodium carbonate and potassium nitrate, followed by leaching, acidification with HCl and the addition of barium chloride, will reveal the presence of *acid-insoluble sulfates* such as barite, or possibly of *organic sulfur*.

The determination of the total sulfur is almost always necessary (this is usually all that is needed for small amounts, i.e., 0.1% or less). A separate determination of either the sulfate or sulfide sulfur will permit the results to be expressed in terms of the latter two forms, assuming that elemental sulfur and organic sulfur are not present. If only the total sulfur content is determined and the form in which the sulfur exists in the sample is not known, then it should be reported as sulfide sulfur only.

In a summation of the analysis of a sulfur-bearing sample it is usually necessary that a correction be made to the analysis total by subtracting the O-equivalent of the S that is present as sulfide if the other elements which form the sulfides, especially iron, are to be reported instead as oxides. No correction is necessary if the sulfur present as sulfate is reported as SO_3. The correction for the sulfide requires a knowledge of the sulfide mineralogy of the sample, unless the sulfur content is less than 0.1%, because of the differing behavior of the sulfides during the acid decomposition of the sample prior to the determination of ferrous iron (it is highly unusual for sulfides other than pyrite and pyrrhotite to be present in significant amounts in silicates which are not ore samples.) Pyrite is scarcely attacked by the hot sulfuric and hydrofluoric acids of the modified Pratt method (Secs. 9.22.1 or 9.22.2) and not at all by the cold Wilson procedure (Sec 9.22.3); it is much more probable that the pyrite will be decomposed completely during the determination of total iron and the ferrous iron of the pyrite will be counted as ferric iron.

$$2\ FeS_2 \equiv Fe_2O_3$$

$$\frac{3O}{4S} = 0.374$$

and thus the oxygen equivalent of the sulfur known to be present as pyrite is the percent sulfur multiplied by 0.374. Pyrrhotite, on the other hand, is readily attacked by acids and the iron will be reported as ferrous iron; the O-equivalent is per cent S $\times$ 0.437.

$$Fe_7S_8 \equiv 7\ FeO$$

$$\frac{7O}{8S} = 0.437$$

Peck[15] recommends that when the amount of sulfur is small and it is not practical to distinguish between individual sulfides, the correction to be subtracted should be per cent S $\times$ 0.5.   It is also better to report the total sulfur as S, rather than to attempt to separate the types.

The problems that may be encountered in the determination of sulfur in rocks and minerals have been treated in detail by Groves,[7] and by Hillebrand et al.[9]; a more recent general coverage of the analytical chemistry of sulfur is also available.[3,11,23]

6.14.2. "SULFIDE" SULFUR

When sulfur is known to be present only as a sulfide, its determination is a relatively simple task.   The sample may be fused with an oxidizing flux such as a mixture of sodium carbonate and potassium nitrate, or decomposed by digestion with a mixture of oxidizing acids and bromine. Both methods result in the oxidation of the sulfide to sulfate.   The fusion method has the advantage of removing a number of elements such as iron and the alkaline earths, but does introduce a large amount of alkali salts into the solution; when there is much sulfide present (e.g. $> 2\%$),[9] however, the fusion method is preferred to acid digestion because of the possibility that some sulfide may escape oxidation and be lost by the latter method.   The acid digestion method has the advantage of removing silica and of introducing no additional material; it is necessary, however, to remove or reduce ferric iron, if much is present, because it will contaminate the barium sulfate when the determination is finished gravimetrically.

The precipitation, filtration and ignition of barium sulfate, whether for the determination of barium or of sulfate, remains a somewhat unsatisfactory procedure in spite of the voluminous literature that has been accumulated on the subject.   A brief summary only will be given here and the following references should be consulted for more details.[1,9,12,23]

1. The final solution should contain only a small excess of HCl (approx. 0.05N) because the solubility of $BaSO_4$ increases with increasing acid concentration.

2. The volume of the final solution should be kept large (e.g., 300 ml), to minimize both the tendency of $BaSO_4$ to adsorb other ions and the precipitation of any SiO_2 present.

3. The precipitation should be made at the boiling point to minimize the tendency of barium sulfate to supersaturate.

4. The precipitate and solution should be digested hot before filtration, or at least allowed to stand for several hours, in order to reduce the amount of coprecipitation that will occur. Hesse first removes colloidal organic matter, which interferes seriously, by collecting it with ferric hydroxide in a preliminary precipitation.[8]

5. Ignition of the precipitate must be done with care. Because $BaSO_4$ is easily reduced by carbon there must be no flame during the burning of the paper, which should be done at a low temperature ($<600°C$) under oxidizing conditions. In the presence of either Si or Fe the precipitate must not be heated too strongly, i.e., $>1000°C$, or some decomposition will occur.

6. The ignited residue should be treated with HF and one drop of concentrated H_2SO_4, reignited, and reweighed to correct for any silica that was carried down with the precipitate. For accurate work the $BaSO_4$ should again be fused with Na_2CO_3 and the precipitation repeated. Dimethylformamide is reported to prevent the precipitation of silica when $BaSO_4$ is precipitated from the solution of a soluble silicate (Azeem, Supp. Ref.).

7. Ferric iron should either be reduced with powdered Al, Fe or Zn (these also will remove Pb and Sb), or removed by extraction or mercury cathode electrolysis; this latter procedure will remove most of the heavy metals present, a particularly useful arrangement if chromium is present as Cr(VI) and which would otherwise coprecipitate as $BaCrO_4$.

The above considerations apply for the most part to the determination of large amounts of sulfur. For most silicates the sulfur content is so low that these precautions are not as critical.

The sample may also be heated at a high temperature (about 1400°C) in a stream of oxygen or air (in this case the sample is mixed with an oxidizing flux) to convert the sulfide to SO_2 or SO_3, or to a mixture of both[2,4,5,17,21,24]; the gases are absorbed in a suitable solution and the determination finished colorimetrically, titrimetrically or gravimetrically.*

It is not usually possible, however, to assume that the sulfur is present only as sulfide and the methods given above could, for the most part, be used to determine the total sulfur content as well. It is common practice to so determine the total sulfur and then to determine, on a separate portion, that part of it which can be put into solution by boiling the sample with dilute HCl (so-called soluble sulfate); the difference between the two values is expressed as sulfide sulfur. It is obvious that this is not entirely satisfactory; insoluble sulfate will be counted as sulfide and care must be taken to avoid the oxidation of soluble sulfides during the acid treatment. Vlisidis (Supp. Ref.) does the acid digestion under an

* Groves[7] heated the sample (after extraction of any free sulphur) in a current of chlorine to produce sulfuryl chloride, SO_2Cl_2, which was absorbed in dilute HCl; after removal of the accompanying ferric iron, the sulfate was precipitated as barium sulfate.

atmosphere of nitrogen, and adds $CdCl_2$ to precipitate any sulfide ion liberated at the same the time; precipitated $BaSO_4$ and any barite present are removed, fused with Na_2CO_3, dissolved in HCl and the Ba(II) precipitated with Na_2SO_4. For the usually low sulfur content of silicates the errors so introduced will be small, but unless there is reason to suspect the presence of more than one form of sulfur, it is better to report the total sulfur as sulfide.[12,15]

The most reliable method is one in which the sulfide ion itself is liberated. Murthy et al.[13,14] digested a mixture of sulfides and sulfates with a mixture of hydriodic and hydrochloric acids in a special apparatus; the H_2S formed by the reaction is swept into a suspension of cadmium hydroxide, the latter is poured into $2N$ acetic acid containing a known amount of iodine, and the excess iodine is titrated with a standard solution of sodium thiosulfate. The reaction is slow but satisfactory for sulfides such as galena and sphalerite; for pyrite and chalcopyrite it is necessary to employ a more concentrated hydriodic acid solution and to add a small globule of mercury to the reaction vessel.[14] A similar precipitation of CdS is used by Smith et al.[19] to determine sulfide sulfur in oil shale in which sulfate and organic sulfur compounds are also present; after the separation of the soluble sulfate, the residue, in a special refluxing flask, is treated with aluminum hydride dissolved in tetrahydrofuran to decompose the sulfides and the hydrogen sulfide formed is passed through a solution of cadmium sulfate. The sulfuric acid formed in the absorption vessel is titrated with sodium hydroxide solution. Any free sulfur present in the sample will be reduced also.

Small amounts of sulfides in glass have been determined indirectly by decomposing a sample with hydrofluoric acid in the presence of silver acetate; the precipitated silver sulfide is filtered, dissolved in H_2SO_4 and HNO_3, and the silver titrated potentiometrically with ammonium thiocyanate.[21] Corrections must be made for any Se, Br, I and Cl present. Satisfactory results for sulfur were also obtained by X-ray fluorescence spectroscopy, using a glass block as sample.

Organic sulfur and free sulfur are two forms of sulfur that are not often encountered, especially in silicates. They are determined as part of the total sulfur content because they are oxidized to sulfate during decomposition of the sample, but for their separate determination such oxidation must be avoided. Groves[7] extracts the free sulfur by refluxing the sample with carbon tetrachloride in a Soxhlet extractor; the extract is evaporated to dryness in a weighed platinum dish and the extraction repeated until the weighings show that all free sulfur has been extracted. Portions of the sample, after the extraction, are used for the determination of sulfide and sulfate sulfur. Smith *et al.*[19] use the dried residue remaining after the removal of sulfate and sulfide sulfur from oil shale for the determination of *organic sulfur*, by fusing it with Eschka mixture (MgO/Na$_2$CO$_3$:1/2), leaching the sinter, acidifying the filtrate after the addition of bromine water, and precipitating the sulfur as barium sulfate.

6.14.3. "SULFATE" SULFUR

Reference has already been made to the determination of sulfate sulfur in the previous section and little more need be added. If it is the only form in which sulfur is present in the sample, whether as a soluble or insoluble sulfate, its determination is a simple matter; fusion with sodium carbonate will insure that all of it will be in solution prior to its determination by one of several possible methods. When sulfides are present certain precautions must be taken to avoid including all, or a portion, of them in the value for sulfate.

The determination of *soluble sulfate* (usually reported as SO$_3$) is accomplished by digesting the sample for a brief period with dilute hydrochloric acid. Water-soluble sulfates, the silicates nosean and häuyne, gypsum, anhydrite and (possibly) celestite will dissolve wholly, while some others may dissolve to a limited extent. Peck[15] boils the sample with water before adding the HCl, in order to expel air which might oxidize the hydrogen sulfide formed when soluble sulfides (e.g., pyrrhotite) are also dissolved; Washington[20] flushed the beaker with a stream of carbon dioxide before boiling the sample with dilute acid, and maintained the passage of the gas during the boiling period.

Some acid-insoluble sulfate minerals can be put into solution by boiling the sample with sodium carbonate (e.g., lead sulfate) or sodium hydroxide (alunite). This is useful when it is necessary to determine the truly acid-soluble sulfates. An initial treatment with sodium hydroxide will extract alunitic sulfate; the residue can then be boiled with dilute HCl to dissolve the gypsum and anhydrite. Peck[15] has pointed out, however, the uncertain state of knowledge regarding the behavior of the sulfide minerals in a boiling basic solution and suggests that, unless the mineralogy of the sample is known and a separation will serve a useful purpose, it is better to confine oneself to the simplest determinations and to allow the petrographer to distribute the sulfur as he sees fit. A simple digestion with HCl, under non-oxidizing conditions, will yield a value for acid-soluble sulfate that will not be greatly in error, provided that the total sulfur content is low.

To obtain *acid-insoluble sulfate*, and thus *total sulfate sulfur*, it is necessary to remove the difficultly soluble sulfides such as pyrite from the residue left over from the determination of acid-soluble sulfate. This latter is digested with a mixture of aqua regia and bromine and the remaining residue is then fused with sodium carbonate to decompose barite and other insoluble sulfates.

Fusion and acid treatment are not the only methods of attack that may be used to solubilize sulfate, although they are usually the simplest and most direct means. Mention has been made of the use of a high-temperature combustion furnace for decomposition of the sulfates to give either SO_2, SO_3 or a mixture of both. It is also possible to reduce the sulfate to sulfide. Keattch[10] determined sulfate in soils by refluxing the sample with hypophosphorous and hydriodic acids and absorbing the H_2S so formed in ammoniacal zinc sulfate solution; the acidified solution is titrated with potassium iodate–potassium iodide. Rafter[16] reduced sulfate by heating it with graphite in a nitrogen atmosphere and precipitated the sulfide as silver sulfide.

6.14.4. TOTAL SULFUR

The decomposition of the sample by acid digestion in an oxidizing environment, to insure the conversion of the sulfur present to sulfate, is the simplest approach but this will include only the acid-soluble sulfate and sulfide; any acid-insoluble minerals will be lost during the determination. With this limitation in mind, however, a rapid determination can be made; Shapiro and Brannock[18] digest the sample with a mixture of aqua regia and carbon tetrachloride, precipitate the sulfate with barium chloride and, after a 10 min period for digestion, filter, ignite and weigh the barium sulfate. Goldich et al.[6] use the filtrate from the R_2O_3 separation for the determination of total sulfur in marl; the sample was first ignited, under oxidizing conditions, to convert all sulfur to sulfate. Although some loss of sulfur occurred during the ignition, a comparison of the values obtained by this rapid method and those obtained by digesting the marls with hydrochloric acid and bromine favored the rapid method in reproducibility and completeness of recovery. Wilson et al.[22] prefer an acid digestion to fusion because the latter approach usually results in the crucible being attacked; they recommend a mixture of aqua regia, HF and $HClO_4$ (in a Teflon dish) with V_2O_5 added to expedite the oxidation of pyrite and any organic matter that may be present. Additional calcium is added when necessary to give at least a two-fold excess over that needed to combine with the sulfate present in order to prevent a loss of SO_3 during the evaporation with perchloric acid.[9] These authors also investigated the effect of any barium present in the sample on the recovery of sulfate and found that a serious loss of sulfate occurs both when the BaO content is around 1%, and when the BaO content is low but the sulfur content is high; when both the BaO and S contents do not exceed 0.2% each, as in most silicate rocks, the loss of sulfate is negligible.

The procedure used by Keattch,[10] which was described briefly in the preceding section, presumably will also serve to determine total sulfur as well.

The fusion of the sample with an alkaline flux, to which an oxidant is added if necessary, is the most widely used procedure for decomposing the sample and rendering the sulfur in the form of a soluble alkaline sulfate. The flux is almost always anhydrous Na_2CO_3, with a small amount of KNO_3 or Na_2O_2 added unless sulfates are known to be the only form of sulfur present in the sample. Na_2O_2[12] and $NaOH$[1] are also used; these two fluxes put more aluminum into solution than does Na_2CO_3 and because of the predilection of $BaSO_4$ for adsorbing and occluding foreign ions, the latter flux is preferred. The fusions are better done in an electric muffle than over a gas flame because of the danger of the sulfur from the gas being picked up by the alkaline flux, and the subsequent leaching of the cake should be done on an electrically heated water or steam bath, or hot plate. Because of the presence of oxides of sulfur in the normal laboratory air, as well as in the reagents used, a determination of the reagent blank should be made at the same time as the sample determination.

The determination of total sulfur by the fusion method can be combined conveniently with the determination, on the same portion of sample, of zirconium, barium, chromium and rare earths.[7,12,20]

The total sulfur can be removed from a sample by burning the latter in a stream of oxygen in a combustion train, at a temperature of 1400–1600°C; by absorbing the SO_2 in a suitable medium the determination can be finished gravimetrically, titrimetrically or colorimetrically.[4,5,21] The high temperature is necessary to insure that all SO_3 is converted to SO_2, and the method is rapid and of general applicability, but it does shorten the life of combustion tubes and requires an expensive high-temperature furnace to obtain such a high temperature. The use of an oxidizing flux such as V_2O_5 permits the employment of a lower temperature, e.g., 900–950°C, and air or nitrogen, rather than oxygen, are used as carrier gases; the sulfur is evolved as SO_3 and determined as such, or is reduced to SO_2 by passing it over heated copper gauze.[2,17] If organic matter is present (e.g., organic sulfur compounds) the ignition products must be completely oxidized by first passing the gases over a heated roll of oxidized copper. Bloomfield[2] used the lower temperature to determine the total sulfur in soils, with satisfactory results; Sen Gupta[17] used a modification of this procedure for rocks, but later found the high temperature procedure to be generally more reliable.

The SO_2 may be determined spectrophotometrically by absorbing it in a medium such as sodium tetrachloromercurate, to which is added pararosaniline hydrochloride and formaldehyde; the purple compound which is formed, pararosaniline methylsulfuric acid,[4] has its absorbance maximum at 560 mμ. The system obeys Beer's law up to 18 μg/500 ml total volume[17] and is thus suited for the determination of microgram amounts of sulfur only. An advantage of the use of sodium tetrachloromercurate as absorbing medium is that the absorbed SO_2 is not readily oxidized by air and so a number of absorbates can be accumulated before the colorimetric determinations are made.

The use of barium chloranilate, $BaC_6Cl_2O_4$, has been proposed for the indirect determination of sulfate but the color system is not sufficiently sensitive for the determination of microgram amounts.

Absorption of the SO_2 in a solution of potassium iodide, starch and HCl followed by titration with potassium iodate,[5,21] or with the addition of a measured excess of KIO_3, followed by titration with sodium thiosulfate, has been proposed.[17] Wilson et al,[24] absorbed the SO_3 formed in a solution of barium chloride and indirectly determined the sulfur by titrating the excess barium with EDTA; an excess of the latter may be added and the unused EDTA back-titrated with a standard magnesium solution.[23]

References

1. Bennett, H., and Hawley, W. G. *Methods of Silicate Analysis*, 2nd ed., 1965. London: Academic Press, pp. 99–100, 286–287.
2. Bloomfield, C. A colorimetric method for determining total sulfur in soils. *Analyst* **87**, 586–589 (1962).
3. Boltz, D. F., (Ed.). *Colorimetric Determination of Nonmetals*, 1958. "Sulfur," by Gordon D. Patterson, Jr. New York: Interscience, pp. 261–308.
4. Burke, K. E., and Davis, C. M. Combustion–spectrophotometric method for determination of trace quantities of sulfur in metals. *Anal. Chem.* **34**, 1747–1751 (1962).
5. Coller, M. E., and Leininger, R. K. Determination of total sulfur content of sedimentary rocks by a combustion method. *Anal. Chem.* **27**, 949–951 (1955).
6. Goldich, S. S., Ingamells, C. O., and Thaemlitz, D. The chemical composition of Minnesota lake marl—comparison of rapid and conventional methods. *Econ. Geol.* **54**, 285–300 (1959).
7. Groves, A. W. *Silicate Analysis*, 2nd ed., 1951. London: George Allen and Unwin, pp. 117–123, 207–208, 238.
8. Hesse, P. R. The effect of colloidal organic matter on the precipitation of

barium sulfate and a modified method for determining soluble sulfate in soils. *Analyst* **82**, 710–712 (1957).

9. Hillebrand, W. F., Lundell, G. E. F., Bright, H. A., and Hoffman, J. I. *Applied Inorganic Analysis*, 2nd ed., 1953. New York: John Wiley & Sons, pp. 711–724, 948–951, 979–980.

10. Keattch, C. J. Estimation and determination of sulfate in soils and other siliceous and calcareous materials. *J. Appl. Chem.* **14**, 218–220 (1964).

11. Kolthoff, I. M., and Elving, P. J., (Eds.). *Treatise on Analytical Chemistry*, Part II, Vol. 7, 1961. "Sulfur," by B. J. Heinrich, M. D. Grimes, and J. E. Puckett. New York: Interscience, pp. 1–135.

12. Kolthoff, I. M., and Sandell, E. B. *Textbook of Quantitative Inorganic Analysis*, 3rd ed., 1952. New York: Macmillan, pp. 322–335.

13. Murthy, A. R. V., Narayan, V. A., and Rao, M. R. A. Determination of sulfide sulfur in minerals. *Analyst* **81**, 373–375 (1956).

14. Murthy, A. R. V., and Sharada, K. Determination of sulfide sulfur in minerals. *Analyst* **85**, 299–300 (1960).

15. Peck, L. C. Systematic analysis of silicates. *U.S. Geol. Surv. Bull.* **1170**, 51–53, 82–83 (1964).

16. Rafter, T. A. Sulfur isotope variations in nature. Part I. The preparation of SO_2 for mass spectrometer examination. *J. Sci. Tech. New Zealand* **38**, 849–857 (1957).

17. Sen Gupta, J. G. Determination of microgram amounts of total sulfur in rocks. *Anal. Chem.* **35**, 1971–1973 (1963).

18. Shapiro, L., and Brannock, W. W. Rapid analysis of silicate, carbonate and phosphate rocks. *U.S. Geol. Surv. Bull.* **1144–A**, A53 (1962).

19. Smith, J. W., Young, N. B., and Lawlor, D. L. Direct determination of sulfur forms in Green River oil shale. *Anal. Chem.* **36**, 618–622 (1964).

20. Washington, H. S. *The Chemical Analysis of Rocks*, 2nd ed., 1910. New York: John Wiley and Sons, pp. 170–175.

21. Williams, J. P., Farncomb, F. J., and Magliocca, T. S. Determination of sulfur in glass. *J. Am. Ceram. Soc.* **40**, 352–354 (1957).

22. Wilson, A. D., Sergeant, G. A., and Lionnel, L. J. The determination of total sulfur in silicate rocks by wet oxidation. *Analyst* **88**, 138–140 (1963).

23. Wilson, C. L., and Wilson, D. W., (Eds.). *Comprehensive Analytical Chemistry*, Vol. 1C (Classical Analysis), 1962. "Sulfur," by L. A. Haddock. Amsterdam: Elsevier, pp. 282–295.

24. Wilson, H. N., Pearson, R. M., and Fitzgerald, D. M. Improvements in the determination of small amounts of sulfur. *J. Appl. Chem.* **4**, 488–496 (1954)

Supplementary References

Azeem, M. Determination of sulfate in presence of soluble silicate. *Analyst* **92**, 115–117 (1967).

Schafer, H. N. S. An improved spectrophotometric method for the determination of sulfate with barium chloranilate as applied to coal ash and related materials. *Anal. Chem.* **39**, 1719–1726 (1967). The determination is made in 80% isopropyl alcohol and the reaction is almost stoichiometric. The blank value is considerably reduced.

Vlisidis, A. C. The determination of sulfate and sulfide sulfur in rocks or minerals. *U.S. Geol. Surv. Bull.* **1214–D**, D1–5 (1966).

6.15. Chlorine*

The chlorine content of most rocks is <0.1 % and rarely exceeds 0.2 %; the higher concentrations will be found in highly sodic, nepheline-bearing rocks and in those containing much sodalite ($Na_8[Al_6Si_6O_{24}]Cl_2$, $Cl \cong 7\%$) and members of the scapolite group (($Na, K, Ca)_4[Al_3(Al, Si)_3 Si_6O_{24}](Cl, F, CO_3, SO_4)$, $Cl \cong 2\%$), minerals in which it is an essential constituent. It is present in many other minerals as well, e.g., chlor-apatite, and as a constituent in fluid inclusions, such as those occurring in quartz. It may be present also as sodium chloride, from contamination of the sample by sea-water or brine, but in this form it is easily removed by leaching the sample with water. Because of the generally low chlorine content of most rocks it is important that the sample, at all stages of the preparation and analysis, be protected from contamination by chlorides present in the laboratory environment, including the atmosphere. The ubiquity of chloride in chemical laboratories makes it essential that a blank determination be run at the same time, and under the same conditions, as the sample.

Bromine and iodine, if present, seldom exceed 0.01% and, as a rule, will accompany chlorine in its reactions and be determined with it without causing a significant error. Their separate determination will be necessary only under very special circumstances and will not be discussed here; further details are available in other sources.[2,9,10,19]

Chlorine may be extracted from silicates by various methods, the choice depending upon the form in which it is present in the sample. As a water-soluble chloride, e.g., NaCl, it is readily extracted by boiling the sample with water and filtering the solution through a fine-textured single or double paper†; the residue is used for the determination of the chlorine present in a water-insoluble form. Brief boiling of the sample with dilute nitric acid will extract the chlorine from acid-soluble minerals

* References for this section will be found on pp. 247, 248.

† Hillebrand et al.[9] warn that a finely divided suspension (SiO_2?) may be obtained which will not settle. If the determination is to be done gravimetrically they recommend that the precipitation as AgCl be continued as usual, the precipitate and suspended matter collected on a filter paper and the paper and contents ignited at a low temperature in a porcelain crucible. Most of the AgCl is reduced to metallic Ag. The residue, after the carbon has disappeared, is evaporated with a few drops of nitric acid; a drop or two of HCl is added, the contents are evaporated to dryness, and the crucible is weighed. The residue is digested with dilute aqueous ammonia to dissolve the AgCl, filtered and the weight of the coprecipitated siliceous matter is obtained by igniting and weighing the insoluble residue.

such as sodalite and chlorapatite; the addition of a few drops of hydrofluoric acid may be necessary to obtain all the acid-soluble chlorine. There is no danger of losing chlorine by volatilization if the boiling is not unduly prolonged (Groves[8] recommends not more than 2 min), nor will gelatinous silica be formed, at least not in sufficient amount to be troublesome. The chlorine present in the scapolite group of minerals is, however, not so easily extracted* and it is necessary to fuse the sample with sodium or potassium carbonate and extract the chlorine by aqueous leaching of the cake. The fusion approach is the most reliable one, unless it is desirable to know the extent to which chlorine is present in an acid-soluble form. Robinson[14] heated the sample, wetted with H_2SO_4 and HNO_3, at 310°C in a combustion tube; the evolved hydrogen chloride is swept out by nitrogen and absorbed in either a dilute solution of sodium carbonate or in deionized water. He applied the method to chloride contents ranging from several per cent down to a few ppm; the majority of the interfering elements are retained in the combustion boat as insoluble sulfates. A similar separation of chlorine can be achieved by steam distillation; Geijer[5] fused amphiboles with NaOH and then distilled the neutralized solution of the cake with H_2SO_4; the condensate is evaporated to small volume and again distilled, this time in a smaller apparatus. Pyrohydrolysis has been used by Caldwell[3] and Gillberg[6]; Greenland and Lovering[7] distilled chlorine from chondritic meteorites at 1700°C under an argon atmosphere, and finally recovered it by diffusion in a special polythene microdiffusion cell.

The gravimetric determination of chlorine as silver chloride is the most widely used procedure,[8,9,11] although it is not too reliable for chlorine concentrations of <0.05%. Only bromine and iodine interfere significantly in the determination and in most rocks this interference is negligible. Precipitation is done by the addition of a small excess of 5% $AgNO_3$ solution to the chloride solution, acidified with HNO_3 and containing about 0.1 ml in excess per 100 ml. The precipitate is coagulated by heating it near the boiling point for one hour and then is allowed to stand overnight. Throughout these steps the precipitated AgCl must be shielded from light, or decomposition to metallic silver will occur (the precipitate develops a purple coloration). The precipitate is collected on a fritted glass crucible or filter paper and the AgCl is dissolved in warm, dilute aqueous ammonia to separate it from any coprecipitated

---

* Washington[17] believed that digestion with HNO_3 and HF would release all of the chlorine and he considered a fusion to be unnecessary.

siliceous matter; the filtrate is acidified with nitric acid and the precipitation procedure repeated.   The precipitate is collected on a fritted glass
crucible, dried at 130–150°C and weighed.

Groves[8] recommends a turbidimetric finish for samples which yield a
precipitate that is obviously not weighable (e.g., <0.05%).   The turbid
solution is *not* heated but is compared, as quickly as possible, with a
solution of $AgNO_3$ to which is added a standard solution of NaCl.

Peck and Tomasi[13] have described a procedure for the routine and
rapid determination of chlorine up to 0.2% which utilizes a turbidimetric
titration also, but with mercuric nitrate as titrant and sodium nitroprusside as the indicator.   The sample is sintered with a mixed flux (Na_2CO_3:
$ZnO:MgCO_3::7:2:1$), leached with water and acidified with nitric acid.
Sodium nitroprusside is added and the solution is titrated in the dark
with $0.01N$ $Hg(NO_3)_2$, in an apparatus which permits comparison of
twin beams of light passed through the sample solution and through a
standard turbid solution (0.4 $\mu g/ml$ AgCl dispersed in a gelatin solution).
When a chloride solution containing sodium nitroprusside is titrated with
$Hg(NO_3)_2$, all chloride must be complexed by the mercury before a near
colloidal precipitate of mercuric nitroprusside will form.   This near colloidal precipitate scatters the light and the end point is marked by
brightening of the sample light beam relative to the reference beam in
the standard turbid solution.   The results obtained are lower than those
obtained by the gravimetric procedure for chlorine contents >0.2%, if
an 80-mesh powder is used for the fusion; for the occasional sample
having Cl >0.2% a more finely ground sample should be used.

Other titrimetric procedures[9,10] which can be used include Volhard's
method, in which the excess silver from the precipitation is titrated with
ammonium or potassium thiocyanate in the presence of ferric ion as indicator, and Mohr's method in which the chloride solution, containing
K_2CrO_4, is titrated with $AgNO_3$, the end point being indicated by the
first appearance of red silver chromate; Geijer[5] used dichlorofluorescein
as indicator for a $AgNO_3$ titration; in the Greenland and Lovering method[7] the diffusing Cl^- is absorbed in KI solution and the final titration is
made with sodium thiosulfate.   Robinson[14] completed his combustion
method by titrating the absorbed hydrogen chloride potentiometrically
with $AgNO_3$ solution; the titration may also be done amperometrically.[15]

There are numerous procedures available for the colorimetric determination of chlorine,[2,10] but they are seldom applied to this determination
in silicates.   Some use has been made of the red complex formed when
mercury(II)thiocyanate is added to a chloride solution in the presence of
Fe(III)[14,16] but the most direct procedure is that of Kuroda and Sandell[12]

for the determination of chlorine contents $<0.05\%$ on as little as 0.1 g of sample, with a precision of $\pm$ 0.005%. The sample is fused with Na_2CO_3, leached with water, the filtrate made acid with HNO_3 and the chlorine precipitated as AgCl; the filtered precipitate is dissolved in dilute aqueous ammonia and sodium sulfide is added to form a yellow-brown, stable silver sulfide sol. No protective colloid is needed if the silver concentration <0.1 mg/ml. The transmittance is measured at about 415 mμ and the relation between silver concentration and extinction is linear up to 100 ppm. Caldwell[3] added $Hg(NO_3)_2$ to the condensate from pyrohydrolysis (weakly ionizing $HgCl_2$ is formed) and determined the excess mercury present colorimetrically with diphenylcarbazone in order to obtain the chlorine by difference.

It is possible to determine chlorine by flame photometry,[4] either directly using the CuCl systems or indirectly by measuring the decrease in luminosity of a standard silver solution due to precipitation of the silver as the chloride; atomic absorption spectroscopy has been used to determine silver in an ammoniacal solution of precipitated AgCl.[18] CaCl bands are suitable for the detection of chlorine in rocks and minerals by emission spectrography.[1] Gillberg[6] collects the chloride (and fluoride) in the condensate from the pyrohydrolysis of micas and amphiboles on an ion exchange resin, irradiates the latter, and finally measures the ^{38}Cl formed by gamma-ray spectrometry.

As in the case of sulfur (Sec. 6.14.1), it is necessary to apply a correction to the analysis summation for the oxygen equivalent of the chlorine present as chloride, if the constituents are reported as oxides.

$$2 \ KCl \equiv K_2O$$

$$\frac{O}{2 \ Cl} = 0.22$$

References

1. Ahrens, L. H., and Taylor, S. R. *Spectrochemical Analysis*, 2nd ed., 1961. Reading, Mass.: Addison-Wesley, pp. 282–283.
2. Boltz, D. F., (Ed.). *Colorimetric Determination of Nonmetals*, 1958. "Chlorine," by D. F. Boltz and W. J. Holland. New York: Interscience, pp. 161–180.
3. Caldwell, V. E. Determination of chloride in glass by phrohydrolysis. *Anal. Chem.* **38**, 1249–1250 (1966).
4. Dean, J. A. *Flame Photometry*, 1960. New York: McGraw-Hill, pp. 262–265.
5. Geijer, P. Halogens in skarn amphiboles. *Arkiv. Mineral. Geol.* **2**, 482–484 (1960).
6. Gillberg, M. Halogens and hydroxyl contents of micas and amphiboles in Swedish granitic rocks. *Geochim. Cosmochim. Acta* **28**, 495–516 (1964).

7. Greenland, L., and Lovering, J. F. Minor and trace element abundances in chondritic meteorites. *Geochim. Cosmochim. Acta* **29**, 848 (1965).
8. Groves, A. W. *Silicate Analysis*, 2nd ed., 1951. London: George Allen and Unwin, pp. 30, 145–147, 277–278.
9. Hillebrand, W. F., Lundell, G. E. F., Bright, H. A., and Hoffman, J. I. *Applied Inorganic Analysis*, 2nd ed., 1953. New York: John Wiley and Sons, pp. 725–736, 936–937.
10. Kolthoff, I. M., and Elving, P. J., (Eds.). *Treatise on Analytical Chemistry*, Part II, Vol. 7, 1961. "The Halogens," by G. W. Armstrong, H. H. Gill and R. F. Rolf. New York: Interscience, pp. 335–424.
11. Kolthoff, I. M., and Sandell, E. B. *Textbook of Quantitative Inorganic Analysis*, 3rd ed., 1952. New York: Macmillan, p. 721.
12. Kuroda, P. K., and Sandell, E. B. Determination of chlorine in silicate rocks. *Anal. Chem.* **22**, 1144–1145 (1950).
13. Peck, L. C., and Tomasi, E. J. Determination of chlorine in silicate rocks. *Anal. Chem.* **31**, 2024–2026 (1959).
14. Robinson, J. W. Rapid determination of chloride in silica or other solids. *Anal. Chim. Acta* **20**, 256–258 (1959).
15. Stock, J. T. *Amperometric Titrations*, 1965. New York: Interscience, pp. 200–208.
16. Unicam Instruments, Ltd. Determination of low concentrations of chlorides. *Method Sheet* **78**, 1 p. (1965).
17. Washington, H. S. *The Chemical Analysis of Rocks*, 2nd ed., 1910. New York: John Wiley & Sons, pp. 21, 175–176.
18. Westerlund-Helmerson, U. The determination of chloride as silver chloride by atomic absorption spectroscopy. *Perkin-Elmer Atomic Absorption Newsletter* **5**, 97 (1966).
19. Wilson, C. L., and Wilson, D. W., (Eds.). *Comprehensive Analytical Chemistry*, Vol. 1C (Classical analysis), 1962. "Chlorine," by L. A. Haddock. Amsterdam: Elsevier, pp. 341–352.

Supplementary References

Huang, W. H., and Johns, W. D. The chlorine and fluorine contents of geochemical standards. *Geochim. Cosmochim. Acta* **31**, 597–602 (1967); Simultaneous determination of fluorine and chlorine in silicate rocks by a rapid spectrophotometric method. *Anal. Chim. Acta* **37**, 508–515 (1967). The sample is fused with sodium carbonate and zinc oxide at 900°C, leached with water, filtered and the filtrate acidified with nitric acid. An aliquot of this is used for the determination of chlorine by the ferric ammonium sulfate–mercuric thiocyanate procedure.
Ogita, H., Nakai, N., and Oana, S. Chlorine in sedimentary rocks. *Geochem. Journal* **1**, 139–148 (1967). Total chlorine is determined colorimetrically with ferric ammonium sulfate and mercuric thiocyanate, following fusion of the sample with Na_2O_2 in a Parr bomb and distillation of the fusion cake with H_2SO_4 to separate the chlorine. Soluble chlorine, obtained by prolonged leaching of the sample in distilled water, is also determined colorimetrically.

6.16. Fluorine*

6.16.1. GENERAL

Like chlorine, fluorine seldom occurs in rocks in an amount exceeding a few tenths of 1%, and is generally much less; it is more abundant in the acid rocks. In such minerals as fluorite and cryolite it is a major constituent, and it is an essential component of fluorapatite and tourmaline and an important minor constituent of micas and amphiboles. Its distribution in rocks and minerals has been summarized recently by Gillberg.[10]

The determination of fluorine is not only of importance to the petrographer but also to the rock analyst, provided that its determination precedes that of silica if the latter is to be done gravimetrically. The problems raised by the presence of more than 2% fluorine have already been discussed in some detail (see Sec. 6.2.1), and it is better to determine the fluorine first, especially if the rock is rich in mica or amphibole, so that the necessary precautions may be taken. A fluorine determination was once a lengthy and difficult procedure but there is no reason now for not determining it if mineralogical or petrographical evidence suggests that it is necessary, and failure to do so can lead to serious error in the results of an analysis.

Qualitative tests for the presence of fluorine are not very satisfactory. The etching or hanging-drop test, which is specific for fluoride ion, requires some experience to be able to estimate the amount involved; the sample is heated with concentrated sulfuric acid and the liberated hydrogen fluoride will etch a wetted glass slide exposed to the vapors, or will dissolve in a drop of water, and in the presence of silica (e.g., glass) will form silicon(IV) fluoride and hydrolyze to gelatinous silica, causing the drop to become cloudy. If the sample is not decomposable by sulfuric acid it must be fused with Na_2CO_3, leached in a small amount of water and the excess SiO_2 separated with ammonium carbonate and/or zinc oxide; a mixture of calcium carbonate and fluoride is then precipitated and used for the test for fluoride. There are also several tests which involve the bleaching of a colored complex of fluoride, but because many anions and cations interfere, it is necessary first to separate the fluoride by volatilization; the additional work required to make the determination a quantitative one is not much more.

As in the case of chlorine it is important that the fluorine determinations be protected from contamination by laboratory reagents such as

* References for this section will be found on pp. 256–258.

hydrofluoric acid.   In particular, bottles of this acid should be removed from the vicinity when alkaline, fluoride-bearing solutions are being evaporated.

Because almost all methods for the determination of fluorine are subject to many interferences, the separation of the fluorine is the first step in the determination.   This is generally achieved by steam distillation, pyrohydrolysis or, more recently, by ion-exchange.   Once the fluorine is separated, the determination may be completed by a variety of titrimetric, spectrophotometric or gravimetric methods.   Bakes and Jeffery (Supp. Ref.) have recently described the determination of fluorine in mill products by neutron activation without prior separation; the ^{16}N produced by the (n, α) reaction is measured.   There has been much written about the determination of fluorine and this has been extensively reviewed recently by Horton,[19] who has covered the literature up to 1959. Because of the availability of this review, it is only necessary here to summarize briefly the methods of separation and some methods of determination now in use, and to mention some papers published since 1959* which deal specifically with the determination of fluorine in rocks and minerals.

As in the case of chlorine and sulfur, it is necessary that a correction be applied for the oxygen equivalent of fluorine when present as fluoride, if the constituents are reported as oxides; the correction is as follows:

$$CaF_2 \equiv CaO$$

$$\frac{O}{2F} = 0.42$$

6.16.2.   SEPARATION OF FLUORINE

(a) By steam distillation

Fluorine is most probably[7,8,24] evolved as a mixture of SiF_4 + HF when the sample, in one form or another, is heated with one or more acids while steam is passed through the mixture, and is collected in the condensate.

1. Some samples may be used without prior treatment but for most silicates it is preferable to fuse the sample with an alkaline flux, remove

* Particular mention must be made, however, of the earlier work of Shell and Craig,[25,26] who made a detailed study of the steam distillation and titrimetric determination of fluorine in fluorosilicates.

the bulk of the silica (and alumina) from the resultant solution* and to steam distill the filtrate directly, or after concentrating it to a desirable volume.† The gelatinous silica that separates on acidification, if the bulk of it is not first removed, will not only retard the recovery of fluorine but will cause undesirable "bumping" of the contents of the distillation flask; removal is usually achieved by precipitating the silica with Zn(II), either by adding it after an alkaline carbonate fusion or, better, by fusing the sample with a mixed alkaline carbonate–zinc oxide flux (see Secs. 6.2.1a, b, and 9.3). However, Blake[4] has reported no difficulty with SiO_2 when a sodium carbonate fusion alone was used; Moyzhess[21] found no interference if the SiO_2 content of the sample was less than 40 mg.

2. There is much disagreement over the choice of acid for the decomposition, as there is also over other procedural details.[28] Sulfuric and perchloric acids are most frequently used, with perchloric being the least troublesome.‡ A higher distillation temperature can be used with sulfuric acid (and with phosphoric acid also), however, thus decreasing the distillation time required for complete recovery of the fluorine, but the presence of these acids in the distillate is undesirable (indeed, the use of sulfuric acid has been definitely not recommended[19]). Because the higher distillation temperature is necessary to ensure the complete recovery of fluorine from some samples, e.g., aluminiferous ones [4,8,16] some analysts prefer to eliminate the harmful volatiles from the distillate by a second distillation at a lower temperature[5,9]; Fox and Jackson[8] have described a simultaneous double distillation in which the uncondensed gas from the first still, at 160°C, is passed through perchloric acid, at 125°C, in a second distilling flask. Other precautions include the use of stills with special splash-traps to remove entrained acid,[16,27] and passing the distillate through an anion-exchange column from which the fluoride is later eluted preferentially.[23] Phosphoric acid is sometimes used alone but more often as a mixture with perchloric or sulfuric acids, usually for the purpose of eliminating the adverse effect of aluminum in solution on

---

* Ingamells[16] describes a rapid method involving direct distillation from H_3PO_4 after fusion of the sample with Na_2O_2; it cannot be used with samples containing other volatile elements which will interfere in the subsequent determination of fluorine.

† Such evaporations must *not* be done in borosilicate vessels but in ones of platinum or vitreous silica.

‡ Ingamells,[16] who prefers H_2SO_4 or H_3PO_4 as the decomposition acid, states that the presence of some $HClO_4$ improves the recovery of the fluorine, possibly because it distills over with the fluorine and minimizes adsorption of fluoride by glass surfaces.

the recovery of fluorine. It is thought that a strong complex is formed between the aluminum and fluoride which reduces the partial pressure of fluorine in the gas phase[8] but opinions differ on the efficacy of phosphoric acid as a releasing agent. Blake[4] investigated the use of both H_2SO_4 and $HClO_4$, alone and mixed with H_3PO_4, for samples rich in aluminum and obtained the best separation of fluorine with an H_2SO_4–H_3PO_4 mixture; Moyzhess[21] tested the evolution of fluorine from a similar acid mixture in the presence of a variety of elements which form stable fluoride complexes, including zirconium and thorium, and found no interference. Others[5,16,24,27] recommend the use of H_3PO_4 in the distilling flask, usually as a complexing agent for aluminum, but Fox and Jackson[7,8] are opposed to its use.

3. The optimum temperature for distillation[5,9,19,22] ranges from 130–150°C, although there is considerable disagreement on the exact temperature to be used. A higher temperature than 150°C is recommended for aluminiferous samples[4] up to 160–170°C if a second distillation is to be made.[5] Ingamells[16] determines the optimum temperature for his special still by experiment, which is that temperature at which no harmful amount of distillation acid is also distilled. Moyzhess[21] distilled solutions containing H_2SO_4 and H_3PO_4 at 140–190°C and found only an insignificant amount of these anions in the distillates collected at 140–150°C, and so chose 140 ± 5°C as his optimum temperature.

4. While most of the fluorine is usually collected in the first 50 ml or so of the distillate, the presence of such elements as aluminum, boron, zirconium or thorium, to mention only a few, dictates that a larger volume must be collected in order to be certain of complete recovery. Thus it is usual to collect 150–250 ml; Ingamells[16] recommends that a second 100 ml of distillate be collected if the fluorine content exceeds 0.5–1%. Peck and Smith[22] collect about 200 ml but recommend that a 3% (relative) correction be added to the results to allow for retention of some fluorine by interfering compounds and adsorption on glass surfaces.

5. Suffice it to say that the apparatus recommended by various analysts ranges from the relatively simple to the very complex and the references cited should be consulted for further details. The apparatus used at the Geological Survey of Canada is described in Sec. 9.32. The apparatus should be thoroughly cleaned between distillations to remove traces of fluoride adsorbed on glass surfaces; hot, 10% Na_2CO_3 solution is an effective cleaning agent.

(b) By pyrohydrolysis

The pyrohydrolytic or pyrolytic method for the evolution of fluorine (and chlorine) is a beautifully simple procedure that is destined to replace the distillation methods now in use. Experience with the procedure at the Geological Survey of Canada has been tentative only and the following comments are based chiefly upon recent pertinent publications. The sample, mixed with an accelerator (catalyst) and flux to break down the structure and facilitate the release of fluorine, is placed in a combustion boat and heated to an elevated temperature in a reactor tube through which passes a fairly rapid stream of either superheated steam or oxygen saturated with water vapor. The gaseous products are swept through either a dilute sodium hydroxide solution or through water, the hydrated fluorine compounds are condensed and adsorbed, and the determination finished titrimetrically or colorimetrically. The boat and reactor tube may be of platinum, nickel, quartz or fused silica and the operating temperature ranges from 800–1000°C. Uranium(IV,VI) oxide, WO_3 and V_2O_5 serve as accelerators and sodium tungstate is often used as a flux. Horton has discussed the various aspects of the procedure in his review.[19]

Bennett and Hawley[2,3] have described their procedure for aluminosilicates in some detail, including the apparatus used. They found that V_2O_5 was the best catalyst and that the minimum temperature for complete recovery of the fluorine was 850°C. Cluley[6] used U_3O_8 as accelerator and heated samples of opal glass to 1000–1050°C to ensure complete release of the fluorine; his apparatus is very similar to that used by Bennett and Hawley.

While no interference is found from silica and phosphate, which remain in the combustion boat, other volatile anions present, such as SO_4^{2-}, NO_3^- and Cl^-, will be carried over and condensed with the fluorine; the method used for the final determination of the fluorine must take into consideration their possible presence, but may be as simple as an acid-base titration.

(c) By ion exchange

Mention has been made of the use of ion exchange by Sergeant[23] to eliminate anionic interferences in the distillate prior to the colorimetric determination of fluorine. The resin, Deacidite FF (chloride form, 100–200 mesh), is placed in a glass column positioned to receive the distillate as it leaves the condenser of the distillation apparatus. The adsorbed fluoride is eluted with $N/10$ sodium acetate and almost 100% recovery is found for up to 600 μg F^-. Glasö[11] preferred ion exchange to distillation as a means for the separation of fluoride but the procedure as described is applicable only to those samples in which the fluoride is present in acid-soluble form, e.g., iron ore, apatite and phosphate rock. The sample, dissolved in warm hydrochloric acid, is passed over Dowex 2-X 10 resin (chloride form, 200–400 mesh) and the fluoride eluted with 10M HCl.

6.16.3. DETERMINATION OF FLUORINE

(a) Titrimetric

The titrimetric determination of fluorine almost always means visual titration of the fluoride-bearing distillate with a solution of thorium nitrate, in the presence of an indicator which again is usually Alizarin Red S (sodium alizarin sulfonate). In spite of the uncertainty of the end point, which requires much practice and perseverance, it is still the most widely used procedure.[28] Horton has also considered the method in detail[19]; numerous variations in operating conditions, including the proper pH, concentration range and composition of the solution to be titrated (i.e., aqueous or alcoholic) have been suggested. In general, a pH of 3.0 is favored (some analysts prefer 3.3–3.4) and this is maintained during the titration (the pH tends to decrease as the titration proceeds) by the use of a buffer, usually sodium monochloracetate, or by the drop-wise addition of dilute sodium hydroxide to maintain the pH at 3.0 as indicated by a pH meter during the titration. The color change at the end point (and the determination of the end point blank is very necessary) is from yellow to pink, as the first excess of thorium forms the thorium lake of alizarin sulfonic acid; further addition of thorium changes the color to rose red. Serious interference is caused by sulfate, phosphate, borate and arsenate.

Other titrimetric procedures have been proposed, including a simple acid-base titration, with phenolphthalein as indicator, following pyrohydrolysis. The thorium nitrate titration may be used with SPADNS, the sodium salt of 2-(p-sulfophenylazo)-1,8-dihydroxynaphthalene-3,6-disulfonic acid, which also forms a thorium lake; the end point color change (crimson to blue violet) is best observed spectrophotometrically. A simple amperometric titration of fluoride ion by thorium nitrate is not possible without recourse to special techniques (Stock, Supp. Ref.); a conductometric end point has been investigated by Israel, Bernas and Yahalom (Supp. Ref.).

(b) Colorimetric

There are many suggested methods for the colorimetric and spectrophotometric determination of fluorine, but only a few of these have been applied to silicates. Methods have been based upon the bleaching effect of fluorine on certain complexes, such as the yellow peroxytitanic ion and the red zirconium Alizarin Red S complex, upon the formation of colored complexes or lakes such as that previously described for the thorium–SPADNS–fluorine system, and upon the liberation of the free colored chromophore by the complexing action of fluoride on a chromophoric metal system such as lanthanum or thorium chloranilate. These methods have been covered in detail in recent texts.[15,19,28]

When fluoride ion is added to a solution of zirconium Eriochrome Cyanine R, the color of the complex is reduced and this has been used for the determination of fluorine in silicate rocks[23] and in iron ore, apatite and phosphate rock[11]; Ingram[17] has described modifications of the thoron (o-(2-hydroxy-3,6-disulfo-1-naphthyazo) benzenearsonic acid disodium salt) bleaching procedure which increases the reliability of the method in the 0–20 μg range.

The red cerous chelate of alizarin complexone (1,2-dihydroxy-anthraquinonyl-3-methylamine-N,N-diacetic acid) changes to blue in the presence of fluoride ion and was used by Jeffery[18] to measure the fluorine content of some reference samples, including G-1 and W-1; the fluoride was first isolated by distillation. The method is relatively free from the interferences which affect other color systems; large amounts of nitrate, chloride and perchlorate, and small amounts of borate, silicate and sulfate, are without adverse effect. The stability and sensitivity of the reagent are enhanced by developing the color in an acetonitrile–water or acetone–water medium.[29] Greater sensitivity is said to be obtained by replacing the cerium in the complex with lanthanum[12]; only borate and silicate interfere in the microdetermination of fluorine (2–25 μg F$^-$) and this can be overcome by collecting larger volumes of distillate. This latter procedure has been adapted for use in an automated apparatus (Chan and Riley, Supp. Ref.).

Peck and Smith[22] have developed a simple colorimetric method that utilizes a zirconium–SPADNS–sulfuric acid solution as the single reagent following distillation of the fluorine, for fluoride contents up to about 1%. The transmittance of the complex is sharply affected by temperature changes and special precautions are taken to minimize this effect. The method is described in more detail in Sec. 9.32.2.

(c) Spectrographic

The quantitative determination of fluorine by emission spectrography usually involves the measurement of the calcium fluoride (CaF) band emission at 5291 Å, although barium or strontium fluoride emission is sometimes used instead.[1,19]   A detailed procedure using CaF emission is given for chondritic meteorites[13]; the silicate–sulfide portion of the sample is mixed with $CaCO_3$ and graphite, after removal of the metallic fraction, and the CaO band at 5473 Å is employed as internal standard. The CaF emission was found to be particularly dependent on the sample matrix and a standard curve was constructed from materials of similar composition to the samples.

References

1. Ahrens, L. H., and Taylor, S. R. *Spectrochemical Analysis*, 2nd ed., 1961. Reading, Mass.: Addison-Wesley, pp. 277–282.
2. Bennett, H., and Hawley, W. G. The determination of small quantities of fluorine by pyrohydrolysis. *Trans. Brit. Ceram. Soc.* **62**, 397–404 (1963).
3. Bennett, H., and Hawley, W. G. *Methods of Silicate Analysis*, 2nd ed., 1965. London: Academic Press, pp. 97–98, 288–294.
4. Blake, H. E., Jr. Fluorine analyses. *U.S. Bur. Mines Rept. Invest.* **6314**, 29 (1963).
5. Boltz, D. F., (Ed.). *Colorimetric Determination of Nonmetals*, 1958. "Fluorine" by S. Megregian. New York: Interscience, pp. 231–259.
6. Cluley, H. J. The rapid determination of fluorine in opal glass by pyrohydrolysis. *Glass Technol.* **2**, 74–78 (1961).
7. Fox, E. J. Discussion of "Steam distillation of fluorine from perchloric acid solution of aluminiferous ores." *Anal. Chem.* **32**, 1530 (1960).
8. Fox, E. J., and Jackson, W. A. Steam distillation of fluorine from perchloric acid solutions of aluminiferous ores. *Anal. Chem.* **31**, 1657–1662 (1959).
9. Geijer, P. Halogens in skarn amphiboles. *Arkiv Mineral. Geol.* **2**, 483–485 (1960).
10. Gillberg, M. Halogens and hydroxyl contents of micas and amphiboles in Swedish granitic rocks. *Geochim. Cosmochim. Acta* **28**, 495–516 (1964).
11. Glasö, O. S. Determination of fluorine in iron ore and apatite. *Anal. Chim. Acta* **28**, 543–550 (1963).
12. Greenhalgh, R., and Riley, J. P. The determination of fluorides in natural waters, with particular reference to sea water. *Anal. Chim. Acta* **25**, 179–188 (1962).
13. Greenland, L., and Lovering, J. F. Minor and trace element abundances in chondritic meteorites. *Geochim. Cosmochim. Acta* **29**, 821–858 (1965).
14. Grillot, H., Béguinot, J., Boucetta, M., Rouquette, C., and Sima, A. Méthodes d'analyse quantitative appliquées aux roches et aux prélèvements de la prospection géochimique. *Mém. Bur. Recherches Géol. Min.* **30**, 51–53 (1964).
15. Hillebrand, W. F., Lundell, G. E. F., Bright, H. A., and Hoffman, J. I. *Applied Inorganic Analysis*, 2nd ed., 1953. New York: John Wiley & Sons, pp. 737–748, 938–948.
16. Ingamells, C. O. The application of an improved steam distillation apparatus to the determination of fluoride in rocks and minerals. *Talanta* **9**, 507–516 (1962).
17. Ingram, B. L. Spectrophotometric determination of fluorine with thoron. *U.S. Geol. Surv. Profess. Papers* **450–E**, E130 (1963).

18. Jeffery, P. G. The fluorine content of some standard samples. *Geochim. Cosmochim. Acta* **26**, 1355–1356 (1962).

19. Kolthoff, I. M., and Elving, P. J., (Eds.). *Treatise on Analytical Chemistry,* Part II, Vol. 7, 1961. "Fluorine," by C. A. Horton. New York: Interscience, pp. 207–334.

20. Kolthoff, I. M., and Sandell, E. B. *Textbook of Quantitative Inorganic Analysis,* 3rd ed., 1952. New York: Macmillan, pp. 721–723.

21. Moyzhess, I. B. Methods of analysis of mineral raw materials. *All-Union Sci. Res. Geol. Inst., Leningrad* **117**, 27–32 (1964).

22. Peck, L. C., and Smith, V. C. Spectrophotometric determination of fluorine in silicate rocks. *Talanta* **11**, 1343–1347 (1964).

23. Sergeant, G. A. Geological Survey of Great Britain. *Rept. of Govern. Chemist, Dept. Sci. Ind. Res., London* **1963**, 61–62 (1964), and private communication.

24. Shell, H. R. Discussion of "Steam distillation of fluorine from perchloric acid solutions of aluminiferous ores." *Anal. Chem.* **32**, 1529–1530 (1960).

25. Shell, H. R., and Craig, R. L. Determination of silica and fluoride in fluorsilicates. *Anal. Chem.* **26**, 996–1001 (1954).

26. Shell, H. R., and Craig, R. L. Synthetic mica investigations: VII. Chemical analysis and calculation to unit formula of fluorsilicates. *U.S. Bur. Mines Rept. Invest.* **5158**, 30 (1956).

27. Wade, M. A., and Yamamura, S. S. Microdetermination of fluoride using an improved distillation procedure. *Anal. Chem.* **37**, 1276–1278 (1965); the apparatus has since been further modified to minimize the carry-over of perchloric acid (private communication).

28. Wilson, C. L., and Wilson, D. W., (Eds.). *Comprehensive Analytical Chemistry,* Vol. 1C (Classical analysis), 1962. "Fluorine," by A. M. G. Macdonald. Amsterdam: Elsevier, pp. 319–340.

29. Yamamura, S. S., Wade, M. A., and Sikes, J. H. Direct spectrophotometric fluoride determination. *Anal. Chem.* **34**, 1308–1312 (1962).

Supplementary References

Bakes, J. M., and Jeffery, P. G. The determination of fluorine by neutron activation. *Analyst* **91**, 216–217 (1966).

Berthelay, J. C. Dosage spectrophotometrique du fluor dans les roches et les minéraux. *Ann. Faculté Sci. Univ. Clermont, Fasc. 11, Géochimie I* **28**, 3–11 (1966).

Chan, K. M., and Riley, J. P. The automatic determination of fluoride in sea water and other natural waters. *Anal. Chim. Acta* **35**, 365–369 (1966).

Evans, W. H., and Sergeant, G. A. The determination of fluorine in rocks and minerals. *Analyst* **92**, 690–694 (1967). The procedure briefly described in the text (Ref. 23) is given in more detail.

Huang, P. M., and Jackson, M. L. Fluorine determination in minerals and rocks. *Am. Mineral.* **52**, 1503–1505 (1967). Distillation from $HClO_4$ at 140–150°C is used to separate fluorine prior to its colorimetric determination.

Huang, W. H., and Johns, W. D. The chlorine and fluorine contents of geochemical standards. *Geochim. Cosmochim. Acta* **31**, 597–602 (1967). See also *Anal. Chim. Acta* **37**, 508–515 (1967). The sample, fused with sodium carbonate and zinc oxide at 900°C, is leached with water, filtered and the filtrate acidified with nitric acid. An aliquot is used for the determination of fluorine by Erio-chrome Cyanine R and zirconium oxychloride.

Israel, Y., Bernas, B., and Yahalom, A. Conductometric end point detection in the titration of fluoride and fluosilicic acid with thorium nitrate. *Anal. Chim. Acta* **36**, 526–529 (1966).

Selig, W. An improved end-point for the determination of fluoride with thorium nitrate. *Analyst* **93**, 118–120 (1968). Methylthymol blue, in a solution buffered with glycine-perchlorate (pH 3.35 ± 0.10), gives a deep-blue color at the end point.

Sen Gupta, J. G. Determination of fluorine in silicate and phosphate rocks, micas and stony meteorites. *Anal. Chim. Acta* **42**, 119–125 (1968). Modifications to the method of Huang and Johns are proposed.

Shapiro, L. A simple and rapid indirect determination of fluorine in minerals and rocks. *U.S. Geol. Surv. Profess. Papers* **575-D**, D233–D235 (1967). A 5 mg (35 mg if F < 1%) sample is treated with H_3PO_4 in a borosilicate test tube and the SiO_2 released by the action of the HF on the glass is determined colorimetrically.

Stock, J. T. *Amperometric Titrations*, 1965. New York: Interscience, pp. 303–309.

Van Loon, J. C. The rapid determination of fluoride in mineral fluorides using a specific ion electrode. *Anal. Letters* **1**, 393–398 (1968). Prior separation of fluoride is not necessary. The sample is fused with NaOH in a silver crucible and the H_2O leach is carefully adjusted to pH 7.0–8.0 with HCl before determination of the F^- using the specific ion electrode and a pH meter in the mV mode. No interference is found from cations, nor from either H^+ or OH^- at pH 7–8.

6.17. Strontium and barium*

6.17.1. GENERAL

The determination of these elements is not usually part of a routine rock analysis, unless they are present in unusually high concentrations, but for highly accurate work their separation and/or determination are necessary if correct values are to be obtained for calcium and magnesium. It is the practice at the Geological Survey to determine these elements by optical emission spectroscopy, together with chromium, nickel, cobalt and vanadium, on most samples submitted for a conventional chemical analysis so that the appropriate precautions may be taken if such are shown to be necessary. Strontium seldom exceeds 0.1% in concentration and probably occurs as a replacement for calcium or potassium in minerals, favoring as a rule the calcium-rich rocks. Barium, on the other hand, is present to a greater extent in potassium-rich rocks, because it readily replaces potassium in a variety of minerals; it is usually below 0.2% in concentration. If forms a rare feldspar, celsian ($BaAl_2Si_2O_8$), and partly replaces the potassium in hyalophane, a variety of orthoclase; it may also be present as barite, $BaSO_4$.

* References for this section will be found on pp. 261, 262.

In the course of the analysis of a silicate rock or mineral, as previously outlined, strontium will accompany calcium almost quantitatively in its reactions (Sec. 6.5.1a). For most purposes it will be sufficient to correct the value obtained for $CaO + SrO$, whether obtained by gravimetric precipitation as oxalate or titrimetrically with EDTA or $KMnO_4$, on the basis of the Sr content found by optical emission spectroscopy; if greater accuracy is desired the strontium may be separated from the calcium (and barium) and determined gravimetrically or it may be determined, without preliminary separation, by flame photometry or other methods (Sec. 6.5.1a). Methods of separation have been discussed in the section on calcium (Sec. 6.5.1a, Refs. 12 and 25).

Barium will separate with the silica during dehydration, if sulfate is also present (Sec. 6.2.1), and will be weighed with the R_2O_3 group residue and counted as Al_2O_3 unless its presence is detected (e.g., the weight of the nonvolatile residue from the silica determination is abnormally high). If present in unusually high concentration, some barium may be precipitated with the R_2O_3 group; it can be recovered from the ignited residue after the pyrosulfate fusion and a correction applied to the Al_2O_3 value. In most analyses the barium content will be low enough (i.e., $<0.5\%$) that it will all pass into the filtrate after the second oxalate precipitation of calcium (and strontium) but if it accompanies the calcium, it will be weighed as $BaCO_3$ (provided that the ignition at 1000°C is not continued longer than 10 min); the residue can be dissolved in HCl and the barium separated as $BaSO_4$ from a volume of solution sufficiently large to keep $CaSO_4$ in solution. If not previously separated, barium will coprecipitate with magnesium during the ammonium phosphate precipitation, unfortunately not quantitatively, and will be weighed as a mixture of $Ba_2P_2O_7$ and $Ba_3(PO_4)_2$; because of these uncertainties in completeness of precipitation and composition it is not possible to coprecipitate the barium with magnesium, as is done for small amounts of calcium, and it must be separated as $BaSO_4$ before the precipitation is made. The small amount of calcium and strontium that escaped precipitation as oxalate will also be recovered with the $BaSO_4$ (Secs. 9.10.1, 9.10.2), and a separation of barium and strontium can be made on the basis of the differential solubility of their respective dry nitrates in fuming nitric acid, or absolute alcohol and absolute ether (see Sec. 6.5.1a); barium is then separated by precipitation as barium chromate and can be weighed as such.

6.17.2. STRONTIUM

Strontium may be determined gravimetrically as the nitrate, after separation from calcium,[7] but it is better to make the final determination as strontium sulfate, as is done following the separation of barium as barium chromate.[7]   Other gravimetric procedures are applicable also[11] but require a preliminary separation of the strontium and have little advantage over the sulfate as a weighing form.   Similarly, titrimetric methods find little application in rock analysis.

A spectrophotometric procedure, using murexide (ammonium purpurate), has been proposed for microgram quantities of strontium[9] but most of the elements associated with Sr (Mg, Ca, Ba, K, Li) interfere and the strontium must first be isolated, which limits the usefulness of the procedure in rock analysis.

Strontium is rapidly and conveniently determined by flame photometry at 460.7 mμ provided that precautions are taken to minimize the depressant effects of other elements present in the solution.[5]   Dinnin[6] has described the use of praseodymium as a releasing agent for strontium in overcoming the depressive interference of associated elements commonly found in silicate rocks and minerals; the addition of an acetic acid solution of 8-hydroxyquinoline to the strontium solution is recommended also for this purpose.[12]   Similarly, flame absorption spectroscopic techniques have been applied to the determination of Sr in geological materials[3,10] and the use of optical emission spectroscopy has been mentioned previously.   The state of the analytical chemistry of strontium, to 1961, has been reviewed by Vinogradov and Ryabchikov,[11] and more recently by Turekian and Bolter (see Kolthoff and Elving, Supp. Ref.).

6.17.3. Barium

Precipitation as $BaSO_4$ from a slightly acid solution containing an excess of sulfate ion is the most widely used procedure for the determination of barium, and the comments made about the conditions of precipitation, filtration and ignition of barium sulfate in a previous section (Sec. 6.14.2) are pertinent here.   Barium may be determined directly on a separate portion of the sample (often combined with the determination of other constituents such as total sulfur, zirconium and the rare earths[7,8]—see Sec. 9.30.1) by decomposing the sample with a mixture of HF and H_2SO_4, or by fusing it with Na_2CO_3, leaching and dissolving the residue in H_2SO_4 to precipitate the barium as $BaSO_4$; in each case it is necessary to fuse the impure $BaSO_4$ with Na_2CO_3 and repeat the leaching and precipitation in order to obtain a pure precipitate.   It may also be recovered at certain stages in the main portion analysis (Sec. 6.17.1); Bennett and Hawley[2] describe the addition of a pyrosulfate fusion of the nonvolatile SiO_2 residue to the silica filtrate to collect all of the barium at the same time.   Precipitation of barium sulfate may, incidentally, be conveniently done in homogeneous solution through the hydrolysis of sulfamic acid or dimethyl sulfate.[13] Nishimura (Supp. Ref.) has investigated the rapid weighing of $BaSO_4$ without ignition; the precipitate, in a sintered-glass filter crucible, is washed with alcohol and ether at room temperature.

If the barium must be separated from appreciable strontium, such as following the separation of barium and strontium from calcium by the nitric acid or absolute alcohol-absolute ether procedures, precipitation of the barium as chromate from a nearly neutral solution offers the best chance of a reasonably good separation. A double precipitation is usually necessary; the precipitation can be made in homogeneous solution containing Ba(II) and CrO_4^- through the hydrolysis of urea,[13] a single precipitation sufficing for the separation of 100 mg barium from 100 mg calcium and 50 mg strontium, but a double precipitation being required when the ratio of barium to strontium is less than 1:2. An investigation[14] of several procedures for this precipitation has shown that for relatively large amounts of barium and strontium, minimum coprecipitation of strontium is obtained if sodium chromate is added to a cold solution buffered at pH = 4.6, followed by boiling and cooling.

Little use is made of colorimetric methods for the determination of barium; Brobst and Ward[4] have described a rapid turbidimetric method for geochemical prospecting.

Trace amounts of barium are best determined by flame emission or optical emission spectroscopy, particularly if calcium and strontium are both present as well. The intensity of the emission of barium in the flame is less sensitive than for calcium and strontium[5] but the spectral sensitivity in optical emission spectroscopy is 1–5 ppm in rocks and minerals.[1] Turekian and Bolter (in Kolthoff and Elving, Supp. Ref.) have recently reviewed the analytical chemistry of barium, with emphasis upon ion-exchange, X-ray fluorescence, isotope dilution and neutron activation methods of determination.

References

1. Ahrens, L. H., and Taylor, S. R. Spectrochemical Analysis, 2nd ed., 1961. Reading, Mass.: Addison-Wesley, pp. 205–209.
2. Bennett, H., and Hawley, W. G. Methods of Silicate Analysis, 2nd ed., 1965. London: Academic Press, pp. 93–94.
3. Billings, G. K., and Adams, J. A. S. The analysis of geological materials by atomic absorption spectrometry. Atomic Absorption Newsletter, No. 23, 1–7 (1964). Perkin-Elmer Corporation, Norwalk, Conn. Part II. Idem., 4, 312–316 (1965).

262   ROCK AND MINERAL ANALYSIS

4. Brobst, D. A., and Ward, F. N.   A turbidimetric test for barium and its geologic application in Arkansas. *Econ. Geol.* **60,** 1020–1040 (1965).
5. Dean, J. A.   *Flame Photometry,* 1960. New York: McGraw-Hill, pp. 203–213.
6. Dinnin, J. I.   Flame photometric determination of strontium with the use of releasing agents. *U.S. Geol. Surv. Profess. Papers* **424–D,** D392–394 (1961).
7. Groves, A. W.   *Silicate Analysis,* 2nd ed., 1951. London: George Allen and Unwin, pp. 63–66, 165.
8. Peck, L. C.   Systematic analysis of silicates. *U.S. Geol. Surv. Bull.* **1170,** 50–51, 80–82 (1964).
9. Russell, D. S., Campbell, J. B., and Berman, S. S.   The spectrophotometric determination of strontium with murexide (ammonium purpurate). *Anal. Chim. Acta* **25,** 81–84 (1961).
10. Trent, D., and Slavin, W.   Factors in the determination of Sr by atomic absorption spectrophotometry with particular reference to ashed biological samples. *Atomic Absorption Newsletter,* No. 22, 1964, 12 pp; Perkin-Elmer Corporation, Norwalk, Conn.   Determination of various metals in silicate samples by atomic absorption spectrophotometry. *Idem.* **3,** 118–125 (1964).
11. Vinogradov, A. P., and Ryabchikov, D. I., (Eds.).   *Detection and Analysis of Rare Elements,* 1961. Moscow: Izdatel'stvo Akademii Nauk SSSR, translation by Israel Program for Scientific Translations, Jerusalem, 1962; pp. 112–134.
12. Voinovitch, I. A., Debras-Guedon, J., and Louvrier, J.   *L'analyse des silicates,* 1962. Paris: Hermann; pp. 347–350, 487.
13. Wilson, C. L., and Wilson, D. W., (Eds.).   *Comprehensive Analytical Chemistry,* Vol. 1C (Classical analysis), 1962. Amsterdam: Elsevier, pp. 83–85.
14. Wonsidler, G. J., and Sprague, R. S.   Determination of strontium coprecipitated with barium by X-ray emission spectroscopy. *Anal. Chim. Acta* **31,** 51–57 (1964).

Supplementary References

Kolthoff, I. M., and Elving, P. J., (Eds.).   *Treatise on Analytical Chemistry,* Part II, Vol. 4, 1966. "Strontium and Barium," by K. K. Turekian and E. Bolter. New York: Interscience, pp. 153–217.
Nishimura, M.   Weighing of barium sulfate precipitate without ignition. *Anal. Chim. Acta* **34,** 246–249 (1966).

6.18. Chromium, nickel and cobalt*

6.18.1. General

For the most part these three elements are present in concentrations not exceeding 0.01–0.05% each but in some rocks, particularly those rich in ferromagnesian minerals, they may attain the status of minor constituents.   Because they can affect the accuracy of the determination of other constituents, especially that of aluminum, it is important that their approximate concentrations be known and a preliminary spectrographic analysis of the sample is a useful first step.

* References for this section will be found on pp. 270, 271.

Chromium is usually found in ultrabasic rocks such as peridotite and pyroxenite, as chromite ($FeCr_2O_4$) or as picotite, the chrome spinel; it may also occur as the chromiferous varieties of many common rock-forming minerals (pyroxene, olivine, talc, serpentine) and the chromian mica, fuchsite, may contain up to 5% Cr_2O_3. *Nickel* may replace magnesium to some extent in some common silicate minerals (olivine, amphibole, mica) and is an important constituent in garnierite ($H_2(Ni, Mn)$ $SiO_4 \cdot n$ H_2O) and some hydrated magnesium silicates; it is also found as a natural alloy with iron, and as the sulfides millerite and pentlandite. *Cobalt* is generally associated with nickel but is much less abundant (Ni : Co $\cong$ 10 : 1). All three elements are commonly found in meteorites; nickel and cobalt are usually in the metal phase alloyed with iron, whereas chromium is found for the most part with the ferromagnesian minerals, or separately as chromite.

In the course of the analysis *chromium* will separate with the ammonia precipitate, and unless a correction is made, will be counted as Al_2O_3; because some oxidation of the chromium to form chromic anhydride will occur during the ignition in air, a correction made for Cr_2O_3 will be too low. If the total iron determination is made on the solution of the pyrosulfate fusion of the R_2O_3 precipitate, the choice of reducing agent is important if chromium is present; reduction of Cr(III) by zinc (i.e., with a Jones reductor) will give a high result for iron (and a low result for aluminum) but $SnCl_2$ does not reduce Cr(III) and no error will be introduced into the iron determination when it is used.

If only a moderate amount of *nickel* is present, such as 0.1% or so, all of it will be found in the filtrate from the magnesium determination provided that double precipitations are made. Larger amounts tend to be distributed throughout the main portion, including a small portion that may separate with the magnesium ammonium phosphate; it is usually safe to assume, however, that the deficiency in the amount of nickel found in the magnesium filtrate, compared with the total nickel content determined on a separate portion, is balanced by that which was separated with the R_2O_3 precipitate and an appropriate correction may be applied to the latter. A more direct approach is to determine the nickel content of the R_2O_3 precipitate in an aliquot of the solution of the pyrosulfate fusion. If much nickel is present it is better to separate it with ammonium sulfide (Sec. 9.9.1) following the precipitation of the R_2O_3 group with aqueous ammonia, in order to prevent it from separating with the calcium and magnesium precipitates in varying amounts; the nickel may then be determined in both the R_2O_3 group and the sulfide precipitate to give the total nickel content of the sample.

Cobalt generally behaves like nickel during the analysis of the main portion but tends to separate to a slight extent with both the R_2O_3 and magnesium precipitates regardless of its concentration in the sample. In most analyses the errors introduced by this are negligible but when the cobalt is present in unusually large concentration it is better to remove it with ammonium sulfide (Sec. 9.9.1) before continuing with the determination of calcium and magnesium.

6.18.2. CHROMIUM

The literature on the analytical chemistry of chromium is very extensive and has been reviewed in detail by Hartford.[8] Chalmers[12] has reviewed gravimetric and titrimetric methods for its determination.

Little use is made of gravimetric methods for the determination of chromium in silicates, chiefly because the quantity of chromium present is too low, but also because titrimetric methods are so reliable and convenient to use. As mentioned in Sec. 6.18.1, chromium will be separated and weighed with the R_2O_3 group but, because some oxidation of Cr(III) to Cr(VI) will occur during ignition of the hydrous oxide in air, the precipitation with aqueous ammonia is not recommended for the direct determination of chromium when it is present in an amount sufficient for such a determination. There are several other precipitants for the hydrous oxide, such as metal carbonates and oxides, urea, bromate–bromide and pyridine but none are of much use in silicate analysis.

For the usually low concentration of chromium present in most silicates a colorimetric method is best choice but for larger amounts, i.e., $>1\%$, a titrimetric method is to be preferred. Preliminary separation of the chromium prior to the titration is seldom necessary. The best-known procedure* involves the oxidation of the chromium to Cr(VI) and reduction of the latter to Cr(III) with Fe(II) and, if necessary, back titration of the excess Fe(II). The oxidation of the chromium is best accomplished by fusing the sample with a mixture of Na_2CO_3 and KNO_3, or with Na_2O_2; the cake is leached in water and the insoluble residue filtered off, leaving the chromium in the filtrate as chromate ion (unless appreciable barium or lead are also present). Decomposition of the sample in fuming perchloric acid will also oxidize the chromium to Cr(VI) but many silicates are not opened up by $HClO_4$ alone and the use of HF may result in some loss of chromium as a volatile fluoride; chlorides

---

* There are many other redox titrimetric methods, as well as complexometric and amperometric ones, but it is only seldom that there is sufficient chromium present to warrant a titrimetric approach and the analyst seeking other, and possibly better, methods should consult a more detailed text.

must be absent during the acid decomposition, for the same reason. It is usually preferable to ensure the complete oxidation of the chromium by a separate procedure, such as boiling with peroxide in alkaline solution or, preferably, by treatment in hot acid solution with permanganate, or with ammonium persulfate and silver nitrate. In most silicate analyses the determination of the chromium, if necessary, will be made on the sulfuric acid solution of the cake resulting from fusion of the ignited R_2O_3 group residue with pyrosulfate and a secondary oxidation of the chromium will be necessary. The reduction by Fe(II) may be done by adding an excess of a solution of ferrous ammonium sulfate and back-titrating the excess Fe(II) with either potassium permanganate or potassium dichromate, or as a direct titration of the Cr(VI) with Fe(II), the end point being determined potentiometrically or by use of an indicator such as ferroin. The only significant interference that is likely to be encountered is that from vanadium and manganese. Vanadium is quantitatively reduced from V(V) to V(IV) by the Fe(II), and because it is reoxidized to V(V) by permanganate, no error is introduced when the excess Fe(II) is back-titrated with $KMnO_4$; when potassium dichromate is used for the back-titration, or when no back-titration is used, the titration represents the sum of the chromium and vanadium present and must be corrected for the vanadium present on the basis of a separate determination of the latter (e.g., optical emission spectroscopy) or by determining it in the same solution by various methods, such as boiling the titrated solution with HNO_3 (about $3N$) for 1 hr, during which time the vanadium, but not the chromium, will be reoxidized and can be re-titrated with Fe(II) and the second titration subtracted from the first one. Manganese is converted to the dioxide when the alkaline peroxide solution is boiled for about 30 min, if oxidation is being done in alkaline solution. In acid solution, if the manganese content is high and a re-oxidation with ammonium persulfate and silver nitrate has been done, the permanganate formed may impart a distinctly reddish color to the dichromate solution, whereas it will be present as the dioxide if $KMnO_4$ was used to oxidize the chromium to Cr(VI); both of these forms are destroyed by boiling the solution with dilute hydrochloric acid, sodium chloride or sodium azide (NaN_3).

The method usually employed for the determination of chromium in silicates is a colorimetric one, involving either the measurement of the yellow CrO_4^{2-} anion, or of the red-violet complex formed by Cr(VI) with s-diphenylcarbazide. The *chromate* method is simple and suitable for an amount of $Cr_2O_3 > 0.01\%$; it readily follows an alkaline fusion of the sample or oxidation of the chromium in a solution having pH >8, and only uranium ($asUO_2^{2+}$) and cerium(IV) interfere by forming yellow complexes also (a yellow color may be due to platinum dissolved during an alkaline fusion in a platinum crucible). The separation of the chromium, as chromate, from other constituents such as iron, calcium and magnesium following an alkaline fusion and leaching of the cake in water is quantitative unless much chromite is present, when a second fusion of the insoluble residue may be necessary. The CrO_4^{2-} anion has a maximum absorbance at 366 mμ and can be measured spectrophotometrically, by visual comparison with a series of standards, or by colorimetric titration; the standard solutions should have about the same pH and salt concentration as the sample solutions, the chromium content of which should be in the range 5–100 ppm. The *diphenylcarbazide* complex is much more sensitive than the chromate anion and is used for determining chromium in solutions containing 0.2–10 μg of Cr(VI); only Mo(VI) gives a similar color but of such low sensitivity that up to 200 ppm molybdenum will not interfere. Both Fe(III) and V(V) form interfering yellow or brown compounds with the reagent but the iron is easily removed and the vanadium color fades rapidly enough that for V : Cr < 10 : 1, it is only necessary to wait about 10 min before measuring the absorbance of the red-violet chromium complex, which has maximum absorbance at 540 mμ. The final solution should have an acidity of 0.05–0.2N, preferably in H_2SO_4 (in HCl the interference from Fe(III) is greater and a separation will be necessary), and it is recommended that the reagent be added as a 0.25% solution in acetone. For further details about both of these colorimetric procedures reference should be made to Sandell[9] or to the review by Hartford.[8] Easton[6] found that when manganese is present in considerable excess of the chromium concentration (e.g., Mn : Cr > 10 : 1) it will interfere with the chromium diphenylcarbazide complex in spite of the addition of alcohol or sodium azide; reduction of the permanganate ion with EDTA (disodium salt) will completely remove the interference. Other colorimetric reagents have been proposed for the determination of chromium but have found little application in silicate analysis; Bennett and Marshall[2,3] used the violet complex formed by Cr(III) and EDTA to determine chromium in chrome ores and chrome-bearing refractories.

Much use has been made of optical emission spectroscopic methods to determine chromium in trace amounts (to 1 ppm) because of the high sensitivity that is attainable. Several spectral lines are suitable, the most sensitive being 4254.35 Å; for larger amounts of chromium (such as in ultrabasic rocks) Cr 2843 is usefully employed. Poor duplication of results will be found when the chromium is present as chromite unless the sample is very finely ground and thoroughly mixed. Further details are given in recent texts.[1,8] Little use is made of flame emission spectroscopy for the determination of chromium, because of the low sensitivity of the method in aqueous solution under normal conditions. Similarly, no application has been made of atomic absorption spectroscopy to the determination of chromium in geological materials. X-ray fluorescence spectroscopy will enable the detection of chromium down to about 50 ppm, but for quantitative determination it lacks the sensitivity of the optical emission spectroscopic and colorimetric methods.

6.18.3. NICKEL

Many reagents have been proposed for the determination of nickel by gravimetric, colorimetric, titrimetric, electrolytic and amperometric methods,[8,9,12] but in silicate analysis only the gravimetric and colorimetric procedures based upon the use of dimethylglyoxime have significant application and only they will be considered here.

The gravimetric dimethylglyoxime method is used for a nickel content >0.1% but because of the bulkiness of the red precipitate (Ni $(C_4H_7O_2N_2)_2$, containing 20.32% Ni) not more than 30–50 mg nickel should be handled at one time.* The precipitation is made in a hot, weakly acid solution containing the reagent by the addition of aqueous ammonia in slight excess; the precipitate thus formed is more easily filtered than that formed in cold solution. Tartaric or citric acid must be present to prevent the precipitation of members of the R_2O_3 group and the solution should contain ammonium salts if magnesium, alkaline earths, manganese or zinc are present in quantity. The reagent is usually added as a 1% solution in ethanol (its solubility in H_2O is only about 0.04 g per 100 ml) or in aqueous ammonia (13:7) and unnecessary excess of the alcoholic solution must be avoided because of the significant solubility of the nickel precipitate in ethanol, especially when the nickel content of the sample is high. The use of alcohol is not necessary if the water-soluble sodium salt of the reagent is used but this latter has the disadvantage of enhancing the interference of copper and cobalt at much lower concentrations than is the case when the dioxime alone is used. Only palladium is also partially precipitated with nickel but cobalt and

---

* By precipitation from homogeneous solution, 100 mg of nickel can be conveniently handled as a coarsely crystalline, compact precipitate; the acidified solution containing nickel, urea and sufficient reagent (1% solution of dimethylglyoxime in 1-propanol) is heated to hydrolyze the urea and raise the pH to the point at which precipitation occurs.[8]

copper tend to coprecipitate if present in large amounts (a double precipitation will eliminate the interference of these two latter elements); iron must be oxidized to Fe(III) because Fe(II) will form a red-colored compound in ammoniacal solution that may cause high results to be obtained. The precipitate, which has a solubility of <0.1 mg per 100 ml of cold water and about the same in a cold 20% alcohol solution containing 0.04 g of the reagent per 100 ml, can be dried at 110–120°C and weighed as nickel dimethylglyoxime; if it is suspected that some excess reagent has coprecipitated with the nickel salt, the excess can be volatilized by drying the precipitate at 150°C to constant weight.

The determination of trace amounts of nickel is conveniently done by utilizing the colored complex formed when dimethylglyoxime is added to a basic solution of nickel which has been previously treated with an oxidizing agent such as bromine or ammonium persulfate. The exact color will depend upon the pH of the solution to which the oxidizing agent is added; if it is neutral or faintly ammoniacal, the complex that forms on addition of the reagent (and aqueous ammonia, if necessary) will be wine-red, but if the original solution is acid, the color will be brown. Sandell[9] prefers this latter choice because the procedure is capable of giving highly reproducible results, provided that constant conditions are maintained and the color intensity is measured (at 445 mμ) within 5–10 min after mixing, so that the adverse effect of changes in the color intensity with time will be minimized. The system is not sensitive to changes in the quantities of the reagents used, and Beer's law holds for up to 5 ppm of nickel. Because the color is developed in a basic medium, citrate or tartrate must be present to prevent the precipitation of the R_2O_3 group constituents such as iron and aluminum; copper and cobalt will interfere only slightly but an appreciable amount of chromium will cause high results. Extraction of nickel dimethylglyoximate into chloroform will separate the nickel from most interfering metals; iron must be in the ferric state and hydroxylamine hydrochloride may be added to prevent manganese from inhibiting the extraction. Cobalt is extracted only slightly and this amount is removed by washing the organic extract with dilute aqueous ammonia. The nickel can be readily returned to an aqueous medium by shaking the organic phase with 0.5–1M HCl.

Much use has been made, of course, of optical emission spectroscopic methods[1] for the determination of traces of nickel. The element is very conveniently determined in this way, with a detection limit of about 2 ppm; the 3414 Å line is most often used but a wide selection of other lines is also available. Similarly, nickel has been determined in silicates by atomic absorption spectroscopy[4,10] with a sensitivity of 2 ppm; a minimum detection limit of 0.1 ppm is reported.

6.18.4. COBALT

As in the case of nickel, there are many methods and reagents which have been proposed for the determination of cobalt,[8,9,12,13] but their use in rock and mineral analysis has been limited. Cobalt is seldom present in more than trace quantity and macro methods, such as precipitation with 1-nitroso-2-naphthol or as potassium hexanitritocobaltate(III) and various titrimetric methods, have little application. It is usually necessary to separate the cobalt from ions such as Fe, Cu and Ni which seriously interfere in its determination, and separation methods such as solvent extraction or ion exchange lend themselves to a subsequent colorimetric determination of the cobalt, which can be done conveniently for amounts as low as 0.1 ppm. Optical emission spectroscopic methods[1] have again been widely used (lower limit about 2 ppm with the 3453 Å line), and much use has already been made of atomic absorption spectroscopy for the determination of traces of cobalt in soil and other geochemical studies.[10]

Among the reagents proposed for the colorimetric determination of cobalt, nitroso-R salt (sodium 1-nitroso-2-naphthol-3,6-disulfonate) is the best-known and most widely used. It forms a soluble red complex with cobalt at a pH of 5.5 ± 0.5, in either a hot acetic acid-sodium acetate, or a cold citrate-phosphate–borate, buffer. The absorbance is measured usually at 525–550 mμ (425 mμ is a more sensitive wavelength but the reagent also absorbs strongly here). The working range of cobalt concentration is generally 1–10 μg and Beer's law is obeyed. Because the cobalt complex is stable in acid solution, the complexes of other heavy metals can be decomposed by boiling the solution with nitric or hydrochloric acid. The citrate–phosphate–borate buffer medium is preferred[9]; the citrate prevents the precipitation of cobalt and other hydroxides. Ammonium salts tend to weaken the colored complex, Fe(III) forms a yellow-brown color in citrate medium if present in appreciable amount, and copper and nickel consume reagent. The interference from copper and nickel is minimized if the reaction is carried out at pH 8 (citrate medium) with excess reagent being added; the addition of potassium fluoride prior to that of the nitroso-R salt will also eliminate this interference, and that of iron(III) as well.[8] The Fe(III) can also be removed by extraction with ether, but if this is necessary, it is as easy to separate the cobalt from all interferences by extracting it from an ammonium citrate medium with dithizone in carbon tetrachloride or other solvents, and then, after destruction of the organic matter, to determine the cobalt colorimetrically; if copper is present in appreciable amount it will be necessary to extract it first with the dithizone-CCl_4 solvent from acid solution. Easton and Lovering[7] used an anion exchange resin to separate a small amount of cobalt from much iron and nickel in the metal phase of chondritic meteorites, before determining the former with nitroso-R salt in a hot acetate medium.

Strongly colored complexes are formed by both 1-nitroso-2-naphthol and 2-nitroso-1-naphthol which can be extracted with either carbon tetrachloride or chloroform; the second of these, 2-nitroso-1-naphthol, is preferred.*   The colored complex is formed at pH 2.5–5, with citrate present to prevent precipitation of metal hydroxides and also to complex Fe(III); reduction of the latter to Fe(II) is prevented by the addition of hydrogen peroxide.   Extraction with chloroform will remove not only cobalt but also nickel and copper; these latter complexes are not stable in strong acid and are removed by washing the organic phase with hydrochloric acid.   Excess reagent is removed by washing with sodium hydroxide.   Clark[5] studied this procedure in detail before applying it to the determination of 0–10 μg of cobalt in soils and rocks; he preferred to develop the colored complex in basic ammonium citrate solution at pH 8.5 and to extract the cobalt complex with isoamyl alcohol.   The interference of Cu(II) and Ni(II), which are also extracted with the cobalt, is eliminated by washing the organic extract with HCl and NaOH (which also remove excess reagent), that of Fe(II) by oxidizing it to Fe(III) with bromine, and that of Mn(II) and Mn(VII) by reducing and complexing them with sodium thiosulfate, which also eliminates the interference of palladium.

Voinovitch et al.[11] recommend the use of isonitrosomalonylguanidine for the colorimetric determination of cobalt in silicates; copper interferes seriously and must be removed.

References

1. Ahrens, L. H., and Taylor, S. R. Spectrochemical Analysis, 2nd ed., 1961. Reading, Mass.: Addison-Wesley, pp. 227–232.
2. Bennett, H., and Hawley, W. G. Methods of Silicate Analysis, 2nd ed., 1965. London: Academic Press, pp. 179–202.
3. Bennett, H., and Marshall, K. The separation and determination of chromium

* Sandell[9] notes the superiority of the extraction of cobalt from silicate solutions with 2-nitroso-1-naphthol over that with dithizone in CCl_4 but cautions about the possible interference of platinum which may be introduced during decomposition of the sample; Clark[5] found no interference from platinum.

sesquioxide in chrome ores and chrome-bearing refractories. *Analyst* **88,** 877–881 (1963).

4. Billings, G. K., and Adams, J. A. S. The analysis of geological materials by atomic absorption spectrometry. *Atomic Absorption Newsletter*, No. 23, 1–7 (1964). Perkin-Elmer Corporation, Norwalk, Conn.

5. Clark, L. J. Cobalt determination in soils and rocks with 2-nitroso-1-naphthol. *Anal. Chem.* **30,** 1153–1156 (1958).

6. Easton, A. J. The determination of chromium in the presence of Mn in rocks and minerals. *Anal. Chim. Acta* **31,** 189–191 (1964).

7. Easton, A. J., and Lovering, J. F. The analysis of chondritic meteorites. *Geochim. Cosmochim. Acta* **27,** 753–767 (1963).

8. Kolthoff, I. M., and Elving, P. J., (Eds.). *Treatise on Analytical Chemistry*, Part II, Vol. 8, 1963. "Chromium," by W. H. Hartford. New York: Interscience, pp. 273–378; *Idem.*, Vol. 2; "Cobalt and Nickel," by J. M. Dale and C. V. Banks, pp. 311–440.

9. Sandell, E. B. *Colorimetric Determination of Traces of Metals*, 3rd ed., 1959. New York: Interscience, pp. 388–401, 409–436, 665–681.

10. Slavin, W. The application of atomic absorption spectroscopy to geochemical prospecting and mining. *Atomic Absorption Newsletter* **4,** 243–254 (1965). Perkin-Elmer Corporation, Norwalk, Conn.

11. Voinovitch, I. A., Debras-Guedon, J., and Louvrier, J. *L'analyse des silicates,* 1962. Paris: Hermann, pp. 385–389.

12. Wilson, C. L., and Wilson, H. N., (Eds.). *Comprehensive Analytical Chemistry,* Vol. 1C (Classical analysis), 1962. "Chromium," by R. A. Chalmers. Amsterdam: Elsevier, pp. 581–588; "Cobalt," by H. Diehl, pp. 656–671; "Nickel" by W. T. Elwell and G. R. Sutcliffe, pp. 672–679.

13. Young, R. S. *Cobalt*, 1960. New York: Reinhold (Am. Chem. Soc. Monograph), pp. 372–398.

Supplementary Reference

Fuge, R., The determination of chromium in rocks and minerals. *Chem. Geol. 2,* 289–296 (1967). A 0.1 g sample is fused with Na_2CO_3 + MgO, leached, filtered and the filtrate made acid with H_2SO_4; chromium is determined as the diphenylcarbazide complex using an Autoanalyzer. Because glass was found to adsorb Cr, all evaporations should be made in polythene vessels and any glassware used should be soaked overnight in concentrated H_2SO_4

6.19. Vanadium*

Vanadium is a common trace constituent of silicate rocks and its concentration, in the basic varieties in particular, may exceed 0.05%. It is more abundant in silicate minerals such as the amphiboles, pyroxenes and micas (e.g., biotite); the vanadium mica roscoelite, $K_2V_4Al_2Si_6O_{20}$ $(OH)_4$, is the most important vanadium silicate. It may also occur as a sulfide, oxide, sulfate and vanadate in accessory minerals in silicate rocks.

* References for this section will be found on pp. 275, 276.

In most routine rock analyses the determination of vanadium can be omitted because its effect upon the accuracy of the values obtained for Al_2O_3 and Fe_2O_3 will be negligible. Its determination should be included in any general optical emission spectroscopic analysis, even if no correction is made to the Al_2O_3 for its presence; at the Geological Survey vanadium is included among the seven trace constituents (Mn, Ti, Cr, Ni, Ba, Sr and V) determined by quantitative optical emission spectroscopy in all samples submitted for conventional chemical analysis.

Vanadium is usually reported as V_2O_3 but it is more likely that it exists in more than one valence form in the host minerals. It is completely oxidized to V(V) during the $Na_2CO_3-Na_2O_2$ fusion of the sample but is reduced to V(IV) during the dehydration of the silica if hydrochloric acid is used as the solubilizing acid. If the combined filtrates are treated with bromine the vanadium is again oxidized to V(V). In any event, virtually all of the vanadium will be found in the R_2O_3 residue following precipitation with aqueous ammonia, provided that both iron and aluminum are present (iron is particularly important.) It is completely oxidized to V(V) during the ignition of the hydrous oxides, and while the vanadium content is reported as V_2O_3 in the final summation, the value for Al_2O_3 is later corrected for the presence of vanadium as V_2O_5 should this be necessary.

The correction of the Al_2O_3 value is thus a relatively simple matter when vanadium is present in only a trace amount. It is unfortunately not so simple when the vanadium content is more than this, because an error may be introduced into the determination of the total iron which will affect not only the latter result but that for Al_2O_3 as well. If the total iron is determined titrimetrically the magnitude of the error will depend upon several factors, such as the method of reduction of the Fe(III), the choice of oxidant, the temperature and acidity of the solution during titration and the oxidation–reduction indicator used (if one is used). For example, reduction of Fe(III) with H_2S, SO_2 or $SnCl_2$ will also reduce V(V) to V(IV) and the latter will be reoxidized to V(V) during titration of the Fe(II) with $KMnO_4$ (but *not* with $K_2Cr_2O_7$) *provided* that the acidity is not too high and the solution is heated to 50°C. Reoxidation of the V(IV) will result in a high value for total iron and a correspondingly low value for Al_2O_3, unless corrections are applied on the basis of a separate determination of vanadium. If the above titration is made at room temperature, in a strongly acid solution and with an oxidation–reduction indicator such as the *o*-phenanthroline-ferrous complex, negligible oxidation of V(IV) will occur. Reduction with a Jones reductor, or simply with pure zinc metal, will reduce the vanadium to V(II) and the errors then introduced into the Fe_2O_3 and Al_2O_3 values, if not corrected, will be much greater than when the other reductants were used. For further details the analyst should consult the references cited.[1,2,5]

Vanadium may also interfere in the colorimetric determination of titanium with hydrogen peroxide or tiron (Sec. 6.7.1), and in the precipitation of phosphorus as the phosphomolybdate (provided that vanadium is present as V(V) and the precipitation is made at an elevated temperature—see Sec. 6.9.1), but generally the effect will be negligible.

A convenient separation of vanadium from such accompanying elements as Fe, Cr(III), Ti, Zr and the rare earths is achieved by fusing the sample with Na_2CO_3 or Na_2O_2 and leaching it with water (H_2O_2 must be added as well to ensure that all vanadium is present as V(V) and to reduce Cr(VI) to Cr(III)). For best results the precipitate should be dissolved and the separation repeated. Gravimetric separations of vanadium with, for example, aqueous ammonia, mercurous nitrate or cupferron, are not selective but serve as a preliminary step before solvent extraction. Mercury cathode electrolysis is a convenient and simple means of removing interfering elements such as iron and chromium, especially when the latter are abundantly present.[3]

The titrimetric determination of vanadium, involving either an oxidation (V(IV) → V(V)) or a reduction (V(V) → V(IV)) in acid solution (e.g., 1:9 H_2SO_4) can be used for a wide range of vanadium concentrations. When interfering elements such as iron, arsenic and antimony are absent it is only necessary to reduce the vanadium to V(IV) with SO_2, expel the excess reductant with CO_2 and titrate the solution with $KMnO_4$ to the first tinge of pink. The titration may be made in the presence of chromium, but because Cr(III) is also slowly oxidized at elevated temperatures, it is necessary to make the titration at room temperature, and the end point is not as sharp (see Sec. 6.18.2 for the joint titration of vanadium and chromium). If the reduction is done with Fe(II) it is then possible to oxidize the excess Fe(II) with ammonium persulfate without reoxidizing V(IV), which is then titrated directly with either $KMnO_4$ or Ce(IV); the titration may be made potentiometrically without first oxidizing the excess Fe(II), and the first inflection will mark the oxidation of the Fe(II), the second (the solution should be heated to 50–60°C unless much chromium is present), that of the oxidation of V(IV). The green color of Fe(III) may be eliminated by adding H_3PO_4 or HF prior to a visual titration.

A reductimetric titration is also conveniently done. The complete oxidation of vanadium to V(V) is ensured by adding $KMnO_4$ to the solution until a permanent pink color is obtained; the excess oxidant is then eliminated by adding sodium nitrite and the reduction products are removed by the addition of sulfamic acid or urea. The titration may be made directly with a standard Fe(II) solution, either visually with an indicator such as barium diphenylamine sulfonate or phenanthroline-ferrous sulfate, or potentiometrically.

When the vanadium content is below 0.1%, the determination should be made colorimetrically and several sensitive methods are available.[2,4] The phosphotungstate method, in which the yellow phosphotungsto-vanadic acid complex is formed, is probably the best known and most widely used procedure, although not the one having the greatest sensitivity (it can be used to determine as little as 10 ppm vanadium, provided that the vanadium is first concentrated by extraction or freed from interfering ions by mercury cathode electrolysis). The strongest color development is obtained when the mole ratio of H_3PO_4 : Na_2WO_4 in the sample solution is in the range 3:1 to 20:1 and the concentration of Na_2WO_4 in the sample solution is $0.01M$ to $0.1M$; recommended[4] optimum concentrations are $0.05M$ H_3PO_4 and $0.025M$ Na_2WO_4. The colored complex may be developed in HNO_3, H_2SO_4 or HCl media and over a wide range of acid concentration but because the color tends first to be noticeably brown at higher acidities (a brown color is formed at lower acidity but is not as apparent) and changes slowly to yellow, the solution must be boiled to hasten the attainment of equilibrium. If the vanadium concentration of the sample solution is very low (<10 μg per ml), this latter step is not necessary because a small amount of brown coloration has only a negligible effect (provided that the acid concentration is not more than $0.5N$ and iron is absent). There are a number of possible interfering ions, such as those which give a color, those which form insoluble or slightly insoluble phosphates or basic salts (K, NH_4, Ti, Zr, Bi, Sb, Sn), and molybdenum(VI) which forms a yellow complex (see Sec. 6.9.2); this latter interference is serious only when molybdenum is present in a relatively high concentration. Ferric iron may be tolerated up to 10 mg/100 ml, but the solution must in this instance always be boiled. The phosphotungstovanadic acid complex is stable for at least 24 hr; the vanadium concentration of the solution used for final measurement should be in the range 10–250 μg.

Mention has already been made of the reddish-brown complex formed when an acid solution of V(V) is treated with hydrogen peroxide (Sec. 6.7.1). The yellow peroxy complex formed by the titanium present can be removed by bleaching it with HF, which will also remove any color due to Fe(III); Mo(VI) will interfere and Cr(VI) forms a blue color which, however, quickly fades if the chromium is present in only small amount. Extraction of chromium as perchromic acid with ethyl acetate is recommended by Fuge (Supp. Ref.) when the concentrations (ppm) Cr: V >500:100 (the ratio may be 10:1 when V>100 ppm, as in ultra-basic rocks); the vanadium, in aqueous solution, is determined colori-metrically by utilizing the catalytic action of vanadium on the oxidation of aniline to aniline black by H_2O_2, under very carefully controlled conditions.

Vanadium forms a magenta-black complex with 8-hydroxyquinoline which can be extracted with chloroform at a pH of 3.5–4.5 and used for the determination of vanadium. Other elements, such as Fe(III), Mo (VI), Cu and Al are also extracted but the vanadium can be separated from most of these by shaking the organic extract with an alkaline solu-tion (pH about 9.5) and transferring the vanadium to the aqueous phase, from which it can be extracted a second time with chloroform. Because some vanadium is lost during the double extraction it is important that the vanadium solutions used for the preparation of the working curve should be carried through the whole extraction procedure.

This extraction procedure may also be utilized for the removal of interfering ions prior to determination of the vanadium by some other means. Because some vanadium(V) is reduced during the extraction it is customary to evaporate the organic extract to dryness and to fuse the residue with sodium carbonate. Vanadium can also be extracted as diethyldithiocarbamate with chloroform at pH 3; aluminum is not extracted but iron and titanium will accompany the vanadium and it is necessary to shake the extract with dilute HNO_3 and H_2O_2 in order to remove the vanadium into the aqueous phase. Chan and Riley (Supp. Ref.) pass the sample through a cation-exchange column which retains the vanadium; it is eluted with dilute H_2O_2 and the final determination of the vanadium is done colorimetrically with diaminobenzidine.

Vanadium has an extensive and highly complex emission spectrum and much use is made of its determination at the trace level by optical emission spectroscopy, using dc arc excitation. This has been discussed in detail by Grady.[2] Some use has also been made of flame emission and atomic absorption spectroscopy; Sachdev et al. (Supp. Ref.) use an or-ganic extract with an oxygen–acetylene flame, but aspirate an aqueous solution directly into a nitrous oxide flame for absorption measurements.

References

1. Hillebrand, W. F., Lundell, G. E. F., Bright, H. A., and Hoffman, J. I. *Applied Inorganic Analysis*, 2nd ed., 1953. New York: John Wiley & Sons, pp. 452–463, 901–905.

2. Kolthoff, I. M., and Elving, P. J., (Eds.). *Treatise on Analytical Chemistry*, Part II, Vol. 8, 1963. "Vanadium," by H. R. Grady. New York: Interscience, pp. 177–272.
3. Page, J. A., Maxwell, J. A., and Graham, R. P. Analytical applications of the mercury electrode. *Analyst* **87**, 245–272 (1962).
4. Sandell, E. B. *Colorimetric Determination of Traces of Metals*, 3rd ed. (rev.), 1959. New York: Interscience, pp. 923–940.
5. Wilson, C. L., and Wilson, D. W., (Eds.). *Comprehensive Analytical Chemistry*, Vol. 1C (Classical Analysis), 1962. "Vanadium," by R. A. Chalmers. Amsterdam: Elsevier, pp. 542–550.

Supplementary References

Chan, K. M., and Riley, J. P. The determination of vanadium in sea and natural waters, biological materials and silicate sediments and rocks. *Anal. Chim. Acta* **34**, 337–345 (1966).
Fuge, R. The spectrophotometric determination of vanadium in rocks. *Anal. Chim. Acta* **37**, 310–315 (1967).
Jeffery, P. G., and Kerr, G. O. The determination of vanadium in silicate rocks and minerals with N-benzoyl-o-tolylhydroxylamine. *Analyst* **92**, 763–765 (1967). The reagent, which is almost specific for vanadium in 4–$8N$ HCl (violet color), is added as a CCl_4 solution and the colored complex is extracted with CCl_4. The interference of Ti is suppressed with NaF.
Sachdev, S. L., Robinson, J. W., and West, P. W. Determination of vanadium by atomic absorption spectrophotometry. *Anal. Chim. Acta* **37**, 12–19 (1967).

6.20. Zirconium and hafnium*

Zirconium is another trace element that is almost always present in silicate rocks but it usually occurs in such a small quantity that failure to determine it will introduce only a slight error into the value for aluminum, if the latter is determined by difference. It usually does not exceed 0.2% ZrO_2, but in very acid rocks and those having a high Na_2O content (for example, nepheline syenite, phonolite, tinguaite), its concentration may exceed 1%; if present as zircon ($ZrSiO_4$) in more than trace amount, it can usually be detected by the optical microscope but a preliminary optical emission spectroscopic examination is the most useful approach. In rocks having high contents of calcium, magnesium and iron, i.e., basic rocks, the Zr concentration is usually very low.

Zircon is the most common zirconium silicate mineral but there are many other more complex ones which have zirconium as a major constituent, such as eudialyte, wöhlerite, elpidite and rosenbuschite.[3,8] Zirconium is frequently associated with Ti, Nb, Ta and the rare earths, as

* References for this section will be found on pp. 281, 282.

in some of the above minerals, but it is also found in the more common amphiboles and in sodium-rich pyroxenes such as acmite.

Hafnium accompanies zirconium at all times, generally in the ratio Hf:Zr $\cong$ 2:100; only in thortveitite, a rare earth–scandium silicate, does the hafnium content exceed that of zirconium.

Because zirconium tends to hydrolyse readily, part of it may accompany the silica during its separation in the main portion analysis (Sec. 9.2 and footnote), and will be found, as ZrO_2, in the residue remaining after the volatilization of silicon with hydrofluoric acid. If the phosphorus content of the sample is sufficiently high, the zirconium may precipitate as the phosphate and accompany the silica in this form, decomposing to ZrO_2 during the ignition of the latter. Any zirconium remaining in the silica filtrate will precipitate during the separation of the hydrous oxides with aqueous ammonia, and all of it will be weighed with the R_2O_3 group and be counted as aluminum unless a correction is made for it. Zirconium will not, as a rule, cause difficulties in the analysis, beyond increasing the weight of the non-silica residue (an unusually large residue is often a good indicator of the presence of more than normal amounts of such elements as zirconium, titanium and phosphorus), but when present in higher than usual concentration and accompanied by sufficient phosphorus to precipitate it as the phosphate, it may cause difficulties in the filtration and ignition steps of the silica determination. It is better in this instance to leach the sodium carbonate fusion cake in water and to process the filtrate and residue separately. Hafnium, of course, accompanies zirconium in its reactions.

There are available several recent and detailed discussions of the analytical chemistry of zirconium and hafnium, particularly those by Hahn,[8] and Elinson[14]; Blumenthal[3] has covered the general chemistry, including the geochemistry. The following is a brief summary of the better known methods for the determination of zirconium and hafnium; for further details about these and other methods the references should be consulted. In the following discussion the methods described are for the determination of zirconium *plus* hafnium, unless specific indication to the contrary is given.

The separation of zirconium and hafnium from most other elements requires a combination of procedures. A separation from Si, Al, Cr, V and Mn is achieved by fusing the sample with Na_2CO_3 and leaching the cake in water. Mercury cathode electrolysis of the acid solution of the water-insoluble residue (H_2SO_4 or $HClO_4$) will remove such heavy metals as Fe, Ni, and Cu without adding anything to the solution. Extraction of the zirconium and hafnium as cupferrates in chloroform will separate them from Be, Al, Mg, Zn and uranium(VI); a useful separation from Al, Fe, Th, U and rare earths (i.e., R_2O_3 group residue) can be made in $6M$ HCl with a $0.5M$ solution of thenoyltrifluoracetylacetonate (TTA) in benzene or xylene. Culkin and Riley[5] used tri-n-butyl phosphate in isooctane to extract zirconium and hafnium from the solution of a silicate rock, together with cerium and thorium; these are all back-extracted into $0.1M$ oxalic acid solution and a separation of the zirconium and hafnium from cerium and thorium is made by ion exchange. Chromatographic separation of hafnium from zirconium has been described, and a separation may also be made by differential extraction using thiocyanic acid in ether, TTA or trifluoroacetylacetone in benzene.[14]

Although titrimetric and colorimetric methods are now generally preferred for the determination of zirconium and hafnium, especially for small amounts, the gravimetric approach has not been entirely neglected.[8,12,14,16] Precipitation with aqueous ammonia serves to separate zirconium and hafnium from a number of elements but it is hardly specific. Diammonium phosphate will precipitate them (Zr + Hf > 1 mg) from strongly acid solution (H_2SO_4, HCl) without interference from Al, Cr, Cu, Cd, Co, Mg, Mn, Ni, Zn and Fe(II) (less than 0.1 g Fe(III) will not interfere). Addition of H_2O_2 will prevent interference from Ti (if < 10 mg) but Nb and Ta must be removed before precipitation of zirconium; a double precipitation is necessary if thorium is also present.[7,12,15,16] The precipitate is ignited to ZrP_2O_7 at 1100–1150°C; if there is doubt about the purity of the residue it should be fused with potassium pyrosulfate and the zirconium reprecipitated as before. The tendency of the precipitate to hydrolyse during washing (ammonium nitrate solution should be used) may cause some PO_4^{3-} to be lost and as a result, the gravimetric factor for the calculation of zirconium is variable; 0.518 ($ZrP_2O_7 \rightarrow ZrO_2$) is suggested as an empirical factor when only small amounts of zirconium are involved.[14] The determination of zirconium can be combined with that of other constituents, e. g., total S, barium and rare earths.[6]

Mandelic acid is recommended[8,16] as a specific precipitant for Zr + Hf (0.2–500 mg) which will quantitatively separate the latter, in about 5N HCl or $HClO_4$, from Al, Cr, Ca, Cu, Fe, Mg, Th, Ti, U, V and Zn; the precipitate is transferred to a fritted-glass crucible, washed with alcohol and ether and dried in a current of air, and weighed as the mandelate. Derivatives of mandelic acid are more sensitive and more convenient to use while yet retaining equivalent selectivity,[16] but because they do not yield precipitates of constant stoichiometric composition[14] the precipitate must be ignited to ZrO_2 at 1000°C. The relative usefulness of mandelic acid and two derivatives, p-bromomandelic and m-nitromandelic acids, for the determination of small amounts of zirconium, has been investigated by Rafiq, Rulfs and Elving[10]; they found p-bromomandelic acid to be the most useful for the determination of 0.1 mg or more in 1–4M HCl, and to have the advantage over the other two precipitants that the washing of the precipitate can be done with hot water alone.

Among other precipitants for zirconium and hafnium are tannin,[12,16] phenylarsonic acid and cupferron. The latter is used by Bennett and Hawley[2] in the analysis of glazes and zirconium-bearing refractories and a mixed precipitate of Fe, Ti and Zr is obtained which is finally ignited to the oxides at 1200°C; the ZrO_2 is determined by difference following the determination of the iron and titanium in the solution of the pyrosulfate fusion of the oxides.

Titrimetric methods for the determination of zirconium (and hafnium) involve either a direct or an indirect titration with EDTA. A colored complex is first formed between zirconium and an organic reagent such as Eriochrome Cyanine, Chromeazurol S, SPADNS, Xylenol Orange or Pyrocatechol Violet at a pH of 1.5–2.5, after which the zirconium (and hafnium) is preferentially removed from the complex by EDTA, with complete discharging of the color at the end point. The results by direct titration have not, however, been too satisfactory because of difficulties that have been attributed to zirconium hydrolysis. Pilkington and Wilson[9] have shown recently that the inconsistencies occurring in direct titrations are attributable to the formation of polynuclear zirconium species ('polymerization') and that these can be overcome by the use of depolymerization procedures before the direct titration. Conversely, a measured amount of EDTA may be added and the excess of the latter back-titrated with a standard solution of bismuth or iron (III). For further details, the reviews by Milner[16] and Elinson[14] should be consulted.

A number of reagents have been proposed for the colorimetric determination of zirconium and hafnium. Alizarin sulfonic acid (Alizarin Red S), in 0.05M HCl or 1.0M $HClO_4$, can be used to determine 1–600 μg Zr^8; the complex is measured at 525 mμ and equilibrium is rapidly reached if the solution is heated for a few minutes in a boiling water bath. Fe(III) must be reduced (mercaptoacetic acid) and the thorium present must not exceed 10 mg. Culkin and Riley[5] use quinalizarin sulfonic acid to form, in not more than 0.5M $HClO_4$, an intense red complex that has about twice the sensitivity of that formed with Alizarin Red S; the maximum color development is reached after 3 hr and does not change for several hours when stabilized with 20% acetone. A linear relationship exists at 565 mμ for 40–140 μg Zr (+Hf); below this range the relationship is non-linear but reproducible. Iron and titanium will interfere but Culkin and Riley separate them from the zirconium by a preliminary solvent extraction-ion exchange procedure. The complex, with aluminum as carrier, is precipitated and hafnium determined by optical emission spectroscopy in the mixed oxides, with a sensitivity of 0.5 ppm Hf if a 2–3 g sample is used.

Sandell[11] measures the fluorescence of the complex formed between zirconium and morin, a method which has the greatest sensitivity of all methods for zirconium (0.01 μg Zr/ml). The sample is fused with Na_2CO_3, the cake leached in water and the insoluble residue dissolved in HCl; Fe(III) is reduced with mercaptoacetic acid, morin is added and the fluorescence of the solution is measured before and after the addition of the disodium salt of EDTA which quenches the fluorescence of Zr but not of such elements as Al, Ga, Sn, Th and uranium. The concentrations of niobium and tantalum should not exceed 0.05%. The method is very sensitive to trace impurities and satisfactory precision is obtained only by strict adherence to the details of the procedure.[5,8]

The use of Xylenol Orange and methylthymol blue has been thoroughly investigated by Cheng,[4] who by means of a masking and demasking approach, has demonstrated the possibility of determining zirconium and hafnium directly in the presence of each other. Zirconium forms an intense red complex with xylenol orange (XO) and methylthymol blue (MTB) in strongly acidic medium (0.8–1.0N), whereas hafnium forms a similar complex in 0.2N HClO$_4$ with XO and at pH 3 with MTB. The maximum absorbances of the Zr and Hf complexes with XO are at 535 mμ and 530 mμ respectively; those for the MTB complexes are at 580 mμ and 570 mμ, but when either Zr or Hf are present in excess a violet color is formed having a maximum absorbance at 595–600 mμ. The complexes are stable for at least 24 hr, and are bleached by EDTA, citrate, phosphate and fluoride ions; the absorbance of the zirconium–XO complex begins to decrease when [SO_4^{2-}] > 2 micromoles, but the hafnium-XO complex will tolerate more sulfate. The formation of the zirconium-XO complex is masked by hydrogen peroxide which also masks slightly the similar Hf complex in 0.2–0.3N HClO$_4$; Cheng overcomes this mixed masking effect by adding sulfate which, since it forms complexes with both Zr and Hf, acts in a demasking sense to eliminate the effect of the hydrogen peroxide on Hf when Zr is also present. The determination of the two elements can thus be done by measuring the total absorbance of the two XO complexes before the addition of both hydrogen peroxide and sulfate, and then measuring the absorbance of the Hf complex alone after the addition of these reagents, to obtain the absorbance of the Zr complex by difference. Cheng suggests a preliminary coprecipitation of zirconium and hafnium, with carrier elements such as aluminum and titanium which do not react with Xylenol Orange in strongly acid medium. Difficulty in obtaining the reagent in sufficiently pure form has been noted.[13]

Elinson[14] discusses the application of several other organic reagents, including Arsenazo I and III, Pyrocatechol Violet and chloranilate. The use of 8-hydroxyquinoline has recently been proposed by van Santen, Schlewitz and Toy.[13]

Zirconium is a difficult element to volatilize completely and the reproducibility of results obtained by optical emission spectroscopy[1] is not always satisfactory. As little as 10 ppm zirconium can be detected, using dc arc excitation and the lines Zr 3438 and 3391, but hafnium does not have a high spectral sensitivity and usually not less than 100 ppm can be detected. Hafnium is thus not readily detected in rocks but requires either the preparation of a zirconium-rich sample fraction or the chemical concentration of the two elements.[5,14] A chemical precipitation, using phenylarsonic acid, is also a useful preliminary step for determining zirconium and hafnium (with niobium and tantalum also) by X-ray fluorescence spectroscopy.[1]

References

1. Ahrens, L. H., and Taylor, S. R. *Spectrochemical Analysis*, 2nd ed., 1961. Reading, Mass.: Addison-Wesley, pp. 211–214.
2. Bennett, H., and Hawley, W. G. *Methods of Silicate Analysis*, 2nd ed., 1965. London: Academic Press, pp. 95, 220.
3. Blumenthal, W. B. *The Chemical Behavior of Zirconium*, 1958. Princeton, N. J.: D. Van Nostrand, pp. 1–13, 201–239.
4. Cheng, K. L. Determination of zirconium and hafnium with xylenol orange and methylthymol blue. *Anal. Chim. Acta* **28**, 41–53 (1963).
5. Culkin, F., and Riley, J. P. The determination of trace amounts of zirconium and hafnium, thorium and cerium in silicate rocks. *Anal. Chim. Acta* **32**, 197–210 (1965).
6. Groves, A. W. *Silicate Analysis*, 2nd ed., 1951. London: George Allen and Unwin, pp. 121–122.
7. Hillebrand, W. F., Lundell, G. E. F., Bright, H. A., and Hoffman, J. I. *Applied Inorganic Analysis*, 2nd ed., 1953. New York: John Wiley & Sons, pp. 564–575.

8. Kolthoff, I. M., and Elving, P. J., (Eds.). *Treatise on Analytical Chemistry*, Part II, Vol. 5, 1961. "Zirconium and Hafnium," by R. B. Hahn. New York: Interscience, pp. 61–138.

9. Pilkington, E. S., and Wilson, W. The influence of polynuclear zirconium species on direct titration of zirconium with EDTA. *Anal. Chim. Acta* **33**, 577–585 (1965).

10. Rafiq, M., Rulfs, C. L., and Elving, P. J. Determination of small amounts of zirconium. I. Gravimetric procedures using mandelic acid and its derivatives. *Talanta* **10**, 696–701 (1963); II. *Idem.* 827–832.

11. Sandell, E. B. *Colorimetric Determination of Traces of Metals*, 3rd ed., 1959. New York: Interscience, pp. 966–981.

12. Schoeller, W. R., and Powell, A. R. *The Analysis of Minerals and Ores of the Rarer Elements*, 3rd ed. (rev.), 1955. New York: Hafner, pp. 134–155.

13. van Santen, R. T., Schlewitz, J. H., and Toy, C. H. The spectrophotometric determination of zirconium with 8-hydroxyquinoline. *Anal. Chim. Acta* **33**, 593–596 (1965).

14. Vinogradov, A. P., and Ryabchikov, D. I., (Eds.). *Detection and Analysis of Rare Elements*, 1961. Moscow: Izdatel'stvo Akademii Nauk SSSR, Translation by Israel Program for Scientific Translations, Jerusalem, 1962; "The Present State of the Analytical Chemistry of Zirconium and Hafnium," by S. V. Elinson, pp. 334–414.

15. Voinovitch, I. A., Debras-Guedon, J., and Louvrier, J. *L'analyse des silicates*, 1962. Paris: Hermann; p. 501.

16. Wilson, C. L., and Wilson, D. W., (Eds.). *Comprehensive Analytical Chemistry*, Vol. 1C (Classical Analysis), 1962. "Zirconium," by G. W. C. Milner. Amsterdam: Elsevier, pp. 507–519.

6.21. Rare earths (yttrium and scandium)*

The determination of these elements is seldom required in rock analysis but not because they are, as their name implies, only rarely present. They are actually widely distributed, albeit in low concentration, in most rocks; yttrium and cerium are about equal in abundance to such common elements as zinc and cobalt, and neodymium and lanthanum have about the same abundance as lead. They are also present as both major and minor constituents of a large number of minerals, including various silicates. The possibility of their presence should not be ignored, since these elements will accompany the R_2O_3 group in the course of an analysis and, unless a correction is made for them, will be counted as aluminum when the latter is determined by difference. The recognition in the sample of such characteristic minerals as allanite, monazite and xenotime is a certain indication of their presence but a preliminary optical emission spectroscopic examination is again the surest and most comprehensive approach.

* References for this section will be found on pp. 287, 288.

Much attention has been paid to these elements recently and the availability of several authoritative accounts of the analytical chemistry of the group,[4,8,9,11,12,13,14] makes it unnecessary to attempt a detailed coverage here. The following is a brief resumé that will be of interest to those concerned only with their occasional determination.

The *rare earths* include scandium, yttrium and lanthanum as well as the fourteen elements having the atomic numbers 58–71 (cerium to lutetium inclusive). This latter group is termed the *lanthanides* or *lanthanons* and with lanthanum (at. no. 57) make up the *lanthanum series* (Ln). These elements are known to occur in more than one hundred minerals, never singly as a major constituent but always together with one or more members of the series; the minerals in which they occur are chiefly phosphates, tantalates, niobates and silicates, minerals which are usually associated with pegmatite dikes and with deposits derived from the weathering of these dikes.

An interesting feature of the rare earths is their tendency towards a fractionation into two groups during crystallization of their host minerals. The separation is only partial but in the silicates, for example, there is a group of minerals, such as cerite, allanite and beckelite, in which the lighter rare earths predominate (cerium group) and another group in which the heavier members (yttrium group) are predominant, such as gadolinite and yttrialite. The rare earths are generally present in the trivalent state, whether as a major constituent or substituting for such elements as calcium, strontium and lead; it is reported that europium and samarium occur in the divalent state in some potassium feldspars.

Scandium is widely distributed and occurs in most rocks in low concentration; it is usually present, in larger amounts, in most rare earth minerals, and is a major constituent of the rare silicate, thortveitite. Scandium, in its chemical properties, resembles yttrium and the lanthanum series in some respects (e.g., the formation of sparingly soluble fluorides and oxalates in weakly acid solutions), but also resembles thorium and zirconium, and even iron and aluminum, in others.

Yttrium, as previously mentioned, is closely associated chemically with the heavier rare earths and can be separated from them only with difficulty. It is also very widespread in most rocks.

The rare earth elements are often accompanied by thorium and, frequently, by such elements as uranium, zirconium and hafnium, actinium, titanium, niobium, tantalum, iron, calcium and aluminum. Further details about their occurrence and distribution are given in the references cited, particularly those by Topp[12] and Eyring,[4] and the recent review of their geochemistry by Fleischer.[6]

The rare earths are precipitated together with other members of the R_2O_3 group when aqueous ammonia is added to bring the filtrate from the silica separation to a pH of 7.0 ± 0.5. The precipitation is not, however, complete; the most basic lanthanides require a higher pH than 7.5, lanthanum may not precipitate to any significant extent and the precipitation of scandium is never quantitative because it in part forms a soluble ammino complex ion with aqueous ammonia.[9] The further addition of ammonia to insure complete precipitation of the lanthanides will also result in the redissolving of some of the hydrous alumina (Sec. 6.4.1a). It is better to do the precipitations as usual and then, if the presence of the rare earths is known or suspected, to add a large excess of aqueous ammonia to the combined filtrates, to let stand cold for an hour or so and to filter any precipitate that forms through a separate small filter paper.

Unless a correction is made for them, the rare earth elements will be counted as Al_2O_3. Unfortunately not all of the rare earths form trivalent oxides—cerium is oxidized during the ignition to CeO_2 and praseodymium and terbium form higher oxides of variable composition; a simple correction will thus be in error if significant amounts of these latter three elements are present. In most rock analyses, however, the error thus introduced will be of negligible proportion.

In most rock and mineral analyses it will be sufficient to determine only the total rare earth content of the sample and this can be conveniently combined with the determination of such other constituents as the total sulfur, barium, chromium and zirconium.[7] The sample (1–2 g) is fused with sodium carbonate and the fusion cake is leached with water; the rare earths are found in the water-insoluble residue (Sec. 9.42). For minerals containing larger amounts of the rare earths, however, acid decomposition is preferred to fusion. Concentrated H_2SO_4, $HClO_4$, HNO_3 and HCl are used, the latter acid being preferred for silicates; hydrofluoric acid is preferred for titanates, niobates and tantalates, since it will, at the same time, give a nearly quantitative separation of the rare earths and thorium from titanium, niobium and tantalum. When acid decomposition is not feasible, however, fusion with alkali pyrosulfate or potassium acid fluoride (KHF_2) should be used.

None of the rare earths are amphoteric and a preliminary separation of most of them from the alkalis and alkaline earths can be obtained by precipitating them with aqueous ammonia. Precipitation as the hydrous oxides is also a useful step in the purification of the rare earths at a later stage, following their separation by some other means. The insolubility of the fluorides of the rare earth elements in dilute acid solution is a useful method of separating them from Ti, Ta, Nb, Zr and PO_4^{3-}, but not from Th and U (IV); scandium will dissolve if ammonium fluoride is also present. The fluoride separation is recommended for the separation of the rare earths from large amounts of iron and aluminum, such as are present in the R_2O_3 group. Because coprecipitation of the alkaline earths will also occur, calcium or strontium may be used as carriers; they can be readily separated later after decomposing the fluorides by fuming with H_2SO_4, $HClO_4$ or, better, with HNO_3 and boric acid.[14] The best method of separation is by precipitation of the rare earths as oxalates, preferably from a hydrochloric acid solution, by which only thorium and uranium(IV) are not separated from them; the precipitation will be incomplete if much uranium or iron is present and the quantity of oxalate added is not sufficient to complex these latter also. Scandium requires a large excess of oxalate to ensure complete precipitation. Calcium may interfere if it is present in sufficient quantity but, again, either calcium or strontium can be used as a carrier, and is easily separated by a subsequent precipitation of the rare earths as hydroxides.

A chlorination–volatilization procedure, at 900°C, has been described for the isolation of the lanthanum series and yttrium from most elements associated with them in uranium and thorium ores and concentrates.[15] A subsequent precipitation with aqueous ammonia separates the rare earth elements from the alkalis and alkaline earths and yields a pure concentrate suitable for analysis. A preliminary extraction of U and Th with tributyl phosphate in carbon tetrachloride is recommended when these elements are present in high concentration.

Scandium can be separated as the thiocyanate by extraction into ether, according to Sandell,[10] and a separation of the rare earths from thorium can be achieved by extracting the latter into such solvents as mesityl oxide or tributyl phosphate (cerium must be in the trivalent state[2]).

Edge and Ahrens[3] have used cation exchange to separate Sc, Y, Nd, Ce and La from hydrochloric acid solutions ($2N$) of granite and diabase (previously decomposed with HF and $HClO_4$). The sample solution is soaked into the top of a cation exchange column (Dowex 50, 8x, 200–400 mesh) previously washed with $3N$ HCl. Elution of the major constituents is done with $3N$ HCl and then $6N$ HCl is used to remove the rare earth elements.

A rough separation between the lighter and heavier rare earths can be achieved[14] by precipitating the lighter ones as double sulfates with an alkali sulfate.

Some chemical methods are available for the quantitative determination of individual rare earths but in general this is better done by physical methods such as optical emission and X-ray fluorescence spectroscopy, and the chemical methods reserved for group separations and determination.

The gravimetric determination of the rare earths has been discussed briefly in the previous section dealing with their separation. Precipitation as oxalate is considered to be the best method, preferably from a hydrochloric acid solution having a rare earth concentration of approximately 1 mg per ml. Both precipitation at pH 2 from a cold solution[12] followed by digestion near the boiling point for 1–2 hr, and precipitation from a hot solution at pH 1, after which the solution is boiled for a few minutes,[14] have been suggested. It is reported[12] that the presence of large quantities of added salts will result in the formation of mixed oxalate precipitates which do not form the proper oxides on ignition; the precipitate should be dissolved, the rare earths separated with aqueous ammonia, redissolved and reprecipitated as oxalate. It is better to ignite the oxalates of the rare earths to oxides than to ignite the hydroxides.

The rare earth elements, in a solution buffered at pH 7 with triethanolamine, can be titrated directly and quantitatively with EDTA, using Eriochrome Black T as indicator, or indirectly, after addition of excess EDTA to the above solution, with a standard zinc solution and Zincon as the indicator.[12] Coulometric titration is also used.[13]

Sandell[10] describes the use of Alizarin Red S to determine the lanthandes and yttrium in a solution from which all constituents of the sample except the alkalis have been removed. The red, colloidal, acetate-buffered solution is stable for at least 30 min and reproducible results are obtained for 10–100 μg in 10 ml, provided that a strict control of the pH is observed. Arsenazo I has been used to determine the total rare earths at a pH of 6.7 and at 570 mμ[13]; the rare earths are first precipitated homogeneously as oxalates with acetonedioxalic acid at pH 0.5–2 on a calcium carrier. The bleaching effect of Ce(IV) on the red iron(II)-phenanthroline complex is used by Culkin and Riley[2] to determine trace amounts of cerium in silicate rocks, following its extraction as Ce(IV) into tri-n-butyl phosphate and separation by cation exchange (see also Sec. 6.20).

The spectra of the rare earth elements obtained by optical emission spectroscopy are extremely complex, with both enhancement and depression caused by interelement interference in the dc arc. A spectrograph having a high dispersion and resolution is necessary, and the arcing should be done in a CO_2 atmosphere to eliminate the interference of cyanogen bands. The group has a high spectral sensitivity, with a detection limit of about 10 ppm for most of them. Yttrium and lanthanum are the elements most commonly reported; if they are absent it is usually safe to assume that the others are absent also. Detailed discussions of the methods and problems are readily available.[1,3,5,13]

X-ray fluorescence spectroscopy appears to be the best method for the determination of yttrium but the spectra of the other rare earths again are subject to much interelement effect, making their determination in anything but simple binary mixtures a complex problem. Woyski and Harris[9] have discussed this problem in detail, as have Fassel[5] and others[13]; methods for both trace and larger amounts are given.[13]

References

1. Ahrens, L. H., and Taylor, S. R. *Spectrochemical Analysis*, 2nd ed., 1961. Reading, Mass.: Addison-Wesley, pp. 218–226.
2. Culkin, F., and Riley, J. P. The determination of trace amounts of zirconium and hafnium, thorium and cerium in silicate rocks. *Anal. Chim. Acta* **32**, 197–210 (1965).
3. Edge, R. A., and Ahrens, L. H. The determination of Sc, Y, Nd, Ce and La in silicate rocks by a combined cation exchange—spectrochemical method. *Anal. Chim. Acta* **26**, 355–362 (1962).
4. Eyring, L., (Ed.). *Progress in the Science and Technology of the Rare Earths*, Vol. I, 1964. New York: Macmillan. "Aspects of the Geochemistry of the Rare Earths," by L. H. Ahrens, pp. 1–29; "La chimie analytiques des terres rares," by J. Loriers, pp. 351–398; "Soviet Research on Analytical Chemistry of the Rare Earths," by D. I. Ryabchikov and V. A. Ryabukhin, pp. 399–415.
5. Fassel, V. A. Analytical spectroscopy of the rare earths. *Anal. Chem.* **32**, 19 A (1960).
6. Fleischer, M. Some aspects of the geochemistry of yttrium and the lanthanides. *Geochim. Cosmochim. Acta* **29**, 755–772 (1965).
7. Groves, A. W. *Silicate Analysis*, 2nd ed., 1951. London: George Allen and Unwin, pp. 121–123.
8. Hillebrand, W. F., Lundell, G. E. F., Bright, H. A., and Hoffman, J. I. *Applied Inorganic Analysis*, 2nd ed. (rev.), 1953. New York: John Wiley & Sons, pp. 547–563.
9. Kolthoff, I. M., and Elving, P. J., (Eds.). *Treatise on Analytical Chemistry*, Part II, Vol. 8, 1963. "The Rare Earths and Rare-Earth Compounds," by Mark M. Woyski and Ray E. Harris. New York: Interscience, pp. 1–146.
10. Sandell, E. B. *Colorimetric Determination of Traces of Metals*, 3rd ed., 1959. New York: Interscience, pp. 742–749.

11. Schoeller, W. R., and Powell, A. R. *The Analysis of Minerals and Ores of the Rarer Elements*, 3rd ed. (rev.), 1955. New York: Hafner, pp. 94–112.
12. Topp, N. E. *Chemistry of the rare-earth elements*, 1965. Amsterdam: Elsevier, 165 pp.
13. Vinogradov, A. P., and Ryabchikov, D. I., (Eds.). *Detection and Analysis of Rare Elements*, 1961. Moscow: Izdatel'svo Akademii Nauk SSSR. Translation by Israel Program for Scientific Translations, Jerusalem, 1962. "The Present State of the Analytical Chemistry of Rare Earth Elements, Scandium and Yttrium," by D. I. Ryabchikov and V. A. Ryabukhin, pp. 135–192.
14. Wilson, C. L., and Wilson, D. W., (Eds.). *Comprehensive Analytical Chemistry*, Vol. 1C (Classical Analysis), 1962. Amsterdam: Elsevier. "Scandium," pp. 423–435, "Yttrium," pp. 436–443, "Lanthanum and the lanthanides," pp. 444–476 and "Cerium," pp. 477–491, by M. L. Salutsky.
15. Zimmerman, J. B., and Ingles, J. C. Isolation of the rare earth elements. *Anal. Chem.* **32**, 241–246 (1960).

Supplementary References

Boyadjieva, R. A method for the enrichment of the rare earth elements in silicate rocks by extraction with tributyl phosphate. *Bull. Geol. Inst. Bulgarian Acad. Sci.* **16**, 45–52 (1967). The rare earths are separated with calcium as fluorides and the F^- expelled by heating with $HClO_4 + HNO_3$. The rare earths are extracted into tributyl phosphate ($Al(NO_3)_3$ is used as salting agent), back-extracted into H_2O, precipitated as oxalates with calcium as collector, and ignited to oxides. Final determination is by X-ray fluorescence spectroscopy.
Brunfelt, A. O., and Steinnes, E. Cerium and europium content of some standard rocks. *Chem. Geol.* **2**, 199–207 (1967). The determinations were made by neutron activation.
Cobb, J. C. Determination of lanthanide distribution in rocks by neutron activation and direct gamma counting. *Anal. Chem.* **39**, 127–131 (1967). No chemical separations are made.
Herrmann, A. G., and Wedepohl, K. H. Die quantitative Bestimmung der Seltenen Erde (La–Lu) und des Yttriums in silicatischen Gesteines. *Z. Anal. Chem.* **225**, 1–13 (1967). The rare earths are concentrated by precipitation as oxalates from 30 g samples, and the final determinations made by a combination of X-ray fluorescence and optical emission spectroscopy.
Hughson, M. R., and Sen Gupta, J. G. A thorian intermediate member of the britholite-apatite series. *Am. Mineral.* **49**, 937–951 (1964). The rare earths are concentrated on a cation exchange resin, eluted with $4N$ HCl, precipitated with aqueous ammonia and ammonium sebacate, and the individual rare earths determined in the ignited oxides by X-ray fluorescence spectroscopy.
Schnetzler, C. C., Thomas, H. H., and Philpotts, J. H. Determination of rare earth elements in rocks and minerals by mass spectrometric, stable isotope dilution technique. *Anal. Chem.* **39**, 1888–1890 (1967). Pr, Tb, Ho and Tm cannot be determined by this method. The rare earths are concentrated by ion exchange from 0.5–1 g sample before being placed, as chlorides, on filaments of mass spectrometer.

6.22. Uranium and thorium*

It is only rarely that the analyst need be concerned about the quantitative determination of uranium and/or thorium as part of a normal silicate analysis. These actinide elements may, however, be present occasionally in concentrations (0.01–0.1%) which are sufficient to cause a significant error in the aluminum value when this latter is obtained by difference, and perhaps in the value for total iron also, if no correction for them is made. It is more likely that there will be a need to determine these elements in trace quantities for geochemical studies. A brief mention of them is thus necessary but in view of the very extensive literature that is extant on the analytical chemistry of these elements, to attempt to do more would be both pointless and foolhardy. Grimaldi (1961),[3] and Ryabchikov and Korchemnaya (1961)[7] have recently reviewed the analytical chemistry of thorium in detail†; Booman and Rein[3] have done the same (1962) for uranium.

These elements are probably more abundant and widespread than is commonly supposed but unfortunately a preliminary optical emission spectrographic analysis is not likely to indicate their presence below 0.01%, because of the relative insensitivity of their arc and spark spectra. Uranium is found in the more siliceous rocks (e.g., 25 μg/g for granite, 0.2 μg/g for basalt); thorium tends to favor the more sodic rocks. Both are commonly associated with Nb, Ta, Zr, Ti, Pb and the rare earths in minerals.

Uranium and thorium will separate with the R_2O_3 group during precipitation with aqueous ammonia; the presence of CO_2 in the ammonia may render incomplete the separation of uranium (it forms a soluble complex with $(NH_4)_2CO_3$) but at the concentrations being considered here, and under the conditions of the precipitation (Sec. 9.6) it is unlikely that any significant loss will occur. Grimaldi[3] states that thorium has a tendency to be occluded with silica and other elements such as

---

* References for this section will be found on pp. 291, 292.

† A detailed study of the determination of submicrogram quantities of thorium, by Sill and Willis,[5,6] has appeared since these reviews. They have developed a fluorimetric method using morin in an alkaline solution of diethylenetriaminepentaacetic acid (DTPA) that has high sensitivity (0.01 μg Th) and a tolerance to most of the common constituents of silicates[5]; errors are caused by Be, Zr and Hf, Ti, U, Al (in presence of citrate), Cr, Nb, Ta and Ba, most of which can be counteracted. The negative interference (quenching effect) of barium is caused by the coprecipitation of thorium with $BaSO_4$ and this is used as an efficient means of quantitatively separating minute quantities of thorium prior to its fluorimetric determination; the precipitate is dissolved in alkaline DTPA.[6]

Nb, Ta and Ti that hydrolyze in acid solution, especially if much phosphate or zirconium are present. Thorium, incidentally, can be separated with the rare earths from the R_2O_3 group residue by precipitation as fluoride or oxalate.

The distribution between aluminum and iron of the error resulting from the uncorrected presence of uranium will depend upon the titrimetric method used to determine iron. If the Fe(III) is reduced with SO_2, H_2S or $SnCl_2$, the U(VI) is not reduced and all of the error will fall upon the aluminum; if zinc is used as reductant the U(VI) will be reduced to something lower than U(IV) and the bulk of the error will fall upon the iron, with a small positive error still occurring in the aluminum value because the Fe_2O_3 titer of $0.1N$ $KMnO_4$ is less than the U_3O_8 titer.

The fluorimetric method given in Sec. 9.43 for the determination of traces of uranium (up to 10 ppm) is in routine use at the Geological Survey of Canada. It was established in its present form by Sydney Abbey, and is based upon the method that has been used for many years in the Mines Branch of the Department.[2] The sample, in a nitric acid solution after complete decomposition by a combination of acid treatment (HF and HNO_3) and Na_2CO_3 fusion, is extracted with ethyl acetate in the presence of a high concentration of aluminum nitrate, to separate and concentrate the uranium. The latter is transferred to an aqueous phase and an aliquot is fused with sodium fluoride under controlled conditions; the fluorescence of the solidified bead, and that of standards and a blank treated in a similar way, is measured with a fluorimeter.

Arsenazo III is one of a series of reagents that are similar in structure to thorin but which differs from the other members in that it forms stable blue complexes with thorium at acidities up to $8N$, and is more sensitive. Much work has been done with the Arsenazo reagents by S. B. Savvin at the V.I. Vernadsky Institute of Geochemistry and Analytical Chemistry in Moscow, U.S.S.R.; the method used at the Geological Survey was developed by Sydney Abbey[1] and it is based upon that described by Savvin. By substituting a perchloric acid medium for the hydrochloric acid one of Savvin, greater sensitivity and stability of the colored complex are achieved. Beer's law is effective over the range 0–25 μg thorium in a final volume of 25 ml.

The sample (usually 0.5 g) is decomposed by sintering with Na_2O_2 in an iron crucible (carbonates are first dissolved in acid and the residue either decomposed by HF and $HClO_4$, or by sintering it with Na_2O_2).

The sinter is leached, filtered and the water-insoluble residue dissolved in HNO_3 and HCl. Calcium is added and the oxalates are precipitated homogeneously with methanolic methyl oxalate; the filtered precipitate and paper are digested with nitric acid to destroy the paper and finally fumed with perchloric acid. If the TiO_2 content of the sample is greater than 1 per cent a second oxalate precipitation is made.* Excess $HClO_4$ is removed, the residue is dissolved in a controlled quantity of HCl, a small amount of ascorbic acid is added to reduce any Fe(III) that may be present, and controlled quantities of $HClO_4$, oxalic acid (to mask Zr and Hf)and reagent are added.

The absorbance of the colored complex (a mixture of the blue thorium complex and the red excess reagent) is measured at 660 mμ. Calcium was found to affect the range of thorium concentration over which Beer' law is effective, as well as positively affecting the absorbance of the thorium complex; it is important that the calcium contents of the sample, standard and blank be as close together as possible.

May and Jenkins[4] have recently described a similar method using Arsenazo III but they prefer to use the HCl medium (2.5–3N) of Savvin, and a series of precipitation steps is used to separate the thorium from other elements, including the rare earths (iodate separation of thorium).

The solvent extraction of thorium is often done with 8-quinolinol (chloroform solution); Goto et al. (Supp. Ref.) have applied it also as a reagent for the spectrophotometric determination of thorium in monazite sand.

References

1. Abbey, S. Determination of thorium in rocks. The Arsenazo III reaction in perchlorate medium. Anal. Chim. Acta 30, 176–187 (1964).
2. Ingles, J. C. Manual of analytical methods for the uranium concentrating plant. Mines Branch Monograph, Dept. Mines and Technical Surv. 866, 1959. Ottawa: Queen's Printer Cat. No. M32–866, Part II, Method U–1, pp. 1–37.
3. Kolthoff, I. M., and Elving, P. J. (Eds.). Treatise on Analytical Chemistry, Part II, Vol. 5, 1961. New York: Interscience. "Thorium," by F. S. Grimaldi, pp. 139–216. Idem., Vol. 9, 1962. "Uranium," by G. L. Booman and J. E. Rein, pp. 1–188.
4. May, Irving, and Jenkins, L. B. Use of Arsenazo III in determination of thorium in rocks and minerals. U.S. Geol. Surv. Profess. Papers 525–D, D192–D195 (1965).

* A very small amount of titanium may be carried down with the oxalate precipitate (at times this may appear as insoluble residue or suspension after dissolution of the oxalate precipitate in HCl and/or $HClO_4$) which will have a negative effect on the absorbance of the thorium, becoming progressively more pronounced with the passage of time.

5. Sill, C. W., and Willis, C. P. Fluorometric determination of submicrogram quantities of thorium. *Anal. Chem.* **34,** 954–964 (1962).
6. Sill, C. W., and Willis, C. P. Precipitation of submicrogram quantities of thorium by barium sulfate and application to fluorometric determination of thorium in mineralogical and biological samples. *Anal. Chem.* **36,** 622–630 (1964).
7. Vinogradov, A. P., and Ryabchikov, D. I. (Eds.). *Detection and Analysis of Rare Elements*, 1961. Moscow: Izdatel'stvo Akademii Nauk SSSR, translation by Israel Program for Scientific Translations, Jerusalem, 1962. "The Present State of the Analytical Chemistry of Thorium," by D. I. Ryabchikov and E. K. Korchemnaya, pp. 415–444.

Supplementary References

Bloxam, T. W. Quantitative determination of uranium and thorium in rocks. *J. Sci. Instruments* **39,** 387–389 (1962). Uranium and thorium in the ppm range are quantitatively determined by gamma ray spectrometry. The sample, 200 g or less, is placed in a special air-tight can which fits over the detector crystal.

Goto, K., Russell, D. S., and Berman, S. S. Extraction and spectrophotometric determination of thorium with 8-quinolinol. *Anal. Chem.* **38,** 493–495 (1966).

Nemodruk, A. A., and Glukhova, L. P. Determination of microamounts of uranium in soils, rocks and minerals, by means of Arsenazo III. *Zh. Anal. Khim.* **21,** 688–693 (1966). The uranium is extracted from a nitric acid solution of the sample with tributyl phosphate in presence of a large amount of ammonium nitrate and a complexing agent (to mask Zr, Th and rare earths), then re-extracted with Arsenazo III solution. Concentration nitric acid, pretreated with urea, is added and the spectrophotometric measurement is made in *exactly* 5 min (formation of colloidal particles of Arsenzao reagent).

Redman, H. N. An improved weighing bottle for use in weighing of hygroscopic materials. *Analyst,* **92,** 584–586 (1967). A platinum crucible is placed in the weighing bottle; a small hole drilled through the neck and lid of the bottle allows equilibration of the contents without removing the lid.

6.23. Niobium and tantalum*

These two elements, usually referred to as the "earth acids," occur only as traces in most silicate rocks and minerals. In igneous rocks the niobium concentration is usually about 20–30 ppm but there is a tendency for it to be higher in syenites, especially in nepheline syenites. The concentration ratio of niobium to tantalum is about 20:1; the average concentration of tantalum in igneous rocks is about 2–5 ppm, with the highest concentration in granites. These elements usually occur in association with iron, manganese, calcium, uranium and the rare earths and are often present as the minerals columbite and tantalite, a gradational series of the niobates and tantalates of iron and manganese.

In the analysis of silicates it is seldom necessary to consider the possible presence of these elements in significant concentrations, but again,

* References for this section will be found on p. 293.

the surest approach is that of a preliminary examination of the sample by optical emission spectroscopy. The detection limit for niobium by this latter method is about 10 ppm but tantalum is not usually detectable at best under 100 ppm, and a preliminary chemical enrichment is preferable.[1] It will usually be satisfactory to estimate the tantalum concentration on the basis of the approximate Nb:Ta ratio mentioned previously.

If the analytical procedures of Sec. 9.2 are followed, most of the niobium and tantalum will separate with the silica and will be part of the residue remaining after treatment of the silica with hydrofluoric acid. They will thus be added to the value obtained for aluminum, if this is obtained by difference. If a recovery of silicon from the R_2O_3 group residue is made as given in Sec. 9.6.1, the earth acids will likely again be separated with the residual silica.

Methods for the determination of these elements will not be discussed here but are readily available in other specialized texts,[1,4,5,6,7] particularly in the recent detailed review by Kallman.[3] Esson[2] has recently described a radiometric–spectrophotometric method for the determination of 0.5–150 ppm niobium in silicate rocks and minerals.

References

1. Ahrens, L. H., and Taylor, S. R. *Spectrochemical Analysis*, 2nd ed., 1961. Reading, Mass.: Addison-Wesley, pp. 260–263.
2. Esson, J. The radiometric-spectrophotometric determination of microgram amounts of niobium in rocks and minerals. *Analyst* **90**, 488–491 (1965).
3. Kolthoff, I. M., and Elving, P. J., (Eds.). *Treatise on Analytical Chemistry*, Part II, Vol. 6 (1964). "Niobium and tantalum," by S. Kallman. New York: Interscience, pp. 177–406.
4. Liebhafsky, H. A., Pfeiffer, H. G., Winslow, E. H., and Zemany, P. D. *X-Ray Absorption and Emission in Analytical Chemistry*, 1960. New York: John Wiley & Sons, 357 pp.
5. Schoeller, W. R., and Powell, A. R. *The Analysis of Minerals and Ores of the Rarer Elements*, 3rd ed. (rev.), 1955. New York: Hafner, pp. 199–227.
6. Vinogradov, A. P., and Ryabchikov, D. I., (Eds.). *Detection and Analysis of Rare Elements*, 1961. Moscow: Izdatel'stvo Akademii Nauk SSSR, translation by Israel Program for Scientific Translations, Jerusalem, 1962. "The present state of the analytical chemistry of tantalum and niobium," by I. P. Alimarin and G. N. Bilimovich, pp. 542–596.
7. Wilson, C. L., and Wilson, H. N., (Eds.). *Comprehensive Analytical Chemistry*, Vol. 1C (classical Analysis), 1962. "Niobium and tantalum," by B. Bagshawe. Amsterdam: Elsevier, pp. 551–569.

Supplementary Reference

Pollock, J. B. A gravimetric method for the determination of mixed oxides (niobium and tantalum pentoxides) in niobium–tantalum minerals. *Analyst* **93**, 93–96 (1968).

6.24. Lead, molybdenum, tungsten and zinc*

These four elements, or "heavy metals," are usually present in silicates as trace constituents only. While they are of much interest geochemically and their determination is a necessary part of most geochemical studies, they are only rarely encountered in significant amounts in ordinary rock and mineral analysis. Because of its geochronological significance, considerably more attention has been paid to the recovery and determination of very low concentrations of lead than to the other three metals.

Lead is a constituent which, in more than trace quantity, can be a source of trouble in silicate analysis and a preliminary examination of the sample by optical emission spectroscopy is a very worthwhile step if its presence is suspected. The danger lies in its possible reduction to the metallic state during fusion and the resultant damage to the platinum crucible or dish, damage which it is not usually possible to repair. Rocks which contain visible sulfide minerals should be regarded with suspicion until the results of an examination by optical microscopy or emission spectroscopy are available.

In the course of a rock analysis, as outlined previously, lead will separate partly with the SiO_2 as $PbCl_2$ during the dehydration step with hydrochloric acid. On ignition of the impure silica, the Pb(II) may be reduced by the carbon of the filter paper and be volatilized, or it may alloy itself with the platinum crucible (reduction of Pb(II) during the initial fusion, which is preferably done in an electric muffle furnace, can be guarded against by maintaining an oxidizing atmosphere within the crucible during the fusion; during ignition of the impure silica the filter paper should be burned off at a low temperature and the paper should not be allowed to catch fire); it may also react to form lead silicate and be weighed as such but, regardless of the initial weighing form, after the volatilization of SiO_2 with HF and H_2SO_4 the Pb(II) will be present as $PbSO_4$ in the final weighing, thus introducing an error into the value for silica. This error will usually be negligible.

That amount of lead which is not separated with the silica will be quantitatively separated with the R_2O_3 group, provided that adequate amounts of iron and aluminum are also present, and will be reported as Al_2O_3 unless a correction is made for it. It is better to separate the lead at the beginning of the analysis, either by using sulfuric acid to dehydrate the SiO_2 and precipitate the lead as $PbSO_4$ or by dehydrating the silica with $HClO_4$ and then separating the lead with hydrogen sul-

* References for this section will be found on pp. 298, 299.

fide (Sec. 6.3). Because of the significant solubility of $PbSO_4$ and the tendency of other elements (Nb, Ta, W, Ba, Sr, Ca) to also be precipitated, the second alternative is the better one.

If lead is present in more than trace concentration, its separation and determination as $PbSO_4$ is the best-known and most commonly used method; because of the solubility of $PbSO_4$, however, a further determination of the residual lead in solution is necessary. This can be done by electrodeposition (in nitric acid solution), by spectrophotometry, by atomic absorption spectroscopy, or by polarography, to name but a few of the possible methods. A disadvantage of the $PbSO_4$ method, in addition to the appreciable solubility of the precipitate, is a tendency of $PbSO_4$ to be contaminated by other elements, and the possibility that difficultly soluble sulfates of other elements such as iron and chromium may be formed. A titrimetric determination of lead with EDTA, both directly and indirectly, is also widely used. These and other macro methods are discussed by Gilbert[7] in his recent (1964) extensive review of the analytical chemistry of lead. Bennett and Hawley,[6] for the analysis of glazes, precipitate the lead in the silica filtrate with hydrogen sulfide and then reprecipitate and determine it as lead chromate. They also describe a special method for lead bisilicate frits in which the filtrate from a single dehydration of SiO_2 (Sec. 9.4.3) is combined with the solution of the nonvolatile residue from the silica determination, and the lead extracted with diethylammonium diethyldithiocarbamate and chloroform and determined gravimetrically as chromate.

For trace concentrations of lead in rocks, the method most extensively applied is that of extraction from basic medium and determination of the lead as the dithizonate in either chloroform or carbon tetrachloride. There are many possible sources of error in this method but detailed studies have led to an understanding of the problems involved and to ways of minimizing them.[4] When lead has to be extracted from solutions containing much calcium (or magnesium) and phosphorus the dithizone procedure fails, but the diethylammonium diethylthiocarbamate extraction (chloroform) can be used in acid medium.

The polarographic method is a very useful means for the determination of trace amounts of lead, and can be conveniently combined with the determination of other trace elements such as copper, bismuth and cadmium.[7] Lead may be determined in either acidic or basic solution.

No data are available for the determination of lead in rocks by atomic absorption spectroscopy; the sensitivity at 2170 Å is given as 0.5 ppm in solution, using a slightly oxidizing oxyacetylene or oxyhydrogen flame.[3] Preliminary extraction of the lead into an organic solvent is recommended.

Optical emission spectroscopy is a convenient way to determine traces of lead, together with many other elements. Ahrens and Taylor[1] describe methods for the analysis of igneous rocks, zircon and feldspar.

Molybdenum, if present only in small amount, will cause no problem in the ordinary rock analysis. Unless it is separated by hydrogen sulfide precipitation at the beginning (Sec. 6.3), it will pass through succeeding steps and be discarded in the filtrate from the separation of magnesium. It is only likely to be encountered in the more siliceous rocks, and mostly likely as a sulfide.

The colorimetric determination of molybdenum with thiocyanate is widely done.[4] In the presence of a reducing agent such as stannous chloride, a reddish-orange product is formed between Mo and CNS^- which can be extracted into isopropyl ether; the usual constituents of igneous rocks do not interfere. If both molybdenum and tungsten are to be determined Sandell[4] recommends the use of dithiol (4-methyl-1,2-dimercaptobenzene (toluene-3,4-dithiol)) when comparable amounts of the two elements are present (see also Stanton and Hardwick, Supp. Ref.). Chan and Riley[9] separate molybdenum by coprecipitation with hydrous MnO_2; the precipitate is dissolved in sulfurous acid and the molybdenum either determined directly with dithiol or, if appreciable Bi, Sb and Sn are also coprecipitated, the molybdenum is first separated by ion-exchange and then determined.

Considerable interest has been shown in the determination of molybdenum by atomic absorption spectroscopy[3,8,10]; Butler and Mathews[8] complex the Mo with ammonium pyrollidine dithiocarbamate (APDC), or with 8-hydroxyquinoline if there is much iron present, and extract it in *n*-amyl methyl ketone for determination in an acetylene–air flame.

The detection limit of molybdenum by optical emission spectroscopy is about 1 ppm (Mo 3170 Å), using cathode excitation.[1]

Tungsten will rarely be encountered by the rock and mineral analyst but, because it may be something of a nuisance if present in appreciable amount, a brief mention should be made of it. It, like molybdenum, occurs with the more siliceous rocks.

Tungsten will separate with the silica (yellow color), but incompletely or not at all if present in its usually very low concentration. That which does separate may be lost by volatilization during the ignition (> 800°C) of the impure SiO_2; an unusually high nonvolatile residue is an indication of its possible presence in the sample. The remainder of the tungsten will be precipitated with the R_2O_3 constituents and will, of course, be counted as Al_2O_3 unless its presence is known (a preliminary optical emission spectroscopic examination will, unfortunately, only detect >100 ppm) and a correction applied. If the reduction of $Fe(III)$ for the titrimetric determination of total iron is done with $SnCl_2$ or zinc metal (but *not* H_2S or SO_2), some of the error will be applied to the value for iron also.

Sandell[4] has discussed the colorimetric determination of tungsten with CNS^- and dithiol; the simultaneous determination of molybdenum and tungsten with dithiol has already been mentioned (see also Chan and Riley, Supp. Ref.).

In ordinary igneous rocks, a chemical enrichment is usually necessary before tungsten can be quantitatively determined by optical emission spectroscopy.[1]

Zinc is a widely distributed constituent of rocks and minerals, commonly as the sulfide (sphalerite) and carbonate (smithsonite). As in the case of lead, a knowledge of its presence in appreciable amount is a useful precaution against possible damage to platinum ware, and optical emission spectroscopy can routinely detect it in concentrations above 100 ppm. In the course of the analysis, if double precipitations are the rule, most of the zinc will pass through into the magnesium filtrate, but some may be caught with R_2O_3 constituents, and some may coprecipitate with magnesium. If preliminary examination of the sample indicates a significantly high concentration of zinc, it should be separated at the beginning of the analysis by precipitation with hydrogen sulfide (Sec. 6.3); usually it is sufficient to separate it by ammonium sulfide precipitation before the precipitation of calcium as the oxalate.

The colorimetric determination of zinc as the dithizonate, following its extraction into chloroform, has been described in detail by Sandell[4] and is a widely used procedure. Mah *et al.* [13] used this color reagent after first separating zinc in rocks and soils by circular paper chromatography.

Although the detection limit of zinc by optical emission spectroscopy is, as mentioned previously, only about 100 ppm, this can be lowered to 1 ppm by a double-arc method.[1] Chemical enrichment is a useful preliminary step. Filby[12] has recently described a neutron activation procedure for the determination of zinc in rocks.

Atomic absorption spectroscopy has proven to be a particularly sensitive and appealing method for the determination of trace quantities of zinc. An oxidizing flame (hydrogen or acetylene, with air or oxygen) is used, usually with an aqueous solution of the sample, and the sensitivity in solution is reported[3] as 0.05 ppm. A study of the procedure was made at the Geological Survey by Sydney Abbey[11] and tests on a variety of rock types showed no significant matrix effects; wide variation in the acidity of the final solution is also without effect. The concentration range (in rocks) covered by Abbey's method is 5–250 ppm; a 0.5 g sample is diluted to 50 ml.

The analytical chemistry of zinc has been reviewed (1961) by Kanzelmeyer.[7]

References

GENERAL

1. Ahrens, L. H., and Taylor, S. R. *Spectrochemical Analysis*, 2nd ed., 1961. Reading, Mass.: Addison-Wesley; "Lead," pp. 242–244; "Molybdenum," pp. 259–206; "Tungsten," pp. 258–259; "Zinc," pp. 264–267.
2. Hillebrand, W. F., Lundell, G. E. F., Bright, H. A., and Hoffman, J. I. *Applied Inorganic Analysis*, 2nd ed. (rev.), 1953. New York: John Wiley & Sons. "Lead," pp. 223–231; "Molybdenum," pp. 302–316, 901–905; "Tungsten," pp. 683–693; "Zinc," pp. 425–438.
3. Robinson, J. W. *Atomic Absorption Spectroscopy*, 1966. New York: Marcel Dekker. "Lead," pp. 131–133; "Molybdenum," pp. 139–140; "Tungsten," p. 156; "Zinc," pp. 157–160.
4. Sandell, E. B. *Colorimetric Determination of Traces of Metals*, 3rd ed. (rev.), 1959. New York: Interscience. "Lead," pp. 555–583; "Molybdenum," pp. 640–664; "Tungsten," pp. 883–899; "Zinc.," pp. 941–965.
5. Wilson, C. L., and Wilson, D. W., (Eds.). *Comprehensive Analytical Chemistry*, Vol. 1C (Classical Analysis), 1962. Amsterdam: Elsevier. "Lead," by J. W. Price, pp. 185–202; "Molybdenum and Tungsten," by R. A. Chalmers, pp. 589–604; "Zinc," by L. Ginsburg, pp. 395–401.

LEAD AND ZINC

6. Bennett, H., and Hawley, W. G. *Methods of Silicate Analysis*, 2nd ed., 1965. London: Academic Press, pp. 90–93.
7. Kolthoff, I. M., and Elving, P. J., (Eds.). *Treatise on Analytical Chemistry*, Part II, Vol. 3, 1961. New York: Interscience. "Zinc," by J. H. Kanzelmeyer, pp. 95–170. *Idem.*, Vol. 6, 1964. "Lead," by T. W. Gilbert, Jr., pp. 69–175.

MOLYBDENUM

8. Butler, L. R. P., and Mathews, P. M. The determination of trace quantities of molybdenum by atomic absorption spectroscopy. *Anal. Chim. Acta* **36**, 319–327 (1966).
9. Chan, K. M., and Riley, J. P. The determination of molybdenum in natural waters, silicates and biological materials. *Anal. Chim. Acta* **36**, 220–229 (1966).
10. David, D. J. The determination of molybdenum by atomic absorption spectrophotometry. *Analyst* **86**, 730–740 (1961).

ZINC

11. Abbey, S. Analysis of rocks and minerals by atomic absorption spectroscopy: I. The determination of magnesium, lithium, zinc and iron. *Geol. Surv. Canada Paper* **67–37**, 1967, 35 pp.
12. Filby, R. H. The determination of zinc in rocks by neutron activation analysis. *Anal. Chim. Acta* **31**, 557–562 (1964).
13. Mah, D. C., Mah, S., and Tupper, W. M. Circular paper chromatographic method for determining trace amounts of Co, Cu, Ni and Zn in rocks and soils. *Can. J. Earth Sci.* **2**, 33–43 (1965).

Supplementary References

Chan, K. M., and Riley, J. P. Determination of tungsten in silicates and natural waters. *Anal. Chim. Acta* **39**, 103–113 (1967). Tungsten is concentrated by coprecipitation with hydrous manganese dioxide. Tungsten and molybdenum are separated from other coprecipitated elements by cation exchange, and from each other by extraction of the molybdenum dithiol complex from a $2.4N$HCl medium containing citric acid. After destruction of the citric acid, tungsten is determined photometrically with dithiol.

Stanton, R. E., and Hardwick, Mrs. A. J. The colorimetric determination of molybdenum in soils and sediments by zinc dithiol. *Analyst* **92**, 387–390 (1967). Interference by tungsten is suppressed by careful control of the time allowed for complex formation.

6.25. Beryllium, boron and copper*

These three elements are grouped together only for convenience in consideration and because the following brief treatment does not warrant a separate section for each one. The writer has had but little occasion to determine these elements and can add nothing to the analytical literature already available; for this reason no methods for their determination are given in Chapter 9 and the reader is referred to the accompanying references for further guidance.

Beryllium is widely distributed in trace amounts in most silicate rocks (about 4–6 ppm average concentration); it is also a major constituent of such silicates as beryl, phenakite and bertrandite. The detection limit by optical emission spectroscopy is about 1–10 ppm, so that its presence can be readily ascertained by the preliminary examination so often suggested in the foregoing pages.

Beryllium behaves like aluminum, except that its hydrous oxide is not amphoteric; it will thus precipitate with the R_2O_3 group and be reported as Al_2O_3 unless an appropriate correction is made for it. Emission spectrographic analysis of the ignited oxides is one convenient

* References for this section will be found on pp. 300–302.

approach; as little as 0.02 ppm Be in solution can be detected by atomic absorption spectroscopy, provided a nitrous oxide flame is used.

Boron is a common trace component (generally much less than 10 ppm) of highly siliceous rocks, occurring usually as tourmaline; axinite and datolite are other common boron-containing silicates.

It is a difficult element to deal with because of its high volatility. It will accompany SiO_2 during the analysis of the main portion and will be volatilized with it, in part at least, when the ignited, impure SiO_2 is treated with HF. This has been discussed in Secs. 6.2.1a and 6.2.1c, and the preliminary removal of the boron with methyl alcohol has been described in Sec. 9.2.3. That part of the boron which remains with the nonvolatile residue from the HF treatment of the impure SiO_2 will precipitate with the R_2O_3 group and will finally be reported as aluminum.

Like beryllium, boron is readily detected by optical emission spectroscopy (about 1–10 ppm); its detection limit by atomic absorption spectroscopy is only about 10 ppm in aqueous solution, using a nitrous oxide flame.

Copper is ubiquitously present as a trace element in most rocks but care is necessary to ensure that its presence is not the result of contamination from brass or bronze screens used during the preparation of the sample. It has high a detection limit by optical emission spectroscopy (about 1 ppm); its sensitivity of determination by atomic absorption spectroscopy is much increased by a preliminary extraction into an organic solvent.

It is seldom that the presence of copper will present a problem in silicate analysis. Groves ("Copper," Ref. 2) found that from 10 to 30% of the total copper tended to accompany the R_2O_3 group and would be reported as Al_2O_3 but that usually no correction was necessary after the second precipitation. If the filtrate from the separation of SiO_2 is treated with H_2S (Sec. 6.3), copper is separated; if not, that part of it which escaped precipitation with the R_2O_3 may be removed by ammonium sulfide treatment of the filtrate before the precipitation of calcium (Sec. 9.9.1). If not removed, a portion of it will precipitate as phosphate with magnesium.

References

BERYLLIUM

1. Ahrens, L. H., and Taylor, S. R. *Spectrochemical Analysis*, 2nd ed., 1961. Reading, Mass.: Addison-Wesley, pp. 204–205.
2. Groves, A. W. *Silicate Analysis*, 2nd ed., 1951. London: George Allen and Unwin, pp. 147–149.

3. Kolthoff, I. M., and Elving, P. J., (Eds.). *Treatise on Analytical Chemistry*, Part II, Vol. 6, 1964. "Beryllium," by B. R. F. Kjellgren, C. W. Schwenzfeier, Jr., and E. S. Melick. New York: Interscience, pp. 1–67.
4. Sandell, E. B. *Colorimetric Determination of Traces of Metals*, 3rd ed., 1959. New York: Interscience, pp. 304–324. Silicates and silicate rocks (Be < 10 ppm)—spectrophotometric determination with morin, following separation of Be with aqueous ammonia in presence of mercaptoacetic acid to complex iron.
5. Schoeller, W. R., and Powell, A. R. *The Analysis of Minerals and Ores of the Rarer Elements*, 3rd ed. (rev.), 1955. New York: Hafner, pp. 54–68. Gravimetric methods, with emphasis upon separation from interfering elements.
6. Vinogradov, A. P., and Ryabchikov, D. I., (Eds.). *Detection and Analysis of Rare Elements*, 1961. Moscow: Izdatel'svo Akademii Nauk SSSR, translation by Israel Program for Scientific Translations, Jerusalem, 1962. "The Present State of the Analytical Chemistry of Beryllium," by V. G. Goryushina, pp. 80–111.
7. Wilson, C. L., and Wilson, D. W., (Eds.). *Comprehensive Analytical Chemistry*, Vol. 1C (Classical Analysis), 1962. "Beryllium," by E. Booth. Amsterdam: Elsevier, pp. 56–64. Gravimetric and titrimetric methods only.

BORON

1. Ahrens, L. H., and Taylor, S. R.   (See under Beryllium); pp. 202–204.
2. Boltz, D. F., (Ed.). *Colorimetric Determination of Nonmetals*, 1958. "Boron," by G. Porter and R. C. Shubert. New York: Interscience, pp. 339–353. Titration with NaOH, in presence of mannitol, for macro amounts (>0.01% B); spectrophotometric determination with quinalizarin (<0.01% B).
3. Hillebrand, W. F., Lundell, G. E. F., Bright, H. A., and Hoffman, J. I. *Applied Inorganic Analysis*, 2nd ed. (rev.), 1953. New York: John Wiley & Sons, pp. 749–765. Titration with NaOH, following separation of boron as methyl borate, in presence of mannitol.
4. Maurel, P., and Ruppli, C.   Application de la méthode des additions successives au dosage, par spectrographie d'émission dans l'arc, du bore en trace dans les roches silicatées. *Compt. Rend.* **253**, 1830–1831 (1961).
5. Morgan, L.   The determination of boron by a pyrohydrolysis technique. *Analyst* **89**, 621–623 (1964). Here applied to various refractories, including alumina insulation rods.
6. Stanton, R. E., and McDonald, A. J.   The colorimetric determination of boron in soils, sediments and rocks with methylene blue. *Analyst* **91**, 775–778 (1966). This method was developed for geochemical studies of material in which boron is present as colemanite (calcium borate). The sample is decomposed with cold $HF-H_2SO_4$, and methylene blue is added to form a complex with fluoborate; the complex is extracted with 1,2-dichloroethane (blue color) and measured at 640 mμ.
7. Voinovitch, I. A., Debras-Guedon, J., and Louvrier, J. *L'analyse des silicates*, 1962. Paris: Hermann; "Boron," pp. 351–355. Titration with NaOH following separation of boron as methyl borate, in presence of mannitol.

COPPER

1. Ahrens, L. H., and Taylor, S. R.   (See under Beryllium); pp. 244–246.
2. Groves, A. W.   (See under Beryllium); pp. 142–145, 157–158.

3. Kolthoff, I. M., and Elving, P. J., (Eds.). *Treatise on Analytical Chemistry*, Part II, Vol. 3, 1961. "Copper," by W. C. Cooper. New York: Interscience, pp. 1–41.
4. Maurel, P., and Ruppli, C. Contribution aux méthodes de dosage du chrome, du cobalt, du cuivre, du nickel et du vanadium dans les roches silicatées. *Bull. Soc. Franç. Minér. Crist.* **84**, 303–376 (1961).
5. Sandell, E. B. (See under Beryllium); pp. 437–470. In silicate rocks, the copper is extracted into 2,2'-biquinoline in isoamyl alcohol.
6. Trent, D. J., and Slavin, W. Determination of various metals in silicate samples by atomic absorption spectroscopy. *Atomic Absorption Newsletter* **9**, 118–125 (1964). Perkin-Elmer Corporation, Norwalk, Conn. The methods are given; Cu is extracted from acid solution (pH 1.5) with ammonium pyrrolidine dithiocarbamate (APDC), or iron is first extracted with methyl isobutyl ketone before Cu is extracted at pH 2.4 with APDC.

CHAPTER 7

CARBONATES

In contrast to the recent, relatively voluminous literature having to do with the analysis of silicate rocks and minerals, very little has been written about the analysis of the carbonates. This in part arises from the fact that carbonates, both rocks and minerals, seldom present analytical problems peculiar to themselves, in part because the detailed analysis required for the relatively more complex silicates is not usually required for the carbonates, but largely it is because the methods and techniques of silicate analysis are generally applicable, with minor modification, to carbonate analysis as well. While some new methods specifically intended for carbonate analysis are discussed in the following pages, the emphasis will be largely upon the use, when applicable, of the silicate methods given in Chapter 9.

The most complete scheme for the analysis of carbonates remains that of Hillebrand et al.[10]; a "condensed" scheme of analysis is included for those constituents stressed in technical analyses (see also Lundell and Hoffman[12]). Similar abbreviated schemes are given in various handbooks of approved methods, such as the one outlined by Hanna.[8] More recent methods of carbonate analysis have been described by Hill et al.,[9] Shapiro and Brannock,[18] Grillot et al.,[6] Bennett and Hawley[2] (dolomites, limestones, magnesites) and Goldich et al.[5] (marls). Ahrens and Taylor[1] present a brief survey of the optical emission spectroscopic determination of trace elements in carbonates.

7.1. Moisture

The discussion of the determination of moisture in Sec. 6.1 is applicable to a similar determination in carbonates. The heating, at 105–110°C, is often a preliminary step before the determination of loss on ignition at 1000°C. If a porcelain crucible is used, the inner glaze should first be removed by igniting the crucible, half-filled with carbonate-rich material, in an electric muffle furnace for 3–4 hr; little weight change should occur after this treatment.

7.2. Loss on ignition

The errors inherent in the value for loss on ignition have been discussed in Sec. 6.12 but in spite of these, its determination can serve a useful purpose, particularly for carbonate rocks. The preferred ignition temperature is about 1000°C and the heating is usually started in a cold muffle furnace, the temperature being gradually raised to prevent the loss of sample by too sudden expulsion of volatiles; about 30–60 min heating at 1000°C is usually sufficient to bring the sample to constant weight. Manheim[13] has discussed this determination in considerable detail.

A method for the determination of carbon dioxide in limestone and dolomite by differential loss on ignition has been described.[9] The loss on ignition for high calcium samples approaches closely the actual carbon dioxide content but the presence of carbonaceous material and other volatiles in less pure samples introduces a large error into the result. By determining the loss on ignition after heating at 550°C for 25 min (CO_2 is retained, other volatiles are evolved), and then heating again for at least 1 hr at 1000°C, the loss in weight due to evolution of carbon dioxide can be calculated. The oxidation of pyrite present in the sample ($>0.2\%$) will introduce a further error but a correction can be applied; the method fails when siderite and magnesite are present.

7.3. Acid-insoluble residue

The determination of the acid-insoluble residue is, by its nature, largely empirical and is of most value when used as a relative measure of the degree of purity of a series of carbonate samples. In most samples the acid-insoluble residue will consist largely of impure silica, particularly when the chief non-carbonate constituent is quartz, but the degree of correlation diminishes with increase in the proportion of silicates only partially resistant to acid attack, as well as of such other impurities as organic matter and pyrite. The amount of acid-insoluble residue obtained will depend upon the acid used, the time and temperature of the digestion and the average grain size of the sample. Quoted figures for acid-insoluble residue are meaningful only when the conditions under which the determinations were made are given; Bisque[3] has suggested 30 min in 150 ml of $3N$ HCl at 50–60°C as being suitable for the treatment of a 1 g sample.

The sample used for the determination of loss on ignition at 1000°C should not be used for the determination of acid-insoluble residue because of the solubilizing effect that the ignition has upon the non-carbonate impurities. Hillebrand et al. (Ref. 10, p. 966) state that a finely ground limestone containing not more than 15% SiO_2 and not more than 6%

R_2O_3 may be converted to a sintered product wholly soluble in hydrochloric acid by heating it for 10–15 min at 1100–1200°C.

Hydrochloric acid is usually selected for the digestion, but because it does have a strong chemical action that will dissolve or disintegrate some non-carbonate constituents (e.g., phosphatic material), acetic or formic acids are sometimes preferred when the preserving of structures is important for subsequent mineralogical and other studies of the residue.[4]

7.4. Silica

The factors affecting the determination, by various methods, of SiO_2 in silicates have been discussed in detail in Sec. 6.2 and apply equally as well to carbonates, bearing in mind the generally much lower SiO_2 content and different matrix of the latter.

The simplest gravimetric approach is one in which the acid-insoluble residue is treated with hydrofluoric and sulfuric acids to obtain the pure SiO_2 by volatilization. This is not recommended by Goldich et al.,[5] who show that an erroneous value may be obtained as a result of the conversion of the associated alkali metals and alkaline earths to sulfates. It also does not allow for the silica dissolved during the acid treatment. If the silica is to be determined by dehydration following acid attack of the sample, a preliminary ignition of the sample, as is done for the determination of the loss on ignition, will insure a more complete decomposition of the silicates present; Hill et al.[9] employ perchloric acid and separate the silica by a single evaporation to copious fumes, followed by gentle refluxing. If a determination of the acid-insoluble residue is done, then it is only necessary to fuse this latter with Na_2CO_3 and to combine the hydrochloric acid solution of the fusion with the filtrate from the acid-insoluble residue determination before continuing with the usual dehydration step. Finally, if the whole sample is to be directly attacked by a flux, this can be reduced to a minimum and the sample sintered with a quantity of anhydrous sodium carbonate equal to one-half or one-quarter of its original weight; this has the advantage of reducing the amount of alkali salts present and simplifies the dehydration and washing of the silica. Prolonged periods of dehydration, or heating at a temperature >110°C should be avoided because of the possibility of the formation of soluble magnesium silicate.[10] A double dehydration is recommended.

The colorimetric methods discussed in Sec. 6.2.3 can be applied to carbonate samples with some modification but the writer is without experience in this respect. Shapiro and Brannock[18] recommend that the sample weight for their solution A be increased to 200 mg for carbonates, the weight of the standard (NBS No. 99, feldspar) remaining the same. It would seem preferable to use a more appropriate standard, such as NBS 1a (argillaceous limestone), instead.

7.5. Aluminum

In most carbonate rocks the content of aluminum is very low (it may, however, be high in the argillaceous types) and it is often, particularly in commercial analyses, not reported separately but is included in the total R_2O_3 oxides, or in the balance of this group remaining after subtraction of the total iron, for which a separate determination is made.

Precipitation of the aluminum together with various other elements has been discussed in Sec. 6.4.1. Precipitation with aqueous ammonia is the most frequent method used; the procedure is the same as that for silicates but is simpler because of the much reduced weight of the mixed oxides that is present. Particular care is necessary, however, to avoid coprecipitation of calcium and magnesium; Bennett and Hawley[2] recommend a triple precipitation, with additional ammonium chloride being added before each treatment with aqueous ammonia. The aluminum content is usually determined by difference; a useful check determination can be conveniently made by emission spectrographic analysis of the ignited mixed oxides.

The colorimetric methods discussed in Sec. 6.4.3 probably apply to carbonates equally as well as to silicates; Shapiro and Brannock[18] (Sec. 9.8.1) merely increase the size of sample taken for the determination of SiO_2 and Al_2O_3 in carbonate rocks to 200 mg. Again, it seems preferable to use a more suitable reference standard than the NBS feldspar (No. 99).

Titrimetric methods have little application here but a procedure for the rapid determination of aluminum (and iron, calcium and magnesium) by EDTA titration has been described.[3]

7.6. Calcium and magnesium

The determinations of calcium and magnesium have been discussed in detail in Secs. 6.5 and 6.6 respectively, and while the concentration ranges of these constituents in carbonates are much wider than those in the silicates, the comments in Secs. 6.5 and 6.6 are fully applicable. Because of this wide variation possible in relative concentrations, the Mg:Ca ratio in a sample is an important consideration in deciding upon the feasibility of a particular method.

For large amounts of calcium a gravimetric finish, following separation of the calcium as oxalate (Sec. 6.5.1a), is recommended.[2,10] Small amounts of calcium, such as in magnesite, are better precipitated together with the magnesium by diabasic ammonium phosphate and later recovered as sulfate from the H_2SO_4 solution of the ignited pyrophosphates.

Titration of the H_2SO_4 solution of the separated calcium oxalate with $KMnO_4$ is a rapid and convenient routine procedure for large amounts of calcium (Sec. 6.5.2); Bennett and Hawley[2] also recommend this for calcium contents less than about 5%. Hill et al.[9] prefer a titration with EDTA, using Calred as indicator (Sec. 6.5.2). Grillot et al.[6] recommend the addition of calcium to bring the Ca:Mg ratio to less than 1:5 for the EDTA titration of calcium in samples having much magnesium; for larger ratios the calcein end point is seriously impaired. For their rapid methods of analysis, Shapiro and Brannock[18] prefer a visual EDTA titration of the total Ca + Mg in carbonates (MgO > 2.5%), using Eriochrome Black T and methyl red as a mixed indicator; the calcium, in another aliquot portion, is separated by a single precipitation as oxalate, the magnesium titrated (photometrically) with EDTA, using Eriochrome Black T alone as indicator, and the calcium content obtained by difference. When the MgO content is < 2.5%, it is determined colorimetrically as the Thiazole Yellow complex and the calcium again obtained by difference.

In magnesites, the MgO content is so high that it is impossible to prevent some precipitation of Mg $(OH)_2$ when ammonia is added, even in the presence of ammonium chloride, and this results in a slow end point (Ref. 2, p. 79). This, together with the possible effect of iron and manganese on the methylthymol blue complexone indicator favored by Bennett and Hawley[2] for the EDTA titration of calcium and magnesium, makes necessary both the use of an indicator having a sharper end point and the prior removal, by solvent extraction under slightly acid conditions with sodium diethyldithiocarbamate, of the iron and manganese. Following the extraction, sufficient EDTA is added to complex about 95% of the magnesium to prevent its precipitation, triethanolamine is then added to complex any aluminum and titanium present, the pH is adjusted to 10–10.5 and the remaining magnesium is titrated with EDTA, using Solochrome Black 6B (Eriochrome Blue Black B) as indicator (red through purple to clear blue). A correction must, of course, be made for any calcium also present.

The quantitative determination of calcite and dolomite in soils and limestone is often very important for agricultural purposes. Frequently

308 ROCK AND MINERAL ANALYSIS

this involves the determination of calcium and/or magnesium and a variety of methods have been suggested for this purpose. These have been summarized recently and a new method involving selective dissolution of calcite in a solution of sodium citrate and sodium dithionite[15,16] has been described.

7.7. Titanium and manganese

The methods for the determination of titanium and manganese in carbonates are the same as those used for silicates and these have been discussed in detail in Sec. 6.7 (Titanium) and Sec. 6.8 (Manganese).

7.8. Phosphorus

See Sec. 6.9 for discussion of the determination of phosphorus in silicates. The phosphorus content of carbonates will generally be low and amenable to a spectrophotometric determination (Sec. 9.19). Shapiro and Brannock[18] employ a double dilution and bracketting standards for samples having $>2.5\%$ P_2O_5, but otherwise the procedure is similar to that (solution B) used for silicates. For limestones with a very low P_2O_5 content, a gravimetric concentration procedure is recommended (Ref. 10, p. 974); the phosphorus is precipitated (from the nitric acid solution of a large sample) as calcium phosphate, dissolved and reprecipitated as ammonium phosphomolybdate.

7.9. The alkali metals

The methods given in Secs. 9.20 and 9.21 for silicates are equally applicable to carbonates with few modifications.

In general, a larger sample weight (about 1 g) should be taken because of the generally lower alkali content of the sample. If decomposition is to be done by the J. Lawrence Smith sintering procedure, the addition of $CaCO_3$ to the sample should not be omitted but the amount may be halved.[10] Acid decomposition is simpler and more frequently used; Hill et al.[9] employ ignition of the residue (see Ref. 23, Secs. 6.10 and 9.21.3) but carry out a precipitation with aqueous ammonia (bromthymol blue) as well in the aqueous leach of the ignited residue, presumably to make the separation as complete as possible (no precautions are taken to avoid losses of the alkali metals through adsorption.)

Bennett and Hawley (Ref. 2, p. 210) recommend the addition of aluminum sulfate (1 g in 250 ml) to solutions of limestones and dolomites, to minimize the interference of calcium in the flame photometric determination of the alkalis.

7.10. Ferrous and total iron

The problems inherent in the determination of ferrous iron have been discussed in considerable detail in Sec. 6.11.1. Because carbonates are much more readily acid-soluble than are silicates, these problems are minimized, and for a pure carbonate the method involves little more than mere decomposition of the sample in acid in an erlenmeyer flask and immediate titration with either $KMnO_4$ or $K_2Cr_2O_7$.

If the sample contains an appreciable amount of silicates, it may be desirable to determine the ferrous (and ferric) iron in the acid-soluble and acid-insoluble portions. The sample is attacked by dilute H_2SO_4 (or dilute HCl if $K_2Cr_2O_7$ is to be used in the titration, and the higher oxides of manganese are absent) in an erlenmeyer flask from which air has been expelled by steam (or by a curent of CO_2), cooled rapidly, and titrated immediately. The solution is filtered, the residue washed thoroughly with water and then washed from the paper into a platinum crucible. The bulk of the water is evaporated and the determination continued as described in Sec. 9.22.1. Total iron is determined in each portion by the method described in Sec. 9.23.1; the titration of both the ferrous and total iron may be done potentiometrically (see Secs. 9.22.2, 9.23.2), but this would involve transferring the contents of the flask, in the case of the acid-soluble portion, to a suitable beaker and thus increasing the possibility of some air-oxidation of the ferrous iron.

Hillebrand et al. (Ref. 10, pp. 975–976) have emphasized the virtual impossibility of obtaining a reliable value for ferrous iron in the presence of carbonaceous matter, so often found in limestones. They recommend that the decomposition be done with cold acid in a carbon dioxide atmosphere (for dolomite some heating will be necessary), that the solution be filtered through an asbestos pad rapidly (and preferably again in a carbon dioxide atmosphere), washed and titrated rapidly ($K_2Cr_2O_7$ is preferable as titrant because it is less affected by the presence of organic matter than is $KMnO_4$). The decomposition time should be kept to a minimum.

7.11. Combined water

The determination of total water in carbonates is made precisely as described for silicates (Sec. 9.25) and is subject to the same errors (Sec. 6.12). In addition, more care is needed to insure complete expulsion of carbon dioxide from the tube after the ignition; the heating should be done gradually so as to avoid losses of water vapor because of too sudden evolution of carbon dioxide from the carbonates. Because carbonaceous matter is more often present in carbonates than in silicates, there is the possibility that values from combined water will tend to be on the high side as a result of the oxidation of the hydrogen of the organic compounds.

A combined micro-method for the determination of water and carbon dioxide in carbonate (and silicate) minerals has been described[17]; the sample is heated to 1100°C and the carbon dioxide and water absorbed in suitable absorbents.

The determination of moisture (H_2O^-) is described in Secs. 7.1 and 10.1. The value so obtained is subtracted from the value for total water to give the percentage of combined water in the sample.

7.12. Carbon

7.12.1. "CARBONATE" CARBON

Methods for the determination of "carbonate" carbon in silicates have been discussed in detail in Sec. 6.13.1 and should be consulted. The determination of carbon dioxide by an acid-evolution gravimetric procedure, and by a volumetric procedure, is given in detail in Secs. 9.26.1 and 9.26.2 respectively; a modification of the volumetric method has recently been described by Hülsemann.[11]

Two methods have been proposed for the determination of CO_2 in limestone and dolomite by acid-base titration.[7] A routine determination can be made by digesting the sample overnight in a measured amount of $0.5N$ HCl, and titrating the excess HCl with NaOH to a bromphenol blue end point. A more rigorous method involves boiling the acid mixture and back-titrating the excess as before to obtain a first approximation of the quantity of HCl required; to a second sample a sufficient quantity of HCl is added to give a back-titration with NaOH of 1–7 ml, the mixture is boiled and the titration made as before. Any dark insoluble matter should be removed by filtration before the NaOH titration; if the sample is Fe(II)-rich, the end point can be improved by adding several grams of NH_4Cl to the solution before the titration.

A combined micro-method for the determination of water and carbon dioxide in carbonate (and silicate) minerals has been described[17] (see Sec. 7.11). Another, but very different, micro-method using a small sample (0.1–0.2 g) involves the acid release of the carbon dioxide within a sorption apparatus where it is swept by helium through a purification train and then through a thermal conductivity cell in which it is measured[19]; very good results are reported for the NBS argillaceous limestone and dolomite and for G-1 (granite) and W-1 (gabbro), and a single determination is said to require only 15 min for completion.

Some rapid and indirect methods for the determination of carbon dioxide have been mentioned in Sec. 6.13.1. Turekian has described a rapid technique for the determination of the carbonate content of deep-sea cores[20]; a 20 mg sample is dissolved in HCl, the solution is buffered with NH_4OH-NH_4Cl, KCN is added and the soluble alkaline earths and magnesium are titrated with EDTA, using Eriochrome Black T as indicator. It is assumed that the total soluble alkaline earths and magnesium are present as carbonates, that all of the carbonate in the sample is bound to the elements titrated, and that no appreciable amount of alkaline earths and magnesium is released from silicates by the hydrochloric acid.

7.12.2. CARBONACEOUS MATTER

This also has been discussed in detail in Sec. 6.13.2 and a combustion-gravimetric method for the determination of total carbon (in silicates) is given in Sec. 9.27. This latter is applicable to carbonates as well but it is preferable to remove first that portion of the carbon present as carbonate; a procedure for this has been suggested in Sec. 9.27 (see footnote) and a more rigorous procedure is given by Hillebrand et al. (Ref. 10, p. 977).

7.13. Sulfur, chlorine, fluorine

The discussions of the determination of these elements in silicates should be consulted. There are no special problems involved in their determination in carbonates.

"Sulfur," see Sec. 6.14; "Chlorine," Sec. 6.15; "Fluorine," Sec. 6.16.

7.14. Strontium and barium

See Sec. 6.17 for a discussion of the separation and the determination of these constituents in silicates. This is directly applicable to carbonates. Oosterom[14] has recently described an X-ray fluorescence spectroscopic method for the determination of strontium and barium (lower limit of detection 5–10 ppm) in carbonates. The sample (20–5 μ particle size) is compared with standards prepared by adding $BaCO_3$ and $SrCO_3$ to calcite or dolomite.

7.15. Other constituents

It is seldom necessary to determine other than those constituents listed previously; when other elements are required, their determination is usually conveniently done by optical emission spectroscopy. The discussions given in Chapter 6 for a number of elements which are present in silicates, as a general rule, in trace concentrations only (Secs. 6.18 to 6.25), are directly applicable to the determination of these constituents in carbonates and should be consulted.

References

1. Ahrens, L. H., and Taylor, S. R. *Spectrochemical Analysis*, 2nd ed., 1961. Reading, Mass.: Addison-Wesley, pp. 297–301.
2. Bennett, H., and Hawley, W. G. *Methods of Silicate Analysis*, 2nd ed., 1965. London: Academic Press, pp. 203–219.
3. Bisque, R. E. Analysis of carbonate rocks for calcium, magnesium, iron and aluminum with EDTA. *J. Sediment. Petrol.* **31**, 113–122 (1961).
4. Gault, H. R., and Weiler, K. A. Studies of carbonate rocks. III. Acetic acid for insoluble residues. *Penn. Acad. Sci. Proc.* **29**, 181–185 (1955).
5. Goldich, S. S., Ingamells, C. O., and Thaemlitz, D. The chemical composition of Minesota lake marl—comparison of rapid and conventional chemical methods. *Econ. Geol.* **54**, 285–300 (1959).
6. Grillot, H., Béguinot, J., Boucetta, M., Rouquette, C., and Sima, A. Méthodes d'analyse quantitative appliquées aux roches et aux prélèvements de la prospection géochimique. *Bur. Rech. Géol. Min.*, *Mém.* **30**, 63–92 (1964).
7. Grimaldi, F. S., Shapiro, L., and Schnepfe, M. Determination of carbon dioxide in limestone and dolomite by acid-base titration. *U.S. Geol. Surv. Profess. Papers* **550-B**, B186–188 (1966).
8. Hanna, W. C. "Portland Cement," in *Standard Methods of Chemical Analysis*, Vol. 11A, 6th ed., F. J. Welcher, Ed., 1963. Princeton, N. J.; D. Van Nostrand, pp. 1053–1076.
9. Hill, W. E., Jr., Waugh, W. N., Galle, O. K., and Runnels, R. T. Methods of chemical analysis for carbonate and silicate rocks. *State Geol. Surv. Kansas Bull.* **152**, Part I, 1–30 (1961).
10. Hillebrand, W. F., Lundell, G. E. F., Bright, H. A., and Hoffman, J. I. *Applied Inorganic Analysis*, 2nd ed., 1953. New York: John Wiley & Sons, pp. 958–986.
11. Hülsemann, J. On the routine analysis of carbonates in unconsolidated sediments. *J. Sediment. Petrol.* **36**, 622–625 (1966).
12. Lundell, G. E. F., and Hoffman, J. I. *Outlines of Methods of Chemical Analysis*, 1938. New York: John Wiley & Sons, pp. 180–183.
13. Manheim, F. Water determination in sediments and sedimentary rocks. *Acta Univ. Stockholm.* **6**, 127–143 (1960).
14. Oosterom, M. G. Strontium and barium in carbonates of crystalline rocks. *Konink. Ned. Akad. Wetenschap. Proc.* **49**, 710–719 (1966).
15. Petersen, G. W., and Chesters, G. Quantitative determination of calcite and dolomite in pure carbonates and limestones. *J. Soil Sci.* **17**, 317–327 (1966).
16. Petersen, G. W., Chesters, G., and Lee, G. B. Quantitative determination of calcite and dolomite in soils. *J. Soil Sci.* **17**, 328–338 (1966).
17. Riley, J. P., and Williams, H. P. The microanalysis of silicate and carbonate minerals. II. Determination of water and carbon dioxide. *Mikrochim. Acta* **4**, 525–535 (1959).

18. Shapiro, L., and Brannock, W. W.   Rapid analysis of silicate, carbonate and phosphate rocks.   *U.S. Geol. Surv. Bull.* **1144–A**, 1962, 56 pp.
19. Thomas, J., Jr., and Hieftje, G. M.   Rapid and precise determination of carbon dioxide from carbonate-containing samples using modified dynamic sorption apparatus.   *Anal. Chem.* **38**, 500–503 (1966).
20. Turekian, K. K.   Rapid technique for determination of carbonate content of deep-sea cores.   *Bull. Am. Assoc. Pet. Geol.* **40**, 2507–2509 (1956).

CHAPTER 8

OTHERS

This chapter is admittedly an attempt to say something about the "grey area" that exists between methods for the analysis of silicates and carbonates (Chapters 6 and 7) and those for the analysis of ore minerals, such as sulfides and oxides, and the more exotic minerals such as titanates, borates and the sulfosalts. Included in this "grey area" are methods for those very complex materials, the meteorites.

The analysis of ores requires the use of many of the methods described in Chapter 9, but because of the different type of major constituent that is encountered in ore analysis (usually present as a minor or trace constituent in silicates and carbonates), these must often be considerably modified to meet the new analytical conditions required. To attempt to suggest possible modifications for even a few ore minerals would be both unsatisfactory and hazardous in view of the writer's limited experience with this type of analysis and the reader should consult a more specialized text. Among the references cited in the preceding chapters there are a few which include methods for the analysis of material other than a silicate or carbonate.[1,13,16]

Many minerals, especially those in which one or more of the major constituents exist in more than one valence state, present exceedingly difficult problems for the analyst, and as has already been noted, their analysis is an individual study in applied analytical research. Unfortunately, the methods are described only too infrequently in the literature, such as those, for example, used for the analysis of simplotite ($CaV_4O_9 \cdot 5H_2O$),[15] wegscheiderite ($Na_2CO_3 \cdot 3NaHCO_3$),[8] the determination of the ammonium ion in buddingtonite ($NH_4AlSi_3O_8 \cdot \frac{1}{2}H_2O$),[7] and the analysis of molybdates and tungstates.[6] It is devoutly to be hoped that more papers which describe new minerals will include a brief description of the methods used to determine the composition of these minerals, such as was done by Donnay, Ingamells and Mason (Supp. Ref.).

The analysis of *chromite* is difficult, both because of the refractory nature of the mineral which defies most of the commonly used methods of decomposition and because of the high concentrations of Fe(II) and Cr which present serious analytical problems in obtaining reliable results. In recent years, two analytical schemes have been proposed

314

for the complete analysis of chromite, and methods suggested for its decomposition (see also Chapter 5).

Bilgrami and Ingamells[2] describe methods for the determination of total H_2O, Fe(II), Cr, Mn, V, Si, Al, Ti, total Fe, Ca, Mg, and Ni, with separate sample portions being used for Mn and V, total H_2O (Penfield method, Sec. 6.12), Fe(II) and Cr, and the remainder. The sample used for the determination of Mn and V is fused with Na_2O_2 in an iron crucible, but that for the Si, Al, Ti, total Fe, Ca, Mg, and Ni is decomposed by prolonged heating and refluxing with perchloric acid. The *silica* is separated and determined as usual, followed by precipitation of the R_2O_3 group with aqueous ammonia, and its ignition and weighing. Sodium hydroxide is used to separate Fe and Ti from the pyrosulfate solution of the R_2O_3 oxides, the precipitate is dissolved in H_2SO_4 and the *titanium* and *total iron* determined colorimetrically (H_2O_2) and by potentiometric titration respectively. Because some *chromium* separates with the R_2O_3 group, it is necessary to determine it colorimetrically in the filtrate from the NaOH precipitation (*vanadium* can also be separated with cupferron and determined at this stage, if not done previously), in order to correct the value for the *alumina* which is obtained by difference. *Magnesium*, together with Mn, Ca and Ti, are precipitated with diammonium phosphate, ignited and weighed as $Mg_2P_2O_7$, which is then dissolved in H_2SO_4; *calcium* is separated as $CaSO_4$ in 80% alcohol solution (Sec. 9.10.2) and determined either gravimetrically (oxalate) or titrimetrically (EDTA), and the alcoholic filtrate is used for the colorimetric determination of *manganese* (periodate) and the gravimetric determination of *nickel* (dimethylglyoxime). In the determination of manganese and vanadium on a separate portion, the sodium peroxide fusion cake is leached with water to separate the *manganese* as a hydrous oxide, which is then dissolved and the determination finished colorimetrically; chromium in the filtrate is reduced with H_2O_2, the *vanadium* is precipitated with cupferron, ignited and fused with $K_2S_2O_7$ and determined colorimetrically as peroxyvanadate. The method described by Seil,[14] in which the sample is decomposed with a mixture of H_3PO_4 and H_2SO_4 in a CO_2 atmosphere and the SO_2 formed from the reaction with Fe(II) is passed through a standard solution of $K_2Cr_2O_7$, is used in a slightly modified form for the determination of *ferrous iron*. The H_3PO_4–H_2SO_4 solution of the sample is used for the titrimetric determination of *chromium*.

Decomposition of chromite with $HClO_4$ is considered by Malhotra and Prasada Rao[10] to result in low values for chromium because of its volatilization as chromyl chloride through reaction with hydrochloric acid formed by decomposition of the perchloric acid. Because they determine chromium on the sample (0.5 g) used for the main portion analysis, they prefer to fuse the sample with anhydrous Na_2CO_3 (4–5 g) in a platinum crucible into which a stream of oxygen is blown during the fusion (see Sec. 5.3.1). The filtrate from the leaching of the fusion cake is used for the determination of *chromium* (oxidized to Cr(VI) by $S_2O_8^{2-}$ and $AgNO_3$) after a double precipitation with aqueous ammonia to separate *aluminum* and *silicon*; these latter are determined gravimetrically. The residue from the leach is used for the determination of *total iron*, *titanium*, *manganese*, *calcium*, *magnesium* and remaining silicon and aluminum, by conventional methods. *Total H_2O* is obtained by the Penfield method (Sec. 6.12), and *nickel* is determined colorimetrically with α-furildioxime on a separate sample portion which is decomposed with perchloric acid and the chromium removed by volatilization as chromyl chloride. Seil's method,[14] again slightly modified, is used to determine ferrous iron.

Phosphates also pose a number of awkward problems for the analyst, particularly in obtaining a reliable value for aluminum. The systematic analysis of phosphates has been given some attention lately and analytical schemes have been proposed; Glover and Phillips[9] also have described micro methods for the determination of fluorine, phosphate, iron and carbonate in fossil bone.

Wilson[17] uses a double precipitation with benzoic acid at pH 3.5 to separate such constituents as aluminum, titanium and iron from the alkaline earths and magnesium. The *aluminum* is then separated from the indeterminate amount of phosphate in the benzoate precipitate and determined by precipitating it with 8-hydroxyquinoline, after a preliminary cupferron extraction to eliminate iron and titanium. *Calcium* is precipitated as oxalate in the filtrate from the benzoate separation at pH 4, and *magnesium* as the phosphate, in the usual manner. A separate sample portion is used to determine *total iron* with 2,2′-bipyridyl; the interference of phosphate (ferric phosphate will precipitate under weakly acid conditions) is overcome by reducing the Fe(III) in acid solution with ascorbic acid, and that of titanium (which forms a yellow complex with ascorbic acid) by developing the ferrous complex in a chloracetate buffer solution at pH 3.0. A separate sample portion is fused with NaOH in a silver crucible and an aliquot of the resulting solution used for the colorimetric determination of *silica* as the molybdenum blue; oxalic acid is used to suppress the interference of phosphate when the silicomolybdic acid is reduced with ascorbic acid. No special difficulties are encountered with the separate determination of manganese and titanium, but methods for other constituents are not given.

Cruft, Ingamells and Muysson[4] stress that errors in the analysis of phosphates stem from the fact that not only are elements such as the rare earths, lithium, strontium and fluorine apt to be ignored by the analyst and to be collected, weighed or titrated as something else, but that they and phosphate also tend to interfere with ordinary analytical procedures. The complete recovery of combined hydrogen (H_2O^+) from phosphates is difficult and a modified Penfield procedure and apparatus is described (see Sec. 6.12) for this purpose. *Phosphate* is determined gravimetrically by precipitation with magnesia mixture, after a preliminary separation as phosphomolybdate (Sec. 9.18); the sample, with added SiO_2, is sintered with anhydrous Na_2CO_3 and Na_2O_2, leached with water and a further recovery of phosphate made by acid decomposition and dehydration of the siliceous residue. The interference of fluorine in the determination of *silica* is overcome by adding a large excess of aluminum chloride to the fusion cake before solution and dehydration with hydrochloric acid; *thorium* and *total rare earths* are also determined in the same sample portion. Another sample portion is used for the determination of main portion constituents, following fusion with Na_2CO_3 and separation of the silica; phosphate is precipitated and removed by adding a calculated excess of zirconyl chloride, a precipitation with aqueous ammonia separates the R_2O_3 group (because Zr and PO_4^{3-} will be present in indeterminate amounts, *aluminum* must be determined directly), and a further precipitation with ammonium persulfate removes Mn and Ce before the normal separation of *calcium* and *magnesium* is done. The *alkali metals* can be determined by standard procedures, except for *lithium* which is best determined on the main portion sample when decomposition is done by a mixture of boric and perchloric acids instead of by fusion.

Chung and Riley[3] have developed a detailed analytical scheme for the analysis of monazite and monazite concentrates. The scheme is divided into two parts: (*1*) the monazite itself, which is that portion of the sample dissolved by treatment with concentrated sulfuric acid, and (*2*) the unattached portion consisting largely of small amounts of associated minerals in the form of inclusions. The sulfuric acid solution is poured through an ion-exchange column and phosphate is washed out with water; three groups are then eluted as follows, (*1*) Al, Ca, Mg, Fe, Pb, Ti and U (Mn and alkali metals) are removed with 1N HCl, (*2*) the rare earths with 3N HCl, and (*3*) thorium, with 3.6N H_2SO_4. Spectrophotometric methods are given for Al, Fe, Ti, Pb, U and Mn, the alkali metals are determined flame photometrically and calcium and magnesium by photometric titration with EDTA. Cerium is separated from the other rare earths by solvent extraction of its nitrate with methyl iso-butyl ketone, and both groups are determined gravimetrically; thorium is determined spectrophotometrically with thorin (1-(o-arsonophenylazo)-2-naphthol-3,6-disulfonic acid sodium salt). Silica, phosphate, tin and chromium are determined on separate portions by photometric methods, following decomposition of the sample by either fusion (Si, Cr), acid treatment (P), or both (Sn).

Two analytical schemes have been offered for the analysis of *chondritic (stony) meteorites* that are very different in their approaches.

Moss *et al.*[11,12] have developed a procedure in which pure dry chlorine is passed over the meteorite sample heated to 310°C in a porcelain boat and then through an absorption train consisting of a dry, conical flask followed by two wash bottles (see also Sec. 5.5.3). The residue in the boat is leached with water, to give an undissolved fraction consisting of all the silicates, phosphates and oxides in the sample, plus a small amount of Ni-rich taenite; the bulk of the metallic phase of the meteorite, together with all of the sulfides, phosphides and carbides, will be represented by the contents of the absorption train, and the solution obtained by leaching the residue in the porcelain boat. If a preliminary magnetic separation, even if incomplete, is made and the two fractions treated as described previously, it is possible then to make a reliable calculation of the distribution of most elements between the metallic, sulfide and oxidized phases. The final determination of the various constituents in the solutions from the absorption train and leaching of the boat residue, and in the unattacked material, is done by various standard methods.

The scheme outlined by Easton and Lovering[5] involves the use of five sample portions of the meteorite. One portion (A) is treated with mercuric-ammonium chloride to extract the macro constituents of the metallic phase (Fe, Co, Ni). Another portion (B) is decomposed with HF and H_2SO_4 and used for the determination of Na, K and minor constituents (C, Cr, Mn, Ti, P); a third portion (C) is also decomposed with HF, after preliminary oxidation with bromine and nitric acid, and the solution used to determine the sulfide content, and Al, Ca and Mg. Fusion of a fourth portion precedes the combined gravimetric-colorimetric determination of SiO_2, and a fifth portion is required for the determination of water and carbon dioxide. Solution A is passed through an anion exchange column and the retained Fe, Co and Ni are then eluted separately; Solution C is passed through a cation exchange column, from which the sulfide, as SO_4^{2-}, is eluted first, followed by the members of the R_2O_3 group, Ca and Mg. The R_2O_3 oxides are separated from calcium and magnesium by a double precipitation with aqueous ammonia. Final determination of these various constituents is by a combination of standard colorimetric, titrimetric and gravimetric methods.

References

1. Bennett, H., and Hawley, W. G. *Methods of Silicate Analysis*, 2nd ed. (rev.)' 1965. London: Academic Press, 334 pp.
2. Bilgrami, S. A., and Ingamells, C. O. Chemical composition of Zhob Valley chromites, West Pakistan. *Am. Mineral.* **45,** 576–590 (1960).
3. Chung, K. S., and Riley, J. P. A scheme for the analysis of monazite and monazite concentrates. *Anal. Chim. Acta* **28,** 1–29 (1963).
4. Cruft, E. F., Ingamells, C. O., and Muysson, J. Chemical analysis and the stoichiometry of apatite. *Geochim. Cosmochim. Acta* **29,** 581–597 (1965).
5. Easton, A. J., and Lovering, J. F. The analysis of chondritic meteorites. *Geochim. Cosmochim. Acta* **27,** 753–767 (1963).
6. Easton, A. J., and Moss, A. A. The analyis of molybdates and tungstates. *Mineralog. Mag.* **35,** 995–1002 (1966).
7. Erd, R. C., White, D. E., Fahey, J. J., and Lee, D. E. Buddingtonite, an ammonium feldspar with zeolitic water. *Am. Mineral.* **49,** 831–850 (1964).
8. Fahey, J. J., and Yorks, K. P. Wegscheiderite ($Na_2CO_3 \cdot 3NaHCO_3$), a new saline mineral from the Green River formation, Wyoming. *Am. Mineral.* **48,** 400–403 (1963).
9. Glover, M. J., and Phillips, G. F. Chemical methods for the dating of fossils. *J. Appl. Chem. (London)* **15,** 570–576 (1965).
10. Malhotra, P. D., and Prasada Rao, G. H. S. V. Investigations on the chemical composition of Indian chromites. *Rec. Geol. Surv. Ind.* **93,** 215–248 (1966).
11. Moss, A. A., Hey, M. H., and Bothwell, D. I. Methods for the chemical analysis of meteorites. I. Siderites. *Mineral. Mag.* **32,** 802–816 (1961).
12. Moss, A. A., Hey, M. H., Elliott, C. J., and Easton, A. J. Methods for the chemical analyis of meteorites. II. The major and some minor constituents of chondrites. *Mineralog. Mag.* **36,** 101–119 (1967).
13. Schoeller, W. R., and Powell, A. R. *The Analysis of Minerals and Ores of the Rarer Elements*, 3rd ed. (rev.), 1955. New York: Hafner, 408 pp.
14. Seil, G. E. Determination of ferrous iron in difficulty soluble materials. *Ind. Eng. Chem. Anal. Ed.* **15,** 189–192 (1943).
15. Thompson, M. E., Roach, C. H., and Meyrowitz, R. Simplotite, a new quadrivalent vanadium mineral from the Colorado Plateau. *Am. Mineral.* **43,** 16–24 (1958).
16. Vinogradov, A. P., and Ryabchikov, D. I., (Eds.). *Detection and Analysis of Rare Elements*, 1961. Moscow: Izdatels'vo Akademii Nauk SSSR, translation by Israel Program for Scientific Translations, Jerusalem, 1962, 744 pp.
17. Wilson, A. D. A scheme for the quantitative analysis of a phosphate rock including the use of a benzoate separation. *Analyst* **88,** 18–25 (1963).

Supplementary References

Bennett, H., and Marshall, K. The separation and determination of chromium sesquioxide in chrome ores and chrome-bearing refractories. *Analyst* **88,** 877–881 (1963). The sample is fused with Na_2CO_3 + H_3BO_3 and dissolved in H_2SO_4, from which the $Cr_2O_7{}^{2-}$ is extracted by liquid ion-exchange resin in $CHCl_3$ (extraction must not be made in direct sunlight because of possible reduction of $Cr_2O_7{}^{2-}$ by resin (See Bennett and Reed, Supp. Ref.). $Cr_2O_7{}^{2-}$ is stripped into aqueous phase with KOH, the Cr (VI) reduced with Na_2SO_3 and determined spectrophotometrically as the complex formed with EDTA.

Bennett, H., and Marshall, K. The analysis of chrome-bearing materials. II. The determination of silica and lime. *Trans. Brit. Ceram. Soc.* **65,** 681–692 (1966). An aliquot of the solution from the separation of $Cr_2O_7{}^{2-}$ (See Bennett and Marshall, 1963) is used for the direct determination of Si as the yellow silicomolybdate complex. Ca is determined in another aliquot by EDTA titration, after removal of interfering elements with cupferron and diethylammonium diethyldithiocarbamate.

Bennett, H., and Reed, R. A. The analysis of chrome-bearing materials. III. The determination of alumina. *Trans. Brit. Ceram. Soc.* **66,** 569–585 (1967). An aliquot of the solution used for the determination of calcium (See Bennett and Marshall, 1966), from which all interfering elements were removed, is treated with excess EDTA and the Al determined by back-titration of excess EDTA with zinc solution.

Donnay, G., Ingamells, C. O., and Mason, B. Buergerite, a new species of tourmaline. *Am. Mineral.* **51,** 198–199 (1966).

Van Loon, J. C., Parissis, C. M., and Kingston, P. W. Analysis of very small samples of sulfide minerals. *Anal. Chim. Acta* **40,** 334–338 (1968). Methods for the determination of S^{2-} (as methylene blue) and Pb, Bi, Cu and Fe (by atomic absorption spectroscopy) in 100 μg–1 mg sample are given.

PART III

Methods of Analysis—Selected Procedures

CHAPTER 9

SILICATES

9.1. Determination of H_2O^- by heating at 105–110°C

Weigh 0.8000 g of sample,* or less if the supply is limited, which has been ground to approximately 150 mesh and thoroughly mixed,† and transfer it to a clean, weighed 30 ml platinum crucible (a porcelain or nickel crucible if this portion is not to be used further). Weigh the covered crucible and sample; the difference in weight of the sample should not exceed 0.2 mg. Place the uncovered crucible in an oven, cover it with a 7 cm diameter filter paper, and heat it at 105–110°C for 1 hr. Transfer the crucible to a desiccator, cover and allow it to cool for 30 min before weighing.

If the loss in weight exceeds 1 mg the heating, cooling and weighing should be repeated until constant weight is obtained. If the loss in weight exceeds 5 mg, the crucible should be heated at a higher temperature, e.g., 125°C, to see if a further loss occurs, indicating the presence of a significant amount of hydrous material.

When a significantly hygroscopic sample is to be weighed it is essential that it be in equilibrium with the atmosphere. Spread the sample on a clean sheet of glazed paper and allow it to stand exposed overnight (a paper canopy should be erected to protect the sample from atmospheric dust), then weigh all portions to be used in the analysis at the same time.

9.2. Gravimetric determination of silica in samples containing $<2\%$ fluorine

9.2.1. BY FUSION WITH SODIUM CARBONATE

If the determination of moisture (Sec. 9.1) was done in a platinum crucible the same sample portion (0.8000 g) can be used for the main

---

* If this is to be used also for the determination of the "main portion" constituents, 0.8000 g is a suitable sample weight. If only the moisture is to be determined, a 1 g sample will be more convenient.

† The use of a magnetized spatula in the weighing of samples, which may result in a concentration of magnetic particles, is not recommended. The narrow nickel and aluminum spatulas are a very convenient means for transferring material from the sample bottle to the weighing container.

portion analysis. If not, weigh 0.8000 g of sample, which has been ground to approximately 150 mesh, and transfer it to a clean, weighed 30 ml platinum crucible. Weigh the covered crucible and sample; the difference in weight of the sample should not exceed 0.2 mg.

Support the platinum crucible, covered, on a clay or silica triangle over a bunsen burner and heat the crucible with a low flame to drive off combined water. Increase the height of the flame until the bottom portion of the crucible is a dull red in color, move back the lid to permit the entrance of air and maintain the bottom of the crucible at a dull red for 5 min. This will oxidize any reducing substances which may be present and which would otherwise attack the crucible during the fusion. Cover the crucible and allow it to cool.

Weigh 3 grams of anhydrous Na_2CO_3 and add about 2 grams of it to the cooled crucible (Caution! avoid loss of sample by dusting). Mix the sample and flux thoroughly with a short length of glass rod, the ends of which have been flame-polished. Add approximately 0.1 g Na_2O_2* and again mix thoroughly. Do not attempt to brush off the glass rod; tap it against the sides of the crucible to loosen adhering particles and "rinse" it in the remainder of the flux. Tap the crucible gently on the bench top to settle the fusion mixture and sprinkle the rest of the flux over the surface to provide a trap for decrepitating particles during the initial heating stages of the fusion step.

Return the covered crucible to the triangle support and begin heating the crucible again with a low flame (see footnote, page 327). After 2 or 3 min, increase the height of the flame so that, at the end of about 5 min, the bottom portion of the crucible is dull red in color. Maintain this temperature for 2 or 3 min and then increase the flame until the crucible bottom becomes a medium red color; heat at this temperature for 5 min and then adjust the height of the flame to give maximum heat. The whole crucible should be a medium red color but at no time should the flame envelop the crucible, nor should the crucible bottom ever come into contact with the blue cone of the burner flame. Continue the heating for 5 min. Heat the outer surface of the lid with another burner, to fuse any particles that may have been spattered on the inner surface of the lid. Quickly transfer the covered crucible to another triangle and heat at the full heat of a meker burner for 10 min, observing the pre-

* Washington (Ref. 42, Sec. 6.2) prefers roasting of the sample to the addition of an oxidant, because the presence of even a small amount of KNO_3 may cause spattering and also attack the crucible. These difficulties are not encountered with Na_2O_2 (see Sec. 5.3.1).

cautions mentioned previously. It is a useful practice, at the start of this final period of heating, to move back the lid, grasp the crucible firmly with platinum-tipped tongs and swirl it to incorporate any unfused material which may be clinging to the sides of the crucible. The amenability of the melt to such swirling will depend upon the nature of the sample; a highly siliceous sample will, on fusion, favor the formation of a viscous liquid but a basic sample will tend to form a sintered or partly liquid mass that will resist swirling.

The fusion is generally complete when the molten mass appears quiescent. An examination of the crucible contents for unattacked particles is not very feasible because the molten mass is seldom clear. At the end of the heating period described previously only chromite and zircon may be unattacked, and further heating at about 1100°C is necessary to insure the decomposition of these minerals. Their possible presence will be indicated, however, in a preliminary semiquantitative spectrographic analysis. Remove the crucible cover and carefully place it, face up, on a polished rock slab. Grasp the crucible with the platinum-tipped tongs, remove it from the flame and rotate the crucible as it cools so that the contents will solidify around its walls. Cover the crucible and allow it to cool. Replace the covered crucible on the triangle and again heat it with the meker burner for no more than 15 sec at a dull red heat; remove it from the flame and allow it to cool to room temperature. The color of the cooled melt may be significant; a green color may indicate the presence of manganate ion which will react to form chlorine when the melt is dissolved in hydrochloric acid and, because this will attack platinum, a few drops of alcohol should be added to the disintegrated melt before the addition of the acid to reduce the Mn(VI) to Mn(II).

Add about 20 ml H_2O to the crucible and, with a glass stirring rod, apply gentle pressure to the cake to loosen it.* Place the crucible lid in a 200 ml capacity, glazed porcelain casserole or in a 250 ml platinum

---

* The brief reheating of the cooled cake serves to melt the layer adjacent to the crucible walls and, because of the contraction of this layer on cooling, water is able to penetrate the space and loosen the cake in the crucible. There are many ways suggested for the removal of the cake but all of them, including the placing of the crucible and lid in a covered dish prior to the addition of acid, favor the accidental loss of material or deformation of the crucible. It is only rarely that the cake, when pressed with a glass rod, will not separate easily and cleanly from the crucible; for some samples the cake can be removed without water, simply by tapping the crucible sharply on the bench top.

dish* and cover it with water.   Police the lid to remove adhering particles and then, by means of platinum-tipped forceps, remove the lid
and rinse it into the casserole.   Carefully transfer the loosened cake
from the crucible to the casserole and police the inside of the crucible to
remove adhering particles, adding all rinsings to the casserole.

Should the cake not separate freely and easily, place the water-filled
crucible on the water bath and heat for about 30 min.   Gently press
the cake with a glass rod and, if it loosens, proceed as described previously.   If not, continue the heating and gentle pressure until the cake
loosens and disintegrates, and transfer it to the casserole.   Add a few
drops of alcohol, if necessary, to reduce Mn(VI).

Support a 50 mm diameter special funnel, whose lower 1 in. or so of
stem has been bent at about 30° angle from the vertical, in a filter stand.
Place the end of the bent stem in the lip of the casserole and cover the
casserole with a watch glass that extends about 1 cm beyond the side of
the casserole.   Place 10 ml 12N HCl in a graduated cylinder and pour
2–3 ml of it into the crucible.   Carefully police the inside of the crucible
and the inner surface of the crucible cover.   Rinse the cover into the
funnel and slowly pour the contents of the crucible into the funnel also†;
rinse out the crucible into the funnel and set the crucible and cover to one
side.   Pour the remaining acid into the funnel and rinse the funnel into
the casserole.   Allow the funnel and casserole to stand undisturbed
until the reaction has ceased, then remove the watch glass and rinse
the bent tip of the funnel into the dish.   Replace the cover and set the
casserole on the steam bath, keeping it covered until no further effervescence is noted; remove the cover, rinse it into the casserole and rinse
down the walls of the casserole.   With a stirring rod, break up any
lumps that remain and evaporate the contents to dryness.‡   At this

* Because of the difficulty in insuring that all of the separated silica has been
recovered from the porcelain casserole it is preferable to use a platinum basin for this
operation.   For routine rock analysis, however, porcelain casseroles are both eco-
nomically and analytically feasible.   A platinum dish should not be used if the
sample is highly ferruginous because the Fe(III), in the presence of HCl, will attack
the platinum:

$$Pt + 4Fe^{3+} + 6Cl^- \rightarrow Pt\ Cl_6{}^{2-} + 4Fe^{2+}$$

† A fleeting pink color which appears when the HCl is first added indicates the
presence of appreciable manganese in the sample.

‡ It is convenient and time-saving to do this evaporation overnight, although it
does not permit the breaking up of the crust that forms during the evaporation and
which tends to prevent the complete expulsion of hydrochloric acid.   It is good
practice in this case to moisten the dried salts with a few drops of dilute HCl (1:1)
and break up the residue with a stirring rod, and then to continue the dehydration.
Rapid dehydration can be obtained with a combination of bottom heating and
surface heating, the latter with an infrared lamp.

stage any unattacked material will be either visible, or detectable as a gritty residue with the stirring rod; if more than a few particles are found, the sample should be discarded and the fusion repeated at a higher temperature.

Ignite the fusion crucible and cover to redness over a meker burner, allowing free access of oxygen to the crucible, and note if it appears stained.* If not, or if there is only a light stain, place the covered crucible in a desiccator, allow it to cool for 30 min and weigh it; this is the weight of the empty crucible after the fusion and the difference between this weight and the original weight is a measure of the degree of attack, if any, that has occured during the fusion. If there is much stain add 10 ml dilute HCl (1:1) to the crucible, cover it and heat it on the water bath for a short time until the stain has disappeared (the acid will usually become very yellow in color). Rinse the solution into the casserole and again ignite the crucible as before. If the stain persists, continue the treatment with acid until it is eliminated. Finally ignite, cool and weigh the crucible as previously described.

The residue, when dry, will be pale yellow to pale brown in color; a golden-yellow color indicates the presence of free hydrochloric acid. Cover the casserole with a watch glass and heat for 1 hr on the water bath, remove it and allow it to cool. A higher temperature of heating should not be used at this stage, nor should the heating be prolonged, because of the danger of a reaction between magnesium chloride and silica to form soluble magnesium silicate, and because the degree of contamination of the SiO_2 by foreign ions is thereby increased.

To the cool residue add 5 ml $12M$ HCl, wetting all of the residue and particularly the ring marking the original level of the liquid in the casserole, from which it is difficult to remove particles of silica. Allow it to stand for 1-2 min and then add approximately 50 ml H_2O, washing down the sides of the casserole and the stirring rod. Cover, and heat the casserole on the water bath until all soluble salts have dissolved, as indicated by an absence of gritty particles. This heating should not be prolonged, in order that as little as possible of the silica will be dissolved, and more water may be added if necessary.

Prepare an 11 cm Whatman No. 41 filter paper in a 65 mm filter funnel and wash it once with the hot, dilute HCl (5:95) which is used as the

---

* The pick-up of iron from the sample by a platinum crucible when the fusion is done over a flame has been demonstrated by Shell (*Anal. Chem.*, **27**, 2006–2007 (1955)). This does not occur when fusions are done in an electric muffle furnace. He also found that the separated silica, if ignited at 1200°C in an electric muffle furnace and in the same crucible in which the flame fusion was made, will in turn pick up the Fe from the crucible.

wash solution in the filtration. Without delay, decant the solution through the paper, catching the filtrate in a 150 ml beaker, and wash the paper and residue once. Wash down the sides of the casserole, stir the silica into the wash liquid and pour it quickly into the funnel. Wash any remaining loose silica into the funnel with a jet of wash solution. Scrub the stirring rod and the inside of the casserole with a rubber policeman, wash them with the dilute HCl and pour the washings into the funnel. Wipe out the inside of the casserole, and the underside of the lip, with a small piece of filter paper and add it to the filter. Finally wash the residue and paper 10 times with the hot HCl (5:95), paying particular attention to the upper edge of the paper; all traces of yellow color should be removed.* Lift one edge of the paper with forceps to drain the funnel stem and cover the funnel with a numbered watch glass (a moistened, numbered filter paper may be fitted over the funnel; on drying it will form a tight cover). Rinse the tip of the funnel into the beaker.

Quantitatively transfer the contents of the beaker to the original casserole and again evaporate the contents to dryness as previously described.† When all traces of HCl have been removed, place the covered casserole in an oven and heat at 105–110°C for 1 hr. Cool, drench the residue with 5 ml 12M HCl, and dissolve the soluble salts in water as before; it is again important that the heating of the solution should not be prolonged and the filtration should be made without delay. Prepare a 9 cm diameter Whatman No. 40 filter paper in a 50 mm funnel, wash it once with warm dilute HCl (5:95) and filter the silica as described previously, catching the filtrate in a 400 ml beaker. Thoroughly police

---

* The separated silica should be white but may be colored by hydrolyzed iron oxides (reddish) or by platinum (grey). Under normal conditions these will do no harm.

† Shell (Sec. 6.2, Ref. 24, p. 138) prefers to use $HClO_4$ for the second dehydration; the advantages include the almost complete precipitation as the metal of any platinum introduced into the solution during the fusion. The use of perchloric acid should be weighed against the additional caution that is necessary when perchlorates are used (Sec. 5.2.5). Dissolve the fusion cake in 10–15 ml $HClO_4$ (72%) in a 250 ml beaker, or add this amount to the HCl filtrate from the first filtration. Add 5 ml HNO_3 (this, of course, excludes the use of platinum if HCl is present) and evaporate to dense fumes of $HClO_4$, adding more $HClO_4$ if much insoluble perchlorate tends to separate out. Cover with a watch glass and heat strongly for 15 min; the acid should reflux down the sides of the beaker. Cool, dilute to about 75 ml and heat to boiling; filter and wash with hot dilute HCl (5:95) as before, but with additional washings to ensure the complete removal of all perchlorate from the residue and paper. Repeat the dehydration if necessary. Ignite, weigh and treat the separated impure SiO_2 with HF and H_2SO_4 as described.

the inside of the casserole and the stirring rod, and add all washings to the filter funnel. Wipe the stirring rod and lip of the casserole with a piece of filter paper and add it to the funnel. Finally wash the paper and residue 10 times with the wash solution, paying particular attention to the upper edge of the paper. Lift the paper to drain the stem of the funnel and rinse the tip of the funnel into the beaker. Cover the beaker and reserve the filtrate for treatment with hydrogen sulfide to remove platinum that has been introduced into the analysis (Sec. 9.5).

Carefully fold the papers in the funnels with platinum-tipped forceps and transfer them to the original platinum crucible. Wipe out the insides of the funnels with a piece of filter paper and add it to the crucible. Place the crucible, with the cover not quite in place, in a cold electric muffle furnace and burn off the paper by allowing the temperature to rise slowly. Ignite the residue at 1000°C for 30 min (the full heat of a meker burner may be used but the furnace is to be preferred), cool the covered crucible in a desiccator for 30 min and weigh it. Return the covered crucible to the muffle furnace for 20 min, cool for 30 min in a desiccator and again weigh it. Continue the ignition for 20 min periods until two subsequent weights differ by no more than 0.0002 g. The final weight is that of the impure silica.

Cautiously moisten the residue with a few drops of water and add 6 drops of dilute H_2SO_4 (1:1). Quickly pour 5 ml hydrofluoric acid (48%) into the crucible and replace the cover. Allow the crucible to stand for 1 or 2 min to permit the initial reaction to take place, then place the covered crucible on a sand bath heated to about 120°C in a fume hood and allow it to stand, covered, for 5 min. Remove the cover (allow any condensate on the inner surface to drop into the crucible) with platinum-tipped tongs and cautiously heat the outer surface of the cover with a low flame to volatilize condensed hydrofluoric acid. Evaporate the contents of the crucible to fumes of SO_3 (raise the temperature of the sand bath as needed), then cautiously evaporate the excess H_2SO_4 over a low flame, holding the open crucible with platinum-tipped tongs. Care is necessary to prevent spattering, particularly if there is much titanium, zirconium or earth acids present; the crucible should be kept in constant motion and localized overheating must be avoided.*

When the evolution of SO_3 fumes has ceased, heat the crucible to dull redness to decompose most of the sulfates that are present. Cool the crucible and gently wipe the outer surface with a damp cloth to

---

* The addition of a small amount of filter paper pulp to the crucible before the volatilization of the sulfuric acid will prevent spattering.

remove adhering sand particles. Ignite the covered crucible in an electric muffle furnace at about 800°C for 5 min, cool and weigh. The difference in weight is that of the pure silica. The nonvolatile residue that remains should not exceed 5 mg, unless relatively large amounts of titanium, zirconium or earth acids are known to be present*; if $>$ 5 mg, the treatment with HF should be repeated and, if the weight remains the same, the analyst should regard the residue with suspicion and seek to account for the unusual. For accurate work the residue should be fused with a small amount of Na_2CO_3, dissolved in HCl and added to the filtrate from the SiO_2†; the crucible should be ignited and reweighed as usual. For routine work it is usually permissible to ignite the precipitate from the aqueous ammonia separation in the crucible containing the nonvolatile residue and to add the weight of the residue to that of the R_2O_3 oxides.

There still remains a small amount of SiO_2 which has escaped recovery. A third dehydration at this point will serve no useful purpose and it is customary to recover this missing silica from the R_2O_3 precipitate by fusion of the ignited oxides with $K_2S_2O_7$, dissolution of the cake in dilute H_2SO_4 and evaporation of the solution to fumes of SO_3. Shell (Sec. 6.2,

* The composition of the nonvolatile residues left after the volatilization of the separated SiO_2 from each of 16 rocks has been determined spectrographically by Rankama (Ref. 35, Sec. 6.2). He found 28 elements (Al, Ba, Be, Ca, Ce, Co, Cr, Fe, Ga, Ge, K, La, Mg, Mn, Mo, Na, Ni, Pb, Pt, Sc, Sn, Sr, Ti, W, V, Y, Zn, Zr) of which Be, Ge, Pb and Sn were enriched in the residue, Cr, Ga, Ni and Zn tended towards enrichment and W and V were impoverished; S, P, Nb and Ta could not be determined but if they were present they would be in this residue. Rankama also found traces of Na and K which were not washed out of the precipitated silica. Appreciable amounts of these alkali and alkaline earth elements, present as a result of insufficient washing, may cause low results to be obtained for silica, because they are first weighed (probably) as an alkaline silicate and secondly, after HF decomposition, as a sulfate; Goldich et al. (Ref. 15, Sec. 6.2) show that an error of almost 1% may occur when the siliceous insoluble residue from marls (about 40% CaO) is treated directly with HF and H_2SO_4 to determine SiO_2. The alkalies also volatilize in part and thus can cause high results to be obtained under some circumstances. Similarly, elements such as Bi, Sb and Sn will change in an indefinite fashion during the ignition and later volatilization of the silica. The volatilization of fluorides other than SiF_6^{2-} (especially titanium) is prevented by the addition of a few drops of H_2SO_4; if phosphorus is present in an amount greater than normal, there is a chance that some may be lost during the fuming with sulfuric acid.

† Shell (Sec. 6.2, Ref. 24, p. 194) suggests that, as a check on the completeness of the original fusion, the nonvolatile matter be fused with Na_2CO_3–$Na_2B_4O_7$, dissolved and dehydrated as usual, the solution filtered into the main filtrate and the residue, if any, ignited, weighed and treated with HF. If a difference of >1 mg is found it suggests that the original decomposition was unsatisfactory.

Ref. 24, p. 136) does not believe that a good recovery is possible in this way and prefers to use a colorimetric method to determine the silica in either the R_2O_3 oxides, or in the filtrate from the separation of the silica, before precipitation of the ammonia group. Andersson (Ref. 2, Sec. 6.2) has performed a very useful service in his determination of the losses of silicon during a silicate analysis, using NBS 99 (feldspar, $SiO_2 = 68.66\%$) for this purpose; he shows that a satisfactory recovery of residual SiO_2 is possible from the R_2O_3 group (Table 9.1).

TABLE 9.1

Recovery of SiO_2 from Main Portion[a]

| % SiO_2 found in | | | | | | | Total |
A	B_1	B_2	C	D_1	D_2	E	SiO_2
68.20	0.48						68.68
68.16	0.51						68.67
67.91		0.60	0.02	0.03		0.13	68.69
67.96		0.59	0.02	0.03		0.06	68.66
		0.73	0.01		0.10	0.01	
		0.78	0.01		0.08	0.01	

[a] After Andersson, p. 256.

A: fusion with Na_2CO_3, single dehydration with HCl, treatment as usual with HF and H_2SO_4.

B_1: spectrophotometric determination of SiO_2 in filtrate after first dehydration and filtration.

B_2: spectrophotometric determination of SiO_2 recovered by a second dehydration of the silica filtrate.

C: spectrophotometric determination of the SiO_2 remaining in the filtrate from a double precipitation of the R_2O_3 group.

D_1: spectrophotometric determination of the SiO_2 recovered when the R_2O_3 oxides were fused with $K_2S_2O_7$, the cake dissolved in dilute H_2SO_4 and filtered at once.

D_2: spectrophotometric determination of the SiO_2 recovered when the H_2SO_4 filtrate from D_1 was evaporated to fumes of SO_3, diluted and filtered.

E: spectrophotometric determination of the SiO_2 not recovered from D_2, with allowance for interference from phosphorus.

A combined gravimetric–spectrophotometric procedure for the determination of silica will be described later (Sec. 9.4.3), in which a single dehydration is followed by the spectrophotometric determination of the residual silica in an aliquot of the filtrate. This method has certain advantages, although it does somewhat complicate subsequent calculations.

On the basis of many recoveries of SiO_2 from the R_2O_3 precipitate an empirical correction has been derived that may be applied in routine work. This correction takes into account the effect of the increasing basicity of the sample on the solubility of silica. The correction that should be added to the SiO_2 value (and deducted from the total R_2O_3 value), is as follows:

$$SiO_2 > \quad 65\%: + 0.10\%$$
$$SiO_2 \quad 55\text{–}65\%: + 0.15\%$$
$$SiO_2 < \quad 55\%: + 0.20\%$$

Conversion Factors

$$SiO_2 \underset{2.1392}{\overset{0.4675}{\rightleftarrows}} Si$$

9.2.2. By fusion with sodium carbonate (minimum flux method)

Mention has been made (Sec. 5.4.2) of the low flux to sample ratio proposed by Finn and Klekotka, and investigated further by Hoffman (who preferred a 2:1 ratio of flux to sample); the fusion, disintegration and dehydration steps are all done in the same 75 ml platinum dish. Besides reducing the number of transfers involved, the procedure reduces considerably the quantity of sodium salts that are introduced. The disintegration of the melt, which forms a smooth flat mass in the bottom of the dish, is not as rapid as is the disintegration of the cake from the crucible and requires more work with a stirring rod, but this is counterbalanced by the reduction in the time required for transfers from one container to another. Bennett and Hawley (Sec. 6.2, Ref. 5, p. 124) have outlined the procedure in detail and the following method is adapted from their description:

Weigh 0.8000 g of sample, ground to approximately 150 mesh (or transfer the portion used for the determination of moisture), into a platinum dish (approximately 3 in. $\times$ 1½ in., effective capacity of 75 ml). Cover the dish and roast the sample as previously described. Cool, add 0.8 g of anhydrous Na_2CO_3 and 0.1 g Na_2O_2, and mix them intimately with the sample; use the glass rod to form the mixture into a charge about 1½ in. in diameter in the center of the dish. Wipe off the glass rod in an additional 0.8 g of Na_2CO_3 and spread this over the surface of the mixture in the dish.

Cover the dish and heat it cautiously over a bunsen burner, gradually raising the temperature over a period of about 10 min until the maximum heat of the burner is obtained. Finally, heat the covered dish in an electric muffle furnace at 1200°C for 15 min, then remove it and allow it to cool.

Cautiously add 20 ml dilute HCl (1:1) under the cover of the dish, 1 ml dilute H_2SO_4 (1:1) and place the covered dish on the water bath.* When the visible reaction has subsided, remove the cover, rinse it into the dish and set it to one side. Carefully press the fused mass with a glass stirring rod to loosen the outer portions of the cake and scrape away insoluble material that tends to cover the surface of the melt and which will slow down its disintegration. Continue the disintegration and evaporation as previously described; all of the fused melt must be detached from the bottom of the dish in order to determine whether or not the fusion has been complete. Finally evaporate the contents to dryness as before.

Cool the dish, drench the residue with 5 ml $12M$ HCl and allow it to stand for 1–2 min. Add 15–20 ml of hot water, and heat the contents until all soluble matter has dissolved, breaking up any lumps with the stirring rod. Filter the contents of the dish as described in Sec. 9.2.1, and wash the separated silica with hot dilute HCl (5:95).

Return the filtrate to the dish and again evaporate to dryness; when the odor of hydrochloric acid can no longer be detected, heat the dish and contents in an oven at 105–110°C for 1 hr. Cool, add 5 ml $12M$ HCl, and carry out the second filtration as previously described. Reserve the filtrate for treatment with H_2S prior to the precipitation of the R_2O_3 group.

Transfer the filter papers and precipitates to a clean, ignited and weighed 30 ml platinum crucible, burn off the paper slowly as in Sec. 9.2.1 and then ignite the residue to constant weight. Treat the impure SiO_2 with HF and H_2SO_4 to obtain the weight of pure silica. The empirical correction for the unrecovered SiO_2 may be applied, or the SiO_2 itself may be recovered from the R_2O_3 group residue.

9.2.3. BY FUSION WITH SODIUM CARBONATE–BORIC ACID

The advantages and disadvantages of borax and boric oxide as fluxes have been discussed in Sec. 5.3.5. The following procedure is based on that of Bennett and Hawley (Sec. 6.2, Ref. 5, p. 125), who employ this mixture for the decomposition of aluminous materials. The boron is removed by volatilization with methyl chloride.

Proceed as described in Sec. 9.2.1. but use a flux mixture of 2.5 g of anhydrous Na_2CO_3 and 1 g of boric acid, with 0.1 g Na_2O_2 added, and heat very slowly at first until frothing has ceased. Finally fuse for 15

---

* The dried residue will not be a crystalline mass like that obtained in Sec. 9.2.1 but will be powdery, and have a tendency to "ball" together because of the H_2SO_4 present; the latter is added to minimize the hydrolysis of titanium salts at this point.

min at 1200°C. Allow to cool and reheat to loosen the cake. Cool and transfer the cake to a porcelain casserole (or platinum dish) with water and HCl as before, and add 1 ml H_2SO_4 (1:1). Ignite the crucible and, if necessary, remove any stain that may be present.

Warm the casserole on the water bath to complete the disintegration of the melt, cool the solution slightly and then add 25 ml of methyl alcohol. Evaporate the solution gently to dryness, breaking up the crust from time to time. When the residue is completely dry, proceed as in Sec. 9.2.1 for the filtration of the separated SiO_2. Transfer the filtrate back to the casserole or dish, add a further 25 ml of methyl alcohol and proceed with the second dehydration, washing down the walls at least once before evaporation is complete in order to remove any boron that may be been deposited there. From here on the procedure is similar to that given previously.

9.3. Gravimetric determination of silica in samples containing $>2\%$ fluorine

9.3.1. By the method of Hoffman and Lundell[20] (See Sec. 6.2)

Weigh 0.5000 g of sample (or a portion such that the amount of fluorine in solution is between 0.01 and 0.1 g) which has been dried at 110°C (or on which the H_2O^- is being concurrently determined) into an ignited and weighed 30 ml platinum crucible, mix with 5 g Na_2CO_3, cover and fuse as described in Sec. 9.2.1. Transfer the cake, with water only, to a 250 ml platinum dish, and heat on the water bath until disintegration is complete; ignite and weigh the crucible and reserve it for later use.

Filter the contents through an 11 cm Whatman No. 41 filter paper into a 600 ml beaker, keeping the bulk of the insoluble material in the dish. With a jet of water transfer the residue from the paper back into the dish, add 50 ml of a 2% solution of Na_2CO_3, and boil for a few minutes.* Filter through the same paper and wash the dish, residue and paper thoroughly with hot water, combining the filtrate and washings with the first filtrate. Reserve the residue (A) for the determination of silica.

To the combined filtrates, which should have a volume of about 300 ml, add a solution of 1 g of zinc oxide in 20 ml dilute HNO_3 (1:9).

---

* For accurate work the residue should be combined with the first precipitate obtained with the zinc oxide solution and again fused with Na_2CO_3, to ensure complete recovery of fluoride. If this is done then the second boiling of the insoluble residue may be omitted.

Boil the solution for 1 min, filter through an 11 cm Whatman No. 41 paper and wash the residue thoroughly with hot water. Reserve the residue (B) for the determination of silica.

Add a few drops of methyl red to the combined filtrates, nearly neutralize with HNO_3 and evaporate to about 200 ml, making sure that the solution remains alkaline during the evaporation. Cool somewhat, and add dilute HNO_3 (1:1) dropwise until a faint pink color just appears. Add 1.0 g zinc oxide dissolved in NH_4OH and $(NH_4)_2CO_3$ (20 ml H_2O, 2 g $(NH_4)_2CO_3$, 2 ml NH_4OH), and boil in a covered dish (preferably a 250 ml platinum dish) until the odor of ammonia can no longer be detected (this usually requires the concentration of the solution to about 50 ml). Add 50 ml of warm water, stir, allow to settle for a few minutes, filter through a Whatman No. 40 paper and wash with cold water. Reserve the residue (C) for the determination of silica. The filtrate may be used for the determination of fluorine (see Sec. 9.32.1).

With a jet of dilute HCl (5:95), wash the residues A, B and C from the papers into a 250 ml platinum dish (preferably the dish in which the last precipitation was made). Ignite the three papers in the platinum crucible used for the fusion and add any residue so obtained to the contents of the dish. Add 25 ml $12M$ HCl and separate the silica by a double dehydration as described in Sec. 9.2.1. Because of the presence of considerable zinc chloride, which is dried only with difficulty in the absence of an appreciable amount of silica, it is advisable to add 10 ml $18M$ H_2SO_4 to the filtrate from the first dehydration and to evaporate the solution to fumes of sulfur trioxide. A small amount of silica (about 1 mg) remains in solution and is recovered as follows: add 0.04 g Al (as $AlCl_3$) and 10 g NH_4Cl to the filtrate from the second dehydration, heat to boiling, precipitate with ammonium hydroxide and filter through a 9 cm Whatman No. 40 paper. Wash the precipitate with a hot solution of NH_4Cl (5:95), dissolve it in 50 ml of dilute H_2SO_4 (1:9) and evaporate to fumes of SO_3. Filter, wash with hot H_2O and add the residue to the silica already obtained.

Ignite the combined silica residues as described in Sec. 9.2.1, in the platinum crucible used previously, and treat the impure silica as usual with HF and H_2SO_4 to obtain the weight of SiO_2.

9.3.2. BY THE ZINC OXIDE METHOD OF SHELL AND CRAIG[40] (SEE SEC. 6.2)

Weigh 0.5000 g of sample, ground to -200 mesh and dried at 110°C (or on which moisture is concurrently being determined) into an ignited and weighed 30 ml capacity platinum crucible and mix thoroughly with 5 g anhydrous Na_2CO_3 and 1.0 g zinc oxide. Fuse as described in Sec.

9.2.1, but heat at just above the melting point of the Na_2CO_3 for 20 min, then fuse at 1050–1100°C for an additional 20 min. If an electric muffle furnace is used, start with a cold furnace and fuse for 30 min at 950–1050°C. It is essential that an oxidizing atmosphere prevail, in order to prevent the reduction of the zinc and its subsequent alloying with the platinum; nickel crucibles may be used for the fusion if it is expected that the atmosphere will be slightly reducing.

When the fusion is complete, allow the crucible to cool. Transfer the fusion cake to a 500 ml stainless steel beaker containing a stainless steel, or nickel, spatula. Add 200 ml cold water and allow to stand until the soluble salts are dissolved, preferably overnight (if the cake sticks to the crucible, the crucible and lid may be left in the beaker overnight). Break up the cake completely, remove and rinse off the crucible and lid if they were placed in the beaker. Add slowly, with stirring, 25 ml of ammoniacal zinc oxide solution (1.0 g ZnO, 1.3 g $(NH_4)_2CO_3$ and 2 ml concentrated NH_4OH dissolved in 10 ml H_2O and diluted to 25 ml).

Heat to boiling with constant swirling over a flame, then cover and set the beaker in an aluminum air bath on a hot plate; this is necessary to prevent bumping caused by the gelatinous nature of the precipitate. Boil for 8–10 min. Remove from the hot plate, let stand for 5 min, and then filter through a 12.5 cm No. 40 Whatman filter paper, catching the filtrate in a 500 ml volumetric flask. Transfer the precipitate to the paper and wash twice with hot water. Transfer the bulk of the precipitate back to the stainless steel beaker with the spatula, then wash the remainder from the paper into the beaker with hot water. Dilute to 60–75 ml and stir thoroughly, then filter again through the same paper into the volumetric flask. Police the beaker and wash the paper and precipitate five times with hot water. This thorough washing is necessary to prevent the entrainment of fluoride by the voluminous precipitate. Reserve the filtrate for the spectrophotometric determination of the residual silica.

Transfer the paper and residue to a 100 ml platinum dish, char the paper carefully and finally burn it off at 650°C, preferably in an electric muffle furnace. Dissolve any residue left in the stainless steel beaker with 25 ml of hot water and 4 ml of $12M$ HCl, and police the beaker, adding the solution and washings to the platinum dish. Wipe the inside of the beaker with a small piece of filter paper and place this in the platinum crucible to be used for the ignition of the silica (the one in which the fusion was done, if a platinum crucible was used).

Add 20 ml $12M$ HCl to the platinum dish and heat the dish on a water bath. Break up all lumps with a glass stirring rod and evaporate to dryness. Acidification, filtration and washing are done as described in Sec. 9.2.1, catching the filtrate in a 400 ml beaker. Add 2–5 ml 15M HNO_3 and 20 ml 72% $HClO_4$ to the filtrate, cover the beaker with a ribbed watch glass and evaporate the solution just to fumes of perchloric acid. Replace the ribbed glass with a closely fitting one and reflux for 20 min, at such a rate that $HClO_4$ is not lost from the beaker. Allow to cool and dilute with 125 ml of hot water. Add a small quantity of filter paper pulp, filter through a 9 cm No. 40 Whatman paper and wash well with hot water, particularly around the edge of the paper to remove all traces of perchlorate.

Combine the papers and precipitate in the platinum crucible and burn off the paper in an electric muffle furnace, with final heating at 1200°C for 30 min to constant weight. Treat the impure silica as usual with HF and H_2SO_4 to obtain the weight of SiO_2.

Dilute the filtrate from the first dehydration to 500 ml and mix well. Determine the residual silica in this filtrate spectrophotometrically as described in Sec. 9.4.3. using an appropriate aliquot, and add it to the gravimetric value obtained previously.

9.4. Colorimetric determination of silica

9.4.1. BY THE METHOD OF SHAPIRO AND BRANNOCK (SOLUTION A)

A scheme of rapid rock analysis to determine the major constituents of silicate rocks was described by L. Shapiro and W. W. Brannock in 1952, and revised in 1956. It was again revised in 1962, and extended to include methods for carbonate and phosphate rocks as well (Ref. 38, Sec. 6.2). These methods, with some modifications, were used at the Geological Survey of Canada for several years to analyze a very large number of silicate and carbonate rocks.

Two solutions which are prepared from separate portions of the sample are the basis for the determination of 10 constituents. Aliquots of one solution, prepared by fusing the sample with NaOH, are used to determine SiO_2 by measurement of the molybdenum blue complex, and Al_2O_3 by that formed with Alizarin Red-S. The second solution is prepared by dissolving the sample in a mixture of $HF-H_2SO_4-HNO_3$; total iron, TiO_2, P_2O_5 and MnO are determined colorimetrically, Na_2O and K_2O by flame photometry, CaO and MgO by titrimetry. Separate portions are used for the determination of FeO, total H_2O, CO_2, F and acid-soluble sulfur. The analytical scheme is presented schematically, with appropriate page references, in the appendix.

This scheme of analysis was designed to give optimum efficiency for a batch operation of about 30 samples. At the Geological Survey of Canada it was preferred to analyze the samples in batches of six and the methods to be given, which differ in some respects from those of Shapiro and Brannock, are based upon this type of operation.

For the determination of SiO_2 and Al_2O_3, in solution A, the batch of 6 samples is increased to 10 by the addition of a reagent blank, two standard samples from which a factor for obtaining the percentage of SiO_2 and Al_2O_3 will be derived, and a laboratory reference sample which serves as a check on the correctness of each set of results. For the constitutents determined in solution B the factors are derived from standard solutions which are carried through the procedure after the preparation of the solution. The laboratory reference sample is also used with solution B.

Solution A, for silicate rocks, is prepared by fusing a 50 mg portion with NaOH in a nickel crucible for about 5 min at a low temperature. The fusion cake is leached with water and the alkaline extract is added to dilute HCl to give an acid solution with the silica present in a monomeric, non-colloidal form. The small sample size is used in order to avoid too great dilutions but it is then very necessary to exercise extreme care in handling the sample and the concentrated leachate prior to dilution, in order to avoid losses. Again, because of the small sample weight used and the high concentration of SiO_2 that is usually present, the sample must be finely ground and homogeneous, or serious errors will occur.

Preparation of Solution A

Reagents

NaOH solution, 15%: weigh 150 g of reagent grade NaOH pellets and transfer them to a 600 ml polyethylene beaker. Add 100 ml H_2O, cover and allow to dissolve, adding more H_2O if needed; stir with a plastic stirring rod, do *not* use a glass one. Dilute to a liter, mix thoroughly and store in a plastic bottle.

Procedure

1. Clean a series of 75 ml nickel crucibles with HCl (1:1) and rinse thoroughly with water. Place a 10 ml portion of 15% NaOH solution in each crucible, and evaporate to dryness on a sand bath (overhead heating with an infrared lamp will speed up this evaporation). Store the crucibles in a desiccator, until ready to use them.

2. Weigh two 50 ± 0.1 mg portions of National Bureau of Standards sample No. 99 (feldspar) and transfer each portion to a nickel crucible containing the NaOH. Similarly weigh a 50 ± 0.1 mg portion of each sample and transfer it to a similar crucible. Retain one crucible and NaOH to serve as a blank.

3. Place each crucible, covered, on a silica triangle over a meker burner and heat it at a dull redness for about 5 min. Excess heating only increases the attack on the crucible, but some samples may require a longer heating time. Remove the crucibles from the heat, uncover each in turn and swirl the melt around the sides of the crucible, allowing it to solidify as a thin shell. Allow to cool, add 50 ml of H_2O to each crucible and allow to stand, covered, overnight.

4. Transfer the contents of the crucibles to 600 ml beakers containing about 300 ml of water and 10 ml 12N HCl; do not allow the alkaline solution to come into contact with the glass but pour it directly into the dilute acid. Police each crucible and rinse thoroughly with water; do not allow the acid solution to touch the crucible. If the solutions are cloudy, heat until clear and then allow to cool.

5. Quantitatively transfer the contents of the beakers to a series of 1-liter volumetric flasks, dilute to volume at room temperature and mix thoroughly. Rinse out a series of plastic bottles with the solutions, and transfer about 100 ml of each to the bottles, stopper them tightly and reserve them for the determination of SiO_2 and Al_2O_3.

Determination of SiO_2

Reagents

Ammonium molybdate solution, 7.5%: dissolve 7.5 g of reagent grade $(NH_4)_6Mo_7O_{24} \cdot 4H_2O$ in 75 ml H_2O, warming gently if necessary. Cool, and add 10 ml 1:1 H_2SO_4, dilute to 100 ml, and mix thoroughly. Store in a plastic bottle; filter the solution before using if a precipitate collects on the bottom of the bottle, and discard it when the solution becomes cloudy.

Tartaric acid solution, 8%: dissolve 40 g of reagent grade tartaric acid in H_2O, dilute to 500 ml and store in a plastic bottle. Prepare a fresh solution when sediment forms.

Reducing solution: dissolve 0.5 g reagent grade anhydrous sodium sulfite in 10 ml of H_2O. Add 0.15 g of 1-amino-2-naphthol-4-sulfonic acid and stir until dissolved. Dissolve 9 g of reagent grade sodium bisulfite in 90 ml of K_2O, and add this solution to the first solution, mix thoroughly and store in a plastic bottle in a dark place. Do not store the solution for more than three days.

Procedure

1. All solutions should be at 20–25°C. Transfer, by pipet, a 10 ml aliquot of each sample solution A, 10 ml of each standard solution A, and 10 ml of the blank solution A to a series of 100 ml volumetric flasks which have been cleaned with 1:1 HCl and rinsed with water.

2. Add, by pipet, 1 ml of the ammonium molybdate solution to each flask, swirling the flasks during the additions. Mix, and allow to stand for 10 min.

3. Add, by pipet, 5 ml of the tartaric acid solution, swirling the flasks during the additions, and mix.

4. Immediately add, by pipet, 1 ml of the reducing solution, swirling the flasks during the additions, and dilute to volume. Mix thoroughly and allow to stand for at least 30 min.

5. Determine the absorbance of each solution at 650 mμ, using the solution prepared from the blank solution A as the reference blank solution.

6. Compute a factor for each of the two standard solutions, and average them:

$$\text{Factor} = \frac{\% \text{ SiO}_2 \text{ of standard}}{\text{absorbance of standard}}$$

7. Compute the % SiO$_2$ in each sample:

$$\% \text{ SiO}_2 = \text{Av. factor} \times \text{absorbance of sample solution}$$

9.4.2. BY THE METHOD OF RINGBOM, AHLERS AND SIITONEN

In the method described in Sec 9.4.1 it is important that the given procedure be followed exactly to minimize the possibility of serious errors in the results. Ringbom *et al.* (Ref. 36, Sec. 6.2) have proposed a method which, although slightly less sensitive, is characterized by greater stability and reproducibility.

The colored complexes, both yellow and blue, which are formed with the beta form of silicomolybdic acid are slightly more sensitive than those formed with the alpha form. Because the beta form is not stable and changes at a varying rate into the alpha form, there will be varying amounts of the latter present and the resultant color of the solution will lack sufficient stability and reproducibility to make possible the attainment of a high precision in analysis. Sensitivity is not as important as stability and reproducibility when the component being determined is a major one; in most analyses it is not necessary to reduce the yellow complex. There is an added advantage, too, in that the alpha form has a higher absorbance than the beta form at shorter wavelengths.

Ringbom *et al.* made a systematic investigation of the formation of the yellow compounds and found the most favorable range of pH to be 2.3–3.9; they made the final measurement at a pH of 3.0–3.7, using a monochloracetic acid–monochloracetate buffer. Although the absorbance values obtained by their procedure are stable and reproducible, they are temperature-dependent (Ringbom *et al.* obtained a temperature coefficient of 0.39%/°C) and the absorbance must either be determined at a constant temperature or a temperature correction must be applied. Because the formation of the complex is accelerated by a rise in temperature, the solution is heated in a boiling water bath for 5–10 min for rapid equilibration and then cooled to room temperature.

Of the possible interferences, the most serious (and likely) is that of phosphorus. This can be minimized by the addition of tartaric acid (a slight fading of the yellow silicomolybdic complex also occurs), but Ringbom *et al.* preferred instead to apply a correction for the effect of the phosphorus (at 390 mμ the absorbance of 1 part P_2O_5 is equivalent to 0.43 part SiO_2).

Because the silicate melt is dissolved in a solution containing EDTA (to avoid the formation of precipitates) its effect upon the nature of the silicomolybdic acid formed was investigated and, for an EDTA concentration $<0.01M$, no significant effect was noticed. The Fe(III)–EDTA complex formed does have an appreciable absorbance, but this can be overcome by using a portion of the unknown solution, to which no ammonium molybdate has been added, as the reference blank.

The results obtained for a series of National Bureau of Standard samples tended to be slightly higher than the accepted value, but the largest error was only 0.3%.

Procedure

Transfer a weighed sample, usually 100–300 mg containing 20–70% SiO_2, to a nickel crucible containing 10 g of molten NaOH and heat for 5–10 min. Cool and dissolve the melt in H_2O in a plastic dish, to which has been added 40 ml of 0.05M solution of the disodium salt of EDTA. Transfer the solution to a tared plastic flask and dilute, on a suitable balance, until the weight is 509.4 g (500.0 ml at 20°C).

Transfer an appropriate aliquot (10–25 ml) to a 50 ml volumetric flask containing 10 ml of a suitable mixture of 2M monochloracetic acid and 2M ammonium monochloracetate (the pH of the final solution must be 3.0–3.7, and the composition of the buffer solution depends on the size of aliquot of the alkaline (approx. 0.5M) sample solution taken), and add 10 ml of ammonium molybdate solution containing 35.3 g $(NH_4)_6Mo_7O_{24} \cdot 4H_2O$ per liter. Immerse the flask in boiling water for

5–10 min, cool to room temperature and dilute to volume. Measure the absorbance of the solution at 390 mμ, using as reference blank a solution prepared in the same way but omitting the addition of the ammonium molybdate and the heating. The color complex is stable for nearly 2 days.

Correct the absorbance to 20°C, using the expression $\Delta A = (20 - t)$ 0.004 A; the correction will be negative for temperatures over 20°C, and positive for those below 20°C.

All reagents should be prepared using redistilled water free from SiO_2 and should be kept in plastic bottles; a blank determination should be made on all reagents.

Ringbom et al., who used a Beckman DU spectrophotometer for measurement of the absorbance, comment as follows on the measurement procedure: "If a Beckman spectrophotometer (or any other one-cell instrument with a 1:10 sensitivity adjustment) is used, it is advantageous to choose the wavelength and silica concentration so that the absorbance is a little above 1.0. A higher sensitivity can then be used, i.e., one scale unit of the transmittance scale corresponds to a relative error of about 0.4% (Ref. 37, Sec. 6.2). As the reproducibility of a measurement is usually about half a scale unit, the error in analyzing silicate rocks is, as a rule, about ± 0.1% SiO_2.

"The absorbance index was determined according to the procedure given above, using a solution of pure SiO_2 in NaOH, but since the absorbance curve is rather steep, the value obtained may depend somewhat on the instrument used. We have used a value $a_{390} = 1.701.10^3$ (20°). It is advisable to adjust the wavelength very carefully using a magnifying glass."

9.4.3. BY THE COMBINED GRAVIMETRIC–COLORIMETRIC METHOD OF JEFFERY AND WILSON

The separation and determination of the bulk of the silica is achieved by a single evaporation with HCl, followed by the usual treatment with HF and H_2SO_4. The filtrate from the separation of the SiO_2 contains from 2 to 8 mg of silica and the latter is determined in an aliquot of the filtrate, which is diluted to 200 ml and made approximately 0.5N in HCl, by the molybdenum blue method.

Use of the yellow silicomolybdic acid complex was abandoned in this instance because of the possible interference of phosphorus, the instability of the color and because it does not obey the Beer-Lambert law.

Jeffery and Wilson (Ref. 23, Sec. 6.2) determined that a pH of 0.8–1.7 was necessary for the quantitative formation of the yellow silicomolybdic acid complex. Ferric iron will interfere both by the precipitation of ferric molybdate, and by oxidizing the molybdenum blue complex; oxalic acid is added to reduce the iron to the ferrous state. The presence of iron reduces the absorbance values of the molybdenum blue solution but not when the measurement is made at 650 mμ. Titanium forms a precipitate with ammonium molybdate but the effect is not significant unless the TiO$_2$ content of the sample greatly exceeds 5%; phosphorus interferes when its concentration exceeds 10% and in this case the photometric method cannot be used.

When iron and/or titanium are present in excess of 20% and 8%, respectively, they should be removed. Cupferron extraction and alkaline fusion procedures are suggested by Jeffery and Wilson for this purpose.

Procedure

The separation and determination of the bulk of the silica are done as described in Sec. 9.2.1 to the end of the first filtration, but add 10 ml of 12M HCl to the dried residue so that the filtrate and washings, diluted to 200 ml in a calibrated volumetric flask, will be approximately 0.5N in HCl. That portion of the silica present in this solution is then determined as follows:

Transfer a 5 ml aliquot of this solution to a 100 ml volumetric flask and add 10 ml water. Add 1 ml ammonium molybdate solution (10% w/v solution in 1N aqueous ammonia) and let stand for exactly 10 min. Add 5 ml of oxalic acid solution (10%), gently swirl to mix the contents and then add, without delay, 2 ml of reducing solution (prepared as described in Sec. 9.4.1, except that Jeffery and Wilson recommend 0.7 g of anhydrous sodium sulfite). Dilute the solution to 100 ml, let stand for 30 min to (preferably) 1 hr, and measure the absorbance of the solution in a 2 cm cell at 650 mμ.

Prepare the standard silica solution as follows: fuse 1 g of pure quartz with 5 g of anhydrous Na$_2$CO$_3$ in a platinum crucible, and extract the melt with water. Filter into a 1000 ml volumetric flask, wash any residue with a 2% (w/v) solution of Na$_2$CO$_3$ and dilute to volume. This gives a stock solution containing approximately 1 mg of SiO$_2$ per ml. Transfer a 5 ml aliquot of this stock solution to a 200 ml volumetric flask, add 10 ml of 12M HCl and dilute to volume. This working solution may be used to prepare a calibration curve but since the absorbance of the molybdenum blue solutions obeys Beer's law, Jeffery and Wilson use instead a single aliquot from which they derive a factor.

Transfer a 5 ml aliquot of the working solution to a 100 ml volumetric flask, dilute to 15 ml, add 1 ml of the ammonium molybdate solution and, after 10 min, 5 ml of the oxalic acid solution and 2 ml of the reducing solution. Dilute to 100 ml and measure the absorbance as previously described; this will correspond to 125 μg of SiO_2.

A comparison of results obtained for a series of rocks, including G-1 and W-1, by other procedures, shows good agreement. Because the classical procedure tends to give low results, the higher values obtained by this new method are preferred.

9.5. Precipitation with hydrogen sulfide in strongly acid solution

The precipitation is usually made in hydrochloric acid solution although platinum is more readily precipitated from sulfuric acid solution (e.g., the solution of the pyrosulfate fusion of the ignited R_2O_3 groups). Cold dilute nitric or perchloric acids may also be used.

Iron may be carried down to some extent by members of the hydrogen sulfide group. Unless the precipitate obtained is very small (i.e., Pt only), it is preferable to dissolve the precipitate, repeat the precipitation, and combine the two filtrates.

The filtrate from the silica separation, in a 400 ml beaker, should have a volume of about 100 ml and contain approximately 2.5 ml 12M HCl (or 1.5 ml 18M H_2SO_4 if the latter acid was used).

Heat the solution to near boiling and, in a fume hood, pass a rapid stream of hydrogen sulfide through it for about 30 min, keeping the beaker covered at all times. If much ferric iron is present, a white precipitate of sulfur will form in addition to the dark-colored sulfides of the members of the hydrogen sulfide group. If only platinum is to be removed, the gassing may be discontinued but otherwise the solution should be diluted to about 200 ml and the stream of gas continued for another 15–30 min. Those elements which will be precipitated are shown in Table 6.2.

The trapping of precipitate inside the gas delivery tube may be minimized by starting the gas flow before the tube is placed in the solution (i.e., prepare the wash solution first by passing the gas through approximately 0.5N acid solution until it is saturated); similarly, the tube should be withdrawn carefully from the sample solution while a slow stream of gas is maintained.

Remove the delivery tube and rinse it, inside and out, with water. Rinse down the walls of the beaker and underside of the watch-glass and allow the solution to stand for several hours, or overnight. Filter through an 11 cm Whatman No. 40 filter paper, catching the filtrate (which should be colorless or nearly so) in a 400 ml beaker, and wash thoroughly with the wash solution. A cloudiness which forms in the filtrate on standing is due to the air oxidation of the sulfide ion to sulfur and can be ignored. The filtration is best done in a fume hood also (hydrogen sulfide is dangerously toxic and, even in small amount, may cause an unpleasant headache to develop).

Carefully evaporate the solution to about 100 ml, rinse down the sides, and add 10 ml of saturated bromine water. Heat to boiling and boil vigorously to expel the last traces of bromine. This must be done to remove sulfide ion and to insure that all of the iron is in the ferric state; if not done, it will cause serious difficulties in the ammonia precipitation to follow. The bromine must in turn be expelled because it attacks the indicators used in adjusting the pH of the solution.

If the precipitate contains only platinum, it may be discarded. If other members of the hydrogen sulfide group are present in appreciable amount it may be advantageous to utilize this precipitate and reference should be made to the sections dealing with these elements.

9.6. Gravimetric determination of aluminum by difference, following precipitation of the R_2O_3 group with aqueous ammonia

If the oxidized filtrate from either the silica determination* or the H_2S separation, which should be in a 400 ml beaker and have a volume of about 200 ml, does not contain sufficient hydrochloric acid to give 5–6 g of NH_4Cl when neutralized with pure aqueous ammonia, add sufficient NH_4Cl in solution (dissolve reagent-grade NH_4Cl in water and filter) to give this amount. Add pure aqueous ammonia (1:1) cautiously from a graduated cylinder until a precipitate forms and then just redissolves.

If the solution is dark-colored, indicating the presence of much iron, there is no point in adding a pH indicator; if the solution is light-colored, add 5 drops of a 0.04% aqueous solution of bromcresol purple (pH 5.2–6.8) or 3 drops of 0.2% methyl red (60% alcoholic solution, pH 4.2–6.3). Heat to boiling, reduce the heat, and add pure aqueous ammonia from a dropping bottle until a permanent precipitate forms, stirring vigorously at all times. The precipitate will appear textureless until near the end point, at which stage it will coagulate. If an indicator was used the supernatant liquid will, at the end point, be bluish-purple

---

* If the dehydrations were done in platinum, add 10 ml saturated bromine water to the filtrate and boil to expel excess bromine.

(bromcresol purple) or yellow (methyl red); if an indicator was not used, the nearness of approach to the end point can be tested for by adding a drop of the indicator to the quiescent solution and noting the color of the drop as it strikes the surface. Allow the precipitate to settle and confirm that the end point has been reached; the solution should smell faintly of ammonia. When precipitation is complete, heat the contents of the beaker to a rolling boil, boil for 1 or 2 min, remove from the heat and allow the precipitate to settle; the color of the supernatant liquid should be as it was before but if the boiling was too prolonged it may be necessary to add 1 or 2 drops of aqueous ammonia and boil the solution again for 1 min. Immediately proceed with the following filtration.

Wash an 11 cm Whatman No. 40 filter paper, in a 75 mm filter funnel, once with freshly prepared, hot 2% NH_4NO_3 wash solution (if the filtrate is eventually to be acidified in a platinum dish, use 2% NH_4Cl instead). Catch the first few milliliters of filtrate in a 150 ml beaker; if the filtrate is cloudy, pour it again through the filter, catching the solution in the original beaker, and repeat these steps until a clear filtrate is obtained. When the filtrate is clear, replace the small beaker with one of 600 ml capacity and filter the remainder of the solution as rapidly as possible, keeping the bulk of the precipitate in the beaker. If the filtrate remains cloudy, add a small amount of paper pulp to the solution and precipitate, stir well and allow to settle before starting the filtration again; usually this will yield a clear filtrate but if it does not, then add sufficient HCl to just redissolve the precipitate, heat the solution to boiling and repeat the precipitation.

Transfer the precipitate to the paper, wash the beaker and stirring rod twice with the wash solution and then wash the paper and precipitate ten times, taking care to wash the precipitate away from the edges of the paper. Wipe the lip of the beaker and the stirring rod with a small piece of filter paper and add this to the funnel.

With the aid of platinum-tipped forceps, remove the paper from the funnel and carefully spread it out on the inside wall of the original beaker. Wash the precipitate from the paper with a jet of hot 5% HCl* and rinse the walls of the beaker and stirring rod also. Place the beaker under the funnel and rinse the funnel with 5% HCl, then with water. Finally wash the paper once or twice with water, fold the paper into a triangle with the forceps, and drape it over the rim of the beaker. Heat the contents of the beaker until the precipitate has dissolved (add more $12M$ HCl if required), dilute to about 150 ml and heat to boiling.

* Washington (Ref. 26, Sec. 6.4) dissolves the precipitate in 10 ml concentrated HNO_3 in order to eliminate the presence of chloride which may cause volatilization of iron during the ignition.

Repeat the precipitation as previously described but, when the end point has been reached, add 1 or 2 drops of aqueous ammonia in excess, and add the filter paper to the solution, shredding it with the stirring rod and forceps and stirring vigorously to macerate it thoroughly.* It may be necessary, if there was much acid retained by the paper, to add a few more drops of aqueous ammonia to the solution. Rinse down the sides of the beaker, heat to boiling and boil for 1 min. Allow the precipitate to settle and filter as before, combining the two filtrates in the 600 ml beaker. Carefully police the beaker and stirring rod, using the NH_4NO_3 wash solution, and wash the paper and precipitate 6–8 times. Wipe off the lip of the beaker and stirring rod with a small piece of filter paper as before. Cover the funnel with a fitted filter paper.

Rinse the face of the watch-glass with hot 5% HCl, catching the washings in the original beaker, and rinse the sides of the beaker. Cover, heat to boiling and boil for 1 min. Add one drop of indicator, neutralize the solution with aqueous ammonia, add a small amount of filter paper pulp and filter through a 7 cm Whatman No. 40 paper, catching the filtrate in the 600 ml beaker. The precipitate recovered in this way is usually small but it is not negligible. Wash it with the NH_4NO_3 solution and discard the washings.

Make the combined filtrate just acid with HCl and evaporate to about 100 ml. Cautiously add concentrated HNO_3 (3 g HNO_3 for each of NH_4Cl), cover and heat on the steam bath until the vigorous reaction and evolution of gas has ended (add more HNO_3 if crystallization begins before the solution has become quiescent); rinse off the cover and sides of the beaker with water, and evaporate the solution to dryness.

Large amounts of NH_4NO_3 can be removed in the same manner, except that HCl instead of HNO_3 is added when the solution, acidified with HNO_3, is reduced in volume. Dissolve the salts in water and add a few drops of HCl (1:1); the solution is now ready for a final recovery of the R_2O_3 group elements or, if this is to be omitted, for the separation of manganese prior to the precipitation of the oxalate group (Sec. 9.9).

If a final recovery of that part of the R_2O_3 group which may have escaped precipitation is to be made, dilute the solution to about 75 ml, add 1 drop of indicator, heat to boiling and neutralize with aqueous ammonia. Add a very small amount of filter paper pulp and filter through the 7 cm paper used previously to catch the precipitate recovered from the walls of the beaker, catching the filtrate in a 400 ml beaker. Wash the precipitate, beaker and paper thoroughly with hot 2% NH_4NO_3 solution.

---

* Groves (Ref. 5, Sec. 6.4) tears off only the used half of the paper and adds it to the solution.

With the platinum-tipped forceps, carefully lift the paper containing the bulk of the R_2O_3 precipitate from the funnel and place it in the platinum crucible used for the determination of silica.   Fold the upper edges of the paper over the precipitate; avoid soiling the sides of the crucible because, after ignition, it is difficult to remove these stains during the pyrosulfate fusion.   Wipe the upper edge of the funnel with a small piece of filter paper, use it similarly to wipe out the second funnel after its paper has been removed and folded on top of the first paper in the crucible, and then add it to the crucible also.   The bulky precipitate should be partly dried, or at least well drained, before being placed in the crucible; if not, there is danger that entrained liquid will boil and some precipitate will be lost by spurting.

Partly cover the crucible and place it in a cold electric muffle furnace. Allow the temperature to rise slowly and ensure that there is free access of air to the furnace during the initial stages of the ignition.

Finally heat at just below 1200°C for 40 min (it will be found convenient to use one muffle furnace for the slow ignition of the precipitate and another one, kept at near 1200°C, for the final heating), cover the crucible, cool and weigh as usual.   When much iron is present it is a useful precaution to transfer the crucible from the muffle furnace to the full heat of a meker burner for 5 min, with the lid displaced to allow free access of air to the crucible, to ensure that oxidation of the iron is complete; there is little likelihood of the ignited Al_2O_3 absorbing water after having been heated at 1200°C.   Repeat the ignition for 20 min periods until constant weight is obtained.   From the total per cent of the R_2O_3 oxides the percentage of Al_2O_3 is obtained by subtracting the percentages of the other constituents of the group, in particular those for total iron (as Fe_2O_3), P_2O_5, TiO_2 and residual SiO_2.

Conversion Factors

$$Al_2O_3 \underset{1.8899}{\overset{0.5291}{\rightleftharpoons}} Al$$

9.6.1. RECOVERY OF RESIDUAL SILICA FROM THE R_2O_3 GROUP ELEMENTS

Unless the silica was determined by the gravimetric–spectrophotometric method of Jeffery and Wilson (Sec. 9.4.3), or an empirical correction is to be made (Sec. 9.2.1), the residual SiO_2 that was scavenged by the R_2O_3 precipitate must be determined, or both the SiO_2 and Al_2O_3 values will be in error.   A separation of the SiO_2 is made by fusing the ignited oxides with potassium pyrosulfate, dissolving the cake in H_2SO_4 (1:9) and heating the solution to fumes of SO_3 to dehydrate the silica.

If the latter evaporation is omitted only a portion of the SiO_2 will be recovered, because heating to fumes of SO_3 is necessary to decompose the alkaline silicates formed during the fusion.

Potassium pyrosulfate fuses at a relatively low temperature and fusions can be made in silica crucibles (see Sec. 5.3.6) to avoid the introduction of platinum into the analysis. Vincent (Ref. 24, Sec. 6.4) considers the platinum derived from the attack of the platinum crucible and dishes used in these first stages of the main portion analysis to be an important source of error, particularly in microanalysis. He recovers the platinum by H_2S precipitation from the sulfuric acid solution of the $K_2S_2O_7$ fusion cake, weighs it and subtracts it from the total weight of platinum lost by the crucible in the various steps from the initial fusion through to the $K_2S_2O_7$ fusion of the ignited oxides; the difference in weight represents the platinum still present in the total weight of the R_2O_3 oxides. However, if an H_2S separation of the platinum is made before the precipitation with aqueous ammonia, and if the ignited oxides are fused in a Vycor crucible as described, the amount of platinum present in the solution of the pyrosulfate fusion cake will be negligible.

Place a 30 ml Vycor crucible upside down over the uncovered platinum crucible containing the ignited R_2O_3 residue, invert the two crucibles and tap the residue into the Vycor crucible. Use a brush to dislodge residue adhering to the walls of the platinum crucible but do not attempt to brush all of the residue into the Vycor crucible. Add 6 g of potassium pyrosulfate (**Caution!** the ignited residue is easily blown out of the crucible) to the Vycor crucible, cover it and place it in a silica triangle over the low flame of a bunsen burner; the flame should be high enough to melt the pyrosulfate but not high enough to decompose it.

While the decomposition of the R_2O_3 residue is in progress, add 1 g $K_2S_2O_7$ to the platinum crucible. Hold the crucible with platinum-tipped tongs and heat it over a low flame until the pyrosulfate melts; increase the heat until the crucible is a dull red in color and fumes of SO_3 are evolved, and continue the heating until the residue left in the crucible has dissolved. Swirl the crucible contents to remove adhering dust and stains from the walls and then allow to cool. Place a small piece of $K_2S_2O_7$ on the undersurface of the crucible cover and heat the cover to dissolve any stains or adhering particles. The whole cleaning process should not take more than 1 or 2 min. Add a few ml H_2O to the cooled crucible, loosen the cake and quantitatively transfer the contents to a 250 ml beaker (or a 100 ml platinum dish, if preferred), policing the crucible if necessary. Clean the cover in the same manner. Add 20 ml H_2SO_4 (1:1) to the beaker if the determination of TiO_2 is to follow the separation of the SiO_2; if it is not, add 2 ml of $18M$ H_2SO_4.

Grasp the Vycor crucible, the contents of which should be molten, with the platinum-tipped tongs and complete the fusion by hand, heating the crucible to a medium to bright red color. As the liquid cools it will become transparent, and specks of undissolved residue are easily seen. When the fusion is complete, swirl the crucible to allow the contents to solidify as a shell while cooling, and allow the covered crucible to cool to room temperature. Add a few ml H_2O to the crucible and, with the policeman, transfer the contents to the 250 ml beaker. Clean the crucible cover also. Heat the beaker on the water bath until the contents have dissolved, then uncover it and evaporate the contents on a hot plate to moderate fumes of SO_3.

Cool the beaker and contents somewhat and dilute the still hot solution to about 50 ml with H_2O. Heat on a water bath until all soluble salts have dissolved; only a few flocks of SiO_2 should remain undissolved but the dissolution may be somewhat prolonged if the sample contains a large amount of iron, and it may be necessary to add more water.

Add a small amount of filter paper pulp, filter the solution through a 7 cm Whatman No. 40 filter paper into a 250 ml beaker and wash the beaker, stirring rod and filter paper with hot water. It is not necessary (usually) to police the beaker. Reserve the filtrate for the determination of titanium (Sec. 9.15.1a) and, if desired, of manganese (Sec. 9.17a) and total iron (Sec. 9.23.3).

Transfer the filter paper with platinum-tipped forceps to a 20 ml weighed platinum crucible and burn off the paper either in an electric muffle furnace as previously described, or over the low flame of a bunsen burner. Finally ignite for 10 min at about 1000°C, cool and weigh as usual. Treat the small residue, which should be white, with 1 drop of H_2SO_4 (1:1) and 1 or 2 ml HF, as previously described, cool and weigh to obtain the weight of silica. If the nonvolatile residue amounts to more than about 0.1 mg, fuse it with a small amount of $K_2S_2O_7$, dissolve the cake and add the solution to the beaker containing the solution for the determination of titanium.

9.7. Titrimetric determination of aluminum

9.7.1. By the method of Milner and Woodhead

A 100 mg sample (for $>5\%$ Al_2O_3) is sintered with sodium peroxide (Procedure I), dissolved and the solution acidified with hydrochloric acid. Sulfur dioxide is passed through the solution to destroy any titanium peroxide complex formed during the sintering. The excess is removed by boiling, and the iron reoxidized with nitric acid. A larger (500 mg) portion (Procedure II) is taken for those samples having <5% Al_2O_3 and decomposed with a mixture of HF and H_2SO_4; the residue is fused with potassium bisulfate, dissolved in HCl and the iron reoxidized with nitric acid. Both solutions are then treated in the same manner for the separation of interfering elements with cupferron and the precipitation of aluminum as the benzoate. The precipitate is dissolved, the aluminum complexed with an excess of EDTA, and the excess backtitrated with standard iron solution, using salicylic acid as indicator. The following is taken from the outline of the method given by Milner and Woodhead. (Ref. 16, Sec. 6.4).

Reagents

Cupferron reagent: dissolve 10 g of cupferron in about 250 ml of cold H_2O with stirring and filter to remove any insoluble matter. Transfer the filtrate to a 1 liter separatory funnel and add HCl (1:1) carefully until the precipitation of the acid salt of cupferron is complete. Immediately add 400 ml pure chloroform and shake to dissolve the precipitate. Transfer the separated organic layer to a suitable flask. Store, if necessary, in a refrigerator but early use of the reagent is recommended.

Ammonium benzoate wash solution: dissolve 10 g of ammonium benzoate in water, with warming. Add 20 ml glacial acetic acid and dilute to 1 liter.

Standard iron solution, 0.1 M: dissolve 5.585 g pure iron in 20 ml concentrated HCl and oxidize with a few ml of concentrated HNO_3. Expel oxides of nitrogen by boiling, cool and dilute to 1 liter.

Ethylenediaminetetraacetic acid solution, $0.1M$: dissolve 37.23 g of the disodium dihydrate salt of EDTA in water and dilute to 1 liter. Determine the exact molarity as follows: pipet 20.0 ml of the standard $0.1M$ iron solution into a 150 ml beaker and add 25.0 ml of the EDTA solution. Dissolve 3 g ammonium acetate in this solution and add aqueous ammonia (1:4) to adjust the pH to 6, using a pH meter. Add 0.2 g salicylic acid and titrate with the standard iron solution, with stirring, to the end point (yellow to brown)

$$\text{Molarity of EDTA} = \frac{(20.0 + \text{back titration (ml)}) \times 0.1}{25.0}$$

Procedure 1: Weigh 100 mg of finely ground sample, which has previously been dried at 110° for approximately 1 hr, into a platinum dish with a capacity of about 40 ml.   Add 2 g of Na_2O_2, mix thoroughly with the sample using a platinum rod, and heat the dish and contents for 30–45 min in a muffle furnace controlled at 450–500°C.   After cooling, place the dish in a 250 ml beaker containing 50 ml of warm distilled water and allow the melt to dissolve.   When dissolving action has ceased, remove the dish with forceps and wash it well with about 50 ml of water, combining the washings with the sample solution.   Add HCl (1:1) dropwise to make the solution just acid.   Place 10 ml concentrated HCl in the platinum dish to dissolve any traces of adhering material, then rinse it into the beaker containing the sample solution.   Pass gaseous sulfur dioxide through the solution for a few minutes to discharge any color resulting from the formation of a titanium peroxy complex.   Boil the solution to drive off the excess sulfur dioxide and then add dropwise the minimum amount of concentrated HNO_3 needed to reoxidize any iron in the sample.   Continue boiling for a short time to drive out oxides of nitrogen and then refrigerate to a temperature of about 0°C.

Procedure II: Weigh 500 mg of finely ground material, which has previously been dried at 110° for approximately 1 hr, into a platinum dish with a capacity of 40 ml.   Moisten the sample with water, then add 5 drops of concentrated H_2SO_4, followed by about 10 ml of HF (48%).   Warm this solution and evaporate to fumes of SO_3 on a sand bath.   Cool slightly, add a further 2–3 ml HF and reheat to fumes of SO_3.   Continue heating until the acid fumes cease to be evolved and then gradually raise the temperature of the dish and contents to about 1000°C, maintaining it at this temperature for about 5 min.   Cool, add 1 g of potassium pyrosulfate and fuse the residue as described in Sec. 9.6.1.   Cool and dissolve the cake in 30 ml of HCl (1:2).   Transfer the solution to a 250 ml beaker, rinse the dish into the beaker, and dilute the solution to about 100 ml.   Boil and oxidize any ferrous iron with nitric acid as in Procedure I, boil to expel oxides of nitrogen and then refrigerate to about 0°C.

Transfer the cooled sample solution to a 250 ml separatory funnel, add 20 ml of cupferron reagent in chloroform and shake gently for about 2 min.   Allow the two layers to separate and transfer the organic layer to a beaker, washing through with 5 ml of chloroform.   Repeat this extraction procedure with a further 20 ml portion of the cupferron reagent and observe the color of the organic layer after allowing the two layers to separate.   If the color has not changed, the first extraction removed the iron, titanium, etc., from the aqueous layer and the cupferron extractions can be discontinued.   If the color has changed, continue extracting with the cupferron reagent but in most cases extraction with two 20 ml portions will be sufficient. Next add 20 ml of pure chloroform to the separatory funnel and shake for about 2 min. Discard the organic layer, washing through with 5 ml of chloroform. Repeat this extraction with a further 20 ml of pure chloroform to remove as much cupferron or its oxidation products as possible from the aqueous solution.

Transfer the aqueous solution to a 400 ml beaker, washing the separatory funnel walls with water, and bring the solution to the boil. Continue boiling to expel the chloroform, and then dilute the solution to a volume of about 200 ml by the addition of distilled water if necessary. Precipitate aluminum benzoate by adding 1 ml of thioglycolic acid (90%), followed by 15 ml of a hot solution containing 1 g of ammonium chloride, 2 g of ammonium acetate and 3 g of ammonium benzoate. Then add ammonium hydroxide (1:1) dropwise to the appearance of a faint, permanent white precipitate and boil the solution at this stage for about 2 min. Adjust the pH of the solution to approximately 4 by adding diluted aqueous ammonia (1:4) and using narrow-range pH papers. Boil the solution for a further 2 min and allow to stand warm for about 30 min, before filtering through a Whatman No. 41H paper. Wash the precipitate with hot ammonium benzoate solution and then wash from the filter paper with hot water into a 250 ml beaker. Dissolve traces of precipitate from the filter paper by pouring 20 ml of hot HCl (1:1) over it, followed by further washing with hot water. Collect all solutions and washings in the same beaker and heat if necessary to dissolve completely the aluminum benzoate precipitate. The volume of the solution should be about 100 ml.

To the aluminum solution, add sufficient of the EDTA solution to give about a 10% excess of reagent over that needed to complex the aluminum and then add aqueous ammonia (1:1) until the solution is just alkaline to methyl red. Boil this solution for 2 min, cool, add 3 g of ammonium acetate and adjust the pH to 6.5 with a direct-reading pH meter by adding dilute aqueous ammonia and acetic acid as needed. Add about 0.2 g of salicylic acid and back titrate the excess of EDTA with the standard iron solution until the brown iron salicylate color just persists. For solutions containing less than 30 mg of aluminum, use a $0.02M$ EDTA solution with $0.02M$ standard iron solution for the back titration. For solutions with more than 30 mg of aluminum, use $0.1M$ solutions of both EDTA and standard iron.

$$1 \text{ ml } 0.1 \text{ M EDTA} = 2.697 \text{ mg Al} = 5.095 \text{ mg Al}_2\text{O}_3$$

9.7.2. By the method of Bennett, Hawley and Eardley

An aliquot of the filtrate from the separation of SiO_2 is treated with cupferron-chloroform to remove Fe, Ti and other constituents, but no further separation of the aluminum is made. EDTA is added in excess and that not needed to complex the Al is back-titrated with standard zinc solution, using an ethanolic solution of dithizone as indicator (Ref. 1, Sec. 6.4).

Reagents

Ammonium acetate: add 120 ml glacial acetic acid to 500 ml H_2O, followed by 74 ml of $15M$ aqueous ammonia. Dilute to 1 liter.

Dithizone indicator solution (0.025%): dissolve 0.0125 g of dithizone in 50 ml ethanol (95%). This solution will keep for about one week.

EDTA solution ($0.05M$): dissolve 18.6125 g of the disodium dihydrate salt of EDTA in 1 liter of water. Standardize against the standard zinc solution.

Zinc solution ($0.05M$): dissolve 3.2690 g of metallic zinc pellets in 10 ml concentrated HCl (the oxide film on the pellets should be removed by acid washing and drying before weighing). Dilute to 1 liter (1 ml $\equiv$ 2.55 mg Al_2O_3).

Procedure

Dilute the hydrochloric acid filtrate from the SiO_2 separation (Sec. 9.2.1) or H_2S precipitation (Sec. 9.5) to 500 ml and transfer an aliquot of 100 ml to a 250 ml separatory funnel. Extract the Fe, Ti and other constituents with cupferron-chloroform as described in Sec. 9.7.1, transferring the aqueous solution to a 500 ml erlenmeyer flask. Add 1 or 2 drops of methyl orange (0.05% aqueous solution), and then add $15M$ aqueous ammonia until just alkaline. Quickly reacidify with concentrated HCl and add 5 or 6 drops in excess. Add sufficient EDTA to provide an excess of a few ml over the expected amount (if a 1 g sample was used for the SiO_2 determination, 1 ml $0.05M$ EDTA $\equiv$ 1.275% Al_2O_3), then add ammonium acetate buffer solution until the indicator turns yellow, followed by 10 ml in excess. Boil the solution for 10 min and cool rapidly.

Add an equal volume of ethanol (95%) and 1–2 ml dithizone indicator solution, and titrate with the $0.05M$ standard zinc solution from a yellow-green to the first appearance of a permanent pink color.

If the EDTA solution added is not exactly $0.05M$, calculate the equivalent volume of exactly $0.05M$ EDTA. If V ml is the volume of

0.05M EDTA added and v ml is the volume of standard zinc solution used in back titration, then

$$\% \ Al_2O_3 = (V - v) \times 0.00255 \times 100/\text{sample weight}.$$

9.8. Colorimetric determination of aluminum

9.8.1. BY THE METHOD OF SHAPIRO AND BRANNOCK (SOLUTION A)

This procedure utilizes the formation of a calcium aluminum alizarin red-S complex and its measurement at 475 mμ, as studied by Parker and Goddard (Ref. 18, Sec. 6.4). The colored complexes which are formed also with iron and titanium will interfere; that from iron is eliminated by the use of potassium ferricyanide and thioglycolic acid as complexing agents for the iron, but an empirical correction must be made for the effect of titanium. For the amount of Al_2O_3 usually encountered in most silicates, the correction is very small.

For samples containing not more than 24% Al_2O_3, a 15 ml aliquot of the solution A, prepared as described in Sec. 9.4.1, is used.

For samples containing >24% Al_2O_3 a 10 ml aliquot is used instead, and 5 ml of the blank solution (Sec. 9.4.1) is also added to the flask in order to ensure a pH and salt concentration comparable to those in the standards.

It was found convenient at the Geological Survey of Canada to mix certain of the reagents in large quantity and thus reduce to four the number of reagent additions that must be made. The dispensing of reagent aliquots by means of automatic pipets was found to reduce materially the time required for the determination, and the concentrations of the reagent solutions were adjusted to make possible the use of 10 ml and 25 ml aliquots.

The following method is that used routinely at the Geological Survey, and is based on that given by Shapiro and Brannock (Ref. 23, Sec. 6.4).

Reagents

Calcium chloride–hydroxlamine hydrochloride mixture: solution A: to 4.0 g reagent grade $CaCO_3$, in a 250 ml beaker, add 50 ml H_2O, cover and add 12N HCl (about 6 ml) until the $CaCO_3$ is dissolved, swirling the solution during the addition of the acid. Boil for 1–2 min to eliminate CO_2, cool and dilute to 1 liter. Solution B: dissolve 8.0 g reagent grade hydroxylamine hydrochloride in H_2O and dilute to 1 liter. Mix A and B and store in a glass bottle.

Potassium ferricyanide solution: dissolve 1.2 g reagent grade potassium ferricyanide in H_2O and dilute to 2 liters. Store in a glass bottle.

This solution will remain stable indefinitely if kept in a dark place; the reservoir and arms of the automatic pipet from which this solution is dispensed should be wrapped to exclude light.

Sodium acetate–acetic acid buffer and thioglycolic acid mixture: solution A: dissolve 140 g of reagent grade sodium acetate in 200 ml H_2O, add 60 ml of glacial acetic acid and dilute to 500 ml. Solution B: Dilute 8 ml of pure thioglycolic acid to 500 ml and mix well. Mix A and B with vigorous shaking and store in a glass bottle.

Alizarin Red-S solution: dissolve 0.4 g of sodium alizarin-3-sulfonate in H_2O and dilute to 2 liters. Mix well and store in a polyethylene bottle. This solution should be aged for at least 24 hr before use.

Procedure

1. All solutions should be at 20–25°C. Transfer by pipet a 15 ml aliquot of each sample solution A, 15 ml of each standard solution A, and 15 ml of the blank solution A to a series of 125 ml dry, stoppered erlenmeyer flasks. The flasks should have been cleaned with HCl (1:1), rinsed with water and then dried.

2. Add, by pipet, 25 ml of the calcium chloride–hydroxylamine hydrochloride mixture to each flask, and mix.

3. Add, by pipet, 25 ml of the potassium ferricyanide solution to each flask. Mix, and allow to stand for 10 min.

4. Add, by pipet, 10 ml of the acetate buffer–thioglycolic acid mixture to each flask. Mix well, and allow to stand for 10 min.

5. Add, by pipet, 25 ml of the Alizarin Red-S solution to each flask. Mix well and allow to stand for exactly 1½ hr.

6. Determine the absorbance of each solution at 475 mμ, using the solution prepared from the blank solution A as the reference blank solution.

7. Compute a factor for each of the two standard solutions and average them:

$$\text{Factor} = \frac{\% \ Al_2O_3 \ \text{of std}}{\text{Av absorbance of std}}$$

8. Compute the apparent $\% \ Al_2O_3$ in each sample:

Apparent $\% \ Al_2O_3 = \text{Factor} \times \text{absorbance of sample solution}$

9. Correct for the interference of TiO_2 as follows:

Apparent % Al_2O_3	Correction % TiO_2
0	0.25
5	0.20
10	0.15
15	0.10
20	0.05

$$\% \ Al_2O_3 = (\text{Apparent} \ \% \ Al_2O_3) - (\text{Correction} \ \% \ TiO_2 \times \% \ TiO_2)$$

9.8.2. BY THE METHODS OF RILEY AND WILLIAMS

The two methods to be described have in common the final measurement of the aluminum as the yellow chloroform extract of the 8-hydroxyquinolate, but differ considerably in the preliminary treatment of the sample solution. The first method (Procedure I) is designed for ordinary silicate rocks and minerals and interfering elements are complexed in various ways prior to the extraction of the aluminum (Ref. 19, Sec. 6.4). For more complex rocks and minerals, however, it is necessary to remove the interfering elements and this is done by means of a preliminary extraction with hydroxyquinaldine and, if necessary, by ion exchange (Procedure II); See Ref. 20, Sec. 6.4.

In procedure I the iron is complexed with 2,2'-dipyridyl and beryllium is added to remove the interference of fluoride. Manganese does not interfere but a small correction must be made for titanium interference. It is essential that the pH of the solution before extraction should be 4.9–5.0.

Because titanium is only partially extracted by 8-hydroxyquinaldine at pH 10, Procedure II includes a preliminary extraction with this reagent at pH 4 which effectively removes the titanium at the start. Zirconium and uranium are not extracted by 8-hydroxyquinaldine; any zirconium present may be complexed with quinalizarin-sulfonic acid but the uranium must be removed, as its anionic acetate complex, on an ion exchange column.

Preparation of Solution B

Procedure

Weigh out exactly 0.5 g of the sample into a 25 ml platinum crucible. Add, from a pipet, 4 ml of perchloric acid and, from a polyethylene measuring cylinder, 15 ml of HF. Heat the covered crucible on the water bath overnight, then remove the cover and evaporate the HF. When no further fumes of HF are visible, heat the crucible with an infrared heating lamp until most of the $HClO_4$ is removed but do not allow the contents to become completely dry. Add 2 ml $HClO_4$ and repeat the evaporation. Add 4 ml $HClO_4$ from a pipet and 15 ml H_2O. Heat the covered crucible on the water bath and stir occasionally with a platinum rod until the cake has dissolved. If it resists dissolution,* transfer it quantitatively to a 150 ml transparent silica flask, dilute to about 100 ml and boil until all has dissolved. Transfer the cold solution to a 500 ml volumetric flask and dilute to volume. Prepare a reagent blank solution B in the same manner, omitting the sample; a fresh reagent blank solution should be prepared each time a new batch of reagents is used.

Determination of Aluminum—Procedure I

Reagents

Sodium acetate solution $(0.5M)$: dissolve 17.0 g of sodium acetate trihydrate in H_2O and dilute to 250 ml.

Dipyridyl solution: dissolve 0.2 g of 2,2'-dipyridyl in 100 ml of $0.2N$ HCl.

Beryllium sulfate: dissolve 4 g $BeSO_4 \cdot 4H_2O$ in H_2O and dilute to 100 ml.

Complexing reagent: mix 4 ml of hydroxylamine hydrochloride (25%, w/v), 50 ml of $0.5N$ sodium acetate, 20 ml of dipyridyl solution and 10 ml of $BeSO_4$ solution. Dilute to 100 ml.

8-hydroxyquinoline reagent (oxine): dissolve 1.25 g of 8-hydroxyquinoline (reagent grade) in 250 ml of reagent grade chloroform, and store the solution in an amber glass bottle in a refrigerator. Reject it if it becomes colored.

Standard aluminum solution: dissolve 0.0529 g of Specpure aluminum (99.99% Al) in a slight excess of dilute HCl and dilute to 1 liter. This solution contains the equivalent of 100 μg Al_2O_3/ml. Alternatively

---

* Riley (Ref. 19, Sec. 6.4) describes two alternate procedures for decomposing resistant minerals. For samples containing oxide minerals such as corundum, rutile and spinel the acid decomposition is followed by separation of the insoluble residue by centrifuging and fusion of the residue with potassium pyrosulfate. When resistant silicate minerals are present the acid-insoluble residue in again separated by centrifuging and then either fused with NaOH or decomposed with HF and $HClO_4$ in a teflon bomb at 150°C, depending upon whether or not the resistant minerals contain alkali metals.

the standard solution may be prepared by dissolving 0.8894 g of ammonium alum (analytical reagent) in H_2O and diluting to 1 liter.

Procedure

Place 10 ml H_2O in a 50 ml stoppered separatory funnel, add from a pipet 1 ml of solution B (0.5 ml if the sample contains $>20\%$ Al_2O_3) and add 10 ml of the complexing reagent. After 5 min add, from a measuring cylinder, 20 ml of the 8-hydroxyquinoline reagent. Tightly stopper the separatory funnel and shake it mechanically for 5–8 min. Run the chloroform layer through a small plug of filter paper (held in the stem of a 2.5 cm funnel) into a dry 25 ml volumetric flask.* Pour 2–3 ml chloroform into the separatory funnel and rotate it gently before allowing the chloroform to pass through the funnel into the flask. Dilute to 25 ml with chloroform. Measure the absorbance of the solution at 410 $m\mu$ in a covered 1 cm cell against chloroform in the compensating cell. Carry out a blank determination in the same fashion on an appropriate volume of the reagent blank solution. Standardize the method by carrying out a determination on the reagent blank solution to which 1 ml of the standard aluminum solution has been added. 100 μg Al_2O_3 give an absorbance of ca. 0.394, after deduction of the reagent blank (ca. 0.014).

Since the aluminum 8-hydroxyquinolate solutions are somewhat light-sensitive, they should be kept in a dark cupboard until they are measured. The extraction should not be made in strong sunlight. Correct for interference by titanium as follows: Let corrected absorbance $= D$, and the observed absorbance (corrected for reagent blank) $= D_0$; then $D = D_0 - $ (vol. of solution B $\times$ % TiO_2 in sample $\times$ 0.0106)

$$\% Al_2O_3 = \frac{10 \times D}{(\text{corrected absorbance of } Al_2O_3 \text{ std} \times \text{vol of solution B used})}$$

Determination of Aluminum—Procedure II

Reagents

Sodium potassium tartrate solution $(0.2M)$: prepare a solution containing 28.2 g of sodium potassium tartrate per 500 ml and add 1 ml of chloroform as a preservative.

* The color of the 8-hydroxyquinoline extract should be pure yellow. If any traces of green are visible, then Fe(III) is being extracted owing to the pH of the solution being too low for the Fe(III) to be reduced to Fe(II) by the hydroxylamine. Such determinations should be repeated using sufficient buffer to bring the solution to the correct pH (4.9–5.0).

Ammonia-ammonium chloride buffer: dissolve 35 g of ammonium chloride in 300 ml of aqueous ammonia (approx. $15M$), dilute to 500 ml with water.

Sodium acetate solution ($0.5M$): as for Procedure I.

Anion exchanger: shake 10 g of Amberlite IRA 400 (Cl) with 100 ml of $2M$ sodium acetate for 20 min. Allow the resin to settle and decant the supernatant liquid. Wash the resin with water by decantation.

Quinalizarin-sulfonic acid (1,2,5,8-tetrahydroxyanthraquinone-3-sulfonic acid): mix 10.52 g of quinalizarin and 6 g boric acid in a round-bottomed flask, and add 37 ml of fuming sulfuric acid (30% SO_3). Fit the flask with a reflux air condenser and heat to 130°C for 6 hr. Pour on crushed ice (150 g), filter through glass wool and partly neutralize with sodium bicarbonate. After 12 hr filter off the precipitated quinalizarin-sulfonic acid and wash it with 20–30 ml of water (yield 63% of theory). The reagent is a 0.02% w/v solution of quinalizarin-sulfonic acid in water.

8-hydroxyquinoline reagent: dissolve 2.5 g of 8-hydroxyquinoline in 250 ml chloroform. Store in an amber glass bottle at 0°C.

8-hydroxyquinaldine reagent: prepare a 1% w/v solution of the reagent in chloroform.

Standard aluminum solution: as for Procedure I.

Procedure

In presence of interfering elements, except U and Zr

Decompose the sample (about 25 mg) with $HClO_4$ and HF as described in the preparation of solutions for Procedure I, but use only 0.5 ml and 4 ml portions of $HClO_4$ and HF, respectively, and make up the final volume to 50 ml (Sec. 6.4 Ref. 20, pp. 808–809).

Transfer 2 ml of this solution (containing not more than 200 μg of Al_2O_3) to a 50 ml separatory funnel containing 10 ml H_2O, and add 10 ml of 1% hydroxyquinaldine reagent. Shake for a few minutes to transfer the organic phase to the aqueous phase, and then adjust the solution to pH 4.0 by the addition of $0.5M$ sodium acetate. Shake the funnel mechanically for 15 min. Withdraw the lower layer and re-extract the aqueous layer with a further 10 ml of chloroform. Discard the chloroform layers, which contain all the titanium from the sample.

Add 1 ml of $0.2M$ sodium potassium tartrate (to prevent the precipitation of aluminum hydroxide) and mix thoroughly. Add 1 ml of the ammonia-ammonium chloride buffer to bring the solution to about pH 10.0. Extract the solution twice with 10 ml portions of 1% hydroxyquinaldine reagent, shaking for 10 min each time. Extract with 5 ml chloroform to remove any traces of the reagent. Discard the extracts and washings.

Add 0.7 ml glacial acetic acid to the aqueous phase remaining in the funnel to reduce the pH of the solution to about 4.5. Extract the solution for 10 min with 20 ml of 1% 8-hydroxyquinoline reagent. Run the chloroform phase through a pledglet of filter paper (held in the stem of a 2.5 cm funnel) into a 25 ml volumetric flask. Pour a few ml of chloroform into the separatory funnel, rotate it gently and add the chloroform to the graduated flask. Make up to 25 ml with chloroform. Measure the absorbance of the extract at 410 mμ in a covered 1 cm cell against a compensating cell containing chloroform. Determine the reagent blank in the same manner and standardize the method with 1 ml of the standard aluminum solution (100 μg Al_2O_3).

The precautions against excessive light, mentioned at the end of Procedure I, should be observed.

In the presence of uranium

Although uranium is not extracted with 8-hydroxyquinaldine, it can be separated from aluminum by ion exchange. Bring the solution to pH 4 by the addition of 0.5M sodium acetate. Pass it through an ion exchange column, about 8 mm in diameter, containing a 12 cm layer of Amberlite IRA (400) in its citrate form. Wash the resin with 10 ml H_2O and combine the percolate and washings. Continue with the extraction of titanium with 8-hydroxyquinaldine at pH 4 as previously described.

In the presence of zirconium

The extraction with 8-hydroxyquinaldine does not remove zirconium but the latter will, however, accompany aluminum into the 8-hydroxy-quinoline extract. Quinalizarin-sulfonic acid forms a strong, mauve-colored lake complex with zirconium and a much weaker one with aluminum; the aluminum, but not the zirconium, is quantitatively extracted from this complex by 8-hydroxyquinoline in chloroform.

Carry out the extraction with 8-hydroxyquinoline as previously described. Add 1 ml of 0.02% quinalizarin-sulfonic acid to the aqueous phase, mix well and then continue with extraction of aluminum, but using 20 ml of a 2% solution of 8-hydroxyquinoline in chloroform instead of the 1% solution normally used. The addition of a few drops of a 0.1% solution of silicone antifoam in chloroform is helpful in preventing the formation of an emulsion during the extraction.

9.9. Removal of manganese before precipitation of the alkaline earth elements

The use of ammonium sulfide (Sec. 6.3) to separate manganese is advantageous if appreciable amounts of zinc, cobalt and nickel are present also, for they are all separated at this stage and can be separated from each other and determined. If only manganese is present in significant amount it is better to use the ammonium persulfate method described in Sec. 9.9.2 because the separation of manganese with ammonium sulfide is incomplete and the remainder must be recovered from the ignited magnesium pyrophosphate.

When a manganese solution containing ammonium persulfate is heated, the colloidal hydrated manganese dioxide precipitate is filtered with difficulty. Peck and Smith (Ref. 30, Sec. 6.5) improved the filterability of the precipitate by gathering it on zirconium hydroxide, and the separation of the manganese is quantitative. Magnesium, but not calcium, coprecipitates appreciably with the hydrated MnO_2; coprecipitation of magnesium is minimized if the precipitation is made from an acid solution but the precipitation of manganese is then incomplete, and it is best to start the precipitation step in acid solution, heat it, make it ammoniacal and heat again. When the $MnO > 2\%$, a double precipitation should be made. The disadvantage later on of the addition of sulfate ion to the solution has been cited in opposition to the use of ammonium persulfate; the danger of coprecipitation of calcium sulfate is reduced to insignificance if a double oxalate precipitation is made, and the advantage gained from the removal of the manganese makes this additional step worth the time expended upon it. Any barium present will be precipitated as the sulfate and removed.

9.9.1. BY AMMONIUM SULFIDE

To the cold filtrate from the R_2O_3 group separation, made slightly acid with HCl and contained in a 250 ml erlenmeyer flask, add 2–3 ml of aqueous ammonia and saturate the solution with hydrogen sulfide gas, observing the precautions given in Sec. 9.5. Add an additional 2–3 ml of aqueous ammonia, fill the flask to the neck with distilled water and stopper it. Allow to stand for 12–24 hr. Add a small amount of filter paper pulp to the flask and then filter the contents through a 9 cm Whatman No. 40 paper. Wash the precipitate thoroughly with a 2% solution of ammonium chloride made just ammoniacal and saturated with hydrogen sulfide. The filtrate is now ready for the separation of calcium and strontium.

The sulfide precipitate may be dissolved in HCl, the chlorides removed by evaporation with H_2SO_4 and the manganese then determined colorimetrically (if less than 1%) as described in Sec. 9.17, or titrimetrically if present in greater amount (Sec. 9.16). If significant amounts of zinc, cobalt and nickel are present they can be recovered from the precipitate prior to the determination of manganese and also determined at this stage (See Sec. 6.5, Ref. 17, pp. 881–882); in most rocks the concentrations of these elements are too low to make this worthwhile.

9.9.2. BY AMMONIUM PERSULFATE

Reagents

Zirconyl chloride solution (5%): dissolve 25 g of zirconyl chloride (octahydrate) in about 200 ml of H_2O containing 5 ml 12N HCl. Allow the solution to stand overnight. Filter through a No. 42 Whatman filter paper into a 500 ml volumetric flask and dilute to volume.

Procedure

Add 1 ml of a 5% solution of zirconyl chloride octahydrate (acidified with HCl) and a small quantity of filter paper pulp to the acidified filtrate from the R_2O_3 group separation, which should have a volume of about 100 ml. Make the solution just ammoniacal to bromcresol purple (0.04% aqueous solution), then make it just barely acid with HCl (1–1).

Add 1 g of ammonium persulfate, stir the solution and heat it on the water bath for 20 min. Add 1 ml aqueous ammonia and heat for a further five min.

Filter at once through a 7 cm Whatman No. 41 paper. Wash the inside of the beaker three times with hot 2% ammonium nitrate solution, and transfer the washings to the paper. Wash the precipitate and paper 10 times with the ammonium nitrate solution.

If a reprecipitation is considered to be necessary, dissolve the precipitate in warm HCl (1:1) containing a few mg of sodium sulfite, dilute to about 100 ml and repeat the precipitation as described above. Combine the filtrates for the separation of the oxalate group. Boil vigorously to decompose excess persulfate.

The colorimetric determination of manganese in the solution of the precipitate cannot be done directly because of the tendency of the zirconium to hydrolyze, but a procedure for its determination at this stage is given in Sec. 9.17c.

Ingamells (Ref. 20, Sec. 6.5) separates manganese by the concurrent addition of $(NH_4)_2S_2O_8$ (25% solution) and aqueous ammonia to the acid filtrate from the R_2O_3 group separation until a final pH of 6–6.5 is attained; the black precipitate is dissolved in HNO_3 and H_2O_2 and reprecipitated, and the determination of manganese is finished titrimetrically. The recovery of the manganese is, however, not quantitative and a further determination of the manganese separated with the magnesium precipitate must be made.

9.10. Gravimetric determination of calcium

9.10.1. BY PRECIPITATION AS OXALATE

To the filtrate obtained in Sec. 9.9.1 or Sec. 9.9.2, which should have a volume of about 200 ml (if manganese was separated with ammonium sulfide the excess sulfide ions must be removed*), add 5 ml 12M HCl and, if necessary, a few drops of either bromcresol purple (0.04% aqueous solution) or methyl orange (0.02% aqueous solution) indicator. Dissolve 3 g $(NH_4)_2C_2O_4 \cdot H_2O$ in 50 ml H_2O, heat to 70–80°C and filter through a 7 cm Whatman No. 40 paper into the sample solution (if several precipitations are to be done use 50 ml aliquots of a hot, filtered 6% solution of ammonium oxalate). Heat the solution to near the boiling point and add aqueous ammonia (1:1) dropwise, while stirring vigorously, until the indicator changes color and then add about 1 ml in excess. Heat to near boiling (but do *not* boil) and then allow to stand without further heating for 2–3 hr with occasional stirring. When <1% calcium is present the precipitate may not appear for several minutes and calcium should not be judged to be absent until the solution has stood for the time recommended.

Filter the solution through a 9 cm Whatman No. 40 paper into an 800 ml beaker, retaining as much of the precipitate in the orginal beaker as is possible.† Wash the precipitate 3 or 4 times, by decantation, with

* After the addition of the 12M HCl boil the solution vigorously to expel H_2S· Add a few ml of saturated bromine water, again boil vigorously to expel the excess bromine and filter, if necessary, before continuing with the addition of the indicator.

† A recovery of any Fe, Al or rare earths which may have escaped precipitation with the R_2O_3 group may be made here but, since the precipitate must be ignited to the oxide (but not weighed), the precipitate must be quantitatively transferred to the paper. Wash and ignite the precipitate and paper as described following the second precipitation. Transfer the ignited residue, with a little water, to a 150 ml beaker, dissolve the residue in dilute HCl (1:1) and precipitate the R_2O_3 group elements as previously described (Sec. 9.6). Filter through a 7 cm Whatman No. 40 paper, wash the paper and precipitate with the hot 2% ammonium nitrate solution and combine the filtrate and washings with those from the first precipitation of calcium. Ignite the precipitate as previously described and add the weight of the residue to that of the main R_2O_3 group.

cold 0.1% ammonium oxalate and pour the washings through the paper. Wash the paper 3 or 4 times with the ammonium oxalate solution. Reserve the filtrate.

Wash down the sides of the original beaker with hot 5% HCl, add 2 ml 12M HCl to the solution and heat to boiling to dissolve the precipitate. Pour the hot solution through the paper, catching the filtrate in a 250 ml beaker; wash the beaker thoroughly with hot 5% HCl, pouring all washings through the paper, and then wash the paper thoroughly with the HCl solution, making sure that all areas of the paper are reached (lift up the inside flap with platinum-tipped tongs and wash the area beneath it). Finally wash the beaker and paper once with water. The final volume should be about 100 ml. Remove and discard the paper and rinse the funnel into the solution once with water.

Add approximately 0.5 g ammonium oxalate, dissolved in a few ml H_2O, to the acid solution, add 2 drops of indicator and heat the solution nearly to boiling. Precipitate the calcium (and strontium) as described previously and allow to stand cold for at least 4 hr, or overnight if convenient. Filter through a 9 cm Whatman No. 40 paper, combining the filtrate with that obtained from the first filtration, and transferring the precipitate quantitatively to the paper. Wash the precipitate and paper 10 times with the cold 0.1% ammonium oxalate solution. Reserve the combined filtrates for the recovery, if necessary, of any alkaline earths that may have escaped precipitation (see p. 366), and for the subsequent separation of magnesium.

Place the loosely folded filter paper in a weighed 25 ml platinum crucible, partly cover it, and burn off the paper at a low temperature in an electric muffle furnace, starting with a cold furnace. Heat the crucible at 1000°C for 30 min (displace the cover slightly at the start to facilitate the escape of carbon dioxide and then cover tightly), then cool for 30 min in a desiccator and weigh rapidly. Reheat at 1000°C for 15 min, cool for 30 min and weigh as rapidly as possible (place proper weights on the balance in advance of the crucible). Continue the heating until constant weight is obtained.*

The oxide should be white in color but occasionally may be light brown or green because of manganese, or greyish-white because of platinum. If the successive separations have been made as described neither of these elements should be present at this juncture. If platinum was not separated as the sulfide after the separation of silica (Sec. 9.5)

---

* In accurate work the ignited and weighed residues should be dissolved, reprecipitated and reignited until a constant weight is obtained.

it may show up as a yellowish to greenish-gray deposit (probably an ammine) on the bottom of the beaker during the evaporations that follow the separation of the R_2O_3 group. That which separates with the calcium oxalate will be reduced to platinum during the ignition and can be removed by dissolving the ignited oxide and reprecipitating the calcium; the direct solution and reprecipitation of the calcium oxalate will not suffice.

If a correction for manganese is considered to be necessary the determination may be made directly on the solution of the ignited oxide, or on the filtrate resulting from the separation of the barium and strontium from calcium in the oxide (Sec. 9.33). Moisten the ignited residue with water, add 5 ml HNO_3 (1:1) and warm to complete solution (if a brownish precipitate (MnO_2) persists, add a few grains of sodium sulfite to reduce it); determine the Mn present (as Mn_3O_4) by the periodate method as given in Sec. 9.17. If the nitric acid filtrate from the separation of strontium and barium from calcium is used, evaporate it to a volume of about 15 ml and then continue with the determination of manganese.

A knowledge of the strontium and barium content of the sample will indicate whether or not a correction must be applied to the ignited residue for the presence of these elements. It is seldom that any barium will be found here, particularly if manganese was separated with ammonium persulfate, and it is permissible to deduct the total SrO content if it is known (e.g., spectrographically), thus simplifying the work required. If a separation is necessary, however, it can be effected by the nitric acid method described under strontium (Sec. 9.33).

The combined filtrates from the oxalate precipitation will contain, in addition to magnesium, most of the barium (unless previously separated as a sulfate) and a small amount of calcium and strontium that escaped precipitation. These latter three elements can be recovered at this stage if so desired, prior to the precipitation of the magnesium.* Evaporate the filtrate to dryness in a platinum or fused silica dish and destroy the ammonium salts by careful heating of the dry residue with a free flame. Dissolve the residue in a small amount of water and separate the sulfates of the alkaline earths as described in Sec. 9.10.2. Reserve the alcoholic filtrate for the precipitation of magnesium and ignite the

---

* Calcium (and strontium) present in the filtrates will, if not removed, be precipitated as phosphates with the magnesium and can be recovered from the ignited pyrophosphate residue. Unfortunately, the solubility of barium phosphate precludes the simultaneous recovery also of barium in this manner and thus a prior separation as sulfate is necessitated. Some calcium will still remain to be recovered from the pyrophosphate residue.

small precipitate, by heating to dull redness for a few minutes, in a small platinum crucible. Fuse the residue with 1 g anhydrous Na_2CO_3 and leach with water; filter and wash the precipitate with cold 1% Na_2CO_3 solution. Discard the filtrate and dissolve the precipitate in warm 5% HCl; precipitate the calcium (and strontium) as the oxalate, ignite and weigh as the oxide. The barium in the filtrate may be determined as described in Sec. 9.33.

Conversion Factors

$$CaO \underset{1.3992}{\overset{0.7147}{\rightleftharpoons}} Ca$$

9.10.2. BY PRECIPITATION AS SULFATE, FOLLOWING COPRECIPITATION WITH MAGNESIUM AMMONIUM PHOSPHATE

This procedure is to be used when the calcium content of the sample is <0.5% and magnesium is preponderant over calcium. Precipitate the calcium and magnesium present in the filtrate from Sec. 9.9.1 or Sec. 9.9.2 with dibasic ammonium phosphate $((NH_4)_2HPO_4)$, as described in Sec. 9.12.1, and ignite to constant weight. Transfer the ignited residue to a 250 ml beaker, add 10 ml water and sufficient dilute H_2SO_4 (1:1) to dissolve the residue, warming if necessary, and give 0.5 ml in excess; dilute to 15 ml. If the residue will not readily dissolve, add 1–2 ml HNO_3, boil, and then evaporate to fumes of sulfur trioxide to expel the excess nitric acid, and dilute to 15 ml. To the cool solution add 70 ml ethyl alcohol (95%); this suffices for 0.3 g of magnesium pyrophosphate and for larger amounts add 15 ml H_2O and 70 ml 95% alcohol for each additional 0.3 g of $Mg_2P_2O_7$ present. Mix thoroughly and allow to stand overnight.

Filter through a 9 cm Whatman No. 42 paper (use a filter cone and moderate suction to speed up filtration) and wash with a 15:70 mixture of water and 95% ethyl alcohol. Allow the paper and precipitate to dry. Reserve the filtrate for the determination of manganese, after evaporation to fumes of sulfur trioxide to remove the alcohol (Sec. 9.17).

Dissolve the precipitate in a small volume of hot 5% HCl, and catch the filtrate in a 150 ml beaker. Precipitate the calcium with ammonium oxalate as described in Sec. 9.10.1 and ignite it to the oxide. Correct the weight of $Mg_2P_2O_7$ for coprecipitated calcium by subtracting the weight of $Ca_3(PO_4)_2$ equivalent to the CaO found.

Conversion Factors

$$CaO \underset{0.5424}{\overset{1.8438}{\rightleftharpoons}} Ca_3(PO_4)_2$$

9.11. Titrimetric determination of calcium

9.11.1. BY POTASSIUM PERMANGANATE

Reagents

Potassium permanganate $(0.1N)$: dissolve 3.2 g of the reagent in a liter of water and allow it to stand for at least 3 days at room temperature. Filter through a sintered-glass filtration funnel and store in an amber, glass-stoppered bottle covered with a dust cap.

Standardize against pure sodium oxalate, arsenious oxide or against National Bureau of Standards limestone (NBS 1a). If arsenious oxide is used proceed as follows: weigh 0.20 g pure arsenious oxide (As_2O_3) into a 250 ml erlenmeyer flask and dissolve in 10 ml $3N$ NaOH solution (free from reducing substances); add 15 ml HCl (1:1), dilute to 50 ml with water, add 1 drop of $0.002M$ KIO_3 (to catalyze the reaction) and titrate with $KMnO_4$ to the first permanent tinge of pink. About 40 ml of $0.1N$ $KMnO_4$ will be required.

Procedure

Precipitate and reprecipitate the calcium as described in Sec. 9.10.1 but filter the precipitate both times through a Gooch or fritted glass filter crucible, and wash the second precipitate with cold water only until free of chlorides and excess ammonium oxalate, keeping the volume of water used as small as possible. Reserve the filtrate for the determination of magnesium. Pour hot dilute H_2SO_4 (1:8) through the filter crucible to dissolve the precipitate, catching the filtrate in a 250 ml filtration flask. Stir the calcium oxalate with a short glass rod, kept in the filter crucible, to facilitate solution of the precipitate and continue to pour the hot acid through the filter crucible until all of the precipitate has dissolved (about 50 ml will be sufficient). Wash the filter crucible once or twice with water, catching the washings in the flask, and rinse down the sides of the flask. Dilute the solution to make it $1–1.5N$ in H_2SO_4 (about 150 ml) and heat to 80°C. Titrate slowly with $0.1N$ $KMnO_4$ to the first permanent pink tinge; 0.07 g CaO will require about 25 ml $0.1N$ $KMnO_4$. The reaction will be very slow at the beginning of the titration but once some manganous ion is present the reaction will be almost instantaneous and the $KMnO_4$ can be added rapidly. If the titration is made slowly and the solution cools below 60°C before the end point is reached, heat the solution again to 80°C before finishing the titration.

If filter paper is used, wash the second precipitate and paper with cold water only, remove the paper from the funnel with platinum-tipped forceps and spread it out on the wall of the original beaker. Pour 50 ml H_2SO_4 (1:8) through the funnel into the beaker, and rinse the funnel with water. Wash the precipitate from the paper with a jet of water and warm the solution to dissolve the calcium oxalate. Keep the filter paper, folded, on the side of the beaker, and out of the liquid. Heat to 80°C and titrate with 0.1N $KMnO_4$ as described, adding the titrant rapidly to the first permanent tinge of pink; add the paper, rinse off the sides of the beaker, macerate the paper and continue the titration to the second end point.

$$\% \ CaO = \frac{ml. \ KMnO_4 \ used \times Normality}{1000} \times \frac{56.08}{2} \times \frac{100}{sample \ wt}$$

9.11.2. By ETHYLENEDIAMINETETRAACETIC ACID (EDTA)

The procedure to be described is one that has been used at the Geological Survey for the determination of calcium and magnesium in a wide range of silicate rocks. An aliquot of solution B, the preparation of which is described below, is treated with aqueous ammonia to precipitate the bulk of the R_2O_3 group elements; traces of heavy metals are first complexed with triethanolamine and then suitable aliquots are titrated with EDTA, using METAB (see Sec. 6.5.2, and second footnote) and Eriochrome Black T as indicators. Aliquots of a standard solution of dolomite (NBS 88), to which are added iron and aluminum in amounts equivalent to those found in an average silicate rock, are carried through the procedure and factors for the final calculations are derived from the volumes of titrant used.

Preparation of Solution B

Procedure

1. Weigh and transfer quantitatively 0.500 g of each sample to a series of 90 ml platinum crucibles.* Moisten the powder with water and cautiously add 10 ml HF (48%); if a visible reaction occurs on first addition of the HF, cover the crucible and add the remainder of the HF when the reaction has diminished. Add 3 ml concentrated H_2SO_4, cover the crucible and place it in a ring of the water bath†; allow it to heat overnight.

---

* Shapiro and Brannock (Sec. 6.5, Ref. 39, p. A4) have replaced the platinum crucibles with Teflon beakers (approximately 50 ml capacity).

† The crucible should project into the water bath about one-half of the height of the crucible. A suitable length of tygon tubing, split lengthwise and inserted around the circumference of the hole, makes a convenient support for the crucible and protects it from the metal of the water bath.

2. Remove the crucible from the water bath, wipe off the exterior and place it on a sand bath. Remove the cover and evaporate the contents to a volume of about 3 ml. Cautiously add 1 ml concentrated HNO_3.

3. Continue the evaporation until copious white fumes of SO_3 are evolved. Cool; if the solution is not colorless or nearly so, add 1 ml concentrated HNO_3 and repeat the heating to fumes. Continue with the addition of nitric acid until all organic matter has been destroyed.

4. Transfer quantitatively the contents of the crucible (police if necessary) to a 400 ml Vycor beaker, dilute to 100 ml and boil until the solution is clear.*

5. Cool and quantitatively transfer the solution (filter if necessary) to a 250 ml volumetric flask; dilute to volume at room temperature and mix thoroughly. Immediately withdraw a 25 ml aliquot for the determination of the alkalies (Sec. 9.21.2). If the alkalies are to be done the same day, transfer the aliquots to 100 ml volumetric flasks; if not, store the aliquots in plastic bottles.

Determination of CaO

Reagents

EDTA solution, 0.005N (approx.): dissolve 1 g of the disodium dihydrate salt of EDTA in water and dilute to 1 liter. Store in a Pyrex bottle and standardize against the standard dolomite solution for each set of samples.

Standard dolomite (CaO + MgO) solution: weigh and transfer 1.000 g NBS dolomite No. 88 to a 250 ml beaker. Cover, add 10 ml H_2O, 20 ml dilute HCl (1:1) and boil for 5 min. Cool, filter, wash the paper and residue thoroughly with water, and dilute the filtrate to 1 liter at room temperature. Store in a plastic bottle.

$$1 \text{ ml} \equiv 0.305 \text{ mg CaO and } 0.215 \text{ mg MgO}$$

---

* A residue may be present which will resist dissolution. If much calcium is present a precipitate of calcium sulfate will form, but will dissolve on continued boiling. A fine white residue is probable $BaSO_4$ and this can be removed by filtration. Gritty particles are indicative of unattacked sample and, if the amount is more than a few grains, the material should be recovered, identified and, if necessary, treated again with HF and H_2SO_4. Hoops (Ref. 19, Sec. 6.5) found andalusite, kyanite, sillimanite and corundum in the residues; in the preparation of solution B these minerals are of petrological interest only. Tourmaline, rutile, garnet (both spessartite and grossularite) and spinels were also identified by Hoops and these, if ignored, will cause low results to be obtained for elements such as iron, titanium, and manganese.

Fe–Al solution (for standardization): dissolve 0.5 g Fe $(NH_4)_2(SO_4)_2 \cdot 6H_2O$ and 1 g $Al(NO_3)_3 \cdot 9H_2O$ in 100 ml dilute HCl (1:99), add 2 ml saturated bromine water and boil to expel excess bromine. Cool and dilute to 200 ml with dilute HCl (1:99). 1 ml contains approximately 4 mg each of Fe and Al.

Ammonium chloride solution, 25%: dissolve 250 g NH_4Cl in water, filter and dilute to 1 liter.

Triethanolamine, 10% (v/v): dilute 10 ml of the reagent to 100 ml with water and mix thoroughly.

Potassium hydroxide solution, 8N: dissolve 89.6 g KOH in water and dilute to 200 ml. Store in a plastic bottle.

Thioglycolic acid, 20% (v/v): dilute 100 ml pure thioglycolic acid to 500 ml with water. Store in a plastic bottle.

Procedure

1. Transfer a 50 ml aliquot of solution B to a 250 ml volumetric flask and dilute to approximately 200 ml. Add 10 ml NH_4Cl solution, one drop of bromcresol purple indicator (0.04% aqueous solution) and precipitate the R_2O_3 group with aqueous ammonia at room temperature; the solution should smell faintly of ammonia. Dilute to volume and mix thoroughly. Allow to stand for 30 min, or until the precipitate has settled.*

2. Transfer a 25 ml aliquot (pipet) of the standard dolomite solution to a 250 ml volumetric flask, dilute to 200 ml and add 10 ml of the Fe–Al solution. Precipitate the R_2O_3 group as described in (1), dilute to volume and mix thoroughly. Allow the precipitate to settle. Aliquots of this solution are used to standardize the EDTA solution in terms of CaO and CaO + MgO.

3. Filter each solution through a dry 9 cm Whatman No. 40 filter paper into a 300 ml, stoppered erlenmeyer flask.

4. Pipet a 50 ml aliquot of the standard solution, and of each sample solution, into 400 ml beakers. Add 1.0 ml of the standard dolomite solution to each sample aliquot known to be low in calcium, to ensure that an adequate volume of titrant will be required.

5. Dilute each aliquot to approximately 200 ml and add 1 ml triethanolamine solution. If appreciable quantities of Pb, Bi, Cd, Zn or

* For routine work a single precipitation will suffice, unless the R_2O_3 group is particularly abundant. If a double precipitation is necessary it is better to make the precipitation in a smaller volume of solution and at boiling temperature. Combine the two filtrates.

372   ROCK AND MINERAL ANALYSIS

Sn are known to be present, add 2 ml of the thioglycolic acid solution. Add 4 ml of KOH solution, stir well, and allow to stand for 1–2 min.

6. Add one METAB indicator pellet, allow it to dissolve completely, and titrate with EDTA solution to a grey end point.

7. The aliquot taken for the titration is equivalent to 5 ml of the original solution. Compute a factor for the EDTA solution from the weight of the CaO in the standard dolomite solution used:

$$\text{Factor} = \frac{0.305 \text{ mg CaO} \times 5}{\text{ml EDTA used for std.}}$$

8. Calculate the percent CaO in the sample:

$$\% \text{ CaO} = \frac{\text{ml EDTA used for sample} \ddagger \times \text{factor} \times 100}{\text{wt of sample used in titration}}$$

9.12. Gravimetric determination of magnesium

9.12.1. BY PRECIPITATION WITH DIBASIC AMMONIUM PHOSPHATE

To the cold (about 10°C) combined filtrates from Sec. 9.10.1 (if the alcoholic filtrate from the recovery of unprecipitated Ca, Sr and Ba is used, first eliminate the alcohol by evaporation), from which the excess ammonium salts have been removed and which have a total volume of about 250 ml, add a filtered solution of dibasic ammonium phosphate $((NH_4)_2HPO_4)$ to give approximately 1 g of the reagent per 100 ml of filtrate with 1 g in excess. Stir and add, with vigorous stirring,* sufficient concentrated aqueous ammonia to make the solution 10% by volume in aqueous ammonia, and continue the stirring until precipitation begins. Allow to stand overnight, preferably in a fume hood.

Filter the solution through a Whatman No. 42 filter paper of appropriate size (depending upon the size of the precipitate, not upon the volume of the filtrate) into an 800 ml beaker, or use a fine porosity, fritted-glass filter funnel with suction and catch the filtrate in a 1-liter filtration flask. Suction may also be used to expedite the filtration through paper, if the paper is supported in the funnel by a platinum

‡ Corrected for the volume of EDTA solution needed to titrate the 1.0 ml standard dolomite solution, if added.

* It is generally advisable to avoid striking the walls of the beaker with the stirring rod during the stirring. The abrasion of the glass surface encourages the growth of tiny crystals of the precipitate on the walls of the beaker. When only a small amount of magnesium is present, however, precipitation can sometimes be initiated in this fashion.

filter cone.*   However the filtration may be done, it is preferable to do it in, or in front of, a fume hood.   The bulk of the precipitate should be kept in the beaker.

Wash the beaker and precipitate twice with 5% aqueous ammonia (v/v) and pour the washings through the filter; wash the precipitate and paper five times with this wash solution, adding all washings to the filtrate.   Reserve the filtrate for subsequent later examination.

Dissolve the precipitate in the beaker in the minimum volume of hot 5% HCl and rinse down the walls of the beaker.   Cover, heat to near the boiling point, and pour the contents of the beaker through the filter, catching the filtrate in a 250 ml beaker, if suction is not used, or in a clean filtration flask.   Wash the paper and funnel with a small amount of hot 5% HCl (raise the inside flap of the paper and wash behind it to dissolve any trapped precipitate), then with water, and remove and discard the paper.   Wash the inside of the funnel once with 5% HCl, and then with water; rinse the tip of the funnel into the solution also.

To the filtrate, in a 250 ml beaker and having a volume of about 100 ml, add approximately 0.1 g $(NH_4)_2HPO_4$ and cool the solution to about 10°C.   Add pure aqueous ammonia dropwise until a precipitate forms; allow this to settle and continue the alternate addition of reagent and settling of the precipitate until precipitation is complete, as evidenced by no formation of a precipitate on the addition of a drop of aqueous ammonia.   Add 10 ml of aqueous ammonia and allow to stand overnight.

Examine the first filtrate for signs of a precipitate.   If none, discard the solution; if a small precipitate is present, decant about three-quarters of the supernatant solution and filter the rest as previously described. Transfer the second precipitate quantitatively to the same paper and wash the beaker(s), precipitate and paper with cold 5% aqueous ammonia as before.   Police the beaker(s) if necessary.   Wash the precipitate and paper ten times with the aqueous ammonia wash solution, and once with water.   Discard the filtrate.

Fold the paper lightly (do not make a tight fold because this makes the burning away of the carbon more difficult) and place it in a weighed,

* It has been found useful at the Geological Survey to clamp four 1 liter filtration flasks in a 6-hole adjustable filter stand (four of the holes were enlarged to accommodate the necks of the flasks) and to connect these flasks, in two parallel series, to an empty 500 ml filtration flask also clamped in the filtration stand.   The smaller flask is in turn connected to a small vacuum pump and the suction applied to each flask is controlled by individual pinchcocks.   This apparatus is easily cleaned, and can be stored without having to dismantle it.

25 ml platinum crucible. Wipe the inside of the funnel with a small piece of filter paper and add this to the crucible. Place the crucible, with the cover drawn back slightly, in a cold electric muffle furnace; allow the temperature to rise to about 450°C and maintain it at this level until all of the carbon is burned off and the residue is greyish-white in color. Do not allow the crucible to become even a dull red before this stage is reached, and at no time allow the contents of the crucible to catch fire. Heat the covered crucible and contents at approximately 1100°C for 30 min, cool in a desiccator for 30 min, and weigh as $Mg_2P_2O_7$. Repeat the ignition, cooling and weighing until constant weight is obtained.

The ignited residue may now be corrected for coprecipitated calcium and manganese, if these were not removed as previously described. The calcium is recovered as the insoluble sulfate (Sec. 9.10.2) and the weight of the calcium, as $Ca_3(PO_4)_2$, is subtracted from the total weight of the residue. Manganese may be determined in the alcoholic filtrate from the calcium recovery, after elimination of the alcohol by evaporation, or directly in the ignited residue, as described in Sec. 9.17. The manganese found is subtracted, as $Mn_2P_2O_7$, from the total weight of the pyrophosphate.

Conversion Factors

$$MgO \xrightarrow[1.6579]{0.6032} Mg$$

$$MgO \xrightarrow[0.3623]{2.7604} Mg_2P_2O_7$$

$$MnO \xrightarrow[0.4998]{2.0007} Mn_2P_2O_7$$

9.12.2. BY PRECIPITATION WITH 8-HYDROXYQUINOLINE (METHOD OF CHALMERS)

The following procedure is essentially that given by Chalmers (Sec. 6.6, Ref. 7, p. 361). To the combined filtrates from the separation of calcium (Sec. 9.10.1), from which excess ammonium salts have been removed and containing not more than 50 mg MgO in a volume of about 100 ml, add 2 g NH_4Cl. Add 0.5 ml o-cresolphthalein indicator (0.02% in ethanol, pH 8.2–9.8) and then sufficient 6N aqueous ammonia to give a violet color (pH 9.5); add 2–3 ml 6N aqueous ammonia in excess. Heat to 60–70°C and from a buret, with vigorous stirring, add a 2.5% solution of 8-hydroxyquinoline in 1N acetic acid until a *small* excess is present as shown by a deep yellow color in the supernatant liquid (50 mg MgO will require about 15 ml of the reagent). Digest the solution and precipitate for 10 min on the steam bath and then filter through a weighed medium-porosity sintered glass crucible.* Wash with about 50 ml of hot water and then dry the crucible and precipitate at 105°C (for the dihydrate) or at 160°C (for the anhydrous compound) for 1 hr, then at 30 min periods until constant weight is attained.

If a second precipitation is to be made the precipitate can be dissolved easily in warm dilute HCl (1:4).

If a titrimetric finish is preferred, using a standard solution of potassium bromate–potassium bromide, see Sec. 9.13.1 for details of the procedure.

9.13. Titrimetric determination of magnesium

9.13.1. By bromometric titration of the 8-hydroxyquinolate

Dissolve the precipitate, obtained as described in Sec. 9.12.2, in about 100 ml warm† dilute HCl (1:4), catching the solution in a 250 ml stoppered erlenmeyer flask. Cool to about 25°C and add 1 g potassium bromide and a few drops of methyl red indicator solution (0.2% in 60% ethanol). Titrate slowly with 0.1N standard potassium bromate solution (2.783 g pure KBrO$_3$ per liter) until the indicator color changes to yellow, at which point a slight excess of bromine is present. Let stand stoppered for 2 min, add 1 g of potassium iodide and titrate the iodine, formed through reaction with the excess bromine, with 0.1N standard solution of sodium thiosulfate (dissolve 25 g Na$_2$S$_2$O$_3$·5H$_2$O in 1 liter of freshly boiled and cooled water and add 0.1 g of Na$_2$CO$_3$; allow to stand for a day before standardizing) until the solution becomes faintly yellow. Add 0.1–0.2 g Thyodene‡ and continue the titration to the disappearance of the blue color.

The titration of the 8-hydroxyquinolate directly with standard potassium bromate may be made potentiometrically.

* A trace of sodium tauroglycocholate added to the solution will reduce the tendency of the precipitate to stick to the beaker (Sec. 6.6, Ref. 17, p. 363).

† Because 8-hydroxyquinoline is volatile at temperatures greater than 60°C, the temperature of the solution should be no higher than that required to effect the solution of the precipitate. It is of interest to note that excess hydroxyquinoline can be removed from a solution by prolonged evaporation alone.

‡ This is a solid reagent available from Fisher Scientific Company that is a convenient stable replacement for the starch solution used as an indicator in iodimetric titrations. It can be added as a solid reagent (≈0.5 g per titration) or as a 0.6% aqueous solution.

9.13.2. By titration with EDTA

The following procedure is the one that has been used at the Geological Survey of Canada. It utilizes an aliquot of solution B (Sec. 9.11.2) and the steps followed are those described in Sec. 9.11.2 up to step 4.

Reagents

In addition to those reagents listed in Sec. 9.11.2, the following are also required:

Monoethanolamine buffer (containing Mg): add 55 ml 12N HCl to 400 ml H_2O and mix thoroughly. Slowly pour 310 ml of monoethanolamine (reagent grade), with stirring, into the acid mixture and cool to room temperature.

Titrate 50.0 ml of standard $MgCl_2$ solution (see below) with standard EDTA solution, using 1 ml of the monoethanolamine buffer and Eriochrome Black T as indicator.

Add 50.0 ml of $MgCl_2$ solution to the exact volume of the EDTA solution required to titrate it in the above step, add this solution to the HCl–monoethanolamine mixture and mix well. Dilute to 1000 ml and store in a plastic bottle.

Magnesium chloride solution (0.01M): dissolve 2.03 g $MgCl_2 \cdot 6H_2O$ in water and dilute to 1000 ml.

Eriochrome Black T (with KCl): triturate 0.2 g Eriochrome Black T and 50 g KCl until the dye is evenly distributed over the surface of the KCl. Store in a glass bottle.

Procedure

Steps *1–3* are as described in Sec. 9.11.2.

4. Pipet a 50 ml aliquot of the standard solution, and of each sample solution, into 400 ml beakers. Add 1.0 ml of the standard dolomite solution to each sample aliquot to ensure that a satisfactory end point will be obtained.

5. Dilute each aliquot to approximately 200 ml and add 1 ml triethanolamine solution. If appreciable quantities of Pb, Bi, Cd, Zn or Sn are known to be present, add 2 ml of thioglycolic acid solution. Add 5 ml (pipet) monoethanolamine buffer solution. Mix thoroughly.

6. Add approximately 0.2 g Eriochrome Black T indicator and titrate with standard EDTA solution to a clear blue end point.

7. The aliquot taken for the titration is equivalent to 5 ml of the origginal solution. Compute a factor for the EDTA solution from the weight of MgO in the standard dolomite solution used:

$$\text{Factor} = \frac{0.215 \text{ mg MgO} \times 5}{(\text{ml std EDTA for CaO} + \text{MgO}) - (\text{ml std EDTA for CaO})}$$

8. Calculate the per cent MgO:

$$\% \text{ MgO} = \frac{(\text{ml EDTA for sample CaO} + \text{MgO*}) - (\text{ml EDTA for sample CaO})}{\text{wt. of sample used in titration}} \times \text{Factor} \times 100$$

9.14. Colorimetric determination of magnesium with Magon sulfonate (method of Abbey and Maxwell)

Reagents

Standard magnesium solution: dissolve 0.060 g Mg ribbon by heating with 5 ml H_2O and 2 ml 12N HCl. Cool, dilute to 1 liter in a volumetric flask and store in a polyethylene bottle. Just before use, transfer a 10 ml aliquot to a 250 ml volumetric flask, add 1 ml H_2SO_4 (1:1) and dilute to volume (1 ml contains 4 µg MgO).

Acid blank solution: dilute 1 ml H_2SO_4 (1:1) to 250 ml.

Acid calcium solution: (*A*) dissolve 0.25 g $CaCO_3$ by heating with 5 ml H_2O and 2 ml 12N HCl. Cool, dilute to 500 ml in a volumetric flask and store in a polyethylene bottle. (*B*) transfer a 10 ml aliquot to a 1000 ml beaker and add, with stirring, 290 ml H_2O, 100 ml 12N HCl and 100 ml 15N HNO_3. Store in a polyethylene bottle. This is the working solution and contains 4 µg Ca per ml.

Reagent solution: weigh 0.015 g Magon sulfonate (sodium 1-azo-2-hydroxy-3-(2,4-dimethylcarboxanilido)-naphthalene-1′-hydroxybenzene-5-sulfonate) into a 150 ml beaker. Add 100 ml 95% ethanol. Cover and dissolve by gentle warming and stirring with a magnetic stirring bar. Prepare fresh daily.

Triethanolamine solution (15%); mix 15 ml of triethanolamine with 60–70 ml H_2O and dilute to 100 ml.

Borax buffer solution (0.08M): dissolve 3.05 g $Na_2B_4O_7 \cdot 10H_2O$ in 80–90 ml H_2O and dilute to 100 ml

* Corrected for the volume of EDTA solution needed to titrate the 1.0 ml standard dolomite solution added.

Procedure

Transfer a 50 ml aliquot (i.e., 100 mg of sample) of solution B (Sec. 9.11.2: Preparation of Solution B) to a 250 ml volumetric flask and dilute to about 200 ml with water. Add 10 ml of 25% w/v NH_4Cl solution and a drop of bromcresol purple indicator (0.04% aqueous solution). Add aqueous ammonia ($14N$) dropwise, with swirling, until the indicator turns bluish-purple, and then one drop in excess. Dilute to volume, mix thoroughly and allow to stand for 30 min. Filter through a dry 9 cm Whatman No. 40 filter paper into a dry 300 ml erlenmeyer flask, and stopper the flask.

Transfer a 2 ml aliquot of the above filtrate (i.e., 0.8 mg of sample) to a 20 ml beaker. Into another 20 ml beaker pipet 2 ml of the acid blank solution, for use as a blank. One or more standards may be prepared, approximating the expected magnesium contents of the samples, by pipetting appropriate volumes of standard magnesium solution into 20 ml beakers (each ml of standard magnesium solution corresponds to 0.5% MgO).

Add 5 ml of acid calcium solution B (20 μg Ca) to each of the 20 ml beakers. Evaporate the solutions to the appearance of SO_3 fumes and continue heating until fumes are no longer evolved. If necessary, warm each beaker gently over a bunsen burner to evaporate any H_2SO_4 which may condense on the beaker walls. Allow to cool. Add 5 ml H_2O to each beaker, cover and warm to dissolve all residue, boiling if necessary. Allow to cool.

Transfer the solutions to 25 ml volumetric flasks, and rinse each beaker twice with 3–5 ml volumes of ethanol. To each flask add 5 ml of reagent solution, 0.5 ml of triethanolamine solution and 1 ml of borax buffer solution. Dilute to volume with ethanol, mix and allow to stand for 1 hr.

Fill a 1 cm absorption cell with water, place it in the spectrophotometer and adjust the instrument to zero absorbance at 510 mμ. Fill another cell with the blank and measure its absorbance immediately. Measure the absorbances of all the standards, then of all the samples, and finally on fresh portions of the blank and standard solutions from the volumetric flasks. Measure each absorbance immediately after filling the absorption cell. Keep the volumetric flasks stoppered.

Subtract the mean blank reading from all other readings. Calculate the MgO content of each sample solution by linear interpolation of absorbances between the two standards giving readings closest to that of the sample.

9.15. Colorimetric determination of titanium

9.15.1. WITH HYDROGEN PEROXIDE

Reagents

Standard TiO_2 solution (1.00 mg/ml): quantitatively transfer 1.013 g of National Bureau of Standards titanium dioxide (No. 154), dried at 105°C, to a 250 ml erlenmeyer flask. Add 10 g of ammonium sulfate and 25 ml concentrated H_2SO_4. Place a short-stemmed glass funnel in the neck of the flask and heat cautiously to incipient boiling while rotating the flask over a free flame. Continue the heating until complete solution has been effected and no unattacked material remains on the wall of the flask. Cool, and rapidly pour the solution into 450 ml of cold water, with vigorous stirring. Rinse the flask with dilute H_2SO_4 (5:95), and add the washings to the beaker. Let stand for 24 hr, then filter through a Whatman No. 42 paper into a 1 liter volumetric flask; wash the paper thoroughly with dilute H_2SO_4 (5:95) and dilute to volume with the dilute acid. Mix thoroughly and then store the solution in either a glass bottle fitted with a greased stopper, or in a polyethylene bottle made air-tight by a polyethylene thimble pressed into the mouth of the bottle and held in place by a screw cap. Remove all aliquots of this solution only with a pipet.

Potassium titanium oxalate $(K_2TiO(C_2O_4) \cdot 2H_2O)$ and potassium fluotitanate $(K_2TiF_6 \cdot H_2O)$ are also used for the preparation of this solution (Ref. 14, Sec. 6.7).

Hydrogen peroxide solution (3%): dilute 10 ml of 30% reagent grade hydrogen peroxide to 100 ml and store in an amber bottle. Such dilute solutions of hydrogen peroxide deteriorate markedly on standing and should not be kept longer than one month. The strength of the reagent should be checked at intervals as follows: pipet a 5 ml aliquot of 3% H_2O_2 into a 250 ml volumetric flask, dilute to volume with water and mix; pipet 25 ml of this solution into a 250 ml erlenmeyer flask, add 10 ml H_2SO_4 (1:5) and titrate to the first permanent pink tinge with $0.1N$ $KMnO_4$ (a fresh solution will require about 8 ml of $KMnO_4$); the first few drops of titrant will decolorize slowly but once a little $Mn(II)$ is present the reaction will proceed rapidly.

Potassium sulfate solution (16%): dissolve 160 g K_2SO_4 in water, filter through a Whatman No. 42 filter paper and dilute to 1 liter. Store in a glass bottle.

Ferric sulfate solution (10 mg Fe_2O_3/ml): dissolve 35.1 g $Fe_2(SO_4)_3 \cdot 9H_2O$ in water containing 25 ml concentrated H_2SO_4, filter and dilute to 1 liter.

Procedure

Preparation of calibration curve

To a series of 12 100 ml volumetric flasks, each containing 25 ml of K_2SO_4 solution and 20 ml dilute H_2SO_4 (1:1), add from a 10 ml semi-micro buret a series of aliquots of the standard TiO_2 solution (0, 0.10, 0.50, 1.00, 2.00, 3.00, 4.00, 5.00, 6.00, 7.00, 8.00, 10.00 ml) to make solutions containing 0–10 mg TiO_2 per 100 ml. Add 5 ml 3% H_2O_2 solution to each flask, dilute to volume at room temperature (20–25°C) and mix thoroughly. Measure the absorbance of each solution in a 1 cm cell at 410 mμ, using distilled water as a blank.* Prepare a calibration curve based upon the average of several readings made on each solution

(*a*) Determination of TiO_2 in the sulfuric acid solution of the R_2O_3 oxides

To the solution of the pyrosulfate fusion of the R_2O_3 group oxides (Sec. 9.6.1), in a 250 ml beaker and 10% in H_2SO_4, add 5 ml of 3% hydrogen peroxide. If the color of the solution is pale amber, transfer it quantitatively to a 100 ml volumetric flask, dilute to volume with water at room temperature and mix well. If the solution is darker than pale amber, transfer it to a 250 ml volumetric flask and add 37 ml K_2SO_4 solution, 30 ml dilute H_2SO_4 (1:1) and 7 ml of 3% hydrogen peroxide; dilute to volume at room temperature with water and mix thoroughly. Measure the absorbance at 410 mμ and obtain the concentration of TiO_2 per 100 ml of solution from the calibration curve.†

$$\% \ TiO_2 = (mg \ TiO_2/1000) \times (ml \ solution/100) \times (100/sample \ wt)$$

Total colorimetric correction. The sulfuric acid solution of the R_2O_3 group residue will usually be colorless to very pale yellow at room temperature but, if more than the usual amount of chromium or vanadium are

---

* Peck (Ref. 11, Sec. 6.7) prefers to use a system in which the per cent TiO_2 is read directly from transmittance tables. Reproducible readings are necessary for a given concentration and a standard solution is used to "set" the spectrophotometer before a measurement is made. The tables are based upon calibration points; one table is for the range 1–10 mg TiO_2 (435 mμ), the other for 10–50 mg TiO_2 (510 mμ). By using a longer wavelength (510 mμ) to increase the range of measurement, it is not necessary to dilute a solution containing more than the equivalent of 1% TiO_2 in order to be able to measure the transmittance at 410 mμ.

† If the determination of total iron is to be made on the solution of the R_2O_3 group, the cell contents and rinsings should be collected in a suitable beaker, together with the remainder of the solution. The solution is then evaporated to small volume to expel all hydrogen peroxide.

present or if the iron content is high, there may be sufficient color to the solution to warrant a total correction rather than one for iron only (see on).* Transfer the solution quantitatively to a 50 ml volumetric flask, dilute to volume with water, and mix. Measure the absorbance at 410 mμ and deduct *one-half* of the absorbance from that obtained for the peroxidized solution if final dilution was to 100 ml. Transfer the cell contents and all rinsings, together with the remainder of the solution, to a 250 ml beaker and continue as previously described.

Iron correction. The correction to be made for the color of ferric sulfate is determined as follows: to a series of 11 100 ml volumetric flasks, containing 25 ml of K_2SO_4 solution and 20 ml dilute H_2SO_4 (1:1), add from a buret 0, 1, 2, 3, 4, 5, 6, 7, 8, 9 and 10 ml of the ferric sulfate solution, corresponding to a range of 0–10% Fe_2O_3 on a 1 g sample. Add 5 ml 3% hydrogen peroxide, dilute to volume at room temperature with water and mix thoroughly. Measure the absorbance at 410 mμ and obtain the equivalent TiO_2 content from the calibration curve; express these values as per cent TiO_2, assuming a 1 g sample. The correction will probably be −0.01 to −0.04% for the range of Fe_2O_3 studied. The need for an iron correction is eliminated if the color due to the ferric sulfate is bleached by the addition of phosphoric acid (2 ml 85% H_3PO_4). Riley (Ref. 13, Sec. 6.7) uses as reagent solution one containing H_2SO_4 (10%), H_3PO_4 (10%), and H_2O_2 (6%), and adds 10 ml to a 40 ml aliquot (50 mg) of his B solution (Sec. 9.8.2) in a 50 ml volumetric flask; absorbance measurements are made at 400 mμ where the absorbance of the H_3PO_4 is minimum.

(*b*) Determination of TiO_2 on a separate portion of sample

Fuse 0.5 g of sample with 2 g of anhydrous Na_2CO_3 and 0.1 g Na_2O_2 as previously described (Sec. 9.2.1). Remove the cold cake to a 250 ml beaker and add 50 ml water; heat on water bath until all soluble salts have dissolved. Filter the solution through a 9 cm Whatman No. 40 paper, retaining as much as possible of the residue in the beaker; wash the paper and residue thoroughly with 1% Na_2CO_3 solution and discard the filtrate and washings (or use for the determination of sulfur (Sec. 9.30.1) and/or chromium (Sec. 9.34.1). Dissolve the precipitate in the beaker in 50 ml dilute H_2SO_4 (1:10) with warming, and pour the contents

---

* Peck (Ref. 11, Sec. 6.7) measures the transmittance of each sample solution *before* adding 0.2 ml 6% H_2O_2 to sample and standards in the spectrophotometric cells; 5 ml of the sample solution and of each of the high and low standards are used. This procedure ensures that all measurements are made at the same temperature.

through the paper until all of the residue on the paper is dissolved.* Filter, if necessary, into a 100 ml volumetric flask and continue with the determination of the TiO_2 as previously described.

Alternatively, the sample may be decomposed with a mixture of HF, HCl and H_2SO_4 but it is very important that the last trace of fluoride ion be expelled before the colorimetric determination of the titanium is made.

If a spectrophotometer is not available, satisfactory results can be obtained with Nessler tubes, or with a colorimeter of the Duboscq type; it is important the the the standards duplicate the composition of the sample as closely as possible. Details of these methods are given by Kolthoff and Sandell (Sec. 6.7, Ref. 6, pp. 619–627).

Conversion Factors

$$TiO_2 \underset{1.6681}{\overset{0.5995}{\longleftarrow\!\!\!-\!\!\!\longrightarrow}} Ti$$

9.15.2. WITH TIRON (METHOD OF RIGG AND WAGENBAUER)

The following procedure is substantially that given by Rigg and Wagenbauer (Ref. 12, Sec. 6.7) which utilizes an aliquot of solution B for TiO_2 contents up to 2%. For higher contents a smaller aliquot should be taken.

Reagents

Standard TiO_2 solution (0.02 mg/ml): weigh 0.1013 g of NBS standard sample No. 154 (TiO_2) into a 25 ml platinum crucible. Add 2 g of sodium bisulfate, cover and heat, cautiously at first, until the sample is completely fused. Cool, and with a jet of water transfer the fusion cake to a 250 ml beaker containing 125 ml H_2SO_4 (1:1); police the crucible with H_2SO_4 (1:1) and heat the contents of the beaker until the cake is completely dissolved. Cool and dilute to 250 ml in a volumetric flask and mix thoroughly. This solution contains 0.40 mg TiO_2 per milliliter. Pipet a 50 ml aliquot of this solution into a 1 liter volumetric flask, add 20 ml H_2SO_4 (1:1) and dilute to volume with water. Mix well and store in a Pyrex bottle. This solution contains 0.02 mg TiO_2 per ml which is equivalent to that in a solution prepared from a sample containing 1.00% TiO_2, as described in Sec. 9.11.2.

---

* Because it is difficult to be sure that all of the hydrous oxides are dissolved, it is better, when the TiO_2 content is known to be more than just a trace, to ash the paper in the crucible used for the fusion, fuse the residue with a small quantity of $K_2S_2O_7$, dissolve the cake in dilute H_2SO_4 (1:10) and add the solution to the main one.

If the standard solution of TiO_2 prepared as described in Sec. 9.15.1a is available, transfer a 20 ml aliquot to a 1 liter volumetric flask, add 20 ml H_2SO_4 (1:1) and dilute to 1 liter. This solution also contains 0.02 mg TiO_2 per ml.

Sodium acetate–acetic acid buffer solution: dissolve 82.0 g $NaC_2H_3O_2$ (or 136.0 g of the trihydrate) in 1 liter of water and add 390 ml of glacial acetic acid. The final pH of the colorimetric solutions utilizing this buffer should be 3.8.

Tiron solution (5%): dissolve 10 g of disodium-1,2-dihydroxybenzene-3,5-disulfonate in water and dilute to 200 ml. Discard the solution at the first sign of any yellow coloration.

Thioglycolic acid solution (20% v/v): dilute 100 ml of reagent grade thioglycolic acid to 500 ml. Mix thoroughly and store in a plastic bottle.

Procedure

1. To a set of 50 ml volumetric flasks add nothing to the first flask (blank), a 5 ml aliquot (pipet) of the standard TiO_2 solution to the second, and a 5 ml aliquot (pipet) of sample solution B to the third.

2. Add 25 ml of the sodium acetate–acetic acid buffer solution (pipet or automatic dispenser) to each flask and mix well.

3. Add, with a pipet, 5 ml of the Tiron solution to each flask and mix by swirling the solution.

4. Add 2 ml of thioglycolic acid solution to each flask and mix well. Dilute to volume with water and mix again.

5. Allow to stand for 1 hr before measurement of the absorbance of the solutions at 380 mμ, using the reagent blank as the reference blank solution.

6. Compute the factor for the standard solution:

$$Factor = 1.00/absorbance \ of \ standard$$

7. Compute the percentage of TiO_2:

$$\% \ TiO_2 = Factor \times absorbance \ of \ sample \ solution$$

9.16. Titrimetric determination of manganese

9.16.1. BY THE POTENTIOMETRIC TITRATION OF $Mn(III)$ WITH POTASSIUM PERMANGANATE (METHOD OF LINGANE AND KARPLUS)

Reagents

Sodium pyrophosphate (ca. 0.3M): prepare as a saturated solution at room temperature (ca. 120 g $Na_4P_2O_7 \cdot 10H_2O$ per liter; it should not be kept longer than 2–3 weeks. Test each new lot of reagent by using it in the titration of a known amount of manganese; if erratic potential readings are observed during the titration the salt should be recrystallized (Refs. 2–3, Sec. 6.8).

Sodium hydroxide $(5M)$: dissolve 20 g in 100 ml water. Test for freedom from reducing substances by adding one drop of $0.02M$ $KMnO_4$ to 10 ml of the NaOH solution; no green color due to manganate should develop. It is better to prepare this reagent freshly as needed.

Potassium permanganate $(0.02M)$: see Sec. 9.11.1 (Reagents). For the oxidation of Mn(III) to Mn(VII), the normality of this solution is $0.08N$.

Decomposition of the Sample

Transfer an appropriately sized sample* to a 50 or 100 ml platinum dish and moisten with water; stir the mixture with a platinum stirring rod.†   Cover the dish and add 10 ml concentrated HNO_3; when any effervescence has ceased, remove and rinse off the cover, and add 10 ml of 48% hydrofluoric acid.   Place the dish on a sand bath and slowly evaporate the contents to near dryness, with occasional stirring.   Cool, moisten with water and add 5 ml HNO_3 and 5 ml HF; mix and again evaporate to dryness on the sand bath.   Cool, add 20 ml HNO_3 (1:1) and again evaporate to dryness; heat the contents of the dish for a further 30 min after the salts appear to be dry.

Cool, add 20 ml (1:1) HNO_3 (previously boiled and cooled to remove oxides of nitrogen) and 10 ml of 5% (saturated) boric acid solution to the dish and break up the residue with the platinum rod.   Cover the dish with a plastic cover and digest the contents on a water bath until solution appears to be complete (1–2 hr); a gritty residue of quartz is often encountered at this stage.   If a brown precipitate is observed (MnO_2), add a few grains of sodium sulfite and stir; the brown precipitate will disappear.

Filter through a 7 cm Whatman No. 40 paper into a 100 ml volumetric flask,‡ and police the dish with water containing a few drops of HNO_3.

---

* The size of sample taken will be governed by the manganese content. At the Geological Survey it is found convenient to decompose a single portion (usually 1 g) for the determination of both manganese and phosphorus.

† These are conveniently made from 3–4 in. of stiff platinum wire (approximately $\frac{1}{16}$ in. diameter). A tight fold, about $\frac{1}{4}$ in. in length, is made at one end to prevent scratching of the dish by the sharp end of the wire, and a loop (approximately $\frac{1}{2}$ in. in length) at the other end provides a convenient handle.

‡ It is well to reserve sets of glassware for this purpose because of the etching effect of the solution.

Wash the dish and paper several times with this wash solution. If the residue on the paper is dark it may contain appreciable manganese and a white non-gritty residue may be titanium phosphate. In either case, if the residue amounts to more than a few grains, it should be ignited in a platinum crucible and fused with 0.5 g anhydrous Na_2CO_3, the cake leached with water, filtered, the filtrate made acid with HNO_3 and added to the solution in the volumetric flask. Dilute to volume and mix thoroughly.

Procedure

Transfer a suitable aliquot of the solution to an etched 150 ml beaker (see previous footnote) and, if necessary, dilute to approximately 50 ml. Boil vigorously to expel gaseous reaction products and add 1 g of urea or sulfamic acid. Add the contents of the beaker quantitatively to 200 ml of sodium pyrophosphate solution (saturated) in a 400 ml beaker. Adjust the pH of the solution to about 6–7 with $5M$ NaOH solution or dilute H_2SO_4 (1:3); indicator paper may be used or the pH can be measured as usual with the pH meter, using a glass electrode and a saturated calomel reference electrode.

Insert the bright platinum and saturated calomel reference electrodes into the solution and support the beaker on a magnetic stirrer. Place a Teflon stirring bar in the solution and, while stirring, slowly add the standard $KMnO_4$ solution ($0.02M$). The potential will rise steadily and the titration can be made with reasonable rapidity to a potential of about 400 mV; when this is reached add the $KMnO_4$ solution dropwise and record the potential and the buret reading after each addition. The end point is marked by a sudden rise of about 100 mV and the exact point at which the maximum change in potential per unit addition of the $KMnO_4$ solution occurred is determined by inspection and, if necessary, interpolation. A blank determination on all of the reagents used should also be made.

$$\% \ Mn = [(\text{Vol. of titrant}\ddagger \times \text{Normality})/1000]$$
$$\times (54.94/1) \times 100/\text{sample wt.}$$

Conversion Factors

$$MnO \underset{1.2913}{\overset{0.7744}{\rightleftharpoons}} Mn$$

‡ Corrected for blank.

9.16.2. By the visual indirect titration of $Mn(III)$ with potassium permanganate (method of Ingamells)

Reagents

Ferrous ammonium sulfate $(0.1N)$: dissolve 9.8 g of Fe $(NH_4SO_4)_2 \cdot 6H_2O$ in water (freshly boiled and cooled) containing 10 ml $18M$ H_2SO_4, and dilute to 250 ml. For best results the solution should be freshly prepared. Standardize against standard $0.1N$ KMnO$_4$ (see Sec. 9.11.1 for preparation) or against $0.1N$ potassium dichromate, using barium diphenylamine sulfonate as indicator.

Barium diphenylamine sulfonate (0.2%): dissolve 0.2 g of barium diphenylamine sulfonate in 100 ml of water.

Procedure

Transfer a suitable aliquot* of the sample solution prepared in Sec. 9.16.1 to a 300 ml wide mouthed erlenmeyer flask and, if necessary, reduce the volume of solution to about 10 ml by evaporation. Add 10 ml 85% H_3PO_4, plus an additional 15 ml for every 40 mg of MnO expected. Add 5 ml concentrated HNO$_3$ and evaporate, without boiling, on a hot plate until the volume of the solution has been reduced approximately to that of the H_3PO_4 used. Add 5–6 g of sodium pyrophosphate (solid reagent) for every 20 ml H_3PO_4 used and continue heating until the lack of condensation on the walls of the flask indicates that the expulsion of HNO$_3$ is virtually complete (the temperature of the non-boiling solution will be about 145°C). Add 1 ml HNO$_3$ (1:1), which has been freed from lower oxides of nitrogen, and continue the heating for 10–15 min during which time the HNO$_3$ will reflux on the walls of the flask; complete removal of the HNO$_3$ is not desirable.

Cool until the flask can be handled and add an amount of cold water equal to about one-quarter of the volume of the solution; mix and cool in running water (the solution should not be allowed to cool to the point of becoming glassy or it will be difficult to dilute it). When cold, dilute with water to about 20% in H_3PO_4, again cool in running water, and titrate the Mn(III) with standard ferrous ammonium sulfate solution, using barium diphenylamine sulfonate as the indicator; the indicator should be added just before the end point, when the pink color of the Mn(III) complex is nearly discharged.

% MnO = [ml titrant × normality × 70.94/10 × sample wt (g)]

* If a $0.1N$ standard ferrous ammonium sulfate solution is used as titrant, then 40 mg MnO will require about 5–6 ml of titrant, and a 10 ml semimicro buret graduated to 0.02 ml should be used. Ingamells (Ref. 1, Sec. 6.8) states that the end point obtained with $0.02N$ titrant is also extremely sharp.

9.17. Colorimetric determination of manganese with potassium periodate

(a) On a separate portion of the sample

If the sample was decomposed as described in Sec. 9.16.1, it is ready for use and the only precaution necessary is to allow for the presence of those elements which form colored solutions, which can be done as described for the determination of titanium (Sec. 9.15.1a—Total colorimetric correction). Otherwise, all reducing substances must be removed or oxidized, and chloride should be eliminated by repeated evaporations of the solution with nitric acid or by fuming with sulfuric acid.

Reagents

Standard manganese solution: (A) Dissolve 0.387 g pure manganese metal in 20 ml hot dilute HNO_3 (1:1). Boil out the oxides of nitrogen, cool and dilute to volume in a 500 ml volumetric flask. Store in a glass-stoppered bottle. This will furnish a solution containing 1.00 mg MnO per milliliter. (B) Transfer a 25 ml aliquot of the above solution to a 250 ml volumetric flask, add 5 ml concentrated HNO_3 and dilute to volume. The titer of this solution is 0.10 mg MnO per milliliter.

Potassium periodate solution (1%): dissolve 10 g of potassium periodate in about 200 ml of HNO_3 (1:1), with warming. Transfer to a 1 liter flask and dilute to volume. Mix and store in a glass-stoppered bottle.

Procedure

Preparation of Calibration Curve

To a series of 100 ml volumetric flasks add, from a 10 ml semimicro buret, aliquots of standard manganese solution B (0.10 mg MnO per ml) to give 0, 0.10, 0.25, 0.50, 0.75, 0.90 and 1.00 mg MnO per 100 ml. Add 20 ml concentrated HNO_3 and 10 ml of potassium periodate solution. Wash down the sides of the flask with water and mix. Immerse the flasks in a boiling water bath for 1 hr, or until color development is complete.

Cool, and add 5 ml of dilute H_3PO_4 (1:1). Dilute to volume, mix well* and measure the absorbance of the permanganate at 525 mμ in a 1 cm cell, using distilled water as a reference blank. Prepare a calibration curve based upon the average of several readings made on each solution.

* Riley (Ref. 5, Sec. 6.8) warns against the use of a polyethylene water bottle for diluting solutions to volume; fading of the colored complex may occur.

Prepare a similar calibration curve for MnO in the range of 1.00–2.50 mg per 100 ml, using 1.00, 1.50, 2.00 and 2.50 ml aliquots of standard manganese solution A (1.00 mg MnO per ml), as described above but add 20 ml of potassium periodate solution to each flask.

Determination of MnO

Transfer a suitable aliquot of the sample solution* to a volumetric flask (the final solution should preferably contain 0.2–2.0 mg MnO per 100 ml; a 50 ml aliquot of the 100 ml sample solution (0.5 g) is routinely used at the Geological Survey for most rocks (0.04–0.40% MnO)) and, for each 100 ml of final volume, add 20 ml concentrated HNO_3† and 10 ml of potassium periodate solution. Wash down the inside of the flask with water, swirl to mix, and immerse the flask in a boiling water bath for 1 hr. Cool, and add 5 ml of dilute H_3PO_4 (1:1) for each 100 ml of final volume. Dilute to the mark, mix, and measure the absorbance of the colored complex at 525 mμ in a 1 cm cell, using distilled water as a reference blank. Obtain the concentration of MnO per 100 ml of solution from the calibration curve, after applying any necessary correction to the absorbance reading (Sec. 9.15.1).

% MnO = (mg MnO/1000 $\times$ (ml solution/100) $\times$ (100/sample wt)

(b) On an aliquot of solution B

Reagents

Acid mixture (H_2SO_4–H_3PO_4): add 200 ml of 85% H_3PO_4 to 500 ml of water in a 1500 ml beaker in a cold water bath. Cool and add slowly, with stirring, 500 ml of concentrated H_2SO_4. Cool and dilute to 2 liters. Store in a glass bottle.

Standard MnO solution (0.02 mg/ml): pipet 20.0 ml of standard manganese solution A (Sec. 9.17a) into a 1 liter volumetric flask and dilute to volume.

Potassium periodate solution (1%): prepare as described in Sec. 9.17a.

---

* This includes the solution prepared as described in Sec. 9.16.1 and the solutions of ignited precipitates obtained in previous stages of the analysis. The ignited CaO (Sec. 9.10.1) and $Mg_2P_2O_7$ (Sec. 9.12.1) are first dissolved in about 5 ml HNO_3 (1:1); the alcoholic filtrate from the separation of Ca from Mg (Sec. 9.10.2) is evaporated to fumes of SO_3 to remove the alcohol, and the sulfuric acid solution of the R_2O_3 group (Sec. 9.6.1) is first freed of hydrogen peroxide by evaporation.

† If the amount of manganese present is known to be of the order of a few micrograms, make the solution 2N in H_2SO_4 instead and add 20 mg $AgNO_3$.

Procedure

1. To a set of 100 ml volumetric flasks add 20 ml of water to the first flask (to serve as reagent blank), 20 ml (pipet) of the standard MnO solution (to serve as standard), and 20 ml aliquots (pipet) of each Solution B (Sec. 9.11.2) to the rest.

2. Add 50 ml of the acid mixture from an automatic dispenser to each flask and mix.

3. Add 20 ml of potassium periodate solution to each flask and mix.

4. Place the flasks in a boiling water bath (near 100°C) and heat for 1 hr.

5. Cool the solutions to room temperature, dilute to volume and mix thoroughly.

6. Measure the absorbance at 525 mμ, using the reagent blank solution.

7. Compute the factor for the standard solution:

Factor = (1.0/absorbance of standard solution)

8. Compute the percentage of MnO:

% MnO = Factor × absorbance of sample solution.

(c) Following the separation of manganese with ammonium persulfate

The following procedure to utilize the hydrous MnO_2–ZrO_2 separated in Sec. 9.9.2 was devised by J.-L. Bouvier of the Geological Survey of Canada.

Procedure

Transfer the filter paper and contents to the 400 ml beaker which was used for the persulfate separation, add 25 ml concentrated HNO_3 and digest on a hot plate until the paper is destroyed; evaporate to dryness. Add 25 ml HNO_3 and 5 ml H_3PO_4 (1:1) and again evaporate, without boiling, until the volume is that of the H_3PO_4 used. Remove the beaker from the hot plate, cover, and immediately add 50 ml water (**Caution!** some spattering will occur).

Add 20 ml concentrated HNO_3 (previously boiled and cooled), 10 ml potassium periodate solution (1%, Sec. 9.17a), a few boiling chips and boil for about 30 min. Cool, transfer the contents to a 100 ml volumetric flask and dilute to volume. Allow to stand for a few minutes.

Transfer some of the solution to a centrifuge tube and centrifuge for 5 min at about 6000 rpm. Transfer some of the supernatant liquid to a 1 cm cell and measure the absorbance at 525 mμ as before.

9.18. Gravimetric determination of phosphorus as magnesium ammonium phosphate, following preliminary separation as ammonium phosphomolybdate

*Reagents**

Ammonium molybdate (5%): dissolve 50 g of $(NH_4)_6Mo_7O_{24} \cdot 4H_2O$ in approximately 500 ml of warm water, let stand for several hours, filter through a Whatman No. 42 filter paper, dilute to 1 liter and store in a polyethylene bottle.

Ammonium nitrate (50%): dissolve 500 g of NH_4NO_3 in about 500 ml of water, let stand for several hours, filter through a Whatman No. 42 filter paper, dilute to 1 liter and store in a glass bottle.

Magnesia reagent: dissolve 50 g of $MgCl_2 \cdot 6H_2O$ and 100 g of NH_4Cl in 500 ml of water, make ammoniacal and let stand overnight. Filter, if necessary, through a Whatman No. 42 filter paper, make just acid with hydrochloric acid, and dilute to 1 liter. Store in a glass bottle.

Procedure

To an aliquot (usually 50 ml of the sample solution, obtained as described in Sec. 9.16.1) in a 150 ml beaker, add 5 ml concentrated nitric acid and 15 ml of ammonium nitrate solution (50%). Heat to boiling, then allow to cool to about 60°C. Add 20 ml of 6% ammonium molybdate solution and stir vigorously until a turbidity appears; if none appears within a few minutes, scratch the bottom of the beaker occasionally with the glass rod until precipitation begins. Keep the solution at about 60°C for 15 min and if the precipitation is heavy, add additional ammonium molybdate reagent. Allow to stand overnight at room temperature.

Filter the solution through a 7 cm Whatman No. 42 filter paper† keeping most of the precipitate in the beaker. Wash the precipitate and beaker five times by decantation with 2% ammonium nitrate solution made slightly acid with nitric acid, pouring the wash solution through the paper, and finally wash the paper five times also. Discard the filtrate and expel excess wash solution from the stem of the funnel. Wipe the underlip of the beaker with a small piece of filter paper and

---

* There is no agreement in the literature on the composition and concentration of these reagents, particularly of the ammonium molybdate.

† If the phosphorus content of the sample appears low enough that a small uncertainty in composition will not cause serious error, the solution may be filtered through a weighed, fritted glass crucible, the precipitate washed first with a 5% solution of ammonium nitrate acidified with nitric acid and then once or twice with water, and weighed after being heated in an oven at 158°C for 1 hr. The gravimetric factor is 0.0378 for P_2O_5.

add it to the funnel. Place the precipitation beaker under the stem of the funnel and dissolve the precipitate on the paper by washing it with a 5% solution of aqueous ammonia containing 2 g of ammonium citrate per 100 ml.* Wash the paper thoroughly with the wash solution, then successively once with water, once with cold 5% HCl, again with water and once more with the aqueous ammonia–ammonium citrate wash solution. Discard the paper and rinse the funnel into the beaker with the ammoniacal solution. Wash down the sides of the beaker with the wash solution and stir until all of the yellow precipitate is dissolved. The volume should be less than 50 ml.

Add dropwise dilute HCl (1:1) until the solution is just acid to methyl red (0.2% in 60% ethanolic solution) and add 10 ml in excess. Add 10 ml of magnesia reagent and then aqueous ammonia dropwise, with vigorous stirring, until the solution is ammoniacal. After a few minutes add 10 ml of the aqueous ammonia, stir vigorously and allow to stand overnight, preferably in a fume hood.†

Filter the solution through a Whatman No. 42 filter paper of appropriate size, keeping as much of the precipitate as possible in the beaker,‡ and wash the beaker and precipitate twice with 5% HCl as described for the second precipitation of magnesium (Sec. 9.12.1), add 1–2 ml of magnesia reagent and precipitate the magnesium ammonium phosphate as before. Filter, wash, ignite and weigh the $Mg_2P_2O_7$ as described in Sec. 9.12.1.

$$\% \ P_2O_5 = Mg_2P_2O_7 \times 0.6377 \times (100/\text{sample wt})$$

Conversion Factors

$$P_2O_5 \underset{2.2912}{\overset{0.4365}{\rightleftharpoons}} P$$

* Kolthoff and Sandell (Sec. 6.9, Ref. 7, p. 381) point out that phosphates of Ti, Fe and similar elements present in the yellow precipitate are not soluble in aqueous ammonia and will remain on the paper unless a complex-forming ion such as citrate is present in the wash solution. A final wash with dilute hydrochloric acid insures that all of the phosphorus passes into the filtrate and the presence of citrate ion in the final solution prevents the precipitation of such elements as Fe and Ti during the subsequent precipitation with ammonium phosphate.

† The preliminary separation of phosphorus with ammonium molybdate may be omitted and the phosphorus precipitated directly with magnesia reagent, if so desired, as follows: to the aliquot from Sec. 9.16.1 in a 150 ml beaker add 5 ml 12N HCl, 2 g of ammonium citrate and 10 ml of magnesia reagent. Complete the precipitation as described above.

‡ For most rocks a single precipitation with magnesia reagent will suffice, particularly if a preliminary separation with ammonium molybdate has been made. In this case, transfer all of the precipitate to the paper and wash, ignite and weigh the $Mg_2P_2O_7$ as described (Sec. 9.12.1).

9.19. Colorimetric determination of phosphorus

9.19.1. As the yellow molybdovanadophosphoric acid complex
(method of Baadsgaard and Sandell)

The sample is decomposed by a mixture of nitric and hydrofluoric acids, followed by evaporation to dryness with HNO_3 to expel most of the fluoride retained in the residue. The residue is then dissolved in colorless HNO_3 and boric acid solution is added to complex any fluoride ion still present. The procedure is given in detail in Sec. 9.16.1. An aliquot is used for the colorimetric determination of phosphorus, based upon the procedure of Baadsgaard and Sandell (Ref. 2, Sec. 6.9).

Reagents

Ammonium molybdate (5%): prepare as described in Sec. 9.18 (Reagents).

Ammonium metavanadate (0.25%): Dissolve 2.5 g of NH_4VO_3 in 500 ml of hot water, cool and add 20 ml concentrated HNO_3. Stand several hours, filter if not clear, and dilute to 1 liter. Store in a glass bottle.

Standard phosphate solution (1.00 mg P_2O_5/ml): dissolve 0.959 g of KH_2PO_4 (recrystallized and dried at 100°C) in water and dilute to 500 ml. Store in a tightly capped polyethylene bottle. This solution may be further diluted as desired.

Procedure

To the solution of the sample prepared as described in Sec. 9.16.1, containing 5 ml concentrated nitric acid, not more than 9 mg P_2O_5 and in a 100 ml volumetric flask,* add by pipet 10.0 ml of ammonium vanadate solution and 20.0 ml of ammonium molybdate solution, in that order. Mix, dilute to volume and mix thoroughly, and then allow to stand at room temperature for 30 min. Measure the absorbance of the yellow complex at 460 mμ in a 1 cm cell, against a blank consisting of the

---

* If a 1.00 g sample is used for the determination of both manganese and phosphorus, a 50 ml aliquot will enable the determination of up to 1.8% P_2O_5; such a high P_2O_5 content would be unusual and the determination should be repeated using a smaller sample weight. The splitting of the solution may be done with a dry pipet or, conveniently, as follows: rinse a clean pipet with the solution and transfer the rinsings to a 150 ml beaker; pipet a 50 ml aliquot into a 100 ml volumetric flask, then rinse the pipet with water into the beaker and combine the remainder of the original solution with it also. The solution in the beaker may be used for the determination of manganese (Secs. 9.16.1, 9.16.2, 9.17), and also as a means of correcting for the absorbance due to the presence of other colored substances (dilute to 100 ml and measure the absorbance at 460 mμ against distilled water as a blank solution).

reagents and concentrated nitric acid in the proportions given previously. Correct the absorbance for that due to other colored substances, such as ferric iron (as described in the footnote) and obtain the concentration of P_2O_5 per 100 ml of solution from a calibration curve.

Preparation of Calibration Curve

To a series of 100 ml volumetric flasks add, from a 10 ml semimicro buret, a series of aliquots of the standard phosphate solution (1.00 mg P_2O_5/ml) to give 0, 0.50, 1.00, 2.00, 3.00, 4.00, 5.00, 6.00, 7.00, 8.00 and 9.00 mg P_2O_5/100 ml. Add 5 ml colorless concentrated HNO_3, dilute to about 50 ml and continue with the determination as previously described. Measure the absorbance of each solution against the solution containing the reagents only as blank. Prepare a calibration curve based upon the average of several readings made on each solution.

9.19.2. As the Yellow Molybdovanadophosphoric Acid Complex (Rapid Method of Shapiro and Brannock)

This procedure utilizes an aliquot of solution B (Sec. 9.11.2) and measures the absorbance of the yellow molybdovanadophosphoric acid complex at 430 mμ in order to minimize the interference of iron. A correction (-0.01% P_2O_5 for every 4% total iron as Fe_2O_3) is also made as an additional compensation for the effect of iron. The method follows closely that given by Shapiro and Brannock (Sec. 6.9, Ref. 11, p. A31).

Reagents

Ammonium molybdovanadate solution: dissolve 1.5 g NH_4VO_3 in 400 ml dilute HNO_3 (1:1). Dissolve 90 g $(NH_4)_6Mo_7O_{24} \cdot 4H_2O$ in 400 ml H_2O. Mix the two solutions and dilute to 1 liter. Store in a polyethylene bottle.

Standard phosphate solution (0.015 mg P_2O_5/ml): weigh and transfer 0.0950 g National Bureau of Standards standard sample No. 56B (phosphate rock) to a 250 ml beaker. Add 80 ml dilute HNO_3 (1:1), cover and digest on the steam bath until solution is essentially complete. Boil vigorously for a few minutes, cool, dilute to about 200 ml and filter through a Whatman No. 40 paper into a 2 liter volumetric flask. Wash the beaker and paper 2 or 3 times with water. Add 60 ml dilute H_2SO_4 (1:1), cool if necessary and dilute to volume. Store in a glass bottle. The P_2O_5 concentration is equivalent to that in solution B of a sample containing 0.75% P_2O_5.

Procedure

1. To a set of 50 ml volumetric flasks add 15.0 ml of water to the first flask (to serve as reagent blank), 15.0 ml of the standard phosphate solution to the second (to serve as standard) and 15.0 ml aliquots of each solution B (Sec. 9.11.2) to the rest.

2. Add, by pipet, 10 ml of the ammonium molybdovanadate reagent to each flask, swirling the solutions during the addition of the reagent. Dilute to volume, mix and allow to stand for at least 5 min.

3. Measure the absorbance of each solution at 430 mμ, using the reagent blank as the reference blank solution.

4. Compute the factor for the standard solution:

$$\text{Factor} = (0.75/\text{absorbance of standard solution})$$

5. Compute the apparent percentage of P_2O_5:

$$\text{apparent } \% \ P_2O_5 = \text{Factor} \times \text{absorbance of sample solution}$$

6. Correct the apparent per cent P_2O_5 for the effect of iron:

$$\%P_2O_5 = \text{apparent } \% \ P_2O_5 - (\% \text{ iron as } Fe_2O_3 \times 0.0025)$$

9.19.3. As the heteropoly blue complex

(a) Rapid method

The rapid colorimetric method given previously (Sec. 9.19.2) was used in the writer's laboratories as part of a scheme for the rapid analysis of silicates based upon the work of Shapiro and Brannock. This scheme has since been replaced by one utilizing a combination of X-ray fluorescence spectroscopic and chemical methods, and the determination of phosphorus is now made in conjunction with the flame photometric determination of sodium (potassium is among 8 elements determined by X-ray fluorescence spectroscopy). A small sample weight (100 mg) is used, of which an aliquot (10 mg) is taken for the phosphorus; the heteropoly blue complex is used because of the need for greater sensitivity occasioned by the small sample. The procedure to be given is based upon that described by Boltz (Ref. 3, Sec. 6.9), for which he suggests an optimum concentration range of 0.1–1.2 ppm phosphorus.

Reagents

Ammonium molybdate (in 10 N H_2SO_4): dissolve 5 g $(NH_4)_6Mo_7O_{24} \cdot 4H_2O$ in about 200 ml 10N H_2SO_4 (dilute 70 ml 36N H_2SO_4 to 250 ml) and dilute to 250 ml with additional 10N H_2SO_4. Store in a polyethylene bottle.

Hydrazine sulfate (0.15%): dissolve 0.15 g of hydrazine sulfate ($N_2H_6SO_4$) in water and dilute to 100 ml.

Procedure

1. To a 0.100 g sample (about 150 mesh) in a 50 ml platinum (or 100 ml teflon) dish add 1 ml H_2O and 1 ml concentrated HNO_3. Evaporate to dryness on a medium temperature sand bath (about 150°C).

2. Add 5ml 12 N HCl and 5 ml HF (48%) and break up the residue with a platinum or Teflon stirring rod. Evaporate to dryness.

3. Add 5 ml HCl, break up the residue as before and again evaporate to dryness.

4. Add 6 drops of 12N HCl to the residue, break up with the rod and add 20 ml H_2O; heat until solution is complete, adding more H_2O if necessary.

5. Transfer (filter, if necessary) the solution to a 100 ml volumetric flask and dilute to volume.

6. Remove a 10 ml aliquot by pipet and transfer it to a 50 ml volumetric flask. Blow out any liquid remaining in the pipet into the 100 ml flask and rinse the outside of the stem and the interior of the pipet with a fine jet of water, catching the rinsings in the 100 ml flask.*

7. Add to the 50 ml flask 5 ml of the ammonium molybdate solution and 2 ml of hydrazine sulfate solution and mix.

8. Dilute to approximately 1/2 in. from the graduation mark with water, stopper the flask and mix thoroughly. Immerse the unstoppered flask in a boiling water bath for ten minutes, remove and cool rapidly in running water. Allow to come to room temperature, dilute to volume and mix thoroughly.

9. Measure the absorbance of the heteropoly blue complex at 830 mμ in a 1 cm cell, using either distilled water or a reagent blank as the reference blank solution. Obtain the concentration of P_2O_5, in micrograms, from a standard curve prepared as described below.

$$\% \ P_2O_5 = (\text{Micrograms } P_2O_5/100)$$

Preparation of standard curve

Transfer 1.00 ml of the standard phosphate solution (Sec. 9.19.1) containing 1.00 mg P_2O_5 per milliliter to a 100 ml volumetric flask, dilute

* There will be a small error introduced into the determination of phosphorus because of the film of water in the pipet. If the inner surface of the pipet is kept clean, and if the pipet is allowed to drain between aliquots (two pipets, of which one is draining while the other is in use, can conveniently be used), the error is very slight.

to volume and mix thoroughly (1 ml = 10 μg P_2O_5).   To a series of 50 ml volumetric flasks add 0, 1, 2, 3, 4, 5, 6, 7, 8, 9 and 10 ml aliquots of this solution, to give a range of P_2O_5 concentration of 0–100 μg and continue with the addition of reagents, heating and final measurement of the absorbance at 830 mμ as previously described.   Plot the average of several readings for each solution against the concentration of P_2O_5 in micrograms per 50 milliliters.

(b) *On the filtrate from the gravimetric determination of SiO$_2$*

This procedure, devised by J.-L. Bouvier of the Geological Survey of Canada, parallels the colorimetric determination of residual silica by the Jeffery-Wilson method in the filtrate from the first dehydration and separation of silica in the main portion analysis (Sec. 9.4.3).

Reagents

Ammonium molybdate: prepare as given in Sec. 9.19.3a (Reagents).
Hydrazine sulfate: prepare as given in Sec. 9.19.3a (Reagents).
Standard phosphate solution (0.01 mg P_2O_5/ml): transfer 1.00 ml of the standard phosphate solution (1.00 mg P_2O_5/ml) prepared as described in Sec. 9. 19.1 (Reagents) to a 100 ml volumetric flask and dilute to volume.   Mix thoroughly and store in a tightly capped polyethylene bottle.

Procedure

Transfer by pipet a 5 ml aliquot of the filtrate from the first dehydration and separation of silica, diluted to 200 ml in a volumetric flask and approximately 0.5N in HCl (Sec. 9.4.3), to a 50 ml volumetric flask. Add 5 ml ammonium molybdate solution, 2 ml hydrazine sulfate solution and dilute to about 40 ml.   Immerse in vigorously boiling water for 10 min, cool and dilute to volume.   Measure the absorbance of the solution at 830 mμ as before.

9.20. Gravimetric-flame photometric determination of sodium and potassium following a J. Lawrence Smith sinter and extraction

Weigh 0.5000 g of sample (approximately 150 mesh) on a tared watch glass or weighing dish and transfer it, without brushing, to an agate mortar, approximately 5 in. in diameter, which is resting on a piece of white, glazed paper. Grind the powder thoroughly.* Weigh 0.50 g of pure ammonium chloride on the same watch glass, and transfer it to the agate mortar in such a way as to rinse off sample powder which may be clinging to the glass. Mix well the sample and NH_4Cl with the pestle but do not grind the mixture or some decomposition of the NH_4Cl will occur. Weigh 6 g of "low alkali" calcium carbonate,† reserve about 0.5 g of it, and add the rest in several portions to the mixture in the agate mortar, mixing thoroughly each time. Use the first portion of calcium carbonate to rinse any remaining sample powder from the watch glass. The mixing should take about 10 min and is a very important step in the procedure; because no fusion is made the reactants must be finely ground and intimately mixed, or else the decomposition will be incomplete. If the sample contains much iron (greater than 10%) partial fusion may occur on heating and there will be difficulty in removing the cake from the crucible; this can be prevented by increasing the proportion of calcium carbonate in the mixture.

Carefully transfer the mixture, using a thin-bladed flexible steel spatula, to either a special, tapering, finger-shaped J. Lawrence Smith platinum or nickel crucible, approximately 8 cm long with a diameter of 1.8 cm at the top and 1.5 cm at the bottom, or to an ordinary 30 ml platinum crucible. It is easier to heat the mixture to the proper temperature, while keeping the upper part of the crucible relatively cool, in the finger crucible than in the ordinary type; because the alkali metals are easily volatilized, it is necessary to have a cooler area in which they can condense. Use a brush to collect the last vestiges of the mixture in the mortar and add the previously reserved 0.5 g of calcium carbonate to the mortar, using it to rinse down the sides of the mortar and the face of the pestle. Settle the mixture in the crucible by tapping the crucible bottom gently on a yielding surface and then cover the mixture with the final portion from the agate mortar. Cap the crucible loosely and insert it through a hole in a square of asbestos board or transite so that two-thirds of the crucible projects through the board. The crucible should be supported at a 45° angle, either by inclining the board or by drilling

---

* Washington (Ref. 31, Sec. 6.10) prefers to grind a portion of the sample first and then to weigh 0.5 g from this, in order to avoid loss of sampling during the grinding. In the writer's experience, no significant error is introduced if the sample is ground after if is weighed.

† The reagents, in particular the calcium carbonate, will contain appreciable amounts of the alkali metals, especially sodium, and it is necessary to run a blank determination whenever a new lot of calcium carbonate is used. The writer uses analyzed reagents, including calcium carbonate marked "low in alkalies," and obtains average blanks of 0.07 and 0.02% Na_2O and K_2O respectively.

the hole at this angle.   Some analysts prefer to cool the upper part of the crucible by various means, such as a wet strip of cloth (Ref. 12, Sec. 6.10) or by substitution of a small crucible filled with water for the crucible cap, but if the heating is cautiously done these additional precautions are unnecessary.   Begin the heating with a small (about $\frac{1}{2}$ in.) flame of a Tirrill burner, set several inches below the crucible bottom. Gradually move the burner until the whole projecting portion of the crucible has been heated with this low flame, and ammonia fumes can no longer be detected at the cap.   The initial reaction is between the calcium carbonate and ammonium chloride, to produce calcium chloride, ammonia, water and carbon dioxide; if the heating is too strongly done, ammonium chloride will be volatilized directly, as seen by white fumes emerging from around the loosely fitting cap.   The initial heating will require about 10–15 min, after which the flame height should be increased gradually until the lower half of the crucible is bright red (a second Tirrill burner, supported at an angle, may be used to provide the desired temperature.)*   Heat for 1 hr at this temperature, and then allow the crucible and contents to cool to room temperature.   Add water from a wash bottle to within about $\frac{1}{2}$ in. of the top of the crucible, replace the cover, and support the crucible in a nearly vertical position for at least 1 hr. During the ignition the various minerals are decomposed to form silicates and aluminates of calcium, magnesium and iron, and alkali chlorides; the excess calcium carbonate is converted to CaO.   The pasty mass now in the crucible is a mixture of these insoluble compounds and a soluble portion containing the alkali chlorides, calcium and magnesium hydroxides, and sulfate formed by oxidation of any sulfides present in the sample.

Carefully transfer the contents of the crucible to a clean 250 ml beaker and rinse the crucible and lid into the beaker.   Fill the crucible with hot water and allow it to stand.   Peck (Ref. 24, Sec. 6.10) prefers to add 0.5 g of $Ba(OH)_2 \cdot 8H_2O$ to the beaker before adding the contents of the crucible; this is to remove any sulfate ion that may be present. Dilute to 50 ml, break up any lumps with a stirring rod, cover the beaker, heat the contents to boiling and boil for 5 min.   Allow the solid material to settle, rinse off the cover, and decant the liquid through a 9 cm No. 41 Whatman filter paper (previously washed with hot HCl (1:5) and H_2O), into a 600 ml beaker.   Add 50 ml of hot water to the 250 ml beaker, crush any lumps with the stirring rod, and again boil the contents for several

---

* Peck (Ref. 24, Sec. 6.10) prefers to heat the crucible at 1100°C for 30 min in an electric furnace of special design.

minutes. Decant the liquid through the same paper as before. Repeat the crushing, boiling and decantation for a third time. Pour the contents of the crucible into the beaker, dilute to 50 ml, and boil vigorously for several minutes. Filter, transferring the solids to the paper, and wash the beaker, residue and paper with hot water until the volume of the filtrate is about 300 ml.* If there is doubt as to the completeness of the decomposition, the sinter residue may be dissolved in hydrochloric acid and the solution examined for unattacked material.

Heat the filtrate to boiling (ignore any white film of calcium carbonate that forms on the surface) and add slowly, with stirring, a freshly prepared, saturated solution of ammonium carbonate until the precipitation of calcium carbonate appears to be complete. Add 5 ml pure aqueous ammonia (to prevent the formation of soluble alkaline earth bicarbonates) and boil for several minutes. Allow the precipitate to settle and test for completeness of precipitation by adding one drop of ammonium carbonate solution; if additional precipitate forms, add more of the ammonium carbonate solution and again boil the solution. Filter through an 11 cm Whatman No. 40 filter paper, previously washed with hot HCl (1:5) and H_2O, into a 500 ml fused silica or platinum dish. Wash the beaker, precipitate and paper thoroughly with hot water. Place the dish on a water bath and evaporate the solution to near dryness.

Groves (Ref. 13, Sec. 6.10) dissolves the calcium carbonate precipitate previously obtained and reprecipitates it, combining the two filtrates. Peck (op. cit.) prefers to digest the precipitate on a water bath for at least 20 min before filtration; the precipitate is then dissolved, reprecipitated, digested and filtered, and the two filtrates are combined. (See also footnote, this page.)

Before the liquid in the dish has evaporated completely, rinse down the sides to collect all of the salts in the bottom of the dish. Evaporate to complete dryness. Remove the dish from the water bath, cover it with a watch glass, and place it on an asbestos gauze supported on a tripod in the fume-hood. Begin heating the covered dish with a small

---

* Some analysts prefer to reprocess the sinter residue at this point and combine the filtrates. Bennett and Hawley (Ref. 2, Sec. 6.10) dry the sinter residue, grind it, mix it with new portions of the reagents and repeat the sintering if the total alkali metal content of the sample is >5%. Because micas are difficult to decompose, Peck (op. cit.) dries the sinter residue from mica samples and resinters it with additional NH_4Cl. The sinter residue may also be mixed with the precipitate of calcium carbonate obtained in the next step in the analysis and the whole reprocessed; this procedure is preferred by the writer but it is only done if the total content of alkali metals is known to be unusually high.

flame and avoid overheating at first, because of the possibility of losing material by decrepitation. Increase the temperature slowly until ammonium salts begin to volatilize and continue heating until all have been expelled. A second burner, hand-held and with a brush-like flame, can profitably be used to volatilize the ammonium salts from the sides of the dish. Finally remove the cover and heat it cautiously to remove the sublimed ammonium salts. At no time should the dish be heated to redness because at this temperature the alkali chlorides will melt and volatilize. A special electric radiator for this volatilization step has been described (Ref. 24, Sec. 6.10).

Dissolve the residue in the silica dish in about 10 ml of water. The solution contains all of the alkali metal chlorides, a small amount of calcium, and sulfate anion if the sample contained sulfides. There may also be a small amount of brown organic matter from the decomposition of the ammonium carbonate; this is of no significance and can be ignored. To the solution add 1 drop of 5% barium chloride solution (to precipitate sulfate) and warm the dish on the water bath. Add 2 drops of aqueous ammonia and 1 drop of saturated ammonium carbonate solution and warm the dish for another 5 min. Carefully filter the solution through a $5\frac{1}{2}$ cm No. 40 Whatman paper, previously washed with HCl (1:5) and water, into a 50 ml platinum dish; this is one of the most difficult steps of the procedure, because the loss of even one drop of the solution will introduce a serious error into the determination. Rinse out the silica dish thoroughly and wash the paper, keeping the volume of liquid below 50 ml. Empty the stem of the funnel into the dish and rinse off the tip. Place the dish on the water bath, set a glass triangle on the dish (this is preferable to glass hooks) and cover with a watch glass (during the evaporation the decomposition of the ammonium carbonate causes effervescence and some losses may occur if the dish is uncovered). After about 30 min, rinse off and remove the triangle and watch glass and evaporate the contents to dryness. When completely dry, place the dish on an asbestos gauze and heat it with a low flame, keeping the dish covered until the danger of decrepitation is over. Remove the cover and heat it with a low flame to volatilize sublimed salts. Grasp the dish securely with tongs and heat the dish over a free flame until all ammonium salts have been expelled, then heat more strongly until the alkali metal chlorides just begin to melt at the point of contact of the dish with the flame (avoid overheating).

Cool and dissolve the residue in a few ml of water. Add 1 drop each of aqueous ammonia and saturated ammonium carbonate solution, and heat on the water bath for a few minutes. There will likely be a small amount of calcium precipitate present as well as some organic matter. Filter as before into a clean 50 ml platinum dish and wash the dish and paper thoroughly with hot water. Evaporate the solution to dryness on the water bath as before but, before evaporation is complete, add 1–2 ml of 5% HCl from a wash bottle to the dish to decompose any carbonate present. When the contents are completely dry, heat the covered dish as before to expel ammonium salts, and then heat it over a free flame. The residue in the dish should be uniformly white; if not, or if the precipitate obtained during the previous treatment with ammonium carbonate and aqueous ammonia was more than just a trace, it is better to repeat the solution, precipitation and evaporation steps once more. Finally, place the dish in a desiccator, cover it and weigh it, uncovered, after 30 min. Repeat the heating and weighing to constant weight. The alkali metal chlorides, especially lithium chloride, may become moist when exposed to air, and the weighing should be done as rapidly as possible.

Dissolve the mixed chlorides in a few ml of water and transfer the solution to a 35 ml porcelain dish. Dry the platinum dish, ignite and weight it; obtain the weight of the mixed chlorides by difference.

Calculate the volume of chloroplatinic acid solution (0.5 g Pt/ml) needed to form chloroplatinates with the mixed chlorides, assuming the weight to be that of NaCl only:

Wt of mixed chlorides $\times$ 34 = ml K_2PtCl_6 solution needed

It is necessary that sufficient reagent be added to convert the NaCl as well as the KCl, because NaCl is insoluble in the 80% ethyl alcohol used to dissolve the sodium chloroplatinate. Add the calculated volume, plus 0.2 ml extra, from a measuring pipet to the solution in the porcelain dish, mix well with a small glass stirring rod and evaporate the contents to near dryness on the water bath. It is important that the solution *not* be evaporated to complete dryness because the solubility of the sodium chloroplatinate in 80% ethyl alcohol is decreased by dehydration of the salt. During the evaporation the chloroplatinate solution must be kept away from ammonia fumes because ammonium chloroplatinate is insoluble and will behave like the potassium salt.

To the cool and slightly moist residue add 5 ml of 80% ethyl alcohol from a wash bottle and break up the crust with the stirring rod. Let stand for a few minutes and then decant the solution (reddish orange) through a 5½ cm Whatman No. 40 filter paper (Peck (Ref. 24, Sec. 6.10) prefers to use a fritted-glass filtering crucible) into a 50 ml beaker, keeping the bulk of the golden-yellow potassium salt in the dish. Repeat the extraction until no trace of the sodium salt remains (if the filtrate is not clear, refilter through the same paper), and wash the paper thoroughly with the 80% alcohol. Set the beaker containing the soluble chloroplatinates to one side.

Allow the dish and residue to dry in the air; dry the funnel and paper in an oven at 105°C. All alcohol must be removed or reduction of the potassium chloroplatinate may occur when it is heated later on. Add 5 ml of water to the dish and heat it on the water bath, stirring to dissolve the potassium salt. Pour the solution through the filter paper, catching the solution in the previously weighed platinum dish. Repeat with several portions of hot water until all of the potassium chloroplatinate has been dissolved from the dish and the paper.

Evaporate the solution to dryness on the water bath, then cover the dish and heat it in an oven at 130°C (or at 100°C if only a small amount is present) for 1 hr. Cool and weigh. Calculate the weight of KCl equivalent to that of the K_2PtCl_6 and obtain the weight of NaCl by difference from the weight of the mixed chlorides:

$$K_2PtCl_6 \times 0.3067 = KCl$$

$$K_2PtCl_6 \times 0.1938 = K_2O$$

$$NaCl \quad \times 0.5302 = Na_2O$$

It has been observed that the accuracy of this determination, apart from analytical skill, depends upon the absence of the other alkali metals in significant concentration; rubidium and cesium will accompany the potassium, and lithium, or that portion of it not lost during the extraction step, will be found with the sodium. For most rocks, and many minerals, these less abundant alkali metals can be safely ignored but for accurate results a correction must be made. This is conveniently done by removing the platinum from the sodium and potassium chloroplatinate fractions with formic acid, and determining all the alkali metals by flame photometry; the correction also includes the determination of the small amount of potassium that is present in the soluble sodium chloroplatinate fraction. The procedure that follows is based upon the method devised by C. O. Ingamells and privately communicated; a detailed description of a similar method for the determination of major and minor alkali metals by differential flame spectrophotometry following acid decomposition and leaching of the sample is available (Ref. 18, Sec. 6.10).

To the alcoholic solution of sodium chloroplatinate add an amount of reagent grade formic acid (made neutral to pH paper with aqueous ammonia) equal to the weight of platinum metal present,

$$\text{Wt NaCl} \times 34 \times 0.05 = \text{g Pt present}$$

and then add 20% in excess. Evaporate the solution to dryness on a water bath. Add a few ml of water and repeat the evaporation a second time, or until all yellow color has disappeared. Add a few ml of water, digest on the water bath and filter the solution through a 5½ cm No. 40 Whatman filter paper, catching the filtrate in a 50 ml platinum dish. Wash the beaker and paper well with hot water. Evaporate the filtrate to dryness and heat the covered dish gently with a low flame to expel excess ammonium formate. Dissolve the residue in a milliliter of 1:1 H_2SO_4 by heating the dish with a low flame. Cool, dissolve in water and filter through a 5½ cm Whatman No. 40 paper into a 50 ml volumetric flask; wash the dish and paper thoroughly with hot water. Cool the flask to room temperature and dilute to volume. Using a flame spectrophotometer with a recording attachment, determine the lithium and potassium present in the solution as described in Sec. 9.21.3.

Dissolve the potassium chloroplatinate residue in a few ml of water and remove the platinum present with formic acid as described previously. Dilute the filtrate from the sulfuric acid treatment to 50 ml and determine the rubidium and cesium present as described in Sec. 9.21.3. Ingamells (Ref. 18, Sec. 6.10) utilizes the considerable enhancement by potassium of the rubidium and cesium emission by adding sufficient potassium to make its content equal to 1000 ppm; similarly, the rubidium content of a 25 ml aliquot of the solution is adjusted to 50 ppm, prior to the determination of cesium (at 852.11 mμ).

Conversion Factors

$$Na_2O \underset{1.3480}{\overset{0.7419}{\rightleftarrows}} Na \qquad\qquad Li_2O \underset{2.1527}{\overset{0.4645}{\rightleftarrows}} Li$$

$$K_2O \underset{1.2046}{\overset{0.8302}{\rightleftarrows}} K \qquad\qquad Rb_2O \underset{1.0936}{\overset{0.9144}{\rightleftarrows}} Rb$$

$$Cs_2O \underset{1.0602}{\overset{0.9432}{\rightleftarrows}} Cs$$

9.21. Determination of sodium, potassium and lithium by flame photometry, without internal standard.

9.21.1. ON A SEPARATE SAMPLE, WITH RADIATION BUFFER

The sample is decomposed by digestion with HNO_3 and $HClO_4$, followed by HF and HCl; the resulting solution, faintly acid with HCl, is diluted to an appropriate volume after the addition of a radiation buffer solution containing aluminum, iron, calcium and magnesium. A preliminary determination of the sodium, potassium and lithium contents of the sample is made flame photometrically; the final determination is made by closely bracketing the sample with standards which contain equivalent amounts of the buffer solution as well as appropriate amounts of the alkali metals being determined. Because the interval between the concentrations of the standards is small, the relationship between concentration and dial reading is assumed to be linear.

Reagents

Na_2O stock solution (1000 ppm): dissolve 1.8860 g NaCl, dried at 105°C, in water in a 1 liter volumetric flask (place a funnel in the neck of the flask, transfer the salt to the funnel and wash it into the flask) and dilute to volume. Mix well and store the solution in a tightly capped, polyethylene bottle.

K_2O stock solution (1000 ppm): dissolve 1.5830 g KCl, dried at 105°C, as described above, dilute to 1000 ml and store in a polyethylene bottle.

Li_2O stock solution (1000 ppm): weigh 2.4729 g Li_2CO_3, dried at 105°C, and transfer it quantitatively to a funnel inserted in the neck of a 500 ml erlenmeyer flask. Wash the powder into the flask and dilute to about 100 ml. Add 4 ml H_2SO_4 (1:1) through the funnel, warm to dissolve the Li_2CO_3 and boil gently to expel carbon dioxide. Cool, transfer the solution quantitatively to a 1000 ml volumetric flask, dilute to volume and store in a polyethylene bottle.

Radiation buffer solution: dissolve 80 g of $Al(NO_3)_3 \cdot 9H_2O$, 23 g of $Fe(NO_3)_3 \cdot 9H_2O$ and 26 g of $MgCl_2 \cdot 6H_2O$ in 200 ml H_2O; add 60 ml of HCl (1:1) to 8 g of $CaCO_3$ in a covered beaker and boil to expel CO_2; combine the two solutions and filter into a 500 ml volumetric flask. Dilute to volume; 10 ml of this solution, diluted to 250 ml, are equivalent to 500 ppm Al and 250 ppm each of Fe, Ca and Mg.

Standard comparison solutions: Standard comparison solutions, covering the range 0 to 50 ppm for each of sodium, potassium, and lithium, are prepared according to Table 9.2. Aliquots of the stock solutions should be added from a 10 ml microburet; it is convenient, if a series of standards is being prepared, to add the radiation buffer solution from a 25 or 50 ml buret. Add 6 drops of HCl (1:1) to each flask before diluting to volume.

TABLE 9.2
Preparation of standard alkali comparison solutions

ppm	ml Stock Solution (1000 ppm)for 100 ml + 4 ml Radiation Buffer	ml Stock Solution (1000 ppm) for 250 ml + 10 ml Radiation Buffer
0	0.0	0.0
1	0.10	0.25
2	0.20	0.50
3	0.30	0.75
4	0.40	1.00
5	0.50	1.25
10	1.00	2.50
15	1.50	3.75
20	2.00	5.00
25	2.50	6.25
30	3.00	7.50
35	3.50	8.75
40	4.00	10.00
45	4.50	11.25
50	5.00	12.50

Procedure

Weigh 0.2500 g of sample, 150 mesh or finer, and transfer it quantitatively to a 50 or 100 ml platinum dish equipped with a platinum stirring rod (Teflon dishes, with Teflon stirring rods, may be used instead of platinum). Moisten the sample with a few drops of water, cover the dish (if carbonate is present) and add 1 ml $HClO_4$ (72%) and 1 ml HNO_3. Mix, and evaporate to dryness on a sand bath, with occasional stirring.* Cool, moisten the residue with a few drops of water, and add 5 ml HCl (12N) and 5 ml HF (48%). Mix thoroughly and allow to evaporate on a sand bath (moderate temperature) to a volume of approximately 5 ml. Cool, add 5 ml HCl and 5 ml HF and evaporate to near dryness, if no unattacked residue remains. If gritty particles are present, repeat the evaporation with HCl and HF for a third time. Cool, add 30 ml of H_2O and 6 drops of HCl (1:1) and place the dishes

* The use of perchloric insures the expulsion of excess nitric acid and nitrates before addition of hydrochloric acid. This precaution is not necessary with Teflon dishes. If platinum dishes are employed and the lack of a special perchloric acid fume-hood precludes the use of the latter acid, the dried residue should be moistened with water, the crust broken up with the stirring rod, the evaporation repeated and the residue baked for 30 min on a medium temperature hot plate.

on a sand bath (moderate temperature) until a clear solution is obtained
(replace evaporated H_2O if necessary). Transfer the solution, filtering
it through a 7 cm No. 40 Whatman filter paper if necessary, to a 250 ml
volumetric flask and wash the dish, and paper, thoroughly with water.
Add 10 ml of the radiation buffer solution, dilute to volume and mix
thoroughly.

If the sample is in limited supply, use only 0.1000 g, and make the
final dilution to 100 ml, adding 4 ml of the radiation buffer solution.*

Make a preliminary direct determination by flame photometry of the
amount of each alkali metal present by aspirating the sample solution
into the flame and obtaining the apparent concentration by reference to
calibration curves prepared from pure solutions of each alkali metal, or
by setting the maximum scale reading to some convenient concentration
(such as 100 scale divisions equivalent to 50 ppm of alkali metal) and
deriving the apparent concentration from the scale reading. For most
samples the Li_2O content will be very low and the instrument should be
set to give the maximum scale reading for 5 ppm Li_2O. The apparent
concentrations of the alkali metals obtained in this fashion are usually
low by about 15% and should be adjusted accordingly.

Prepare appropriate standard solutions which bracket the apparent
concentration of each alkali metal previously measured, preferably by
±5 ppm (see under *Reagents* for preparation), and contain an equivalent
amount of the other alkali metals and the correct amount of radiation
buffer.

The procedure to be followed in the flame photometric measurements
will be governed by the instrument used. With the Perkin-Elmer Model
146 Flame Photometer with a compressed air-propane flame, without
internal standard, used at the Geological Survey, the instrument is set

---

* A simplified version of this procedure for the determination of Na_2O has been
utilized in the combined X-ray fluorescence–chemical rapid methods of analysis now
in use at the Geological Survey. To the 90 ml (plus rinsings) of solution left in the
100 ml volumetric flask (Sec. 9.19.3a), add 4 ml of radiation buffer solution and
dilute to volume. Determine the Na_2O concentration flame photometrically as
previously described but without the use of bracketing standards; set the maximum
scale reading (100) equal to 50 ppm Na_2O (using a standard solution that also contains 4 ml of radiation buffer per 100 ml) and obtain the concentration directly in
ppm:

$$\text{ppm} = \frac{\text{scale reading}}{2}$$

$$\% \ Na_2O = \frac{\text{ppm}}{9}$$

for maximum response with the desired alkali metal standard solution and then all standard and sample solutions are read in descending numerical order; the zero and maximum scale readings are checked and the standard and sample solutions are read in reverse (ascending) numerical order. The zero and maximum scale readings are again checked and the standard and sample solutions are read once more in descending order. The three readings are averaged and the concentration of the unknown sample is determined by interpolation as follows. Let

X = concn. of alkali metal (M_2O) in sample

Y = scale reading for sample

X_1 = concn. of M_2O in low standard

Y_1 = scale reading for low standard

X_2 = concn. of K_2O in high standard

Y_2 = scale reading for high standard

Then

$$X = \left\{\frac{X_2 - X_1}{Y_2 - Y_1} \times Y - Y_1\right\} + X_1$$

and

$$\% \ M_2O = \frac{X}{10^6} \times \text{volume} \times \frac{100}{\text{sample wt}}$$

For a 0.2500 g sample diluted to 250 ml, or a 0.1000 g sample diluted to 100 ml, the concentration of the alkali metal is given by:

$$\% \ M_2O = X/10$$

It is imperative that a blank determination be made frequently (preferably with each batch of samples, which should include also a laboratory control sample) and a correction made.

The maximum concentration of alkali metal to be measured should not exceed 50 ppm. If the sample solution contains more than this concentration, dilute the sample solution and add an appropriate volume of radiation buffer.

The flame photometric measurements should preferably be made on the same day that the solutions are prepared because of the tendency for iron and aluminum to hydrolyze on standing and thus affect the flow rate of the aspirator. If a delay is necessary, the filtered solution from the acid decomposition step should be stored in a polyethylene container until needed.

9.21.2.   ON AN ALIQUOT OF SOLUTION B, WITH RADIATION BUFFER (SODIUM AND POTASSIUM)

The flame photometric determination of sodium and potassium is usually a part of the various schemes of rapid rock analysis that have been proposed. Shapiro and Brannock (Ref. 28, Sec. 6.10) use a Perkin-Elmer instrument with a modified atomizer that eliminates the need for rinsing and flushing that is required for the funnel type atomizer; the procedure with bracketing standards described in Sec. 9.21.1 is employed but with an internal standard (lithium) rather than a radiation buffer. A 25 ml aliquot of Solution B (Sec. 9.11.2) is diluted to 100 ml; a similar procedure is described in detail by Rigg and Wagenbauer (Ref. 26, Sec. 6.10). Riley (Ref. 27, Sec. 6.10) uses a 2.5 ml aliquot of his Solution B (Sec. 9.8.2) and passes it through an ion-exchange resin column to remove Al, Fe and Ti; final measurement is made with an EEL flame photometer.

Procedure

1. Transfer the 25 ml aliquot of solution B reserved for the determination of the alkali metals (Sec. 9.11.2, Preparation of Solution B, step 5) from the polyethylene bottle to a 100 ml volumetric flask. Add 4 ml of radiation buffer (see Sec. 9.21.1, *Reagents*) and dilute to volume.

2. Make a preliminary flame photometric determination of sodium and potassium (Sec. 9.21.1, *Procedure*).

3. Prepare the appropriate bracketing standards, as described in Sec. 9.21.1 (add 0.2 ml H_2SO_4 (36N) per 100 ml, instead of HCl) or select the appropriate ones for those previously prepared (Table 9.3). Repeat the flame photometric determination and calculate the concentration (in ppm) of alkali metal (as oxide) in the sample solution as before (X).

$$\% \ M_2O = 5(X)$$

It will be found convenient to prepare a series of stock standard solutions to cover the following range of Na_2O and K_2O concentrations using the (1000 ppm) stock solutions of Na_2O and K_2O (see Sec.9.21.1). Measure the aliquots of the stock solutions (Table 9.3) into a 500 ml volumetric flask, add 20 ml of radiation buffer solution and 2 ml of H_2SO_4 (1:1), and dilute to volume. Store in tightly capped polyethylene bottles.

9.21.3.   DETERMINATION OF POTASSIUM BY A "NEUTRAL LEACH" METHOD

In an effort to improve the accuracy of the determination of potassium in micas to be used for the potassium–argon method of geological age

TABLE 9.3

Preparation of stock standard solutions of sodium and potassium

Na$_2$O		K$_2$O		Na$_2$O		K$_2$O	
ppm	ml	ppm	ml	ppm	ml	ppm	ml
0	0	0	0	30	15.0	0	0
0	0	5	2.5	30	15.0	5	2.5
0	0	10	5.0	30	15.0	10	5.0
0	0	20	10.0	30	15.0	20	10.0
0	0	30	15.0	30	15.0	30	15.0
0	0	40	20.0	30	15.0	40	20.0
0	0	50	25.0	30	15.0	50	25.0
5	2.5	0	0	40	20.0	0	0
5	2.5	5	2.5	40	20.0	5	2.5
5	2.5	10	5.0	40	20.0	10	5.0
5	2.5	20	10.0	40	20.0	20	10.0
5	2.5	30	15.0	40	20.0	30	15.0
5	2.5	40	20.0	40	20.0	40	20.0
5	2.5	50	25.0	40	20.0	50	25.0
10	5.0	0	0	50	25.0	0	0
10	5.0	5	2.5	50	25.0	5	2.5
10	5.0	10	5.0	50	25.0	10	5.0
10	5.0	20	10.0	50	25.0	20	10.0
10	5.0	30	15.0	50	25.0	30	15.0
10	5.0	40	20.0	50	25.0	40	20.0
10	5.0	50	25.0	50	25.0	50	25.0
20	10.0	0	0				
20	10.0	5	2.5				
20	10.0	10	5.0				
20	10.0	20	10.0				
20	10.0	30	15.0				
20	10.0	40	20.0				
20	10.0	50	25.0				

determination, a flame photometric method which reduced known sources of error to a minimum was developed by Sydney Abbey of the Geological Survey of Canada (Ref. 1, Sec. 6.10). The sample, following decomposition with sulfuric and hydrofluoric acids, is ignited and the R$_2$O$_3$ group converted to insoluble oxides; the alkaline earths remain as slightly soluble sulfates. Leaching of the ignited residue with H$_2$O and filtration produces a neutral solution of the alkali metals and magnesium. The addition of magnesium, prior to the ignition and leaching steps, prevents the adsorption of potassium by the R_2O_3 oxides (magnesium is preferentially adsorbed), and any possible effect of magnesium on the emission of potassium radiation is overcome by adding an equivalent amount of magnesium to the standards also.* As a further means of improving the accuracy of the determination, a refined flame photometry technique was devised to minimize instrumental drift. A summary of the method is given here but the original paper should be consulted for supplementary details.

Equipment

A Beckman Model DU Spectrophotometer, with a Model 9200 Flame Attachment, Model 4300 Photomultiplier Unit, Model 92300 Spectral Energy Recording Attachment, a Varian Model G-10 Recorder (10 mV span, one second response) and Model 23700 Beckman A.C. Power Supply, was used. The fuel was acetylene at a pressure of 4 lb/sq in. with oxygen at a pressure of 10 lb/sq in.

Procedure

Chemical Treatment

Weigh and quantitatively transfer duplicate 100 mg samples into 100 ml platinum dishes provided with platinum covers. Add 5 ml H_2O, 2 ml H_2SO_4 (1:1) and 5 ml HF. Cover the dishes and heat on a steam bath to decompose the samples. Remove and rinse the covers, add 2–3 ml HF and evaporate the contents to dryness. Cautiously heat the dishes with a low flame until heavy SO_3 fumes are evolved, then allow to cool. Rinse down the sides of the dishes with a few ml of water, and warm on the steam bath until the solution is clear (if any undecomposed sample is visible at this point, the treatment with HF should be repeated).

Add sufficient magnesium sulfate solution to provide 60 mg of magnesium, mix and evaporate to dryness. Heat on a sand bath until no more SO_3 fumes are evolved, then heat cautiously with a meker burner to the reappearance of SO_3 fumes and finally heat strongly with the full flame until all fuming ceases. Allow to cool.

Moisten the residue with about 10 ml H_2O, cover and warm on the steam bath. Remove and rinse the cover, break up the residue with a

---

* Magnesium is the cause of the depression of potassium radiation in the determination of low concentrations (0.005–0.1%) of potassium in stony meteorites; a correction, based upon an investigation of the depressive effect, is applied (Ref. 8, Sec. 6.10).

rubber-tipped glass rod, rinse off the rod, and evaporate the slurry to dryness. Repeat the ignition to expel any SO_3 trapped in the residue from the first ignition. Allow to cool.

Add 25 ml H_2O, cover and warm for 30 min at least on a low temperature hot plate. Decant through a prewashed 9 cm No. 42 Whatman filter paper into a 200 ml volumetric flask. Repeat the leaching and decanting twice, using 10 ml portions of water and 10 min warming periods. Finally transfer the residue to the paper and wash it 3 or 4 times with water. Allow to cool.

Dilute to volume, mix well, and transfer to a polyethylene bottle. At this concentration, 50 ppm of a constituent in the solution corresponds to 10% of that constituent in the original sample.

Calibration

Prepare neutral sulfate solutions of K, Rb, Li and Na, each containing 100 ppm of the alkali metal, to serve as standards. Pipet 0, 10, 20–50 ml of the 100 ppm K standard into 100 ml volumetric flasks and dilute to volume, giving a series of standards containing 0, 10, 20–50 ppm. Using the instrument on manual "Null Reading," with the red-sensitive phototube and the 10,000 megohm resistor, aspirate the 50 ppm K standard into the flame and adjust the wavelength for peak emission at or near 760 mμ. Then adjust sensitivity to give a transmittance of 100 with a 0.05 mm slit. Without further adjustment, take readings for all of the K standards in descending order of concentration, repeat in ascending order, then again in descending and ascending order, giving four readings in all for each standard. Determine the mean reading and average deviation for each standard, and reject all readings which differ from the mean by $>1\frac{1}{2}\times$ the average deviation. Multiply each mean value by the factor required to convert the mean reading of the 50 ppm standard to 100, and plot the adjusted transmittance against ppm potassium.

Repeat the entire procedure using a 0.05 mm slit at 780 mμ for Rb, a 0.03 mm slit at 671 mμ for Li, and a 0.02 mm slit at 589 mμ for sodium. Proceed exactly as for K, but adjust the 50 ppm Rb standard to a transmittance of only 50, and use the photomultiplier tube, with the 22-megohm resistor, for the Na readings.

Preliminary Analysis

Prepare a standard solution containing 50 ppm K, 10 ppm each of Rb, Li, and Na, and 300 ppm magnesium. Aspirate this standard and set the wavelength between 775 and 770 mμ with the red-sensitive phototube and a 0.05 mm slit.   Using the recording system, open the shutter and adjust the dark current control to give a recorded reading close to zero. Scan across the 766 mμ peak, stop the scan at maximum response, and adjust the sensitivity to make the peak read about 95.   A setting of "30" is a suitable scanning speed.

Without changing the sensitivity setting, set the wavelength at 790 mμ, scan across the Rb and K peaks and stop the scan at about 755. Set the wavelength at 675, scan across the Li peak, and stop at about 665. Switch to the photomultiplier tube, reduce the slit to about 0.015 mm, set wavelength at 592, scan across the Na peak and stop at about 585 mμ. A typical standard trace is shown in Fig. 9.1.

Switch to the red-sensitive phototube, increase slit to 0.05 mm, return wavelength to 790 mμ, and repeat the entire procedure with one of the duplicate solutions of each sample.   From the recorder trace determine the net K reading (peak minus adjacent zero readings) for each sample. Correct each reading by the factor required to convert the standard reading to its calibration value, and refer the resulting corrected reading to the "pure alkali metal" calibration curve.   Similarly calculate the approximate content of each of the other alkali metals detected in the sample.

Final Analysis

Prepare new standard solutions, approximating the composition of each sample solution, plus 300 ppm magnesium.   Set the instrument as for the potassium calibration given previously, and aspirate the standard corresponding to the first sample in the series.   Set the transmittance dial to read the value on the calibration curve corresponding to the K content of the standard, and adjust the sensitivity dial to zero the null galvanometer.

Without changing any settings, take a reading for each sample solution, preceding and following each reading by a reading for the corresponding standard.   Take all readings in an uninterrupted sequence with as uniform an interval as possible between readings.   Then repeat the entire series, with the order of the samples reversed.

Multiply each standard reading by the factor required to convert it to the value corresponding to its K concentration on the calibration curve.   Multiply each sample reading by the factors used on both the preceding and following standard reading, giving a total of four corrected readings for the two readings on each duplicate of each sample.   Refer the corrected readings to the calibration curve, obtaining ppm K in each sample solution.   Divide by five, giving per cent K in the original

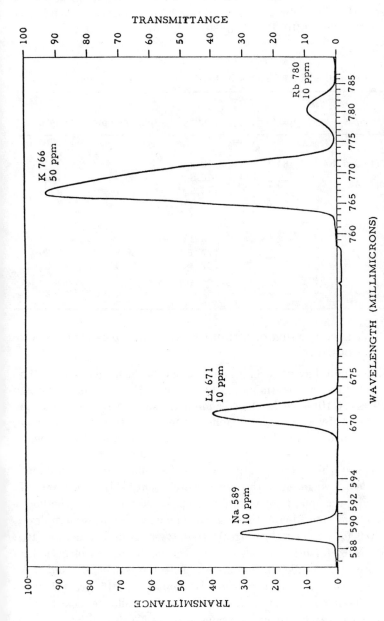

Fig. 9.1. Typical recorder trace of preliminary run with a standard solution containing 50 ppm K and 10 ppm each of Rb, Li and Na. Reproduced with permission of *Chemistry in Canada*.

sample.   Report the mean of the four values for each duplicate.   The procedure for the calibration of results on a typical mica sample is shown in Table 9.4 (Sec. 6.10, Ref. 1, p. 39).

TABLE 9.4

Calculation of the final results for the determination of potassium

Sample	Reading	Corrected reading	K in solution, ppm	K in sample, %
Initial run				
Std (40 ppm K)	83.1	83.2		
Sample A	76.7	76.8	36.4	7.28
		77.3	36.7	7.34
Std	82.6	83.2		
Reverse Run				
Std	84.0	83.2		
Sample A	77.8	77.0	36.5	7.30
		77.6	36.9	7.38
Std	83.4	83.2		
			Mean	7.32

9.21.4.   DETERMINATION OF LITHIUM BY THE METHOD OF ELLESTAD AND HORSTMAN

Neither the J. Lawrence Smith method nor the "neutral leach" method are suitable for the determination of lithium.   Ellestad and Horstman (Ref. 9, Sec. 6.10) have developed a method that involves decomposition of the sample by acid digestion, followed by precipitation of aluminum, iron, calcium and neutralization of free H_2SO_4 with basic lead carbonate, and the flame photometric determination of lithium in the filtrate, using standards to which appropriate amounts of sodium and potassium have been added.   Magnesium, the only other element likely to be present in the filtrate, was found not to interfere up to 360 ppm.   The method is sensitive to 5 ppm Li_2O in the original sample, with a maximum deviation of 0.0005% Li_2O in the range 0.001–0.03%.   A similar method (Ref. 17, Sec. 6.10) for lithium uses calcium carbonate instead of basic lead carbonate for the precipitation step because the former is found to be a more consistently reliable reagent; Horstman (Ref. 16, Sec. 6.10) also used $CaCO_3$ in a later method for Li, Rb and Cs in silicate rocks because the basic lead carbonate led to serious losses of Na, K and Rb by occlusion and adsorption on the lead sulfate precipitate.

The following is essentially a summary of the method as given in the original paper. A Beckman Model DU Spectrophotometer, equipped with a special flame attachment for an air–natural gas flame, was used; corrections were made for the positive continuum effect (background blank on a solution containing Na and K equal to that of the sample solution) and the radiation interference effect (Na and K), which is negative.

Reagents

Standard Li_2SO_4 solution (approx $0.1N$ or 700 ppm Li): weigh approximately 8.5 g $Li(OH) \cdot H_2O$ (purified by recrystallization if necessary), dissolve in 300 ml distilled water, and neutralize with a slight excess of $H_2SO_4(4N)$, using methyl red indicator. Boil several minutes and make just alkaline to methyl red with dilute lithium hydroxide solution. Filter, and dilute to 2 liters. To standardize, pipet 25 ml aliquots into 50 ml platinum dishes, acidify with a drop of H_2SO_4 $(4N)$ and evaporate to dryness. Bring the residue to constant weight by repeated heating to a dull red heat over a low flame. Calculate the normality of the solution from the weight of lithium sulfate found.

Basic lead carbonate (solid reagent): the quality of the reagent with regard to the alkali metal content is very important and some brands of this analytical reagent have too high a content of sodium to be of use.

Procedure

Weigh a 0.5 g sample of finely ground material and transfer it to a 50 ml platinum dish. Moisten with water and add 0.4 ml of concentrated H_2SO_4, followed by 15 ml of HF and 1 drop of nitric acid. Heat below the boiling point until fumes of SO_3 appear (1–2 hr). Cool, moisten with 2 ml H_2O and 0.1 ml H_2SO_4 $(36N)$ and again evaporate to fumes. Make a third evaporation with 5 ml of H_2O to insure complete removal of fluoride. Add 25–30 ml of H_2O and heat, with occasional stirring, until all soluble material has dissolved. Transfer to a 150 ml beaker and dilute to about 70 ml.

Heat to near boiling and add solid basic lead carbonate, with stirring, until the solution is alkaline to methyl red, keeping the beaker covered as much as possible during this addition. Boil several minutes, rinse down and stir well. Filter through an 11 cm Whatman No. 1 filter paper into a 150 ml beaker and wash the residue and paper thoroughly with hot water until the volume of the filtrate and washings is about 125 ml. Evaporate the solution to about 30 ml, cool and filter through a 9 cm Whatman No. 40 paper into a 50 ml volumetric flask. Wash the residue and paper with hot water, cool and dilute to volume.

Complete the determination of lithium flame photometrically by first determining the approximate concentration of lithium, sodium and potassium and then comparing the sample solution with standard solutions containing approximately equivalent amounts of Na and K, and bracketing the approximate lithium content.

9.22. Titrimetric determination of ferrous iron

9.22.1. VISUAL TITRATION, FOLLOWING HOT ACID DECOMPOSITION OF THE SAMPLE (MODIFIED PRATT METHOD)

Reagents

Boric acid solution, saturated (5%): dissolve, with stirring, 100 g of boric acid in 1 liter of hot water, cool and dilute to 2 liters.

Standard potassium dichromate solution (0.1N): weigh 9.810 g of finely ground primary standard grade $K_2Cr_2O_7$ into a weighing bottle, and dry in an oven at 110°C for 2 hr or more.   Weigh the bottle and contents and pour the reagent into a small funnel placed in the neck of a 2 liter volumetric flask; reweigh the bottle to obtain the true weight of $K_2Cr_2O_7$ used.   Wash the powder into the flask with a jet of water, rinse and remove the funnel, and wash down the neck of the flask.   When the reagent is dissolved, dilute to volume at 20°C and mix thoroughly. This solution will serve as a primary standard,

$$N = \frac{\text{wt } K_2Cr_2O_7 \text{ (g)}}{2 \times 4.9035 \text{ g}}$$

but a check on the normality may be made by using an aliquot of the standard iron solution (see below) and following the procedure given in Sec. 9.22.1 or Sec. 9.22.2; 10 ml of the Fe solution requires approximately 10 ml of 0.1N $K_2Cr_2O_7$.   For convenience in use, the oxidizing capacity of the potassium dichromate solution may be expressed as milliequivalent weights of FeO and Fe_2O_3, i.e., for 0.1000N $K_2Cr_2O_7$,

$$FeO \equiv 0.007184 \text{ g/ml}$$

$$Fe_2O_3 \equiv 0.007985 \text{ g/ml}$$

Standard iron solution (approx 5 mg Fe/ml): weigh accurately about 1.3 g of pure iron, as wire or chips (wash in ether to remove oily film and allow to dry in air), and transfer to a 150 ml beaker.   Add 50 ml HCl (1:1), cover and warm on a steam bath to complete solution.   Cool and transfer quantitatively to a 250 ml volumetric flask.   Dilute to volume at 20°C, mix thoroughly and transfer to a 250 ml glass bottle with a greased stopper. Take aliquots only with a pipet.

Procedure

Weigh 0.500 g of sample (100 mesh) and transfer it quantitatively to a 45 ml platinum crucible having a tightly fitting cover, in the center of which is a hole 2 mm in diameter. Add about 1 ml of water and swirl the crucible to distribute the sample over the bottom to prevent caking, and add 2 or 3 drops of H_2SO_4 (1:1) to decompose any carbonates present; cover and allow to stand until all reaction has ceased.

To 10 ml of water in a 50 ml Pt dish add 5 ml H_2SO_4 (18M) and 5 ml HF. Place the covered crucible on a silica triangle suspended firmly over a bunsen burner having a low flame (see description of special stand in Sec. 4.2.9 and Figs. 4.3 and 4.4), slide the cover to one side, quickly (*and carefully*) add the hot acid mixture from the Pt dish, replace the cover and immediately begin brushing the sides and cover of the crucible with the flame of a second burner until the contents are boiling and steam escapes through the hole in the cover. Adjust the height of the flame of the first burner (about ½in.) so that the contents of the crucible boil gently, and continue the heating for 10 min; the heating period should not be prolonged (use an automatic timer) nor should the temperature be sufficiently high to evaporate the water to the point at which the hot concentrated H_2SO_4 begins to oxidize the ferrous iron.

To 200 ml of water in a 600 ml beaker add 50 ml of 5% (saturated) boric acid solution,* 5 ml H_2SO_4 (18M) and 5 ml 85% H_3PO_4, and mix well.

At the conclusion of the heating period at once grasp the crucible firmly with platinum-shod crucible tongs, press the cover firmly on the crucible with a glass stirring rod, and swiftly submerge the crucible below the surface of the acid solution in the 600 ml beaker; do not allow more than the Pt shoes of the tongs to touch the acid solution. At once dislodge the cover from the crucible with the stirring rod and stir to mix the contents of the crucible and beaker; do not stir so vigorously as to draw air into the liquid but make sure that all of the crucible contents have been washed out into the beaker and that all soluble material has dissolved. Peck (Ref. 25, Sec. 6.11) prefers to remove and rinse off the crucible and cover at this point.

* Graff and Langmyhr (Ref. 9, Sec. 6.11) have shown that aluminum trichloride is more effective than boric acid in complexing fluoride ion (see also reference to work of Shell (Sec. 6.2).

Titrate the solution immediately with $N/10$ KMnO$_4$ (see Sec. 9.11.1) to the first appearance of a permanent pink tinge, or with $N/10$ K$_2$Cr$_2$O$_7$ (add 6 drops of 0.2% barium diphenylamine sulfonate (Sec. 9.16.2) to serve as indicator and titrate to the first appearance of a permanent blue-violet color). When the end point is reached, swirl the crucible in the beaker to insure that all of the sample solution has been fully titrated. For accurate work, an indicator blank should be determined and a correction made.

$$\% \text{ FeO} = \frac{\text{ml of titrant} \times \text{normality}}{1000} \times \frac{71.85}{1} \times \frac{100}{\text{sample wt (g)}}$$

Carefully remove the crucible cover with platinum-tipped forceps and rinse it into the beaker. By means of the stirring rod, remove the crucible and rinse it, inside and out, into the beaker. Examine the residue, if any, in the beaker with a hand lens; usually this residue will consist of white, or greyish white grains of quartz which resist decomposition and which can be ignored.

The presence of brassy yellow grains of pyrite in the residue should be noted so that a correction to the FeO value, based upon the amount of sulfur present as sulfide, may be made later on if desired. This will also indicate the potentially unreliable nature of the determination, because of the possible presence in the sample of acid-decomposable sulfides such as pyrrhotite.

If there are more than a few grains of red to black undecomposed material in the residue, a second decomposition and titration should be made as follows: decant carefully the supernatant liquid from the beaker, retaining as much of the residue as possible. Wash the beaker and residue once with water and again decant the liquid. Rinse the residue into a small agate mortar, decant the liquid, and grind the residue until no grittiness remains. Wash the slurry into the original crucible and repeat the decomposition, transfer and titration as before. Add the volume of titrant used to that used for the first titration. If some dark unattacked material remains still undecomposed it is likely chromite, and further acid attack is of no avail. Unfortunately, because of the variable nature of chromite, it is not feasible to add a correction based on the known chromium content of the sample.

Conversion Factors:
$$\text{FeO} \underset{1.2865}{\overset{0.7773}{\rightleftarrows}} \text{Fe}$$

$$\text{Fe}_2\text{O}_3 \underset{1.4297}{\overset{0.6994}{\rightleftarrows}} \text{Fe} \qquad \text{Fe}_2\text{O}_3 \underset{1.1113}{\overset{0.8998}{\rightleftarrows}} \text{FeO}$$

9.22.2. POTENTIOMETRIC TITRATION, FOLLOWING HOT ACID DECOMPOS-
ITION OF THE SAMPLE (MODIFIED PRATT METHOD)

The procedure up to the point of titration is the same as that de-
cribed in Sec. 9.22.1, except that it is not necessary to add the 5 ml of
85% phosphoric acid to the solution in the beaker.

Immerse the platinum and saturated calomel reference electrodes of a
potentiometric apparatus (see Sec. 4.3 and Fig. 4.3) in the solution, add
a plastic-covered stirring bar and place the beaker on a magnetic stirrer
(be careful that the electrodes are kept clear of the stirring bar and avoid
the formation of a vortex by too rapid stirring). The initial potential
of the system will usually be in the neighborhood of 400 mV. Titrate
with $N/10$ $K_2Cr_2O_7$ until a sharp increase in the potential indicates that
all of the Fe(II) has been oxidized. The changes in potential will be
gradual and steady until a reading of about 500 mV is reached, after
which they become more pronounced and the titrant should be added
at the rate of 1 or 2 drops at one time, allowing sufficient time between
the additions for equilibrium to be reached; the abrupt rise in the potential (50–100 mV), marking the end point, usually occurs at 600–650
mV.

Remove the crucible and cover, rinse them into the solution, and examine the residue for undecomposed material, as previously described.
Retain the solution(s), if desired, for the determination of total Fe as
given in Sec. 9.23.2.

9.22.3. VISUAL TITRATION, FOLLOWING THE MODIFIED COLD ACID DE-
COMPOSITION METHOD OF WILSON (Ref. 40, Sec. 6.11)

Reagents

Ammonium metavanadate solution (approx. $0.1N$): dissolve 10 g
reagent grade NH_4VO_3 in 110 ml H_2SO_4 (1:1) and dilute to 1 liter.
Store in a glass bottle or in the reservoir of an automatic pipet.

Acid mixture ($H_3PO_4:H_2SO_4:H_2O$ = 1:2:2): add cautiously, with
stirring, 400 ml of concentrated H_2SO_4 to 400 ml of water in a 1500 ml
beaker which is cooled in a water bath. Add 200 ml of concentrated
H_3PO_4, and mix well.

Boric acid solution: see Sec. 9.22.1 (*Reagents*) for preparation.

Ferrous ammonium sulfate solution ($0.05N$): see Sec. 9.16.2 (*Reagents*) for preparation, but dilute to 1 liter.

Barium diphenylamine sulfonate solution (0.2%): see Sec. 9.16.2 (*Reagents*) for preparation.

Standard potassium dichromate solution (0.05N): prepare as described in Sec. 9.22.1 (*Reagents*), but use only 4.905 g of $K_2Cr_2O_7$.

Procedure

1. Weigh 0.200 g of sample (if FeO is likely to be greater than 10%, use a 0.100 g sample) and transfer it to a 60 ml plastic vial.*

Fig. 9.2. Oscillating device to facilitate sample decomposition in the Wilson cold method for the determination of Fe(II). The 50 ml polythene flasks are stoppered with polyethylene thimbles during the decomposition. Photograph by Photographic Section, Geological Survey.

* A special oscillating device to accomodate 30 determinations has been designed by J. M. Valenzuela of the Geological Survey of Canada to facilitate the rapid and total decomposition of the sample (Fig. 9.2). The sample and acid are contained in 50 ml polythene erlenmeyer flasks which are placed in 1 in. holes drilled in a 1½ in. thick plywood board. The holder is fastened to the top of the oscillator of an oscillating hot plate.

2. To each vial, and to 4 additional vials which serve as blanks, add from an automatic pipet 5 ml ammonium vanadate solution.

3. Swirl gently to form a uniform slurry and add, from a plastic graduated cylinder, 10 ± 1 ml of concentrated hydrofluoric acid.

4. Cover the vials (not tightly) and allow to stand overnight in a fumehood, or until the sample is completely decomposed (absence of gritty particles.)

5. To each vial add 10 ml of the sulfuric–phosphoric acid mixture.

6. Pour the contents of the vial into 100 ml of boric acid solution in a 400 ml beaker; rinse out the vial, using a rubber policeman if necessary, with an additional 100 ml of boric acid solution and add the rinsings to the 400 ml beaker.

7. Stir the contents of the beaker to effect complete solution of the contents of the vial and add, from an automatic pipet, 10 ml of ferrous ammonium sulfate solution.

8. Add 1 ml of barium diphenylamine sulfonate indicator and titrate to a grey end point with standard potassium dichromate solution, using a 10 ml semimicro or equivalent type of buret.

9. % FeO =

$$\frac{\text{ml for sample} - \text{ml for blank*}}{1000} \times N \times \frac{71.85}{1} \times \frac{100}{\text{sample wt (g)}}$$

9.23. Titrimetric determination of total iron

9.23.1. VISUAL TITRATION, USING THE SOLUTION FROM THE FERROUS IRON DETERMINATION

Reagents

Stannous chloride solution (5%): dissolve, with warming, 2.5 g $SnCl_2 \cdot 2H_2O$ in 15 ml 12M HCl and add 35 ml of water. Store in a glass dropping-bottle; this solution should be freshly prepared.

Mercuric chloride solution (5%): dissolve 25 g of $HgCl_2$ in 500 ml of water and store in a glass bottle.

Zimmermann-Reinhardt solution (for use with $KMnO_4$): dissolve 70 g manganous sulfate in 500 ml H_2O; add, with stirring, 125 ml of 18M H_2SO_4 and 125 ml of 85% H_3PO_4, and dilute to 1 liter.

---

* Average of 4 values.

Procedure

To the filtrate(s) from the ferrous iron determination (Sec. 9.22.1 or Sec. 9.22.2) add 5 ml 12N HCl and evaporate to less than 50 ml. The solution should be distinctly yellow in color; if not, add dropwise a 2% solution of KMnO$_4$ until it is yellow. Heat the solution to boiling and carefully add dropwise 5% stannous chloride solution, with vigorous stirring, until the yellow color has been discharged,* and then add 1 or 2 drops more. Cool the solution to room temperature (not over 25°C) and add, all at once, 10 ml of 5% HgCl$_2$ solution; a small, white, silky precipitate should be formed.† Allow to stand for 2 or 3 min only.

If the titration is to be made with KMnO$_4$, add 300 ml of H$_2$O and 25 ml of Zimmermann-Reinhardt solution. Titrate slowly with 0.1N KMnO$_4$ (Sec. 9.11.1) to a faint, but permanent, pink coloration.

% Total iron (as Fe$_2$O$_3$) =

$$\frac{\text{ml titrant} \times \text{normality}}{1000} \times \frac{159.68}{2} \times \frac{100}{\text{sample wt (g)}}$$

If K$_2$Cr$_2$O$_7$ is to be used instead of KMnO$_4$, add 200 ml H$_2$O to the solution, 5 ml 85% H$_3$PO$_4$ and 6 drops of 2% barium diphenylamine sulfonate (Sec. 9.16.2) as indicator. Titrate slowly with 0.1N K$_2$Cr$_2$O$_7$ (Sec. 9.22.1) to the first tinge of purplish blue. The calculations are the same as those given above.

For accurate results an indicator blank should be determined for both the KMnO$_4$ and K$_2$Cr$_2$O$_7$ titrations.

9.23.2. POTENTIOMETRIC TITRATION USING THE SOLUTION FROM THE FERROUS IRON DETERMINATION

To the filtrate(s) from Sec. 9.22.2 add 25 ml of 12M HCl and evaporate to about 100 ml. Heat to boiling and reduce the Fe(III) with 5% SnCl$_2$, as in Sec. 9.23.1, again adding 1 or 2 drops in excess.‡

* If the titration of the Fe(II) was made previously with K$_2$Cr$_2$O$_7$ the presence of Cr(III) ions will impart a light green color to the solution. There is no difficulty, however, in recognizing the disappearance of the yellow color of the ferric ion.

† Only the minimum quantity of stannous chloride must be added because of the danger of reducing the mercuric chloride, added to oxidize the excess stannous ion, to elemental Hg instead of HgCl. This is indicated by the formation of a grey or black precipitate, in which case the determination should be discarded because the elemental Hg will be oxidized by either KMnO$_4$ or K$_2$Cr$_2$O$_7$. The white precipitate, on the other hand, should be small in amount because the presence of too much HgCl is also undesirable.

‡ The reduction of the Fe(III) may conveniently be done with the electrodes immersed in the solution; the potential of the Sn(IV)–Sn(II) couple, under these conditions, is usually about 100 mV.

Immerse the bright Pt electrode and the saturated calomel electrode in the solution, add a magnetic stirring bar and place the beaker on a magnetic stirrer. The initial potential should be about 100 mV. Add $0.1N$ $K_2Cr_2O_7$ very cautiously, to oxidize the excess Sn(II); the end point should be indicated by a sudden jump in potential of about 50 mV, in the neighborhood of a reading of 300 mV. Note the volume of titrant used to reach this first end point. Continue with the titration as described in Sec. 9.22.2 until the end point is reached. If the titration is not satisfactory, the reduction and titration can be repeated.

The volume of titrant needed to oxidize the Fe(II) is the total volume used, less that required to reach the first end point. The calculations are similar to those given previously.

9.23.3. VISUAL OR POTENTIOMETRIC TITRATION, USING THE SOLUTION OF THE R_2O_3 OXIDES

If the sulfuric acid solution of the pyrosulfate fusion of the R_2O_3 group residue (Sec. 9.6.1) was used for the colorimetric determination of TiO_2 (Sec. 9.15.1a, see first footnote), the solution should have been evaporated to a small volume to expel all hydrogen peroxide. Pass a stream of H_2S through the hot solution, which should have a volume of about 100 ml, as described in Sec. 9.5 and continue with the filtration and washing of the platinum precipitate. Evaporate the solution to small volume to expel all traces of hydrogen sulfide, and dilute to about 100 ml. Continue with the visual or potentiometric titration of the iron, as described in Sec. 9.23.1 and Sec. 9.23.2.

9.23.4. VISUAL OR POTENTIOMETRIC TITRATION, USING A SEPARATE PORTION OF THE SAMPLE

If the sample is known to be free of refractory iron-bearing minerals, transfer 0.5 g of sample (or less if the Fe content is high) to a 50 ml platinum dish and roast the sample (Sec. 9.2.1) if sulfides are likely to be present. Moisten the sample with water and add 5 ml $12N$ HCl; stir with a platinum wire stirring rod and add 5 ml HF. Heat on a medium temperature sand bath until the sample is completely decomposed (do not allow the contents to become dry). Cool, add 15 ml of 5% boric acid solution (Sec. 9.22.1) to the dish, and stir. Transfer the contents quantitatively to a 250 ml beaker, rinsing the dish with 10 ml HCl (1:1).

Should the sample not be sufficiently decomposable by acid alone, fuse it with 2 g Na_2CO_3 (Sec. 9.2.1), using either the whole sample after roasting, or the residue obtained by preliminary leaching of the sample with 10 ml HCl (1:1) and filtration, washing and ignition of the acid-insoluble material.* Transfer the cake to a 250 ml beaker and dissolve it in 10 ml HCl (1:1), or combine it with the acid soluble portion if a preliminary leaching was done. Continue with the determination of the iron by visual or potentiometric titration as described in Sec. 9.23.1 and Sec. 9.23.2.

9.24. Colorimetric determination of total iron with o-phenanthroline

9.24.1. USING AN ALIQUOT OF SOLUTION B

Reagents

Hydroxylamine hydrochloride (10%): dissolve 50 g $NH_2OH \cdot HCl$ in water and dilute to 500 ml.

Orthophenanthroline solution (0.1%): dissolve 0.5 g o-phenanthroline monohydrate in water with warming, and dilute to 500 ml. Keep solution out of sunlight.

Sodium citrate solution (10%): dissolve 50 g sodium citrate (2H₂O) in water and dilute to 500 ml.

Standard iron solution (0.2 mg Fe_2O_3/ml): weigh 0.4910 g $FeSO_4 \cdot (NH_4)_2SO_4 \cdot 6H_2O$ and transfer it by means of a small funnel to a 500 ml volumetric flask. Rinse the funnel and the neck of the flask and add about 300 ml H_2O, swirling to dissolve the salts. Add 16 ml H_2SO_4 (1:1) and 10 ml concentrated HNO_3, swirl, dilute to volume and mix thoroughly. Store in a glass bottle.

Procedure

The procedure is essentially that given by Shapiro and Brannock (Ref. 32, Sec. 6.11) and is part of one of the analytical schemes for routine rock analysis used at the Geological Survey of Canada.

1. To a set of 100 ml volumetric flasks, add nothing to the first flask (to serve as blank), pipet 5 ml of the standard iron solution into the second flask and 5 ml of the sample solution B (Sec. 9.11.2, Preparation of Solution B), equivalent to 10 mg sample, to the third.

2. Add, from a graduated cylinder, 5 ml of 10% hydroxylamine hydrochloride solution to each flask, swirl to mix, and allow to stand for 10 min.

3. Pipet 10 ml of the 0.1% o-phenanthroline solution into each flask and swirl to mix.

* If the insoluble, gritty residue remaining after the leaching with HCl is white (quartz?) it is very likely that a subsequent fusion of this residue will not be necessary.

4. Pipet 10 ml of 10% sodium citrate solution into each flask, and swirl to mix.

5. Dilute to volume, mix thoroughly, and allow to stand for 1 hr.

6. Measure the absorbance of the reddish-orange o-phenanthroline ferrous complex at 560 mμ, using the reagent blank as the reference blank solution.

7. Compute a factor for the standard solution:

$$\text{Factor} = \frac{10}{\text{Absorbance of std solution}}$$

8. Compute the percent total Fe, as Fe_2O_3:

% Total Fe (as Fe_2O_3) = Factor $\times$ absorbance of sample solution.

9.24.2. USING A SEPARATE PORTION OF THE SAMPLE*

Reagents

Orthophenanthroline solution (0.25%): dissolve 0.25 g of o-phenanthroline monohydrate in water, warming if necessary; cool and dilute to 100 ml. Keep the solution out of sunlight.

Hydroxylamine hydrochloride solution (10%): prepare as previously described (Sec. 9.24.1).

Sodium acetate solution (25%): dissolve 62.5 g sodium citrate (trihydrate) in water and dilute to 250 ml.

Standard iron solution (0.01 mg/ml): this solution is best prepared by dilution of an aliquot of a more concentrated solution; pipet a 1.00 ml aliquot of the standard iron solution (5 mg/ml) described previously (Sec. 9.22.1) quantitatively into a 500 ml volumetric flask, add 4 ml 12N HCl, dilute to volume and mix thoroughly. The solution is 0.1M in HCl.

Procedure

The method to be used to ensure the complete decomposition of the sample will depend upon whether or not the iron-bearing minerals are decomposed by acid alone.

Treat a 0.5 g sample (or less if the iron content is high), in a 50 ml Pt dish, as described in Sec. 9.23.4 (roasting of the sample will probably not be necessary) and heat on a medium temperature sand bath until

---

* This method is used for samples having a total iron content of 0.1–1%. For samples containing less than 0.1% Fe, such as glass sand, a large sample should be taken and the decomposition procedure modified to suit the sample size.

decomposition is complete, or further acid treatment is obviously futile. Do not allow the contents of the dish to become dry. Add 15 ml of 5% boric acid solution to the dish, stir, filter through a Whatman No. 40 paper into a 250 ml beaker and wash with hot 5% HCl. Ignite the residue (if any) in a 10 ml Pt crucible and fuse it with a small amount of anhydrous Na_2CO_3 (Sec. 9.2.1). Dissolve the cake in the acid filtrate in the 250 ml beaker.

Alternatively, the sample may be fused with Na_2CO_3 and the silica separated by double dehydration as usual (Sec. 9.2.1). Sandell (Sec. 6.11, Ref. 30, p. 546) has described a similar procedure for glass sand. Shell (Ref. 33, Sec. 6.11) prefers to fuse the sample with a mixture of sodium carbonate and sodium borate in a silver crucible, and to determine the iron colorimetrically with o-phenanthroline without preliminary separation; the AgCl that precipitates when the cake is dissolved in HCl in the crucible must first, however, be removed; he has warned about the possible loss of iron to a platinum crucible during sodium carbonate fusion (Sec. 9.2.1, see footnote, p. 327).

If the silica is separated by dehydration and then volatilized as usual with HF and H_2SO_4, any nonvolatile residue should be fused with a small amount of Na_2CO_3, the cake dissolved in dilute HCl, and the solution added to the main filtrate.

Transfer the sample solution to a 100 ml flask, dilute to volume and mix well. With a pipet transfer a 10.0 ml aliquot of the sample solution to a 25 ml volumetric flask. Add 1.0 ml of 10% hydroxylamine hydrochloride solution and adjust the pH to 3–6 with sodium acetate solution (determine the amount required for a similar aliquot portion to which has been added an appropriate indicator). Add 2.0 ml of 0.25% o-phenanthroline solution, swirl to mix and dilute to volume. Mix well and allow to stand for 5–10 min; measure the absorbance of the solution at 508 mμ.

Obtain the iron content of the sample from a calibration curve prepared as follows: to a series of 25 ml volumetric flasks add aliquot portions of the standard iron solution (0.01 mg Fe per ml) to cover the range 0.00–0.20 mg Fe per 25 ml, and continue with the determination as previously described. Use the reagent blank solution as the zero reference solution.

9.25. Determination of water by the modified Penfield method

The method to be described is in routine use at the Geological Survey in both the conventional and rapid analytical schemes. It makes use of a simple water tube without a center bulb and the apparatus for supporting and cooling the tube during the ignition that was devised by Serge Courville (Ref. 3, Sec. 6.12). The elimination of carbon dioxide is achieved by displacing the air in the tube, after drawing off the fused end, with a closely fitting glass rod which is then weighed with the tube; this is the procedure used by Sandell in his micro method (Ref. 22, Sec. 6.12). It is convenient to make the determinations in groups of 8; the tubes, with condensed water and inserted glass rod, are accumulated on a rack in a refrigerator until all 8 are ready for the next step, that of coming to equilibrium in the balance room prior to the first weighing. A supply of the water tubes, cleaned with cleaning solution and thoroughly rinsed with distilled water, is kept in an oven having a temperature of about 130°C.

Apparatus

Penfield tube: this is made from borosilicate tubing having an outside diameter of 9 mm and walls of 1 mm in thickness, overall length of 250 mm and with a bulb at one end having a diameter of approximately 2.5 cm. The walls of the bulb must be of uniform thickness and the open end must be free of any "lip" which reduces the size of the opening.

Tube support: as pictured in Figs. 4.5 and 4.6, it consists of a heavy plywood support having a space at one side of the base to accommodate the bottom of a bunsen burner during the initial stages of the heating. The heat shield is made of two $\frac{1}{4}$in. pieces of asbestos board screwed to the front vertical support; a single sheet of $\frac{1}{2}$in. asbestos board would suffice as well. The overflow tray is made of plastic in order to minimize "sweating" of the dish when the cold water drips into it. The cooling tray is made of aluminum and the rubber supports are made from rubber stoppers cemented to the two trays. The height of these supports should be such that, with the overflow tray in position, the notches in the rim of the cooling tray are level with the bottoms of the two slots in the vertical supports of the tray holder. The open end of the tube should have a slight downward inclination to prevent any condensed water from running back into the hot portion of the tube.

Funnel: this is a long-stemmed funnel which, when supported, will reach nearly to the bottom of the sample tube and which can be easily and rapidly inserted and withdrawn.

Capillary plug: heat and draw a length of capillary tubing to a short taper such that, when cut off and covered with a piece of thin rubber tubing, it will fit tightly into the open end of the water tube.

Tube cap: insert a short piece of glass rod into a $\frac{1}{2}$in. length of firm rubber tubing having an internal diameter such as to enable it to fit easily but firmly over the open end of the water tube.

Displacement rod: this is a 20 cm long piece of 6–7 mm o.d. glass rod, fire-polished at both ends, which will slide easily into the water tube. When not in use these rods should be stored in used water tubes to keep them clean.

Reagents

Lead oxide flux: place a quantity of PbO (litharge) in a 90 ml nickel crucible and heat it with a moderate flame of a bunsen burner for 30 min, stirring frequently with a glass rod and crushing any lumps which form.* Cool in a desiccator and transfer to a tightly sealed bottle which is kept in a desiccator until needed.

Procedure

Place a clean water tube, which has been dried for at least 2 hr at a minimum of 110°C, in a horizontal support, insert a narrow glass tube into it and, by means of suction, draw a gentle current of air through the water tube until it is in equilibrium with room conditions. Support the long-stemmed funnel (which must be dry) in a vertical stand and insert it into the water tube so that the end of the stem is at about the center of the bulb.

Weigh and transfer 1.000 g of sample (or less if the water content is likely to be greater than 1%) to a small porcelain crucible and add 2 g of flux. Mix with a spatula and transfer by means of the funnel to the bulb of the water tube. Brush any remaining powder into the funnel and brush the walls of the funnel; expedite the transfer of powder to the bulb by tapping the rim and stem of the funnel with the spatula. Add about 0.5 g of flux to the crucible, rinse the crucible with it and add it to the funnel.† Lift the funnel so that the lower end of the stem is at the entrance to the bulb and again tap the funnel to dislodge any powder

---

* Peck (Ref. 16, Sec. 6.12) prefers to ignite the flux at 800°C in an electric muffle furnace for 1 hr. The sintered material is crushed and ground in an agate mortar to −20 mesh before it is bottled.

† When the sample is known or suspected to contain appreciable fluorine it is necessary to add CaO to retain the latter, as follows: weigh 0.5 g reagent grade calcium carbonate into an ignited and weighed, covered porcelain crucible, and heat at 900°C for 30 min; cool in a desiccator and weigh rapidly (the weight of CaO should be near to 0.28 g); store, covered, in a desiccator until needed and add it, rapidly, as the final step in the loading of the water tube (do not try to remove dust adhering to funnel).

clinging to the stem walls.   Withdraw the funnel and close the tube with a capillary plug.

Carefully raise the tube to a horizontal position (no powder must escape from the bulb) and wrap the center 3 to 4 in. of the tube spirally with a 6 in. strip of wet cloth (about ½ in. wide). Place the tube in the cooling tray of the tube support, which should contain crushed ice and water, so that the cloth wrapping touches the cold water and the bulb projects about ¾ in. beyond the heat shield.

Place a bunsen burner in position so that a small flame is beneath the bulb. Rotate the tube slowly to prevent caking of the sample, and gradually increase the burner flame until the full heat of the burner is attained and the flame envelops the bulb (this must be done slowly over a period of 5 min in order to avoid explosive expulsion of volatiles with the possibility of some water being carried out of the tube).   Commence heating the bulb with a brush-like flame of an oxygen–propane hand torch, while steadily rotating the tube, and gradually increase the heat of the flame until the walls of the bulb begin to collapse (the sample and flux should be a molten mixture by now).   Rotate the tube rapidly enough to prevent the bulb from sagging and heat it as strongly as possible; collapse the walls of the bulb about the fused sample, starting at the bottom of the bulb and working towards the orifice (this reduces the possibility of the fused material blocking the orifice and causing a rupture of the bulb before the ignition is finished).   Finally, using a sharp flame, fuse the tube at the junction with the bulb, and with tongs, draw off the fused end and discard it.   Heat the sealed end of the tube briefly to round any sharp projections.

Keep the tube in a horizontal position and cautiously wipe the hot end with a wet cloth to cool it somewhat before pushing the wet wrapping strip over the hot end of the tube.   Place the end of the tube in the ice bath, pat dry the other end with a lint-free cloth, remove the capillary plug and quickly insert a clean, dry displacement rod; if there is a space remaining at the upper end of the tube, displace the air in it once or twice with the end of another displacement rod.   Close the tube with a tube cap and carefully dry the outside of the tube by patting it (do not rub) with a lint-free cloth.*   Place the tube in a horizontal rack and allow it

---

* The development by friction of a charge of static electricity on the surface of the tube can cause a serious error in the determination. Sandell (Ref. 22, Sec. 6.12) recommends the use of a lint-free cloth which is stored under a moist atmosphere in a desiccator. At the Geological Survey it was found useful to keep a mounted ionizing unit in the balance chamber to dispel static electrical charges from the water tubes and to pass this, or an ionizing brush, lightly over the tube before weighing.

to stand for 20 min near the balance before removing the tube cap and weighing the tube. The weighing is best done by supporting the tube on a wire saddle of known weight.

Carefully remove the displacement rod and dry it by heating it cautiously with a low flame; dry the water tube in the same manner (they may be left overnight in an oven having a temperature of at least 110°C if this is more convenient). Cool the rod and tube, while drawing a gentle current of air through the latter. Return the rod to the tube, wipe the tube with a damp cloth and again pat dry. Place the tube on a horizontal rack, allow to stand for 20 min near the balance, and weigh as before. The difference between the two weights is that of the water in the sample. A blank should be run at regular intervals and always when a new batch of flux is used.

Conversion Factors

$$H_2O \underset{8.9357}{\overset{0.1119}{\rightleftharpoons}} H$$

9.26. Determination of "carbonate" carbon

9.26.1. BY AN ACID EVOLUTION–GRAVIMETRIC METHOD.

Apparatus

The apparatus used at the Geological Survey of Canada for the acid evolution–gravimetric determination of "carbonate" carbon, as part of both the rapid and conventional schemes of rock analysis, is of a standard design in that the CO_2, liberated by treatment of the sample with hot acid in a flask fitted with a condenser, is purified of unwanted volatile constituents by passage through absorbing media and is finally collected in a weighed absorption tube for the final gravimetric determination. The apparatus is modified, however, to permit the simultaneous running of two determinations, and use is made of a commercially available purification and absorption train.

The dual apparatus is shown in Fig. 9.3.* The reaction vessels are 150 ml flat-bottomed, wide-mouthed extraction flasks fitted with Graham condensers to remove most of the water from the gas, and 50 ml cylindrical separatory funnels for the addition of acid to the samples. Air

* Each of the decomposition flask, condenser and dropping funnel units have since been replaced by a single glass apparatus which requires only minor modifications to standard items.

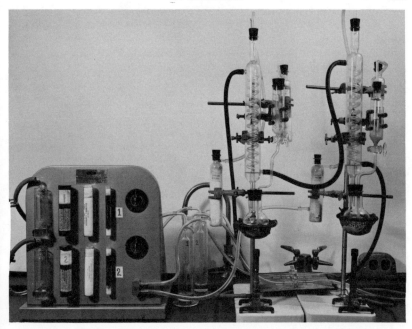

Fig. 9.3.   Dual apparatus for the acid evolution-gravimetric determination of "carbonate" carbon, using a Burrell Carbotrane.   Photograph by Photographic Section, Geological Survey.

enters the system by the way of each dropping funnel after first passing through a small tower containing soda-asbestos (Ascarite) which removes carbon dioxide.   After leaving the condenser the gas passes through a tower containing anhydrous magnesium perchlorate and then into the Burrell Carbotrane,* which carries two separate purification and absorption systems.   The gas passes in order through a cartridge containing granulated MnO_2 (Sulsorber C) which removes the oxides of sulfur, a cartridge containing anhydrous magnesium perchlorate (Exsorber C) which removes the last traces of water vapor, and then through a weighed cartridge (Disorber C) containing soda-asbestos and anhydrous magnesium perchlorate in which the CO_2 is absorbed.   The bubble trap contains a silicone oil and serves both to indicate the rate of flow

* Available from the Burrell Corporation, Pittsburgh, Pa.   In order to accommodate the Carbotrane to a vacuum system, instead of a pressure system as originally intended, it is necessary to modify the bubble trap (Burrell Catalog No. 24–923–04) by adding a right angle glass delivery tube to its top, at right angles to the outlet which plugs into the Carbotrane, for connection to a vacuum pump.

and to prevent any entry of CO_2 or moisture into the Disorber C cartridge should the systems accidentally "back up."

The glass inlet and outlet of each cartridge plug into rubber sleeves on the ends of rigid connecting copper tubing; the admittedly large mass of the Disorber C cartridge (about 100 g) is counterbalanced by the convenience of handling and ease of replacement.

Procedure

Weigh and transfer a 1.00 g sample (the sample weight used will depend on the amount of CO_2 expected to be present and will vary from 0.2 g for limestones to as much as 5 g for silicate rocks containing less than 0.1% CO_2) to the reaction vessel, and add about 35 ml of water which has been freshly boiled and cooled to remove dissolved carbon dioxide. Connect the flask firmly to the condenser and separatory funnel, start a flow of cooling water through the condenser* and open the stopcock of the separatory funnel to admit a slow current of CO_2-free air. Flush the system in this manner for 10 min. A glass delivery tube having a right angle bend at each end is inserted in the Carbotrane in place of the Disorber C cartridge.

Close the stopcock of the separatory funnel and add to the funnel 30 ml of HCl (1:1). Remove the glass delivery tube from the Carbotrane and insert the weighed Disorber C cartridge. Open the stopcock of the separatory funnel and allow the acid to flow slowly into the flask.

After all of the acid has been added, and any visible reaction has ceased, heat the contents of the flask to boiling and boil for 2 min.† Remove the source of heat, reduce the flow of air to approximately one bubble per sec and continue the flow for 15 min.

Close the stopcock of the separatory funnel and also the one connecting the bubble trap to the vacuum line; remove the Disorber C cartridge, seal the outlet and inlet tubes with a connecting piece of rubber tubing, and place it in the balance room. Allow it to stand for 5 min, and weigh. The increase in weight is that of the carbon dioxide from the "carbonate" carbon present in the sample.

---

* It will simplify connections if the condensers are in series.

† Peck (Ref. 18, Sec. 6.13) uses a special absorption train that permits the addition of the acid to the hot slurry of the sample and water while air is, at the same time, being drawn through the system. This allows a better control of the rate of addition of the acid so that excessive buildup of pressure is avoided and a sudden, violent evolution of gas when the sample and acid mixture are heated will not occur. Some carbonates will react with the acid only at higher temperatures.

A blank determination should be made daily, or more often if humid conditions prevail, and a suitable correction applied to subsequent determinations.

Conversion Factors

$$CO_2 \underset{3.6644}{\overset{0.2729}{\rightleftharpoons}} C$$

9.26.2. BY A VOLUMETRIC METHOD

In this procedure the carbon dioxide is liberated by the action of hot acid on the sample in a closed system and the measured volume of air and carbon dioxide is passed through a solution of potassium hydroxide; the CO_2 is absorbed and is determined as the difference between the measured volumes before and after passage through the KOH solution. This method is essentially that described by Goldich et al., (Ref. 8, Sec. 6.13), and modified by Shapiro and Brannock (Ref. 22, Sec. 6.13) for use in their rapid analysis scheme; they recommend its use for samples containing more than 6% carbon dioxide. The need to correct for temperature and pressure conditions is eliminated if standards are run with each batch of samples and the factor used in calculations is based upon these. Very satisfactory results have been obtained at the Geological Survey of Canada for samples containing as little as 0.3% CO_2; the method is simple in operation and rapid, each determination taking about 15 min. The presence of other volatile constituents which are absorbed by KOH will, unless they are removed, cause serious error in the determination.

Apparatus

The apparatus is shown in Fig. 9.4 and consists of: (*1*) a 50 ml Pyrex filtration flask connected by a tightly fitting rubber stopper to a small (8 in.) Liebig condenser, and with the side arm connected by a length of tygon tubing to a 500 ml separatory funnel; (*2*) the condenser is connected by a piece of right angle glass tubing to one outlet of a three-way stopcock joined to a 100 ml buret enclosed in a water jacket in which a thermometer is also suspended; a second outlet of the stopcock is joined to a gas-bubbling pipet having two compartments and a three-way stopcock, and the lower part of the buret is connected by a length of tygon tubing to a leveling bottle, supported on a levelling clamp.

Reagents

Hydrochloric acid–sodium chloride solution: to 1500 ml water add 500 ml of 12*M* HCl and saturate the dilute acid (1:3) with sodium chloride. Add a few milligrams of methyl red indicator to the mixture.

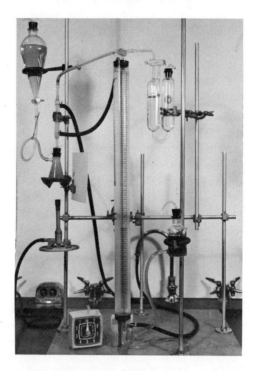

Fig. 9.4. Apparatus for the volumetric determination of "carbonate" carbon.
Photograph by Photographic Section, Geological Survey.

Potassium hydroxide solution (50% w/v): dissolve 1000 g of KOH
in water in a plastic beaker, cool, dilute to 2 liters and store in a plastic
bottle.

Procedure

1. Fill the separatory funnel with the acid solution and allow it to flow
into the connecting tygon tubing until all air is expelled.   Add sufficient
acid solution to the leveling bottle to fill the buret when the bottle is
raised to the level of the stopcock.   Add sufficient solution (about 200 ml)
to the gas pipet, and adjust the level in the absorption compartment with
the leveling bottle, so that there is sufficient KOH to accommodate the
passage through it of 100 ml of gas.   Mark this level.

2. Turn the stopcock of the gas pipet so that gas does *not* bubble
through the solution and open the connection between the gas pipet and
buret by turning the buret stopcock to the appropriate position; by
raising and lowering the leveling bottle, adjust the level of the KOH so-

lution to the reference mark on the absorption compartment. Close
the stopcock of the gas pipet.

3. Open the connection between the buret and the condenser by
turning the buret stopcock to the appropriate position and, by raising the leveling bottle, fill the buret to the stopcock opening and close
the buret stopcock. Lower the leveling bottle to the bottom of the
buret. The system is now ready for use.

4. Weigh 0.1000 g of reagent grade $CaCO_3$* into the flask and attach
the flask firmly to the condenser; connect the side arm of the flask to the
tubing from the separatory funnel (the acid solution should just fill the
side arm) and open the connection between the evolution system and
the buret (the level of the liquid in the buret will drop slightly).

5. Open the stopcock of the separatory funnel and allow 10 ml of
the acid solution to enter the flask. Heat the contents of the flask
cautiously to near boiling, then boil for 3 min and discontinue the heating.

6. Open the stopcock of the separatory funnel and allow the acid solution to flow into the evolution system until the liquid just reaches the
buret stopcock; close the buret stopcock. Raise the leveling bottle
until the levels in the buret and in the bottle are the same and record the
buret reading. Note the temperature inside the water jacket at the
time of the reading.

7. Turn the buret and gas pipet stopcocks to their proper settings
and, by raising the leveling bottle, force the gas from the buret through
the KOH solution. Lower the leveling bottle, after turning the gas
pipet stopcock to the appropriate setting, until the gas has been drawn
into the buret again. Repeat the passage through the KOH twice more.
After the third passage adjust the level of the KOH solution to the
reference level marked on the absorption compartment and close the
buret stopcock. Adjust the height of the leveling bottle so that the
heights of the liquid in the buret and bottle are the same, and record
the buret reading. The difference between the two readings is that of
the volume of CO_2 released from the sample. Note the temperature,
which should not have changed by more than 1°C.

8. Repeat the procedure (steps 1–7) with a sample of suitable size. At
the end of a group of samples, run another 0.1000 g portion of $CaCO_3$ and
average the results obtained for the two standards.

* The volume of heated air and carbon dioxide which is forced into the buret from
the system described here during the period of boiling is between 95 and 100 ml; the
volume before absorption is approximately 95 ml.

9. Compute the factor for the standard:

$$Factor = (44.01/\text{Av volume of } CO_2 \text{ obtained})$$

10. Compute the percentage of CO_2 in the sample:

$$\% \ CO_2 = \text{volume of } CO_2 \times Factor \times (0.1000/\text{sample wt.})$$

9.27. Determination of total carbon by a combustion-gravimetric method

Apparatus

The combustion, purification and absorption train is shown in Fig. 9.5. The combustion is done in a high temperature tube, $1\frac{1}{2}$ in. $\times$ $1\frac{1}{4}$ in. $\times$ 36 in. and having a reduced end, which is heated to 1000°C in a split, cylindrical furnace capable of continuous operation at about 1200°C. The reduced end of the combustion tube is connected to a cartridge containing granulated MnO_2 (Sulsorber C from a Burrell

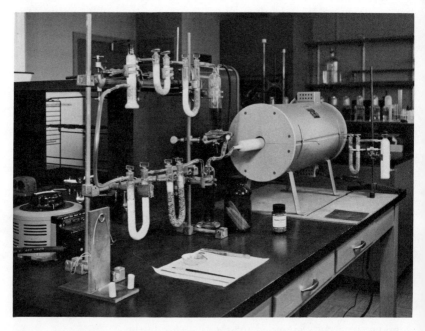

Fig. 9.5. Apparatus for the combustion-gravimetric determination of total carbon. Photograph by Photographic Section, Geological Survey. Note: the absorption train shown here has since been replaced by a Burrell Carbotrane as shown in Fig. 9.3.

Carbotrane, (Sec. 9.26.1)) for the removal of SO_2 and chlorine, which in turn is followed by a Fleming purifying jar containing a mixture of chromic and phosphoric acids (add concentrated chromic acid to 85% phosphoric acid until the solution is a deep reddish-orange) which serves both as an indicator of the rate of flow of gas through the train and as a trap for SO_3 and oxides of nitrogen. A U-tube containing pumice impregnated with ferrous sulfate (the pumice is soaked in 1:3 H_2SO_4 saturated with ferrous sulfate and dried at 110°C) serves as a trap for any of the chromic–phosphoric acid solution that might be carried out of the Fleming jar by the gas. A U-tube packed with anhydrous magnesium perchlorate removes any water vapor from the gas, which then passes through a weighed (about 50 g) Stetser-Norton carbon dioxide absorption bulb containing a lower layer of soda-asbestos (Mikhobite or Ascarite) and an upper layer of anhydrous magnesium perchlorate to absorb any water produced by the absorption of carbon dioxide; a small weighed U-tube also containing soda-asbestos and anhydrous magnesium perchlorate follows the absorption bulb and serves to catch any CO_2 not absorbed in the latter. A U-tube filled with anhydrous magnesium perchlorate and a second purifying jar containing chromic–phosphoric acid solution protect the weighed absorption tubes from contamination in the event of a "backing up" of the system. The purifying jar also serves as a flow rate indicator and, with the first bubbler, as a check on the absence of leaks in the vacuum system.

Air is drawn through the system at a rate of about 3 bubbles per sec by means of suction; it is first made CO_2-free by passing it through a U-tube containing soda-asbestos. The U-tube is connected to the hollow stem of a high temperature, double baffle heat reflector which in turn is connected to the wide end of the combustion tube by a tightly fitting rubber stopper. The temperature of the furnace is regulated by a variable rheostat.

The sample, mixed with an oxidizing flux (reagent grade vanadium pentoxide), is burned in a ceramic boat (95 mm long) which is used only once. A small wooden stand is a useful device for carrying the weighed carbon dioxide absorption vessels to and from the balance room.

All connections in the train are made with tygon tubing, except the connections to, and between, the two weighed carbon dioxide absorption vessels which are here of soft rubber for ease of handling. Closely fitting pieces of heavy rubber tubing, sealed with Teflon plugs at one end, are used to close off the outlets and inlets of the absorption vessels when they are removed from the train; they are removed during the weighing. A piece of glass tubing is inserted in their place in the train during the period of weighing so that flushing of the system may be continued in preparation for the next determination.

Procedure

Weigh a sample of suitable size such that not less than 10 mg of CO_2 will be produced during the combustion,* add 0.50 g of reagent grade vanadium pentoxide and mix well. Carefully spread the mixture on the bottom of a clean ceramic boat (the boats should first be ignited in a furnace for several hours and then stored in a sealed glass jar until needed; they should be handled with forceps, or finger cots).

Insert the weighed carbon dioxide absorption vessels into the train (close off the system on either side of them during this operation), then quickly insert the boat in the mouth of the combustion tube, push it into the hot central portion with a steel rod, insert the rubber stopper and open the system to suction. Continue the heating for 30 min.

Close off the system on either side of the absorption vessels and disconnect them from the line; close the stopcocks of the U-tube and seal the outlet and inlet of the Stetser-Norton bulb with the rubber plugs. Place the tubes on a support and allow them to stand for 10 min in the balance room. Before weighing, remove the plugs from the absorption bulb and briefly open the stopcocks of the U-tube. Insert the glass tube in the train in place of the absorption vessels and flush the system for 10 min before beginning another determination.

A blank determination should be made daily; because of the complexity of the train these blanks tend to be rather high but should not be much more than 1 mg.

9.28. Gravimetric determination of "sulfide" sulfur by acid decomposition and oxidation

This procedure is used when the sample is known to contain an appreciable amount of sulfur in the form of sulfide minerals, so that the determination of "sulfide" sulfur as the difference between total sulfur and "sulfate" sulfur is not likely to be satisfactory. As discussed in Sec. 6.14.1, the procedure will determine the sulfur also present as acid-soluble sulfate (and as elemental and, to some extent, as organic sulfur as well, although this combination will rarely be encountered) but for most silicates the error will be negligible if all of the sulfur is reported as "sulfide" sulfur. The correction to be made for equivalent oxygen in

---

* If the sample contains much "carbonate" carbon, it may be preferable to remove the latter by treatment with acid; the insoluble residue is filtered on an asbestos pad, dried and the pad and residue burned in the usual way (Refs. 3, 5, 11, 25, Sec. 6.13).

the summation of the analysis will depend upon whether the sulfide mineral, if known, is pyrite or pyrrhotite (see Sec. 6.14).

Reagents

Barium chloride solution (5%): dissolve 25 g of $BaCl_2 \cdot 2H_2O$ in water, filter and dilute to 500 ml in a glass-stoppered bottle.

Procedure

Transfer a 1.00 g sample (80–100 mesh) to a dry 250 ml beaker, cover and cautiously add (in a fume-hood!) 5 ml of a mixture of bromine and carbon tetrachloride (2 parts by volume of bromine to 3 parts of carbon tetrachloride). Allow to stand for 15 min at room temperature, with occasional swirling to ensure complete wetting of the sample by the mixture.

Add, through the lip of the beaker, 10 ml of concentrated nitric acid and let stand at room temperature for 15 min, again with occasional swirling of the contents. Place the beaker on a low temperature ($<100°C$) hot plate or on the covered surface of a water bath and heat until all visible reaction has ceased and most of the bromine has been expelled. Increase the temperature of the hot plate, or heat the beaker directly on the water bath, remove the cover, and evaporate the contents to dryness. Cool, add 10 ml of $12N$ HCl, and again evaporate to dryness. Heat the dry residue at $100°C$ for 30 min to dehydrate any dissolved silica.

Drench the residue with 2 ml $12N$ HCl, let it stand for a minute or so, and then add 50 ml of water. Heat to boiling and boil for 5 min. Cool for 5 min and then cautiously add about 0.5 g of reagent grade powdered zinc or aluminum to reduce the ferric iron; warm the mixture if the reaction is too slow, and stir from time to time.* Filter the solution through a 9 cm Whatman No. 40 paper into a 600 ml beaker and wash the paper and residue thoroughly with hot water. Dilute the filtrate to at least 300 ml in volume, and add 2 ml of $12N$ HCl.

Heat to boiling and add dropwise, by pipet, 25 ml of hot 5% $BaCl_2$ solution. Stir vigorously during the addition of the precipitant. Digest the solution and precipitate on the water bath, with occasional stirring, for 1 hr and then allow to stand overnight.

Filter the solution through a 9 cm Whatman No. 42 paper ·by first

* The determination can be extended at this point to include the total sulfur content by fusing the filtered, washed and ignited ($<600°C$) residue as described in Sec. 9.30.1. The filtrate from the leaching of the fusion cake, made acid with HCl, is combined with that obtained by acid digestion prior to the reduction of the ferric iron.

decanting as much of the supernatant liquid as possible through the paper, keeping most of the precipitate in the beaker. Discard this filtrate before transferring the precipitate to the paper with a stream of hot water. If the filtrate appears cloudy, pour it again through the paper and repeat this procedure until the filtrate is perfectly clear. Police the beaker and stirring rod carefully and pour all rinsings through the paper. Finally, wash the paper and precipitate ten times with small amounts of hot water.

Fold the paper around the precipitate with platinum-tipped forceps and place it in a weighed 25 ml covered platinum crucible. Wipe out the funnel and the tips of the forceps with a small piece of filter paper and add it to the crucible. Loosely cover the crucible, place it in a cold electric muffle furnace (or over the low flame of a Tirrill burner) and then burn off the paper at the lowest temperature possible ($<600°C$). Do not allow the crucible to become red, and do not allow the contents to catch fire. When the paper has been completely charred, displace the cover to one side, raise the temperature until the crucible is dull red, and burn off the carbon. When the residue is white, ignite it at about 900°C for 15 min. Cool in a desiccator for 30 min and weigh. Repeat the ignition, cooling and weighing to constant weight.

If the residue is colored it should be fused with a small amount of Na_2CO_3, the leachate filtered and made acid with HCl, and the precipitation repeated. This will, unfortunately, not remove any chromium which will be present as chromate and will coprecipitate as barium chromate (the ignited residue will show a green coloration). If a significant error is likely to be introduced by the presence of chromium it will have to be removed by solvent extraction, mercury cathode electrolysis, or by passage through a cation-exchange resin.

If it is suspected that some silica may have separated with the barium sulfate, the residue should be treated with a drop of sulfuric acid and 1 or 2 ml of hydrofluoric acid, evaporated to dryness, reignited and reweighed.

$$\% \text{ S} = (\text{wt of BaSO}_4 \text{ (g)}/\text{sample wt (g)}) \times 0.1374 \times 100$$

Conversion Factors

$$SO_2 \underset{1.9981}{\overset{0.5005}{\rightleftarrows}} S \qquad\qquad BaSO_4 \underset{7.2807}{\overset{0.1374}{\rightleftarrows}} S$$

$$SO_3 \underset{2.4971}{\overset{0.4004}{\rightleftarrows}} S \qquad\qquad BaSO_4 \underset{3.6438}{\overset{0.2744}{\rightleftarrows}} SO_2$$

$$BaSO_4 \underset{2.9156}{\overset{0.3430}{\rightleftarrows}} SO_3$$

$$BaSO_4 \underset{2.4299}{\overset{0.4115}{\rightleftarrows}} SO_4$$

9.29. Gravimetric determination of acid-soluble "sulfate" sulfur

This procedure presupposes that any hydrogen sulfide formed by decomposition of acid-soluble sulfide minerals also present in the sample is expelled from the solution without being oxidized; the sample is first boiled with water to remove dissolved oxygen before the addition of non-oxidizing acid. Any sulfur that is precipitated during the decomposition of sulfides will either be filtered off before precipitation of the sulfate, or will be burned off during the ignition step.

Procedure

Transfer a 1.00 g sample (80–100 mesh) to a 250 ml beaker. Add about 75 ml of water to the beaker, swirl to make a slurry, cover and heat to a gentle boil on an electric hot plate. Boil gently for 1 or 2 min and then add, through the lip of the beaker, 20 ml of dilute HCl (1:1); the solution should not stop boiling. Continue the gentle boiling for 15 min.

Allow to cool slightly and filter the solution through a 9 cm Whatman No. 42 paper (paper pulp added to the solution in the beaker will help to ensure a clear filtrate) into a 600 ml beaker.* Wash the beaker and residue several times with hot 5% HCl, pouring the washings through the filter paper, and wash the paper and residue several times with the hot, dilute HCl. Discard the paper and residue, and rinse the inside of the funnel, and the outside of the tip, into the beaker.

Add 2 or 3 drops of bromcresol purple (0.04% aqueous solution) or methyl red (0.2% ethanolic solution) to the filtrate and neutralize the solution with pure aqueous ammonia. Make the solution just acid with dilute HCl (1:1), then add 4 ml in excess. Dilute the solution to 300 ml at least.

Continue with the precipitation, filtration, ignition and weighing of the barium sulfate as described in Sec. 9.28.

$$\% \ SO_3 = (\text{wt of BaSO}_4 \ (g)/\text{sample wt. (g)}) \times 0.3430 \times 100$$

9.30. Determination of total sulfur

9.30.1 GRAVIMETRICALLY, FOLLOWING FUSION WITH SODIUM CARBONATE

Transfer 1.00 g of sample (80–100 mesh) to a 25 ml platinum crucible. Add 3 g of anhydrous Na_2CO_3 and 0.2 g Na_2O_2, and mix thoroughly with

---

* Hesse (Ref. 8, Sec. 6.14) has shown that the presence of colloidal organic matter in soils can cause serious errors in the determination of total sulfate; he employs a precipitation of ferric hydroxide to scavenge the colloidal matter prior to the precipitation with barium chloride.

a small glass rod.   Wipe the rod free of particles by rubbing it in about 1 g of Na_2CO_3 on a filter paper, and use this flux to cover the mixture in the crucible.

Place the covered crucible in a cold electric muffle furnace (or at the edge of a heated one, with the door open) and rapidly raise the temperature of the muffle to 1000–1050°C (or gradually insert the crucible into the hot furnace).   Allow to heat at this temperature for 15 min.

Remove the crucible from the furnace, quickly remove the cover (**Caution:** there may be a blob of molten material on the underside!), and swirl the contents of the crucible, if possible, to distribute the cooling melt around the crucible walls.   Place the crucible on a silica triangle supported on a tripod and replace the cover.

Heat the crucible cover to redness for 1 min with a Tirrill burner, to fuse any sample material that may have been spattered on the underside of the cover.   Heat the crucible to dull redness for 15 sec and allow it to cool thoroughly.

Add a few ml of H_2O to the crucible and loosen the cake with a stirring rod (warm the crucible on the water bath if necessary).   Transfer the cake and rinsings to a 250 ml beaker* and police the inside of the crucible and cover with a rubber policeman.   It is not necessary to remove any stain from the crucible, except to clean it for future use.   Dilute to 50 ml, add a few ml of ethyl alcohol, and digest on an electrically heated water bath until all soluble material is in solution, breaking up lumps with a stirring rod when necessary.

Filter the solution into a 600 ml beaker (see footnote) through a 9 cm Whatman No. 40 paper.   Wash the residue and beaker several times with cold 1% Na_2CO_3 solution and pour the rinsings through the paper. It is not necessary to transfer the residue quantitatively to the paper unless the determination of such constituents of the residue as barium, zirconium or the rare earths is planned (Sec. 9.33).   Wash the paper and contents five times with the sodium carbonate solution.   The filtrate can be used for the determination of chromium by colorimetric comparison with standard potassium chromate (see Sec. 9.34.1) prior to the determination of sulfate.

* It is convenient to reserve a set of beakers, cover glasses and stirring rods for this purpose.   They will soon become etched by the strongly alkaline solutions. This glassware, incidentally, should *never* be cleaned with cleaning solution but only with hot dilute HCl or HNO_3.

Similarly, a set of 600 ml beakers, covers and stirring rods should be reserved for the precipitation steps.   Acidification of the filtrate from the leach step usually causes the glassware to become clouded.

Add 5 ml of saturated bromine water to the filtrate and dilute to 300 ml. Cover, and add dilute HCl (1:1) cautiously through the lip of the beaker until the solution is acid (a drop of indicator will last long enough to show when this stage has been reached). Heat to boiling and boil for a few minutes to expel carbon dioxide and bromine. Cool and neutralize the solution with aqueous ammonia, then add 2 ml of 12N HCl.

Continue with the precipitation, filtration, ignition and weighing* of the barium sulfate as previously described (see Sec. 9.28). The total sulfur is usually reported as per cent S.

9.30.2. TITRIMETRICALLY, FOLLOWING HIGH TEMPERATURE COMBUSTION OF THE SAMPLE

The sample, mixed with tin, is burned in a current of oxygen in a combustion tube heated to 1450°C. The sulfur dioxide formed by the combustion is absorbed in a known excess of potassium iodate–iodide solution containing starch, and the excess KIO$_3$ is back-titrated with standard sodium thiosulfate solution.

This procedure, which is less subject to error than is the low temperature (900–950°C) method using V$_2$O$_5$ as oxidant formerly used at the Geological Survey of Canada (Sen Gupta, Ref. 17, Sec. 6.14), is applicable to both large and small amounts of sulfur. A spectrophotometric finish, using sodium tetrachloromercurate to absorb the SO$_2$ which is then reacted with pararosaniline hydrochloride and formaldehyde, has also been described (op. cit.). This is suitable only for microgram amounts of sulfur.

Apparatus

Furnace: the furnace must be capable at least of reaching and maintaining a temperature of 1450°C. One capable of holding two combustion tubes, enabling dual operation, will be found useful.

Combustion Tubes: these should be 30 in. long, 1½ in. o.d., high temperature tubes with one tapered end. The tubes should be placed in the furnace, the air gaps at the two ends closed with quartz wool and left in position until it is necessary to clean them or replace them.†

* Fusion with an alkaline flux puts the silica into solution also and for accurate results the BaSO$_4$ should be treated with HF and H$_2$SO$_4$.

† A residue accumulates in the cool portion of the combustion tube after about three weeks of continuous operation and will reduce the yield of SO$_2$. This can be burned out by moving the cooler end into the hot zone and heating it for 30 min in a stream of oxygen.

Receivers: two borosilicate glass receivers, of the type described by Sen Gupta (op. cit.), are very convenient to use. They are connected to the tapered end of the combustion tube with short pieces of plastic tubing through a short glass tube which is constricted in the middle and packed at both ends with glass wool. The glass wool should be replaced daily.

Combustion boats: any type of disposable high temperature ceramic boat, having a negligible sulfur content, may be used.

Reagents

Potassium iodate solution, primary standard: weigh 1.0703 g of primary standard grade KIO_3, dried at 120°C for 1 hr and cooled in a desiccator, and transfer it quantitatively to a 250 ml beaker. Add about 100 ml of water and stir until the reagent is dissolved. Quantitatively transfer the solution to a 1 liter volumetric flask and dilute to volume (room temperature <25°C). Store in a dark place.

Sodium thiosulfate solution: dissolve 1.5 g of $Na_2S_2O_3 \cdot 5H_2O$ in water and dilute to 1 liter. Determine the equivalence of this solution in terms of 1 ml of the KIO_3 stock solution by running a series of blank determinations through the whole procedure.

Potassium iodide–starch solution: make 0.45 g of soluble starch into a paste with water in a 100 ml beaker; add 100 ml of boiling water, stir to dissolve, boil for 1 min and cool rapidly in running water to room temperature. Add 4.5 g KI, mix well and store in a glass-stoppered bottle. Discard the solution at the first sign of a yellow coloration (usually stable for up to one week).

Hydrochloric acid, 0.2N: dilute 34 ml of 12N HCl to 2 liters.

Procedure

The following steps should be done about 30 min before beginning the determination. Set the furnace temperature at 1450°C and connect the receivers to the tapered end of the combustion tube, via the glass filter tube. Insert a rubber stopper fitted with a heat deflector into the mouth of the combustion tube and pass a stream of oxygen (purified by passing it through a Fleming purifying jar containing concentrated H_2SO_4 and then through a U-tube packed with glass wool, ascarite and anhydrous magnesium perchlorate (Anhydrone)). Connect the exit end of the receivers to a second purifying jar containing water which serves to indicate the flow rate of the oxygen through the system. Flush the combustion tube with a moderate stream of oxygen for 30 min.

Transfer 10–200 mg of sample (200 mesh), depending upon the sulfur content (S > 5%, 50 mg; S, 1–5%, 100 mg; S < 1%, 200 mg) to a new combustion boat and spread it evenly over the bottom. Cover the sample with 1 g of pure, granulated (20–30 mesh) tin, and place a ceramic cover over the boat to prevent spattering of hot material on the tube walls.

Disconnect the receivers from the system and add to each, by means of a pipet, 25 ml of 0.2N HCl. Add 2 to 5 ml of the KI–starch solution, depending upon the expected sulfur content, to the first receiver and 1 ml to the second one. Add, from a buret, a known excess of KIO_3 primary standard solution to the first receiver (3 ml for up to 1% S, 5 ml for up to 2% S, 10 ml for up to 4% S and so on, *for a 100 mg sample*). Connect the receivers into the system.

Adjust the oxygen flow rate to approximately 1.5 liters per min and, without delay, push the boat and contents into the hot zone of the tube and quickly replace the stopper. After a few seconds the combustion of the sample and tin will cause the gas flow at the exit end to diminish; increase the oxygen flow so that the flow rate is uniform in both receivers. A white cloud in the receivers, and the condensation of moisture in the glass tube or plastic tubing connections, may be observed at this stage, and the flow of oxygen should be continued (at a moderate rate) until the white cloud and/or moisture are no longer seen (about 2–4 min).

When combustion and absorption are complete, disconnect the receivers and transfer the contents to a 250 ml erlenmeyer flask, washing the receivers with water and adding the rinsings to the flask also. Titrate the excess iodine to a colorless end point with the standardized sodium thiosulfate solution. If 1 ml $KIO_3 \equiv X$ ml $Na_2S_2O_3$, then:

$$\% \text{ S} = (\text{ml } Na_2S_2O_3 \text{ in final titration} \times 0.481 \times 100 /$$
$$X \times \text{sample wt (mg)}$$

For routine determinations using the solutions described here it will be found that the ratio

$$(0.481/X \text{ ml } Na_2S_2O_3) \cong 0.1$$

and, for a 100 mg sample,

$$\% \text{ S} = \text{ml } Na_2S_2O_3 \text{ in final titration}/10$$

9.31. Gravimetric determination of chlorine

9.31.1. ACID-SOLUBLE CHLORINE

Reagents

Silver nitrate solution (0.2 N): dissolve 3.4 g $AgNO_3$ in 100 ml of water containing a few drops of nitric acid and store in an amber, glass-stoppered bottle.

Procedure

To a 1.0 g sample (80 mesh), in a covered 250 ml beaker, slowly add 50 ml of dilute nitric acid (1:19). Heat to boiling and boil for 2-5 min but no longer. Allow to settle and then decant the supernatant liquid through a 9 cm No. 42 Whatman filter paper, prewashed with dilute HNO_3 (1:99), into a 150 ml beaker; if the supernatant liquid appears turbid it may be necessary to use a double filter paper to ensure a clear filtrate. Wash the residue several times with the dilute HNO_3 (1:99), decanting the wash water through the paper, and finally wash the paper five times with the wash solution; the final volume should be about 75 ml. The residue may be transferred quantitatively to the paper (after removing the beaker containing the filtrate) and used for the determination of acid-insoluble chlorine as described under the determination of total chlorine (Sec. 9.31.2).

Add 3 ml of 0.2N $AgNO_3$ solution and heat to boiling, keeping the beaker out of direct sunlight as much as possible. For most samples the solution will become opalescent at this stage, without visible coagulation of the AgCl. Surround the beaker with a shield of dark paper and allow to stand overnight in a cupboard (away from containers of HCl or NH_4Cl).

Decant the supernatant liquid through a 7 cm Whatman No. 42 filter paper or a fine porosity fritted-glass filter crucible, both washed free of chloride with dilute HNO_3 (1:99), and wash the precipitate, if possible, two or three times by decantation. Wash down the sides of the beaker with 2–3 ml of dilute aqueous ammonia (1:1) and pour the solution through the filter to dissolve any silver chloride present, catching the filtrate in a 150 ml beaker. Wash the original beaker and the paper once with the dilute aqueous ammonia, and then with about 10 ml of water. Dilute to about 25 ml with water and acidify with 1:1 HNO_3, using methyl orange (0.02% aqueous solution) as indicator; add 1 ml in excess.

Add a few drops of 0.2N $AgNO_3$ solution, heat to boiling, cover with a dark shield and allow to stand for 1–2 hr. Filter on a weighed fine porosity fritted-glass filter crucible, wash five times with dilute HNO_3 (1:99) and dry at 130–150°C to constant weight.

A blank determination should be made at the same time as the sample analysis, following all steps of the procedure. The precipitation step will seldom yield much beyond a faint opalescence, the final value of which does not usually exceed 0.01% Cl.

Conversion Factors

$$AgCl \underset{4.0426}{\overset{0.2474}{\rightleftarrows}} Cl$$

9.31.2. TOTAL CHLORINE

Mix 1.0 g of sample (80 mesh), or use the residue from Sec. 9.31.1, with 5 g of anhydrous Na_2CO_3 (low in chlorine) in a 25 ml platinum crucible and fuse as described in Sec. 9.2.1. Transfer the fusion cake to a 150 ml beaker and digest near the boiling point with 50 ml of water until the cake has completely disintegrated and all soluble matter has dissolved. Add 2–3 ml of ethyl alcohol to reduce any manganate present.

Decant the solution through a 9 cm No. 42 Whatman filter paper, previously washed with H_2O, and wash the residue several times with small portions of 1% Na_2CO_3 solution, decanting the wash solutions through the paper. Finally wash the paper five times with the wash solution, and discard the paper and residue.

Cover the beaker and carefully add dilute HNO_3 (1:1) until the solution is acid to methyl orange, and add 1 ml in excess.* Continue with the determination of the chlorine as silver chloride as described in Sec. 9.31.1. The value obtained is that of the total chlorine content of the sample, unless the residue alone from the acid-soluble chlorine determination was used for the fusion, in which case the final figure represents the acid-insoluble chloride only.

9.32. Determination of fluorine after its separation by steam distillation

9.32.1. TITRIMETRIC

This procedure, which involves the fusion of the sample with sodium carbonate, the leaching of the cake and separation of the bulk of the silica

---

* If the acidified filtrate is not clear, allow to stand overnight, heat to boiling and add aqueous ammonia to excess; filter, wash the precipitate with hot water, cool the filtrate, make acid with dilute HNO_3 (1:1), and proceed with the determination as previously described (Ref. 5, Sec. 6.15).

with acid zinc perchlorate, the steam distillation of the fluoride-bearing filtrate from a perchloric acid solution and titration of aliquots of the distillate with thorium nitrate, using sodium alizarin sulfonate as indicator, was adapted by R. B. Ellestad for use in the Rock Analysis Laboratory at the University of Minnesota, where the writer became familiar with it and later introduced it into the laboratories of the Geological Survey of Canada. The details of the method are also given by Kolthoff and Sandell (Ref. 20, Sec. 6.16).

Apparatus

The two-unit distillation apparatus used at the Geological Survey has no special features and is based upon that described by Shell and Craig (Sec. 6.16, Ref. 26, p. 5)

A three-necked, 2 liter, round-bottomed boiling flask serves as the steam generator for two sets of distillation apparatus; the third neck is fitted with an adapter to which is joined a piece of rubber tubing closed by a screw clamp, and this is used to regulate the supply of steam to the distillation flasks. The latter are round-bottomed flasks of 250 ml capacity having a side arm; the center neck is fitted with an adapter which holds the steam delivery tube and the thermometer.* The neck of the side arm is connected by adapters to a Liebig condenser; an 8 oz polyethylene bottle is used to catch the distillate. An "anti-bump" rod is placed in each distillation flask and serves also to reduce the attack on the steam delivery tube by the acid mixture. Heating of the steam generating and distillation flasks is done by electric heating mantles or by Tirrill burners; the latter make possible a closer control of the temperature, provided that the flame is protected by a flame guard.

Reagents

Standard thorium nitrate solution (0.02N): dissolve 2.76 g of reagent grade $Th(NO_3)_4 \cdot 4H_2O$ in water, add 0.4 ml concentrated HNO_3 and dilute to 1 liter.

Standard fluoride solution (0.000339 g F/ml): dry about 0.2 g of reagent grade NaF by heating it in a small platinum crucible at a low

---

* The distillation apparatus is available from Ace Glass, Vineland, N.J., U.S.A., as item 6431 (Fluorine Determination Apparatus). The separatory funnel is replaced, after addition of the acid, by a No. 7640 joint (₮ 10–30) which has a delivery tube 5 mm in outside diameter; it is necessary to join another length of 5 mm Pyrex tubing to this in order to extend the steam delivery tube to near the bottom of the flask, and to make a right angle bend in the upper tube to facilitate connection of the tube to the steam generating flask.

red heat; dissolve 0.1500 g of the dried salt in water and dilute to 200 ml in a volumetric flask.

Sodium alizarin sulfonate (0.05%): dissolve 0.05 g of the indicator in 100 ml of water.

Sodium monochloracetate–monochloracetic acid buffer solution (2 M): dissolve 18.9 g of reagent grade monochloracetic acid in water and dilute to 50 ml; neutralize 25 ml of this solution by adding dropwise concentrated sodium hydroxide solution, using phenolphthalein as indicator. Add the remaining 25 ml of monochloracetic acid and dilute to 100 ml. This solution should not be kept longer than 3 to 4 weeks; beyond this it will cause deterioration of the quality of the end point of the titration.

Acid zinc perchlorate solution: dissolve 10 g of zinc oxide in 112 ml of 70% perchloric acid and add 50 ml of water.

Procedure

Standardization of thorium nitrate solution

Pipet several pairs of various volumes (e.g., 0.50, 1.00 and 2.00 ml) of the standard fluoride solution into 50 ml beakers. Add sufficient water to bring the total volume of each to 15 ml, and add 15 ml of 95% ethyl alcohol. Add 3 drops of the sodium alizarin sulfonate indicator and 0.30 ml of the buffer solution. Titrate each with the standard thorium nitrate solution, using a 5 or 10 ml buret, to a faint pink end point matching a reference solution prepared as follows: dilute 2.0 ml of the 0.02N standard thorium nitrate solution to 100 ml and add 0.50 ml of this 0.0004N solution (equivalent to 0.01 ml of the 0.02N solution) to 15.0 ml of water, 15.0 ml of 95% ethyl alcohol, 3 drops of indicator and 0.3 ml of buffer solution in a 50 ml beaker. The color of this reference solution is equivalent to a 0.01 ml excess of the 0.02N thorium nitrate solution and this reagent blank should be deducted from each titration whose end point is matched to this color. Calculate the F equivalent of 1 ml of the standard thorium nitrate.

Procedure for silicates

Mix 0.5 g of sample (finely ground) with 3 g of anhydrous sodium carbonate in a 25 ml platinum crucible, and fuse as quickly as possible, with only a few minutes over the meker burner. Transfer the cake to a 250 ml beaker and leach it in 40 ml of water; *do not add alcohol to reduce manganate ion.* Filter through a 9 cm Whatman No. 40 filter paper into a 250 ml beaker and wash paper and residue with 2% sodium carbonate solution.   Rinse the residue back into the original beaker, boil briefly with 20–30 ml of sodium carbonate solution, filter through the same paper and continue washing the paper and residue until the total volume is about 100 ml.   Discard the residue.

To this cold solution add, with stirring, 8 ml of the acid zinc perchlorate solution.   Heat to boiling and boil for 1 min.*   Filter through an 11 cm Whatman No. 40 paper and wash well with hot water.   Rinse the precipitate back into the original beaker, stir well with a small quantity of hot water and again filter.   Discard the precipitate.   Concentrate the combined filtrates to 10–15 ml by heating the uncovered beaker on a low temperature hot plate (do *not* boil); care is needed during this evaporation to avoid loss of sample by "bumping."

Transfer the solution to the distillation flask† by means of a long-stemmed funnel, using a minimum amount of water for rinsing.   Place a "boiling tube" in the flask, and connect the flask into the apparatus but use a separatory funnel temporarily in place of the steam inlet tube.   Apply heat to the steam-generating flask, close the steam inlet tubes with screw clamps but leave the vent open to the atmosphere. Start cooling water circulating through the condenser and place an 8 oz polyethylene bottle at the condenser outlet.

Close the stopcock of the separatory funnel and add to it 25 ml of 70% perchloric acid.   Allow the acid to flow steadily into the flask, re-move the funnel and immediately insert the steam inlet tube.

Heat the distillation flask with the full flame of a bunsen burner. When the temperature in the flask reaches 140°C, open the clamp on the steam inlet tube and close the vent of the steam generating flask, the contents of which should be boiling vigorously (boiling chips should be placed in the flask to maintain steady boiling).

Maintain the temperature in the distillation flask at 135–140°C by regulation of the rate of steam generation and the height of the flame under the distillation flask, and continue the distillation until about 150 ml of distillate have collected; this includes the 15–20 ml of excess water that is distilled before the maximum distillation temperature is reached.   Keep the distillate alkaline to phenolphthalein (2 drops) dur-

---

* This precipitates, as zinc silicate, the bulk of the silica that was leached from the fusion cake.

† Samples which are acid-soluble, such as phosphates, may be added directly to the flask without first being fused but the temperature of the distillation must be kept low enough to prevent any carryover of phosphoric acid into the distillate; it is preferable that the distillate be concentrated and redistilled.   Such unfused samples must *not* contain appreciable amounts of organic matter.

ing the distillation by adding 0.1N sodium hydroxide solution as needed (about 2–3 ml).

When the distillation is complete, remove the flame from the distillation flask and allow the temperature to fall below 130°C. Carefully close the steam inlet tube so that no "suck-back" occurs in the steam generation flask. Disconnect the condenser, hold it vertical and rinse the center tube with a small amount of water, combining the rinsings with the distillate.

Transfer the alkaline distillate to a 250 ml platinum dish and evaporate the solution on a water bath to about 15 ml.* Carefully transfer the solution to a 25 ml volumetric flask, cool to room temperature and dilute to volume. Pipet a 10 ml aliquot into a 50 ml beaker and add 0.2N HCl dropwise until the pink color disappears. Add 3 drops of the sodium alizarin sulfonate indicator (which will show its reddish violet alkaline color) and continue to add the drops of 0.2N HCl until the indicator turns yellow. Add 0.3 ml of the buffer solution, dilute to 15 ml with water and add 15 ml of 95% ethyl alcohol. Titrate with the standard thorium nitrate solution to a shade of pink that matches the color of the reference solution, and deduct 0.01 ml as titration blank. Repeat the titration on a second aliquot as a check; if the first titration was unusually large it is preferable to use a smaller aliquot for the second titration and, because the result of the first titration is known, it is possible to add the proper amount of water to give a final volume of 30 ml, thus increasing the accuracy of the titration. Calculate the per cent fluorine in the sample.

Clean the glass apparatus thoroughly with hot 10% Na_2CO_3 between distillations.

9.32.2. SPECTROPHOTOMETRIC

This method, for use *only* with silicates containing 0.01–1.00% F, was adapted by Sydney Abbey from the method described by Peck and Smith (Ref. 22, Sec. 6.16) for use in the Geological Survey laboratories. Slight changes were made in the preparation of the reagents and the final calculation is based upon interpolation between two standards, one of which is the blank, instead of using a calibration curve as described by Peck and Smith.

* It is very important, at this stage, to keep the area free of fumes of HF. Because this method is almost micro in scale, a serious error can be introduced by contamination.

Apparatus

The two-unit distillation apparatus used is that described in Sec. 9.32.1.   A more compact two-unit apparatus is shown in detail by Peck and Smith (op. cit.).

Reagents

Mixed flux: mix thoroughly 350 g anhydrous sodium carbonate, 100 g zinc oxide and 50 g basic magnesium carbonate.   A blank determination should be made on each batch of flux.

Stock sodium fluoride solution (1.00 mg F/ml): Peck and Smith (op. cit.) describe the preparation of this reagent as follows: "dissolve 10 g of reagent-grade NaF in 225 ml of water, add a few ml of HF and an equal volume of methanol.   Collect the salt on paper, wash it with methanol and air dry.   Fuse 3 g of the salt, pour the melt into a platinum dish and cool.   Crush the NaF and quickly weigh 1.105 g.   Dissolve the weighed portion in water and filter it into a 500 ml volumetric flask.   Dilute to the mark and mix (a solution so prepared assayed 0.996 mg of F/ml when analyzed by the lead chlorofluoride method)."   The solution should be stored in a polyethylene bottle.*

Standard sodium fluoride solution (0.100 mg F/ml): dilute 50 ml of the stock NaF solution (1.00 mg F/ml) to 500 ml in a volumetric flask. Store in a polyethylene bottle.†

Zirconium sulfate–sulfuric acid solution: dissolve 5.8 g of zirconium sulfate tetrahydrate in 500 ml H_2SO_4 (1:1) and 400 ml of water.   Filter through a Whatman No. 42 paper into a 1 liter volumetric flask, and dilute to volume.   Determine the ZrO_2 content by evaporating an aliquot of the solution in a platinum dish, ignite and weigh to constant weight as the oxide.

SPADNS solution: dissolve 16 g of SPADNS in 400 ml of water and filter through a No. 42 Whatman paper, covered with filter paper pulp,† into a 1 liter volumetric flask.   Dilute to volume and mix well.

---

* This solution may also be prepared as follows: dissolve 1.1 g reagent grade sodium fluoride in water and dilute to 500 ml in a volumetric flask.   Mix well and store in a polyethylene bottle.   Standardize this solution by pipetting a suitable aliquot into a platinum dish and treating it with H_2SO_4 until all fluoride and excess H_2SO_4 have been expelled by heating.   Weigh as sodium sulfate.

† If the stock sodium fluoride solution does not contain 1.00 mg F/ml then a volume of solution containing 50.0 mg F should be taken for dilution to 500 ml (0.100 mg F/ml).

† For the preparation of the SPADNS solution, Peck and Smith (op. cit.) recommend filtration through a Fiberglas pre-filter and a membrane filter having 0.45 μ pores.

Zirconium–SPADNS reagent: measure a volume of zirconium sulfate–sulfuric acid solution containing 0.75 g ZrO_2 into a 500 ml graduated cylinder. Dilute to 400 ml with H_2SO_4 (1:3), then to 500 ml with water. Transfer to a 2 liter volumetric flask, add 500 ml of SPADNS solution and dilute to volume.

Procedure

Weigh 5.0 g of mixed flux into a 30 ml platinum crucible, add 0.500 g of sample (100 mesh) and mix thoroughly with a glass rod. Cover the crucible and heat it at 900°C for 30 min in an electric furnace. Allow the crucible to cool.

Add water to the crucible, loosen the cake with a stirring rod and transfer it to a 50 ml beaker. Bring the volume to about 30 ml, break up the cake as much as possible and allow to stand overnight.

Crush the cake to a powder with the stirring rod and decant the solution through a 7 cm Whatman No. 42 paper into the distillation flask. Wash the residue in the beaker twice by decantation with hot 1% sodium carbonate solution, transfer the residue to the paper and wash the paper and contents five times with the sodium carbonate solution. Discard the residue.

Place a "boiling tube" in the flask and connect the flask into the distillation apparatus, but use a separatory funnel, temporarily, in place of the steam inlet tube. Apply heat to the steam generating flask, leaving the vent open to the atmosphere but closing the steam inlet tubes with screw clamps. Start cooling water circulating through the condenser and place an 8 oz polyethylene bottle, with approximate calibration marks at the 75 and 200 ml levels, at the outlet of the condenser.

Close the stopcock of the separatory funnel and add 25 ml H_2SO_4 (1:1) to the funnel. Carefully open the stopcock and allow the acid to flow into the flask, maintaining a steady effervescence. Close the stopcock. As soon as the effervescence ceases, remove the funnel and immediately insert the steam inlet tube.

Heat the distillation flask with the full flame of a bunsen burner. When the temperature in the flask reaches 130°C, open the clamp on the steam inlet tube close the vent of the steam generating flask, the contents of which should be boiling vigorously and steadily. When the temperature in the distillation flask reaches 135°C, adjust the burner beneath it to maintain a temperature between 135–145°C. About 75 ml of distillate should have collected by this time, which should be about 25 min after distillation begins.

Continue the distillation until another 125 ml of distillate have collected, which should take about 45 min. Remove the burner from beneath the distillation flask and allow the temperature to drop to below 130°C, then carefully close the steam inlet tube without permitting "suck-back" into the steam generating flask. Disconnect the distillation flask and allow it to cool.*

Allow the distillate to attain room temperature, transfer it to a 250 ml volumetric flask, dilute to volume and mix well. Pipet a 50 ml aliquot of this solution into a 100 ml volumetric flask. (If the sample is known to contain more than 0.20% F, use a smaller aliquot.)

Prepare a reference standard (0.200 mg F) by pipetting 2.00 ml of the standard fluoride solution (0.100 mg F/ml) into a 100 ml volumetric flask. Prepare a reference blank by adding 75 ml of water to another 100 ml flask. Dilute the sample and reference standard solutions to 75 ml also.

To the sample, reference standard and blank add 10 ml of the zirconium–SPADNS reagent, dilute to volume, mix well and allow to stand for 1 hr at room temperature. Transfer aliquots of the solutions to 1 cm cells of a Beckman Model B spectrophotometer, adjust to zero absorbance for the reference standard and measure the absorbance of the sample and blank.

If a 50 ml aliquot of the 250 ml distillate is used (i.e., a 100 mg sample) then the per cent fluorine in the sample is given by:

$$\% \text{ F} = 0.2[(B - X)/B] \times (10\dagger/9)$$

where B = absorbance of blank, X = absorbance of sample.

9.33. Gravimetric determination of barium (as chromate) and strontium (as sulfate)

Fuse 1.00 g of sample with Na_2CO_3 and Na_2O_2, leach the cake, filter and wash the insoluble residue as described in the gravimetric procedure for the determination of total sulfur (Sec. 9.30.1). The residue will contain the barium and strontium, together with calcium, zirconium, rare earths and possible other constitutents.

---

* After every distillation, the distillation flask must be washed thoroughly and freed of adsorbed fluoride by heating a dilute solution of sodium carbonate in it.

† This is an arbitrary factor used to correct for the small amount of fluoride that is lost by adsorption on the glass, coprecipitation or retention in the acid mixture. Peck and Smith (op. cit.) recommend the addition of a 3% correction to the result.

Dissolve the residue in 5 ml of H_2SO_4 (1:1), catching the solution in a 150 ml beaker, and add a few grains of sodium sulfite to reduce manganese compounds. Reserve the filter paper. Dilute the solution to 40–50 ml, add an equal volume of 95% ethyl alcohol and allow to stand overnight.

Filter through a 9 cm Whatman No. 40 paper, refiltering through the same paper if the filtrate is cloudy. Wash the beaker and precipitate with 50% ethyl alcohol containing a few drops of H_2SO_4. Place the paper, together with the paper used for filtering the leachate, in a 25 ml Pt crucible (the one used for the original fusion will serve without first cleaning it) and ignite at about 600°C in an electric muffle furnace until the carbon has been burned off.

Pulverize the residue in the crucible with a glass rod, mix it with 2–3 g Na_2CO_3 and 0.1 g Na_2O_2 and again fuse as before. Do not spread the fusion on the walls of the crucible before cooling it. Remove the fusion cake, transfer it to a 100 ml beaker (see footnote, Sec. 9.30.1), dilute to about 25 ml and digest on the water bath until all soluble material has dissolved. Filter through a 7 cm Whatman No. 42 paper and wash the residue thoroughly with 1% Na_2CO_3 solution to remove all sulfate.

Wash the residue into the 100 ml beaker, without removing the paper from the funnel. Place the 100 ml beaker beneath the funnel and wash the paper with hot 5% HCl to dissolve the alkaline earth carbonates. Heat the solution in the beaker to boiling and boil to remove all CO_2. Add 1–2 ml of saturated bromine water, boil the solution to expel excess bromine, and precipitate any R_2O_3 group elements that may be present with aqueous ammonia, using bromcresol purple (0.04% aqueous solution) as indicator. Filter through a 7 cm Whatman No. 40 paper and wash the precipitate and paper with hot water. Discard the precipitate.

Precipitation of barium as chromate

Make the filtrate, which should have a volume of 50–75 ml, just acid with HCl* and add 10 ml of a 30 % solution of ammonium acetate. Heat to boiling and add 5 ml of a 10% solution of ammonium dichromate.† Stir well and let stand overnight. Filter through a 7 cm What-

* The determination of barium as chromate or as sulfate may be done at this stage using the solution obtained from the separation of Ca, Sr and Ba as sulfates prior to the precipitation of magnesium (Sec. 9.10.1). The sulfate precipitate was fused with Na_2CO_3, leached in water, the carbonates dissolved and the Ca and Sr precipitated as oxalates.

† Wonsidler and Sprague (Ref. 14, Sec. 6.17) recommend the precipitation of $BaCrO_4$ by slow addition of a sodium dichromate solution to a cold, buffered (pH 4.6) barium solution, following by boiling and cooling, on the basis of convenience and effectiveness of separation.

man No. 42 paper and wash well with cold water. Reserve the filtrate. Place the paper in a weighed, small porcelain crucible, burn off the paper at a low temperature and then ignite and weigh the precipitate as $BaCrO_4$. Greenish spots may be visible after the first weighing, owing to reduction of the chromate to $Cr(III)$ during the ignition; these should disappear on further ignition to constant weight. If much strontium is known to be present the precipitated $BaCrO_4$ should be dissolved in dilute HCl and a second precipitation made.

Conversion Factors

$$BaCrO_4 \underset{1.8446}{\overset{0.5421}{\rightleftarrows}} Ba$$

$$BaCrO_4 \underset{1.6521}{\overset{0.6053}{\rightleftarrows}} BaO$$

$$BaO \underset{1.1165}{\overset{0.8957}{\rightleftarrows}} Ba$$

Separation of strontium (and calcium) from the filtrate

Heat the filtrate to boiling and add an excess of aqueous ammonia, followed by a large excess of a saturated solution of ammonium carbonate. Allow to cool, filter through a 7 cm Whatman No. 40 paper, and wash the precipitate 3 times with 5% aqueous ammonia. Discard the filtrate. Dissolve the carbonate in a small amount of dilute nitric acid and repeat the precipitation, filtration and washing. Dissolve the pure carbonates in a small amount of dilute nitric acid, catching the solution in a 50 ml beaker (refilter if necessary to remove filter paper fibers). Place the beaker on a sand bath and carefully evaporate the solution to dryness. When thoroughly dry, heat the beaker and residue in an oven at 160°C for one hour.

To dry nitrates* add 20 ml of nitric acid (Sp. Gr. 1.44) and carefully and thoroughly crush the solid material to a fine powder with a short glass rod. The calcium nitrate will dissolve and the strontium nitrate, if the SrO present is greater than 0.02%, will be visible as a faint turbidity. Alternately stir the contents and then allow to stand for a few minutes, until all of the calcium nitrate appears to have dissolved. Decant the solution through a weighed (if the Sr is to be weighed as the nitrate) fine porosity fritted-glass filter crucible; if some

* The ignited CaO (plus SrO) obtained from the oxalate precipitation (Sec. 9.10.1) may be dissolved in nitric acid, evaporated to dryness, dried at 160°C for 1 hr and the Sr separated as described.

water is placed in the filtration flask to dilute the acid filtrate, and a strong suction is maintained, the attack on the rubber crucible holder will be minimized.

Add 10 ml of nitric acid (Sp. Gr. 1.44) to the beaker and repeat the alternate stirring and settling. Transfer the contents of the beaker quantitatively to the filter crucible and wash the residue 10 times with nitric acid (Sp. Gr. 1.44) to remove all traces of calcium nitrate. Dry the crucible on the water bath to remove excess nitric acid.

The strontium may be determined gravimetrically as strontium nitrate after drying the crucible and contents in an oven at 160°C for 1 hr. It is preferable to obtain the final weight of the filter crucible by dissolving the strontium nitrate in a small amount of hot water, drying (at 160°C) and reweighing the crucible; this latter weight is used in the calculations.

It is better to make the final determination of the strontium as sulfate. Dissolve the strontium nitrate in a minimum amount of hot water, catching the solution in a 100 ml beaker. Add 0.5 ml H_2SO_4 (1:1) and evaporate the contents as much as possible on the water bath. Dilute to about 15 ml, add an equal volume of 95% ethyl alcohol and allow to stand overnight. Filter through a 7 cm Whatman No. 42 filter paper and wash the residue with 50% ethyl alcohol. Dry the paper, transfer to a small platinum or porcelain crucible, burn off the paper at about 600°C and ignite the residue at a low red heat (about 800°C). Weigh as $SrSO_4$.

Conversion Factors

$$SrSO_4 \underset{1.7726}{\overset{0.5642}{\rightleftarrows}} SrO$$

$$SrO \underset{1.1826}{\overset{0.8456}{\rightleftarrows}} Sr$$

9.34. Colorimetric determination of chromium

9.34.1. AS CHROMATE

Reagents

Standard chromium solution (0.0001 g Cr_2O_3 per ml): dissolve 0.2838 g $K_2Cr_2O_7$ in distilled water free of reducing substances and dilute to 1 liter in a volumetric flask.

Procedure*

Weigh and quantitatively transfer 1.00–2.00 g of 100 mesh sample (containing 0.01–0.5% Cr_2O_3) to a 30 ml platinum crucible. Ignite the crucible at low red heat, with the cover displaced, as given in Sec. 9.2.1, and allow to cool. Add 5 g anhydrous Na_2CO_3 and 0.1 g Na_2O_2 and mix thoroughly; cover the mixture with 1 g Na_2CO_3. Fuse the mixture as described in Sec. 9.2.1 (do not prolong the fusion unnecessarily in order to minimize attack on the crucible; platinum imparts a yellow color to the solution), cool and loosen the cake.† Transfer the cake to a 150 ml (etched) beaker, add 25 ml water, and heat it on the water bath until the cake has completely disintegrated and all soluble material has dissolved.‡ Add about 1 ml of alcohol if manganese is likely to be present, and heat near the boiling point until all manganate has been reduced.

Filter through a 9 cm Whatman No. 40 filter paper, previously washed with hot concentrated Na_2CO_3 solution to remove substances which may impart a yellow color to the filtrate, and wash the precipitate and paper with hot 1% Na_2CO_3 solution, keeping the volume below 50 ml. The residue may be used for the determination of other elements (see footnote this section).

If the determination of the chromium is to be done by colorimetric titration, collect the filtrate and washings in a Nessler tube of appropriate size (or dilute to suitable volume in a volumetric flask and take an aliquot). To a similar volume of water in another Nessler tube add 6 g Na_2CO_3 and stir to dissolve. Add the standard $K_2Cr_2O_7$ solution (0.1 mg Cr_2O_3 per ml) from a 10 ml buret in small increments until a satisfactory visual match is obtained between the colors of the two solutions. If the volume of the standard solution required is less than 10% of the total volume of the comparison solution, no correction for the effect of increased volume will be necessary.

% Cr_2O_3 = (ml Std solution × 0.0001/1000) × 1.461 ×

(100/Sample wt (g))

---

* Do *not* clean glassware with chromic acid cleaning solution. Rinse once with hot, dilute HCl (1:3) and then rinse thoroughly with distilled water.

† If it is necessary to determine chromium (and vanadium) in the ignited residue of the R_2O_3 group, fuse the latter with Na_2CO_3 + Na_2O_2 as described above. The insoluble residue from the water-leach of the cake will contain the iron and titanium (and nickel) for subsequent determination.

‡ Any black, gritty material in the insoluble residue is likely chromite and it must be recovered by a second fusion of either the whole insoluble residue, or of the material remaining after the insoluble residue is treated with dilute acid.

The visual colorimetric titration method is rapid, sensitive and simple to carry out, but if preferred, the determination may be made by visual comparison using a colorimeter such as the Duboscq, or by measuring the absorbance of the solutions at 370 mμ. The filtrate and washings are collected in a volumetric flask of suitable size and diluted to volume. The standard solutions used should be similar to the sample in alkalinity and salt content.

The filtrate containing the chromate can also be used for the colorimetric determination of vanadium (see Sec. 9.40).

Conversion Factors

$$Cr_2O_3 \underset{1.4614}{\overset{0.6843}{\rightleftarrows}} Cr$$

9.34.2. WITH s-DIPHENYLCARBAZIDE

Reagents

Sulfuric acid (6N): dilute 25 ml of 36N H$_2$SO$_4$ to 150 ml, heat to near boiling and add dilute KMnO$_4$ solution until a faint pink color remains. Cool, and store in a glass-stoppered bottle.

s-diphenylcarbazide solution (0.25%): dissolve 0.125 g of reagent in 50 ml of reagent grade acetone which has been made faintly acidic with dilute (1:19) H$_2$SO$_4$. It is better to prepare this reagent freshly as needed.

8-hydroxyquinoline solution (2.5%): dissolve 0.25 g of 8-hydroxyquinoline in 10 ml of 2N acetic acid (approx 1 ml of glacial acetic acid diluted to 10 ml).

Standard chromium solution (10 γ Cr$_2$O$_3$/ml): dilute, shortly before use, 10.0 ml of the standard solution from Sec. 9.34.1 to 100 ml in a volumetric flask and mix well. Use distilled water that has been freed of reducing substances.

*Procedure**

A direct determination of chromium can be made provided that the vanadium content of the sample is less than 10 times that of the chromium. If [V] > 10 [Cr], the vanadium must first be removed by extraction with 8-hydroxyquinoline in chloroform. The procedures given are those described by Sandell (Ref. 9, Sec. 6.18).

* Do *not* clean glassware with chromic acid cleaning solution; rince once with hot dilute HCl (1:3) and then thoroughly with distilled water.

Mix 0.25 to 0.5 g of 100 mesh sample, containing 0.0001–0.05% Cr_2O_3, with 1.2–2.5 g anhydrous Na_2CO_3 in a platinum crucible and fuse, leach and filter as described in Sec. 9.34.1 (see also footnote†, Sec. 9.34.1). It may be preferable to use a suitable aliquot of the solution of a larger sample which has been prepared by fusion with Na_2CO_3 and Na_2O_2 (Sec. 9.34.1); in this case it is necessary to reoxidize the chromium after acidification of the solution.

If the chromium is to be determined directly ([V] < 10[Cr]), collect the filtrate and washings in a 25 or 50 ml volumetric flask. Cool and add 6N H_2SO_4 cautiously to neutralize the Na_2CO_3, and then add an excess sufficient to make the solution 0.2N in H_2SO_4 when the solution is diluted to volume.* Swirl the solution carefully to remove dissolved carbon dioxide, add 1 or ml of 0.25% diphenylcarbazide solution and dilute to 25 or 50 ml with water that has been freed of reducing substances. Mix and measure the absorbance of the solution at 540 mμ after 10 min. Prepare a working curve covering the range 0.2–10 γ Cr_2O_3 by adding appropriate aliquots of the standard chromium solution (10 γ Cr_2O_3 per ml) to a series of 25 ml volumetric flasks each containing 15 ml of water and 1 g of Na_2CO_3; add the 6N H_2SO_4 and reagent as described previously, dilute to volume and measure the absorbance after 10 min.

The measurement may also be made with a Duboscq colorimeter against a standard containing about 0.5 ppm Cr_2O_3 in the final volume.

If a separation of vanadium is to be made ([V] > 10 [Cr]), dilute the filtrate and washings to volume and transfer an aliquot representing 0.1–0.25 g of sample to a small separatory funnel. Add from a buret exactly the volume of 6N H_2SO_4 needed to neutralize the solution (as found by titrating a similar aliquot containing methyl orange indicator (0.02% aqueous solution) until the color just changes from a pure yellow)† and swirl the solution to expel dissolved carbon dioxide. Add 0.1 ml 2.5% 8-hydroxyquinoline solution in 2N acetic acid and extract with two 3 ml portions of pure chloroform, shaking 30–60 sec each time. The chloroform extracts may be discarded, or may be run into a 25 ml

* If Na_2O_2 was used in the fusion of the sample, then reoxidation of the chromium is necessary to insure that only Cr(VI) is present. To the acidified filtrate in a 150 ml beaker add 1.0 ml of 1% silver nitrate solution and 5 ml of freshly prepared 10% ammonium persulfate solution; boil for about 10 min, cool and filter to remove any precipitate (e.g., AgCl) that may have formed. Continue with the addition of the diphenylcarbazide as described.

† If the acidified solution is to be treated with ammonium persulfate and silver nitrate (as described in above footnote), the methyl orange indicator may be added to the sample solution and the titration made directly with 6N H_2SO_4, because the color of the indicator will be destroyed during the subsequent reoxidation.

platinum crucible and used for the determination of vanadium (see Sec. 9.40). Add another 0.1 ml of the 8-hydroxyquinoline solution and repeat the extractions with chloroform, the last portion of which should be virtually colorless or showing only a faint yellow color due to the reagent itself. Filter the aqueous layer through a 5.5 cm Whatman No. 40 paper into a 25 or 50 ml volumetric flask and wash the separatory funnel and filter with a few small portions of water. Make the solution 0.2N in H_2SO_4 and continue with the colorimetric determination (including the reoxidation step (see footnote this section) if this is necessary) as described previously. The absorbance may be measured at once.

$$\% \ Cr_2O_3 = (\gamma \ Cr_2O_3/10^6) \times (100/\text{Sample wt (g)})$$

9.35. Titrimetric determination of chromium

When the chromium content is greater than about 1%, it can be conveniently determined by titration. Only manganese and vanadium interfere and the interference of both is easily overcome. In the following procedure a 0.5–1 g sample (depending on the chromium content) is used, requiring a volume of $N/10$ titrant ranging from about 3 to 20 ml; if the amount of sample available is insufficient, a smaller sample may be taken and either a more dilute titrant used, or the titration made with a semimicro buret.

Reagents

Ferrous ammonium sulfate standard solution ($N/10$): prepare as described in Sec. 9.16.2 (*Reagents*).

Potassium permanganate standard solution ($N/10$): prepare as described in Sec. 9.11.1 (*Reagents*).

Potassium dichromate standard solution ($N/10$): prepare as described in Sec. 9.22.1 (*Reagents*).

Procedure

Weigh and transfer a 0.5–1.0 g sample (depending on the amount of chromium present) to a 30 ml platinum crucible (see Sec. 9.34.1, footnote†) and ignite the sample at low red heat as described in Sec. 9.2.1. Fuse the sample with $Na_2CO_3 + Na_2O_2$, leach the cake (omit the alcohol), filter and wash the residue as described in Sec. 9.34.1, catching the filtrate and washings in a 400 ml beaker (see Sec. 9.34.1, footnote‡). Cover the beaker and cautiously but rapidly add H_2SO_4 (1:1) through the lip, swirling the solution after each addition, until the solution is slightly acid; dilute to about 250 ml and add 15 ml concentrated H_2SO_4 and 3 ml concentrated HNO_3. Heat the solution to boiling and boil until all carbon dioxide is expelled. Add 5 ml 0.5% $AgNO_3$ solution and 20 ml of freshly prepared 10% ammonium persulfate solution, and boil the solution for 10 min to complete the oxidation of the chromium and to destroy the excess persulfate. If manganese is present, either previously known or as visible $KMnO_4$ or oxides of manganese, add 5 ml 5% sodium nitrite (or 5 ml 1:3 HCl if $KMnO_4$ is not to be used) dropwise to reduce it to Mn(II), and boil the solution for a further 5 min to destroy the excess of the reducing agent (the presence of AgCl does not interfere in the subsequent titration.) If vanadium is known to be absent, or if a correction is to be applied as the result of separate determination of the vanadium, the solution may be titrated potentiometrically with $0.1N$ ferrous ammonium sulfate which has been standardized against a measured volume of $0.1N$ $K_2Cr_2O_7$ primary standard solution carried through the procedure beginning with the addition of sulfuric acid; the titration may also be made visually with ferroin (o-phenanthroline ferrous complex) as indicator.

$$1 \text{ ml } 0.1N \text{ ferrous ammonium sulfate} \equiv 1.734 \text{ mg Cr}$$

If vanadium is present and is to be determined in the same solution, or is to be prevented from interfering with the determination of chromium, a different approach is necessary. Add a measured excess of standard ferrous ammonium sulfate solution $(0.1N)$ to the solution and back titrate the excess with standard $KMnO_4$ solution $(0.1N)$ to a permanent pink end point lasting at least two min (to ensure oxidation of the vanadium). Standardize the ferrous ammonium sulfate solution against the standard $KMnO_4$.

The above solution can also be used for the determination of vanadium by reduction with ferrous sulfate in the presence of ferroin as indicator.

9.36. Gravimetric determination of nickel with dimethylglyoxime

Reagents

Dimethylglyoxime solution (1% in ethanol): dissolve 1 g of the reagent in 100 ml ethanol and store in a tightly stoppered glass bottle.

Procedure

Weigh and transfer 1.00–2.00 g of 100 mesh sample (containing not more than 50 mg Ni) to a 50 ml platinum dish, moisten with water and

add 1 ml concentrated HNO₃, 10 ml H₂SO₄ (1:1) and 5 ml HF.* Heat
on a sand bath of moderate temperature, stirring frequently with a short
platinum rod, until copious fumes of SO₃ are evolved. Cool, rinse down
the walls of the dish with 2–3 ml H₂O and examine the contents for undecomposed sample. If gritty material is present, add 2–3 ml HF and
repeat the heating to fumes of SO₃. When the dissolution of the sample
appears to be complete, rinse down the walls of the dish with 2–3 ml
H₂O and evaporate to SO₃ fumes; cool, again rinse down the walls and
repeat the evaporation to fumes. Cool, add 10 ml H₂O and 2 ml 12M
HCl, and heat on a water bath for 15 min to dissolve any insoluble calcium sulfate that may have formed. There may be left at this point a
very small residue of undecomposed material in the dish; if it is white it
may safely be ignored but if it contains black grains it should be filtered,
washed, ignited, fused with 0.5–1 g of Na₂CO₃ and the cake dissolved
in dilute HCl and added to the main solution.

Transfer the solution to a 250 ml beaker and dilute to about 100 ml.
Dissolve 2–3 g of tartaric acid in the solution.† Neutralize the solution
carefully (litmus paper) with aqueous ammonia and then make it slightly
acid with dilute (1:1) HCl or H₂SO₄. Add 10 ml of 1% dimethylglyoxime solution, heat to near boiling point, and add dilute (1:1) aqueous
ammonia dropwise until an excess of a few drops is present. Allow the
precipitate to settle and test the solution for completeness of precipitation by adding a few drops of the reagent to the supernatant liquid.
Allow to stand overnight.

Filter through a fritted glass crucible of medium porosity (do not apply
suction until a layer of precipitate has formed on top of the filter pad,
and then apply it cautiously at first) and wash with cold water.

Dissolve the precipitate in a small amount of hot, dilute (1:1) nitric
acid and collect the filtrate and washings in a 150 ml beaker. Add 2 ml

* The sample may also be decomposed by fusion with Na₂CO₃ + Na₂O₂ (Sec.
9.34.1) and the nickel determined in the insoluble residue remaining after filtration
of the alkaline leach solution of the fusion cake; this applies also to the ignited
R₂O₃ residue (see Sec. 9.34.1, footnote†) but this latter will usually be put into
solution by potassium pyrosulfate fusion prior to the colorimetric determination of
titanium and an aliquot of this solution, following expulsion of hydrogen peroxide,
may be used for the determination of nickel.

† Fe, Cr, Al and small amounts of manganese do not interfere when the precipitation is made from ammoniacal tartrate solution. Cu(II) will interfere and
must be reduced to Cu(I) with Na₂SO₃. Cobalt interferes in that it consumes
reagent and additional reagent must be added as compensation or the Co must be
complexed with cyanide. In most rocks, however, the interference of Co will be
negligible.

of H_2SO_4 (1:1) and evaporate as far as possible on a water bath, then heat to SO_3 fumes.   Cool, rinse down the walls of the beaker and repeat the evaporation to fumes of SO_3 to ensure removal of all HNO_3.   Cool and dilute to 25–30 ml (filter if necessary).

Add a few grains of tartaric acid, neutralize the solution with dilute (1:1) aqueous ammonia and then make just acid with dilute (1:1) H_2SO_4 or HCl.   Add 10 ml of the alcoholic dimethylglyoxime reagent, heat to near boiling and add dilute aqueous ammonia until the solution is just alkaline (litmus paper or methyl red indicator), then add 2–3 drops in excess.   Allow to stand cold for 1–2 hr and then filter through a weighed, fritted-glass crucible and wash with cold water as before.   Dry at 110–120°C for 1 hr (or at 150°C if some coprecipitation of the reagent has occurred) to constant weight.

$$\% \text{ NiO} = \text{wt red ppt (g)} \times 0.2585 \times \frac{100}{\text{sample wt (g)}}$$

Conversion Factors

$$\text{Ni}(C_4H_7N_2O_2)_2 \underset{4.9227}{\overset{0.2031}{\rightleftharpoons}} \text{Ni}$$

$$\text{Ni}(C_4H_7N_2O_2)_2 \underset{3.8682}{\overset{0.2585}{\rightleftharpoons}} \text{NiO}$$

$$\text{NiO} \underset{1.2726}{\overset{0.7858}{\rightleftharpoons}} \text{Ni}$$

9.37. Colorimetric determination of nickel with dimethylglyoxime

The following method is that given by E. B. Sandell (Sec. 6.18, Ref. 9).

Reagents

Dimethylglyoxime solution (1% in ethanol): prepare as given in Sec. 9.36.

Sodium citrate (10%): dissolve 50 g $Na_3C_6H_5O_7 \cdot 5H_2O$ in 500 ml water.

Chloroform: reagent grade.

Bromine water: saturated.

Standard nickel solution (0.10 mg/ml): dissolve 0.1000 g pure nickel metal in 20 ml 1:1 HCl (or use 0.405 g uneffloresced crystals of $NiCl_2 \cdot 6H_2O$) and dilute to 1 liter in a volumetric flask (0.1N in HCl).   For preparation of calibration curve dilute 10.0 ml of this solution, plus 1 ml 12N HCl, to 100 ml (10 γ Ni/ml).

Procedure

Weigh and quantitatively transfer 0.500 g of 100 mesh sample (containing not more than 0.1% Ni) to a 50 ml platinum dish, moisten with 1 ml H_2O and add 0.5 ml concentrated HNO_3, 3 ml 1:1 H_2SO_4 and 5 ml HF. Decompose the sample and expel excess HF and HNO_3 by heating to fumes of SO_3 as described in Sec. 9.36. Cool, add 6 ml dilute HCl (1:5) and heat the dish on the water bath for 15 min, or until all soluble material has dissolved. Add 10 ml of sodium citrate solution (10%), heat for 10 min on the water bath, then allow to cool. Neutralize the solution (use litmus paper) with aqueous ammonia (1:1), add 1 ml in excess, and then filter any insoluble material through a 5.5 cm No. 40 Whatman paper, catching the filtrate in a 50 ml volumetric flask. Wash the residue and paper with ammoniacal sodium citrate solution (1%), and dilute the filtrate to volume.*

Transfer a 25 ml aliquot of the above solution for an acid rock (or a 10 ml aliquot for a basic rock) to a small separatory funnel and add 3 ml of the 1% alcoholic dimethylglyoxime reagent. Extract the red nickel compound with 3 portions of 3–4 ml each of chloroform, shaking each portion vigorously for 30 sec. Shake the combined extracts with 5 ml aqueous ammonia (1:50) for 15 sec and draw off the organic layer without removing any of the aqueous phase with it. Shake the aqueous layer with 1–2 ml of chloroform to recover suspended drops of $CHCl_3$ and combine the organic portion with the previous one.

Shake the combined organic extracts vigorously for 1 min with two 5 ml portions of 0.5M HCl (for final dilution to 25 ml; if final dilution is to be 10 ml, use smaller portions) and transfer the aqueous layer to a 25 ml volumetric flask. Dilute to approximately 20 ml and add 1 ml saturated bromine water and 2 ml of concentrated aqueous ammonia. Cool to less than 30°C, if necessary, add 1 ml of 1% alcoholic dimethylglyoxime solution, dilute to volume, stand for 5 min and measure the absorbance of the brownish complex at 445 mµ. Obtain the concentration of nickel in the sample by reference to a working curve prepared from aliquots of the standard nickel solution (10 γ Ni/ml) to cover the range 0–5 ppm Ni;

* The residue, if any, will most likely be a small amount of $CaSO_4$. Return the residue to a small beaker and heat it with 0.25 ml HCl, 1 ml 10% sodium citrate and 25 ml H_2O until a clear solution is obtained. Cool, make just ammoniacal, dilute to 50 ml in a volumetric flask and reserve for a separate extraction of nickel from a 10 ml aliquot, combining the organic extracts with those from the main portion. If the residue refuses to dissolve it may be fused with a small amount of Na_2CO_3 and the cake dissolved in HCl and 1 ml of 10% sodium citrate as described above.

the acidity of the nickel solutions used to establish the curve should be the same as that of the sample solution ($0.25N$) when diluted to 20 ml prior to the addition of bromine, aqueous ammonia and reagent.

9.38. Colorimetric determination of cobalt

9.38.1. WITH NITROSO-R SALT, FOLLOWING DITHIZONE EXTRACTION OF COBALT

This procedure is described by E. B. Sandell (Sec. 6.18, Ref. 9) and utilizes an aliquot of the sample solution prepared for the colorimetric determination of nickel (Sec. 9.37).

Reagents

Nitroso-R salt solution (0.2%): dissolve 0.2 g of reagent in water and store in a glass-stoppered bottle; the solution is stable for several months if kept in the dark.

Dithizone (0.01% in carbon tetrachloride): dissolve 0.01 g dithizone in 100 ml of reagent grade carbon tetrachloride. Store in a glass-stoppered bottle in a cool dark place.

Standard cobalt solution (0.01 mg/ml): dissolve 0.4037 g $CoCl_2 \cdot 6H_2O$ (analyzed reagent) in water, add 1 ml $12N$ HCl, and dilute to 1 liter in a volumetric flask. For preparation of the calibration curve dilute 10.0 ml of this solution, plus 2–3 drops of $12N$ HCl, to 100 ml (10 γ Co/ml).

Procedure

Treat 0.500 g of 100 mesh sample (containing 10–100 ppm Co) as described in Sec. 9.37 (including footnote), finishing with dilution of filtrate and washings to volume in a 50 ml volumetric flask.

Transfer a 10 ml aliquot (use larger aliquot for Co < 10 ppm in original sample) of the above solution to a small separatory funnel, add 0.2 ml aqueous ammonia (1:1) and shake vigorously for 1 min with 5 ml of 0.01% dithizone in carbon tetrachloride. Draw off the CCl_4 layer into a second separatory funnel. Shake the aqueous phase with 2–3 ml of dithizone; draw off the CCl_4 layer into the second funnel and repeat the extraction with additional portions of dithizone, if necessary, until the last portion shows only a green or brownish-green color after shaking for 1 min. Extract a 10 ml aliquot of the solution of any insoluble residue found (Sec. 9.37, and footnote) in the same fashion.

Wash the combined organic extracts with 5 ml of water and draw off the CCl_4 layer into a small (preferably 25 ml) erlemeyer flask. Evaporate the CCl_4, add 0.25 ml $18M$ H_2SO_4 and 0.5 ml 72% $HClO_4$, and heat at about 200°C until the liquid is colorless. Fume off the H_2SO_4 but do not overheat the residue. Cool and dissolve the latter in 0.25 ml HCl (1:10) and 5 ml water, heating to boiling if necessary. Add exactly 0.5 ml of the 0.2% Nitroso-R solution and 1.0 g of sodium acetate ($3H_2O$); the pH should be close to 5.5 at this point. Boil for 1 min, add 1.0 ml of 15M HNO_3 and again boil for 1 min. Cool to room temperature in the dark, transfer to a 10 ml volumetric flask and dilute to volume. Measure the absorbance at 500 mμ and obtain the concentration of cobalt by reference to a working curve prepared from a series of aliquots of the standard cobalt solution (10 γ Co per ml) covering the range 0–1 ppm Co. A blank must be carried throughout the whole procedure.

Conversion Factors

$$CoO \underset{1.2715}{\overset{0.7865}{\rightleftharpoons}} Co$$

9.38.2. WITH 2-NITROSO-1-NAPHTHOL, FOLLOWING EXTRACTION INTO ISOAMYL ACETATE

This method has been proposed by L. J. Clark (Sec. 6.18, Ref. 5). It requires no preliminary separation of interfering metals and the transmittance of the isoamyl acetate extract of the red cobalt complex is measured directly.

Reagents

Water: redistill in an all-borosilicate glass apparatus.

Phenolphthalein indicator solution (1.0%): dissolve 1.0 g indicator powder in 100 ml 95% ethyl alcohol.

Ammonium citrate solution (20%): dissolve 100 g diammonium citrate, $(NH_4)_2HC_6H_5O_7$, in 350 ml H_2O. Add 5 drops 1% alcoholic solution of phenolphthalein, and then add aqueous ammonia until the solution is pink in color. Dilute to 500 ml.

Isoamyl acetate: reagent grade, boiling range 137–142°C.

2-Nitroso-1-naphthol solution (0.04%): dissolve 0.04 g of reagent in 8 drops of 1N sodium hydroxide and 1 ml water, and dilute to 100 ml with water.

Sodium thiosulfate solution (10%): dissolve 10 g of $Na_2S_2O_3 \cdot 5H_2O$ in water and dilute to 100 ml.

Standard cobalt solution (1 γ Co per ml): prepare a solution containing 0.10 mg Co per ml as described in Sec. 9.38.1 (*Reagents*); dilute 10 ml of this solution to 1000 ml.

Procedure

Decompose a suitably sized sample by treatment with acids (as in Sec. 9.36) or by fusion with $Na_2CO_3 + Na_2O_2$ and separation of silica by dehydration (Sec. 9.2.1), such that the final solution is $0.7N$ in hydrochloric acid and a 10 ml aliquot is equivalent to 0.2 g of sample containing 0–10 γ cobalt.

Transfer a 10 ml aliquot of the final solution to a 125 ml separatory funnel and oxidize any ferrous iron present by adding 0.5 ml saturated bromine water. Add 10 ml of ammonium citrate solution (20%), 1 ml of sodium thiosulfate solution (to reduce $Mn(II)$ and/or $Mn(VII)$), and 1 drop of phenolphthalein indicator, then add aqueous ammonia until a distinct pink color appears.

Pipet 2 ml of 2-nitroso-1-naphthol solution (0.04%) into the separatory funnel. Add 5 ml isoamyl acetate, shake the mixture for 1 min and then let stand for 1 hr to allow the layers to clear (the acetate (upper) layer is usually amber to red, the aqueous (lower) layer is yellow). Draw off and discard the aqueous layer.

Wash the acetate layer in four steps to eliminate interference of Cu and Ni: (*1*) add 5 ml $1N$ HCl, shake for 0.5 min and let stand 2 min; repeat the shaking operation three times, then discard the aqueous layer; (*2*) add 5 ml of $1N$ NaOH, shake in 2 intervals as above and discard the aqueous layer; (*3*) repeat the alkaline washing with a second 5 ml portion of $1N$ NaOH; (*4*) wash the acetate layer with 5 ml of $1N$ HCl in two shakings, and discard the aqueous layer.

Filter the organic layer through a pledget of clinical absorbent cotton into a cuvette (12 $\times$ 75 mm) and measure the transmittance at 530 mμ, using isoamyl acetate as the reference liquid. Obtain the cobalt concentration of the solution from a calibration curve prepared from a series of aliquots of the standard cobalt solution (1 γ Co per ml) over the range 0–10 γ cobalt.

9.39. Colorimetric determination of vanadium as phosphotungstovanadic acid

Reagents

Sodium tungstate solution ($0.5M$): dissolve 16.5 g $Na_2WO_4 \cdot 2H_2O$ in 100 ml of water.

Standard vanadium solution (1.00 mg V per ml): dissolve 1.785 g pure V_2O_5 (previously ignited at 500°C) in a slight excess of sodium hydroxide, make just slightly acid with sulfuric acid and dilute to 1 liter. If desired, the solution may be prepared by dissolving 2.296 g of reagent grade ammonium metavanadate, NH_4VO_3, in water. The vanadium concentration may be checked by reducing the vanadium to V(IV) with SO_2 and titrating it with $KMnO_4$ as follows: to a suitable aliquot add sufficient dilute H_2SO_4 to give an acid concentration of 2:98, heat to boiling and add dropwise a solution of $KMnO_4$ until the H_2SO_4 solution is pink; pass a rapid stream of SO_2 through the solution for 5-10 min and follow this with a rapid stream of carbon dioxide to expel excess SO_2 (test for completeness by bubbling the stream through a dilute acid solution of $KMnO_4$); cool to 60–80°C and titrate with $0.1N$ $KMnO_4$ (Sec. 9.11.1); 1 ml $0.1N$ $KMnO_4$ $\equiv$ 0.0051 g V.

Dilute an aliquot of the above standard solution to give 0.01 mg V per ml and use suitable aliquots of this solution to prepare a working curve, carrying them through the extraction procedure.

Procedure

Take a suitable aliquot of the sodium carbonate filtrate obtained in Sec. 9.34.1 for the colorimetric determination of chromium as chromate, or by s-diphenylcarbazide (Sec. 9.34.2), equivalent to 0.1–0.25 g of sample (containing 10–250 γ V) and extract the vanadium with a 2.5% acetic acid (1:8) solution of 8-hydroxyquinoline and chloroform as described in Sec. 9.34.2, transferring the organic extracts to a 25 ml platinum crucible.*

Add 0.1 g anhydrous Na_2CO_3 to the crucible, evaporate the chloroform at low temperature, and then heat more strongly to destroy organic matter. Finally fuse the contents of the crucible and heat for 1–2 min to ensure conversion of all vanadium to vanadate. Dissolve the melt in about 5 ml of water in a small beaker and filter, if necessary, through a sintered glass crucible. Transfer the solution to a 10 ml volumetric flask or, if a visual comparison or colorimetric titration is to be made, to a 25 ml Nessler tube.

Add 1.0 ml of $4N$ H_2SO_4, 1.0 ml of H_3PO_4 (1:2) and 0.5 ml of $0.5M$ Na_2WO_4 solution. Mix, and if a volumetric flask is being used, dilute to volume. Measure the absorbance at 400 mμ against a blank solution in the reference cell or, if visual comparison is to be made, compare with standards that have been carried through the extraction procedure.

---

* If chromium is absent, and the vanadium content is not too small, the extraction may be omitted, although the results tend to be 5–10% low (Sec. 6.19, Ref. 4). The solution should be heated to boiling after the addition of the reagents and cooled before final dilution to volume.

Report as V_2O_3 (but correct the Al_2O_3 value obtained from the R_2O_3 group for an equivalent amount of V_2O_5).

Conversion Factors

$$V_2O_3 \underset{1.4711}{\overset{0.6797}{\rightleftharpoons}} V$$

$$V_2O_5 \underset{1.7851}{\overset{0.5602}{\rightleftharpoons}} V$$

9.40. Titrimetric determination of vanadium

9.40.1. WITH POTASSIUM PERMANGANATE

Reagents

Phenanthroline-ferrous sulfate ($0.025M$): dissolve 1.485 g of phenanthroline monohydrate in 100 ml of $0.025M$ $FeSO_4$ solution.

Procedure

If the vanadium is obtained in a solution free from iron and containing only a small amount of chromium, the titration may be made following SO_2 reduction as described in Sec. 9.39 (*Reagents*); because chromium is slowly oxidized by $KMnO_4$ the solution should be cooled to room temperature before the titration is made, and the final end point should not fade for at least 1 min with continuous stirring.

If iron is present, advantage may be taken of the oxidation of Fe(II), but not of V(IV), by ammonium persulfate (see Sec. 6.19). Chromium (III) is without effect and the determination of vanadium may follow that of chromium in the same solution (Sec. 9.35): make the solution 10% in H_2SO_4, decolorize any Fe(III) present by adding 5 ml dilute (1:1) H_3PO_4, add 1 drop of $0.025M$ phenanthroline-ferrous sulfate, and then add $0.1N$ ferrous ammonium sulfate solution (Sec. 9.35, *Reagents*) until the color of the indicator changes sharply from blue to brownish green, and add 3–5 ml in excess; add 8 ml of freshly prepared 15% ammonium persulfate solution, stir for 1 min, then titrate with $0.1N$ $KMnO_4$ to an end point that persists for at least 1 min with vigorous stirring.

$$1 \text{ ml } 0.1N \text{ } KMnO_4 \equiv 0.0051 \text{ g V}$$

9.40.2. WITH FERROUS AMMONIUM SULFATE

Procedure

To the solution obtained from the titrimetric determination of chromium with $KMnO_4$ (Sec. 9.35) add $0.5M$ sodium nitrite dropwise until the pink color is discharged.   Immediately add 5 g of urea and 100 ml $10M$ H_2SO_4, and dilute to 200 ml.   Add 1 drop of $0.025M$ phenanthroline ferrous sulfate and titrate with $0.1N$ ferrous ammonium sulfate until the blue color of the indicator changes sharply to brownish green.   The titration may also be made potentiometrically.

1 ml $0.1N$ Ferrous ammonium sulfate $\equiv$ 0.0051 g V

9.41. Gravimetric determination of zirconium (and hafnium)

9.41.1. WITH DIAMMONIUM PHOSPHATE

This procedure is suitable for the determination of as much as 50 mg of zirconium (preferably <15 mg).   Fusion with sodium carbonate serves to separate the zirconium from chromium, vanadium, aluminum and silicon and electrolysis of the solution of the residue at a mercury cathode is a convenient way of eliminating excessive amounts of iron which may be present (see Ref. 3, Sec. 6.19).   Hydrogen peroxide will prevent the interference of titanium in moderate amount (<10 mg) but niobium and tantalum will coprecipitate and must first be separated if they are present in significant quantities, either by precipitating them with tannin from an oxalate solution while zirconium remains in the filtrate (Ref. 12, Sec. 6.20), or by separating the zirconium by solvent extraction (Ref. 16, Sec. 6.20).   For most rocks and minerals the only interfering elements will be iron(III) and titanium (i.e., in the R_2O_3 group residue), but a preliminary spectrographic analysis is a valuable source of such information.

Procedure

For most rocks it will be necessary to use a 1 or 2 g portion.   Fuse the sample strongly with Na_2CO_3 and Na_2O_2 in a platinum crucible, leach in water and filter, as given in Sec. 9.34.1 (or use the water-insoluble residue from the separation of chromium.)   Wash the residue from the filter paper into the beaker in which the original leaching was done and reserve the paper.   Add 2 ml $12N$ HCl to the beaker and heat it until all soluble residue is dissolved.   Filter through a small Whatman No. 40 filter paper into a 150 ml beaker (the filtrate should be clear).   Reserve the paper.   Dilute the filtrate according to the estimated amount of zirconium present, i.e., from about 30 ml for 1 mg to 100 ml for 50 mg. Add sufficient $12N$ HCl to make the final solution about $1N$ in acid, then add 5 ml 3% H_2O_2.   Warm the solution to about 50°C, add 3 g diammonium phosphate $((NH_4)_2HPO_4)$ dissolved in 15 ml warm water and filtered, stir and digest at 40–50°C for not less than 2 hr, adding more H_2O_2 if the yellow color (Ti complex) fades because of decomposition of the peroxide. If appreciable precipitate has formed after 2 hr, stir in a small amount of filter paper pulp and allow to settle. If there is no precipitate or only a very slight one, let the solution stand cold overnight, or longer if necessary.

Ignite the two reserved papers in the crucible used for the fusion and fuse well with a small quantity of Na_2CO_3. Leach the cake in a few ml of water, filter through a small Whatman No. 42 filter paper and wash the residue with 1% sodium carbonate solution. Discard the filtrate. Return the paper and residue to the platinum crucible, ignite as usual and fuse the residue with a small amount of $K_2S_2O_7$. Transfer the cake to a 100 ml beaker, add 3–4 ml concentrated H_2SO_4, dilute to about 40 ml and warm to complete solution (if a residue remains ($CaSO_4$?), filter and discard it). Continue with the precipitation of the zirconium as described before, beginning with the addition of H_2O_2.

Filter both precipitates through a single small Whatman No. 42 filter paper and wash thoroughly with 100–300 ml of cool 5% ammonium nitrate solution. The filtrate will contain any rare earth elements that were present (see Sec. 9.42).

Transfer the paper and precipitate to a weighed 25 ml platinum crucible, dry and slowly ignite the paper and residue, observing the precautions given for the ignition of $Mg_2P_2O_7$ (Sec. 9.12.1), with final ignition at 800–1000°C. Weigh as ZrP_2O_7 and repeat the ignition to constant weight.*

$$ZrO_2 = ZrP_2O_7 \text{ (g)} \times 0.4647 \times \frac{100}{\text{sample wt (g)}}\dagger$$

Conversion Factors

$$ZrO_2 \underset{1.3508}{\overset{0.7403}{\rightleftarrows}} Zr$$

$$HfO_2 \underset{1.1792}{\overset{0.8481}{\rightleftarrows}} Hf$$

* If it is suspected that the ZrP_2O_7 is not pure, fuse it with a little Na_2CO_3, followed by leaching, filtering and fusion of the precipitate with $K_2S_2O_7$ and reprecipitation from H_2SO_4 solution as described previously for the residue from the original Na_2CO_3 fusion.

† Because of possible loss of phosphate as a result of partial hydrolysis of the precipitate during the washing, an empirical factor of 0.518 for ZrO_2/ZrP_2O_7 has been suggested for the determination of only small amounts of zirconium (Sec. 6.20, Ref. 14).

Hahn (Sec. 6.20, Ref. 8) has suggested that 0.4667 should be used as the factor for ZrO_2/ZrP_2O_7 if one assumes that the hafnium content of the zirconium precipitate is about 2%.

9.41.2. WITH p-BROMOMANDELIC ACID

This method, for a small amount of zirconium (0.1 mg or more), has been proposed by Rafiq, Rulfs and Elving (Sec. 6.20, Ref. 10).

Reagents

p-bromomandelic acid solution (1.5%): dissolve 1.5 g of reagent in warm water, filter and dilute to 100 ml.

Procedure

The precipitation is made in a solution of the sample from which silicon has been removed and which is about 2.5N in hydrochloric acid. This may be obtained by fusing a 0.5–1 g portion of finely ground sample with Na_2CO_3 and Na_2O_2 (Sec. 9.2.1) or with Na_2CO_3 and H_3BO_3 (Sec. 9.2.3) and removing the silica by dehydration of an HCl solution of the cake (any residue left after volatilization of the silicon must be fused with Na_2CO_3, dissolved in HCl and combined with the main filtrate), or the water-insoluble residue remaining after the separation of chromium as chromate (and silicon and aluminum also) in Sec. 9.34.1 may be dissolved in HCl (any acid-insoluble residue should be fused a second time with a small amount of Na_2CO_3 and the cake dissolved in HCl).

Heat the solution, which should have a volume of about 30 ml and be about 2.5N in HCl, to 50–60°C and add dropwise, with constant stirring, 30–35 ml of 1.5% p-bromomandelic acid. Digest for 15 min at 80–85°C, then let stand at room temperature for 3–4 hr or overnight if the precipitate is small. Filter through a small Whatman No. 40 filter paper and wash with 20–25 ml of hot, distilled water.

Transfer the paper and precipitate to a 25 ml weighed platinum crucible, place it in a cold electric muffle furnace and allow the temperature to rise slowly until the paper has been ashed. Ignite the crucible and contents at 900–1000°C and weigh the residue as ZrO_2. Repeat the ignition and weighing to constant weight.

9.42. Gravimetric determination of total rare earths as oxalates, following hydroxide and fluoride separations

The separation and determination of the rare earths in rocks and most minerals is conveniently done subsequent to the separation of zirconium (Sec. 9.41.1) as the phosphate. A sample of 1–2 g is fused with Na_2CO_3, the cake leached in water and filtered and the filtrate used for the determination of chromium (Sec. 9.34.1), vanadium (Sec. 9.40), and total sulfur (Sec. 9.30.1); the water-insoluble residue is used for the determination of barium and strontium (Sec. 9.33), and zirconium (and hafnium). The sulfuric acid solution of the R_2O_3 group, after fusing it with pyrosulfate, can also be used. A smaller sample should be used when necessary, to give a final rare earth concentration of not more than 50 mg.

To the filtrate from the precipitation of zirconium as the phosphate (Sec. 9.41.1), which should have a volume of about 50 ml, add an excess of a concentrated solution of freshly prepared NaOH to precipitate the rare earths together with Fe and Ti; any Si and Al present will be dissolved by the excess base. Filter the solution through a Whatman 41H hardened paper contained in a polyethylene funnel and wash the precipitate and paper several times with hot water. Discard the filtrate and washings.

Dissolve the precipitate in a minimum volume of hot 5% HCl, catching the filtrate in a 50 ml platinum dish, and retain the paper in the funnel. Evaporate to near dryness on a boiling water bath. Dilute to 30 ml with H_2O, add 15 ml 48% HF, stir, and let stand for at least 3 hr, or preferably, overnight. Filter the solution through the first filter paper into a polyethylene beaker and wash the dish, paper and the rare earth fluorides with 20 ml of 5% HF. Discard the filtrate.

Carefully transfer the paper and precipitate to a 25 ml platinum crucible, ash the paper and ignite the residue. Add 1–2 ml of $18M$ H_2SO_4 and cautiously evaporate to dryness to decompose the fluorides.

Dissolve the residue in 1–2 ml dilute HCl (1:1) and transfer the solution to a 100 ml beaker. Precipitate the rare earths by adding an excess of aqueous ammonia and filter through a 7 cm Whatman No. 40 paper. Wash twice with hot water, then dissolve the precipitate in hot dilute HCl (5%), catching the filtrate in a 100 ml beaker.

Dilute the solution to about 50 ml and make it $0.1N$ in HCl. Heat to boiling and add slowly, with rapid stirring, about 25% of its volume of a hot saturated solution of oxalic acid. Boil gently for a few minutes, cool to room temperature and let stand for several hours, preferably overnight. Filter through a small Whatman No. 40 filter paper and wash first with a saturated solution of oxalic acid, then with water. Transfer the paper to a small weighed platinum crucible and ignite at 800–900°C to the oxides. The latter are trivalent (i.e., Ln_2O_3) except for cerium (CeO_2), and praeseodymium and terbium (approximately Pr_6O_{11} and Tb_4O_7).

If a double precipitation is necessary, the precipitate should first be ignited to the oxide before being dissolved in HCl. It is preferable, incidentally, to ignite the oxalates to oxide, rather than to ignite the hydroxides.

9.43. Fluorimetric determination of traces of uranium, using sodium fluoride fusion

This method was developed by Sydney Abbey for the determination of 0–10 ppm uranium in silicate rocks, for which it is required that the entire sample be decomposed. It will be necessary to modify the procedure where the sample is not a silicate rock, where only the HF-soluble or acid-leachable portion of the sample is of interest, or where the expected uranium content is not in the 0–10 ppm range.

Special Equipment*

Fluorimetry dishes: platinum, approximately 18 mm wide and 3 mm deep. Several sets, of 24 dishes each, will be needed. Store under distilled water.

Pipet (for measuring 0.10 ml aliquots): a 100 μl microchemical pipet equipped with syringe attachment.

Aluminum trays: $5\frac{1}{2}$ in. $\times$ $3\frac{1}{2}$ in. $\times$ $\frac{5}{32}$ in., containing 24 holes (6 rows of 4 each) each $\frac{5}{8}$ in. in diameter.

Burner: a Fletcher radial flame burner, modified by the insertion of a loose roll of bronze screen wire (16 mesh) into the burner barrel (to diffuse gas) and the placing of a nichrome wire screen over the burner cap (to support the dishes).

Hood: it is necessary to modify a standard fume hood by lining it with fire-brick and installing a flame baffle, consisting of two sheets of heavy $\frac{1}{4}$ in. mesh wire screening in the upper portion (to protect exhaust fan). The hood window is replaced by a panel of asbestos board having a mica window (8 in. $\times$ 8 in.).

Fluorimeter: Jarrell-Ash Galvanek-Morrison Fluorimeter Mark V or equivalent, equipped for reflectance measurement.

Reagents

Aluminum nitrate salting solution: transfer 5 lb of reagent grade aluminum nitrate, having a low uranium content, to a 4 liter beaker. Add 150 ml H_2O, cover, place the beaker on a combined hot plate and magnetic stirrer, and heat, with stirring, until solution is complete. Adjust the concentration by evaporation or dilution to give a solution which boils at 130°C, and transfer the hot solution to a 3-neck, round-bottom 2 liter flask. Place the flask in a heating mantle controlled by a Variac, insert a reflux condenser in the middle neck, a thermometer in a

---

* The equipment used in the foregoing method has been described in detail by Ingles (Sec. 6.22, Ref. 2) and only a brief description is given here.

ROCK AND MINERAL ANALYSIS

well in one side neck and stopper the other neck (solution is withdrawn from this neck when required). Store the solution at 80°C and heat to 110°C before using.

Aluminum nitrate wash solution: dilute 100 ml of the above solution with 73 ml H₂O and 4 ml concentrated HNO₃.

Standard uranium solutions: dissolve 211 mg UO₂(NO₃)₂·6H₂O in water, add 50 ml concentrated nitric acid and dilute to volume in a 1 liter volumetric flask to give a solution containing *1 mg U/ml*. Dilute 1.0 ml of this solution to 1 liter with dilute HNO₃ (1:19) to give a solution containing *1 ug U/ml*.

Sodium fluoride pellets (98% NaF, 2% LiF): prepared pellets, each containing 0.588 g NaF and 0.012 g LiF, can be obtained. Those supplied by Analoids (Ridsdale and Co., Middlesbrough, England), in 1000-pellet lots, have been found satisfactory.

Procedure

Decomposition

Weigh 1.00 gm of sample into a porcelain crucible and ignite over a meker burner for 15–20 min. Cool, break up with a spatula and repeat ignition for about 10 min. Cool. Transfer to a 100 ml Teflon dish, rinsing the crucible with a little water.

With the first batch of samples, prepare a *blank*, containing 10 ml of HNO₃ (1:19) and a *standard*, containing 10 µg uranium, in separate 100 ml Teflon dishes and process these in the same way as the samples. Aliquots of the final solutions prepared from these may be used with subsequent batches of samples, provided that there is no change in the reagents and procedure used.

To blank, standard and samples add 5 ml concentrated nitric acid and 10 ml HF. Cover and warm on a steam bath for several hours (preferably overnight).

Rinse off and remove the cover. Add a little more HF and evaporate the contents to dryness. Take up with 5 ml HNO₃ and 5–10 ml H₂O, break up the residue with a rubber-tipped rod and again evaporate to dryness. Repeat the dissolution and evaporation twice more.

Add 5 ml nitric acid and about 25 ml water. Break up the residue with a rubber-tipped rod, cover the dish, and warm on a steam bath to dissolve soluble matter.

Filter hot on a 9 cm Whatman No. 42 paper, receiving the filtrate in a 250 ml beaker. Wash with warm nitric acid (1:19).

Evaporate the filtrate to a low volume, transfer to a 30 ml beaker and evaporate to < 10 ml. Set aside for eventual combination with dissolved residue.

Transfer the filter and residue to a 10 ml platinum crucible. Place the loosely covered crucible in a cold muffle furnace and raise the temperature slowly until all paper is burned off.

To the cooled crucible add 0.5 ml H_2SO_4 (1:1) and about 2 ml HF. Evaporate until SO_3 fumes are no longer evolved. Ignite briefly to expel traces of acid. Add 0.5 g sodium carbonate and fuse for 15–20 min on a meker flame. Allow to cool.

Add about 5 ml water and 2–3 drops ethanol to the crucible. Cover and warm gently to dissolve the sodium carbonate. Transfer the suspension to a 50 ml beaker and set aside. Add 1 ml nitric acid and 4 ml water to the crucible, cover and warm to dissolve soluble matter. Cool.

Carefully add the contents of the crucible to the contents of the 50 ml beaker and warm to dissolve soluble matter and expel carbon dioxide. Evaporate to less than 20 ml if necessary. Transfer this mixture to the 30 ml beaker containing the evaporated filtrate.

Evaporate to less than 10 ml, add one drop hydrogen peroxide (30%), cover and warm until the color of the titanium complex is destroyed and no more oxygen is evolved.

Extraction

Add 13 ml of the salting solution and swirl to mix. Pour into a 25 ml graduated cylinder and measure volume (V ml). Pour into a 60 ml separatory funnel, and rinse graduate and beaker into funnel with (23 − V) ml of water. Allow to cool.

Add 20 ml ethyl acetate, close the funnel with a polyethylene stopper, and shake for 45–60 sec. Allow phases to separate.

Remove and discard aqueous phase, leaving any intersurficial turbidity in the funnel. Add 10 ml aluminum nitrate wash, shake and separate as before.

Rinse out the funnel stem with water, discarding the rinsings. Drain the ethyl acetate phase into a 50 ml beaker. Rinse out the funnel with 7 ml H_2O and drain into the same 50 ml beaker.

Allow the ethyl acetate to evaporate at room temperature (preferably overnight). If the volume, after evaporation, is still well over 10 ml, warm gently to evaporate further.*   Cool, and dilute to volume in a 10 ml volumetric flask.

Sodium fluoride fusion

Place 19 platinum fluorimetry dishes on the perforated aluminum tray and dry under an infrared lamp.

For three samples (1,2,3), a standard (S) and blank (B), pipet 0.10 ml aliquots on to the dishes in the following order (0 represents an empty dish):

```
1  2  3  3
S  B  S  2
S  0  1  B
1  3  2  1
S  3  2
```

Evaporate the aliquots to dryness under an infrared lamp, and transfer to the grid of the fusion burner in the following pattern:

```
   1  2  3
3  S  B  S
2  S  0  1  B
1  3  2  1
   S  3  2
```

Place a flux pellet on each dish except that designated as zero.   Ignite the burner and adjust conditions to give fusion within 90 sec, and keep the beads molten for 120 sec longer.   Turn off the burner and spray steam over the cooling beads until they no longer glow red.

Return the dishes to the aluminum tray and measure fluorescence of all beads between 15 and 60 min after the end of the fusion.   Calculate the mean fluorescence reading for each sample, for the standard and for the blank.

$$\text{ppm U in sample} = 10 \cdot (x - B/S - B)$$

where x = mean fluorescence reading for the sample, S = mean fluorescence reading for the standard, B = mean fluorescence reading for the blank.

* With some samples a reaction may take place in which the solution turns a brownish color and nitrogen oxides are evolved. *On rare occasions this reaction can be quite vigorous, to the point of causing an explosion.* The beaker should therefore be removed from the source of heat as soon as the solution begins to darken. The reaction will usually sustain itself without further heating. After it subsides the beaker should be *cautiously* warmed to ensure that the reaction is complete and if necessary, to reduce the volume.

After fluorimetry, rinse the fusion dishes in warm, running water to remove the beads. Boil the dishes for 30 min in fresh concentrated HCl and rinse in cold water. Repeat boiling and rinsing, finally rinsing with distilled water. Return each set of dishes to the appropriate storage bottle of distilled water.

Conversion Factors

$$U_3O_8 \underset{1.1792}{\overset{0.8480}{\rightleftharpoons}} U$$

9.44. Colorimetric determination of traces of thorium with Arsenazo III in perchlorate medium

This method was developed by Sydney Abbey (Sec. 6.22, Ref. 1), for use in the Geological Survey of Canada, and the following outline is taken directly from the above published reference.

Reagents

Sodium hydroxide (500 g/1): dissolve 250 g in about 325 ml of water. Allow to stand overnight. Decant and dilute to 500 ml. Store in polyethylene. For the *alkaline wash solution* dilute 5 ml of this solution to 500 ml.

Methyloxalate solution: heat oxalic acid crystals at 100°C overnight and cool in a desiccator. Break up the crust, heat at the same temperature 1 hr longer and again cool in a desiccator. Dissolve 100 g of the anhydrous oxalic acid in 250 ml of methanol. Allow to stand at least 3 days and filter immediately before using.

Calcium nitrate solution (10 mg CaO per ml): weigh 9 g of calcium carbonate into a large beaker. Cover with water and dissolve by adding a slight excess of nitric acid. Evaporate to dryness. Take up with a small volume of water and again evaporate to dryness. Dissolve in about 450 ml of water and adjust the pH to the green color of bromophenol blue indicator. Dilute to 500 ml.

Oxalic acid complexing solution: dissolve 40 g of oxalic acid crystals in about 40 ml of hot water. Filter and dilute to 500 ml. For the oxalic acid wash solution, dilute 60 ml of this solution to 480 ml.

Perchloric acid (4:1): mix 400 ml of perchloric acid (70%) with 100 ml of water.

Arsenazo III solution: dissolve 50 mg of arsenazo III in about 90 ml of water and dilute to 100 ml.

Standard thorium solution (5 μg Th per ml): prepare a concentrated solution, containing about 100 μg of thorium per ml, by dissolving thorium nitrate in water, adding sufficient nitric acid to give a final normality of $1N$. Standardize gravimetrically by evaporating aliquots to dryness and carefully igniting to oxide. Dilute an aliquot to give a working solution of 5 μg of thorium per ml, in $1N$ nitric acid.

Procedure

Samples should be finely ground, preferably to -200 mesh. Up to 4 samples can be conveniently handled in one batch.

For *silicate samples*, weigh 0.5 g of sample into a porcelain crucible. Ignite over a meker burner for 15 min, allow to cool, break up lumps with a small spatula and ignite again for 10 min. Allow to cool.

Set up 2 blanks by weighing 5 g of sodium peroxide into each of two 50 ml iron crucibles. Mix each sample with 4 g of sodium peroxide in the porcelain crucible. Line the bottom of a 50 ml iron crucible with about 0.5 g of sodium peroxide, and add the sample-peroxide mixture. "Rinse" the porcelain crucible with about 0.5 g of sodium peroxide, and use the "rinsings" to cover the charge in the iron crucible. Cover all of the iron crucibles and heat in a muffle furnace at 460°C for 1 hr. Remove from the heat and allow to cool to room temperature.

Remove the crucible covers, wipe off the outside of each crucible with clean tissue, and place in a 250 ml beaker. Add 50 ml of water to the crucible and cover the beaker immediately with a watch glass. When the reaction subsides, rinse down the material spattered on the watch glass and the walls of the beaker. Rinse off and remove the crucible. Pipet 5 ml of standard thorium solution into one of the blanks, which then serves as a check standard. Dilute all of the suspensions to about 125 ml and boil for 15 min.

Filter on a Whatman No. 42, 9 cm filter paper and wash with hot alkaline wash solution. Discard the filtrate. Place the original beaker under the filter funnel, and dissolve the residue on the paper with four 5 ml portions of hot $2N$ nitric acid, followed by 3 similar portions of hot $2N$ hydrochloric acid, and 2 more of hot $2N$ nitric acid. Finally, wash the paper 3 or 4 times with water.

If any silica remains on the paper, ignite it in a small platinum crucible, treat with a few drops of perchloric acid and a few ml of hydrofluoric acid and evaporate three times to perchloric fumes. Take up with a little water and combine with the main solution in the 250 ml beaker. (If any insoluble residue remains at this point, it should be separated and sintered with sodium peroxide.) The solution is now ready for the oxalate precipitation.

If the sample is a *carbonate*, raise the temperature slowly in the initial ignition. After the second ignition, rinse the cooled sample into a 150 ml beaker with a little water. Add 5 ml of hydrochloric acid, cover and warm to dissolve, adding a little nitric acid if necessary. Evaporate to dryness and bake to separate silica as usual. Take up the dry residue with hot hydrochloric acid (1:1), warm to dissolve soluble matter, and filter into a 250 ml beaker. Wash with hot hydrochloric acid (1:10). Ignite the silica and treat with perchloric and hydrofluoric acids as described above. While the silica is being treated separately, evaporate the filtrate to dryness. Add 10 ml of nitric acid, boil, and again take to dryness. Add 5 ml of nitric acid and a little water, and warm to dissolve. The solution is now ready for oxalate precipitation.

Add sufficient calcium nitrate solution to bring the CaO content of the solution to about 150 mg, 10 ml of hydrogen peroxide (30%) and sufficient water to bring the volume to 125 ml. Cover the beaker and heat on a hot plate set at low heat. Remove the cover and add sodium hydroxide solution (500 g/1) dropwise from a polypropylene pipet, with stirring, until the pH is just greater than 3.8 (this can be checked with short-range pH paper; it comes quite close to the point where hydrous ferric oxide begins to precipitate). Add 15 ml of freshly filtered methyl oxalate solution, stir and allow to digest uncovered on the low temperature hot plate for 30 min. Remove from the heat, add a little paper pulp, and again add the concentrated sodium hydroxide solution to pH 2. Allow to stand 1 hr longer. Add 5 ml of the calcium nitrate solution, dropwise, with stirring, and allow the solution to stand for another hour. Filter on a Whatman No. 42, 9 cm filter paper, covered with a thin layer of filter paper pulp. Wash with cold oxalic acid wash solution.

Place the filter and precipitate in the original beaker and add 25 ml of nitric acid. Break up the paper, rinse down the walls of the beaker, and rinse off and remove the stirring rod. Cover the beaker with a watch glass and digest on a hot plate set at low heat until a clear solution is obtained (preferably overnight).

(If there is any reason to suspect the presence of more than 1% TiO_2 in the original sample, the solution must be evaporated to dryness and taken up with dilute nitric acid. The entire oxalate precipitation procedure is then repeated, but no calcium is added.)

Rinse off and remove the watch glass, add 5 ml of perchloric acid, cover with a ribbed watch glass and heat on a sand bath until perchloric fumes are evolved. Allow to cool, rinse down the watch glass and the walls of the beaker, and again heat on the sand bath, this time until all perchloric fumes are expelled. Allow to cool, add 1 ml of hydrochloric acid, rinse down the watch glass and the walls of the beaker, and evaporate to dryness on the sand bath. Allow to cool, add 1 ml of hydrochloric acid and 5 ml of water. Swirl to dissolve. Add a few crystals of ascorbic acid and swirl to decolorize. Pour into a dry 25 ml volumetric flask. Rinse the beaker with 10 ml of perchloric acid (4:1) and pour the rinsings into the volumetric flask. Add 5 ml of oxalic acid complexing solution directly to the flask, and swirl to mix. Rinse the beaker with 2 ml of water and add the rinsings to the flask. Add 1 ml of arsenazo III solution and dilute to volume. Mix well before measuring the absorbance in a 1 cm absorption cell at 660 mμ.

Calibration

Into six 150 ml beakers, measure 0,1,2,3,4,5 ml of standard thorium solution (i.e., 0–25 μg of thorium). Add 20 ml of calcium nitrate solution and 5 ml of perchloric acid to each. Cover with ribbed watch glasses and continue as in the last paragraph above. Plot absorbance against μg of thorium.

Conversion Factors

$$\text{ThO}_2 \underset{1.1379}{\overset{0.8788}{\rightleftarrows}} \text{Th}$$

9.45. Colorimetric determination of traces of lead with dithizone

The following method is that given by Sandell (Sec. 6.24, Ref. 4, pp. 572–575) in which the lead is first isolated by extraction as the dithizonate with carbon tetrachloride from a weakly basic solution of the sample, returned to an aqueous solution by shaking the organic extracts with dilute hydrochloric acid, and finally determined by a monocolor standard series procedure, again as the dithizonate after re-extraction with carbon tetrachloride.

Reagents

Perchloric acid (72%): a redistilled grade should be used, unless the possible presence of a few ppm of Pb is of no consequence.

Hydrofluoric acid (48%): as for perchloric acid.

Hydrochloric acid: as for perchloric acid; prepare as 1:1 and 1:500 solutions.

Dithizone in carbon tetrachloride (0.01% (w/v)): dissolve 0.01 g reagent in 100 ml CCl$_4$ and store in a dark, glass-stoppered bottle; the presence of a small amount of diazone (yellow oxidation product) is usually of no consequence; shake a 0.01% solution of the reagent in CCl$_4$ with 1:100 metal-free aqueous ammonia—no more than a faint yellow color in the organic layer indicates that the reagent may be used without further purification, unless only a minute trace of lead is being sought (see Sandell, Ref. 4, pp. 170–171 for purification step).

Dithizone in carbon tetrachloride (0.001% (w/v)): prepare shortly before use by dilution of the previous 0.01% solution. This solution will usually show a progressive loss in strength (if the 0.01% solution is kept cool and in the dark it will show little change over a considerable period of time) and its strength should be checked periodically.

Sodium or ammonium citrate solution (10%): dissolve 10 g of analytical reagent in 100 ml metal-free water; make the solution slightly ammoniacal (1:200 in aqueous ammonia) and remove lead and other reacting metals by extracting the solution with successive portions of 0.01% dithizone solution until the organic layer is colorless, or shows only a slight tinge of pink.

Potassium cyanide solution (5%): dissolve 5 g of analytical reagent in 100 ml metal-free water; no pink color should be observed in the organic layer when 1 ml is diluted with 2 ml metal-free water and shaken with 1–2 ml of dithizone solution, 0.001%.

Hydroxylamine hydrochloride solution (20%): dissolve 20 g of reagent in 100 ml metal-free water; if necessary, lead can be removed by dithizone extraction after adjusting the pH to 9–10 with aqueous ammonia.

Procedure

Decomposition of sample

Weigh 0.50 g of 100 mesh sample into a 50 ml platinum dish and add 2–3 ml H$_2$O, 1.0 ml HClO$_4$ and 5–6 ml HF. Evaporate to dryness, with occasional stirring; add 4 ml HF, repeat the evaporation and then heat at a higher temperature to drive off the perchloric acid. Cool, add a few ml H$_2$O and 1 ml HClO$_4$, and again evaporate to dryness and expel the excess HClO$_4$. Cool, add 2 ml 1:1 HCl, 5 ml H$_2$O and warm, if necessary, to bring all soluble material into solution. Add 5 ml of the 10% citrate solution, cool and add aqueous ammonia until the solution is basic to litmus paper, and then 0.25–0.3 ml in excess. Let stand if necessary, and then filter through a 5.5 cm Whatman No. 40 paper into a small beaker, transfer any residue from the dish to the paper, and wash the paper and residue 3 or 4 times with 1 ml portions of water containing a drop of aqueous ammonia and citrate solution. Reserve the filtrate (I).

Ignite the paper and contents in a 25 ml platinum crucible, at a low temperature.   Add 0.1–0.2 g Na_2CO_3 to the residue and fuse as usual. Add a few ml H_2O to the cooled melt, warm on a steam bath, and filter through a 5.5 cm paper as before, keeping as much as possible of the residue in the crucible.   Wash paper and residue with about 10 ml H_2O, and reserve the filtrate (II).   Ignite the paper and residue in the platinum crucible at a low temperature, cool, add 1 ml HF and 2–3 drops of $HClO_4$ and evaporate to dryness to expel the $HClO_4$.   Add a few drops of water, 1–2 drops of $HClO_4$ and again heat to expel the excess acid. Add 1 ml HCl (1:1), 2–3 ml H_2O, and warm to dissolve the residue (any insoluble material should be filtered, ignited and re-fused with Na_2CO_3). Add 1 ml citrate solution, neutralize with aqueous ammonia and add 3–4 drops in excess (III).

First extraction

Transfer solution I to a small separatory funnel, add 1 ml each of potassium cyanide and hydroxylamine hydrochloride solutions, 5 ml of 0.01% dithizone solution and shake for 30 sec.   If the dithizone is green, shake for another 30 sec.   Allow the phase to separate and draw off the organic layer into another separatory funnel.   Unless the CCl_4 layer is distinctly green, continue the extraction with successive portions of dithizone until the last portion remains green after shaking for 30 sec.

Repeat the above extraction with solution III, using 1–2 ml portions of dithizone, and combine the extracts with those from solution I.   Extract solution II with 2 ml portions of dithizone solution after adding 0.5 ml citrate solution and dilute HCl until an intermediate greenish color is obtained with thymol blue indicator.   Combine the extracts with those obtained previously and wash with 2–3 ml of water; separate the phases, shake the wash water with 1 ml dithizone solution and add this to the main CCl_4 extracts.   These latter must be free from droplets of iron-containing solution.

Return to aqueous phase

Shake the combined CCl_4 extracts with 10 ml HCl (1:500) for 1 min. Draw off the CCl_4 into another separatory funnel and shake vigorously with a fresh 10 ml portion of HCl (1:500).   Combine the two acid extracts, add a few drops of CCl_4 and draw off to remove any colored droplets of CCl_4, transfer to a 25 ml volumetric flask and dilute to volume with HCl (1:500).

Transfer a 5 or 10 ml aliquot of the solution to a separatory funnel, add 0.1 ml citrate solution, 0.1 ml aqueous ammonia, 1 ml potassium cyanide solution (2 ml for a 10 ml aliquot), and shake vigorously with 2.0–3.0 ml of 0.001% dithizone solution (added from a buret). Allow the phases to separate and run the organic layer through a plug of glass wool in a small funnel into a 1 cm cell and measure the transmittance at 520 mμ. Obtain the concentration of Pb from a standard curve prepared under the same conditions (0–2 γ Pb) using aliquots of a standard lead solution prepared from lead nitrate (1:100 HNO$_3$). It is imperative that a blank determination be carried through the entire procedure.

Conversion Factors

$$PbO \underset{1.0772}{\overset{0.9283}{\rightleftharpoons}} Pb$$

$$PbO_2 \underset{1.1544}{\overset{0.8662}{\rightleftharpoons}} Pb$$

9.46. Colorimetric determination of molybdenum (thiocyanate–stannous chloride method)

The procedure to be described is that given by Sandell (Sec. 6.24, Ref. 4, pp. 654–655, Procedure A) for use when only molybdenum is to be determined. The molybdenum thiocyanate is extracted with isopropyl ether from the acidified water leach of a sodium carbonate fusion of the sample and the absorbance measured at 460 mμ.

Reagents

Potassium thiocyanate solution (10%): dissolve 10 g of KCNS in 100 ml water.

Stannous chloride solution (10%): dissolve 10 g of SnCl$_2 \cdot$2H$_2$O in 100 ml 1N HCl and filter if necessary. Prepare fresh weekly.

Ferrous ammonium sulfate solution (1%): dissolve 1 g of reagent in 100 ml 0.2 N sulfuric acid.

Isopropyl ether: reagent-grade; shake with stannous chloride solution before use.

Standard molybdenum solution (0.0100% Mo): dissolve 0.075 g of pure MoO$_3$ in a few ml of dilute sodium hydroxide, dilute with water to about 50 ml, make slightly acidic with HCl and dilute to 500 ml in a volumetric flask. Prepare 0.001% standard Mo solution by diluting the 0.01% solution with 0.1N HCl, preferably weekly.

Procedure

Fuse 0.5 g of finely ground sample with 3 g of Na$_2$CO$_3$ in a 25 ml platinum crucible as described in Sec. 9.30.1 and leach the cake in a

150 ml beaker with 25 ml H_2O and a few drops of ethyl alcohol. Filter through a 7 cm Whatman No. 40 paper into a 125 ml erlenmeyer flask and wash the paper and residue with hot water (retention of Mo in the leached residue is negligible). Keep the volume of the filtrate and washings to about 50 ml.

Swirl the solution in the flask and rapidly add 10 ml 12N HCl (if added too slowly a precipitate may form which will redissolve only with difficulty). Shake to liberate CO_2 and cool to 20–25°C.

Transfer the solution to a separatory funnel* and add 1 ml of 1% ferrous ammonium sulfate solution (the presence of a small amount of iron (or copper) increases the color intensity of the molybdenum thiocyanate complex), 3 ml 10% KCNS solution, and 2 ml of 10% $SnCl_2$ solution, mixing after each addition. After 30–45 sec add from a buret exactly 5 ml of isopropyl ether (previously shaken with stannous chloride solution), shake vigorously for 30 sec, allow to separate, and draw off and discard the aqueous phase. Run the organic extract, through a plug of glass wool if necessary, into a 1 or 2 cm cell and measure the absorbance of the complex at 460 mμ. A blank determination should be carried through the entire procedure.

Prepare a standard curve by carrying aliquots (1–50 γ Mo) of the standard molybdenum solution through the method (omit the fusion but add the aliquots to 25 ml water containing 3 g sodium carbonate.

Conversion Factors

$$MoO_3 \xrightleftharpoons[1.5003]{0.6666} Mo$$

* Chromium, vanadium, tungsten and phosphorus, if present in more than the usual amounts found in silicate rocks, will interfere in the colorimetric determination of Mo as thiocyanate. Molybdenum can be separated from these by precipitation as a sulfide (antimony is added as carrier) before proceeding with the colorimetric determination (Sec. 6.24, Ref. 4, pp. 655–657, Procedure B).

CHAPTER 10

CARBONATES

10.1. Determination of H_2O^- by heating at 105–110°C

Transfer a 1.000 g sample, ground to about 150 mesh and thoroughly mixed, into a weighed, covered crucible.* Reweigh the covered crucible and sample; the difference in the weight of the sample should not exceed 0.2 mg. Continue with the determination of moisture as described in Sec. 9.1. The possible presence of organic matter may make it difficult to obtain a constant weight, in which case the weight loss after a single heating should be arbitrarily taken as representing the H_2O^- in the sample.

10.2. Determination of loss on ignition

Place the crucible and sample used for the determination of moisture, uncovered, in a cold, electric muffle furnace and allow the temperature to rise slowly. When the temperature reaches 550°C, cover the crucible and continue the heating to 1000°C. Heat at this temperature for 30 min, cool in a desiccator and weigh. Repeat the ignition at 1000°C until constant weight is obtained. Calculate the weight loss at 1000°C.

10.3. Determination of acid-insoluble residue

To a 1.000 g sample (about 150 mesh) in a 250 ml beaker, add 25 ml of cold water, cover and, while swirling, add 5 ml 12M HCl. When vigorous effervescence has ceased, add 20 ml of hot water and heat the covered mixture quickly to near boiling. Digest for 15 min at a temperature just below boiling. Filter at once through a 7 cm Whatman No. 40 paper into a 250 ml beaker and wash the paper and residue thoroughly with hot dilute HCl (5:95). The filtrate may be used for the determination of SO_3, or combined with the solution obtained by fusion of the

* If the same sample is to be used for the determination of loss on ignition at 1000°C and the ignited residue fused directly with Na_2CO_3, a platinum crucible should be used. If no fusion is contemplated, a porcelain crucible may be used instead, but the crucible should be pretreated to remove the inner glaze, as described in Sec. 7.1.

acid-insoluble residue and solution of the cake in dilute HCl, for the subsquent determination of SiO_2.

Transfer the paper and residue to a weighed 25 ml platinum crucible (or pretreated (Sec. 10.1, footnote) weighed porcelain crucible if the residue is not to be treated further). Burn off the paper at a low temperature and then ignite the residue at 900–1000°C for 15 min, cool in a desiccator and weigh. Repeat the ignition to constant weight.

10.4. Determination of silica

10.4.1. GRAVIMETRICALLY, BY DEHYDRATION WITH HYDROCHLORIC ACID*

To the platinum crucible containing the sample from the determination of H_2O^- (Sec. 10.1), or the loss on ignition (Sec. 10.2), add 0.5 g anhydrous Na_2CO_3 and mix thoroughly. Heat the covered crucible over a low flame for a few minutes (see Sec. 9.2.1) and then for 10 min at the full heat of a meker burner (or in an electric muffle furnace at 875–900°C). Cool, tap the crucible sharply on the bench top to loosen the sinter cake, and transfer it to a porcelain casserole or platinum dish and continue with the determination of SiO_2 as given in Sec. 9.2.1 or Sec. 9.4.3. If preferred, the "minimum-flux" method (Sec. 9.2.2) may be applied to the ignited residue from the determination of loss on ignition, or to that from the determination of H_2O^- after a preliminary ignition of the dried sample; two 0.5 g portions of Na_2CO_3 should suffice and the second portion should be used to "rinse" out the crucible.

If a determination of the acid-insoluble residue was done (Sec. 10.3), fuse the residue in a platinum crucible with 4 times its weight of anhydrous Na_2CO_3 (Sec. 9.2.1), dissolve the fusion cake in HCl, combine it with the filtrate from the separation of the acid-insoluble residue and continue as given in Sec. 9.2.1 or Sec. 9.4.3.

10.4.2. GRAVIMETRICALLY, BY DEHYDRATION WITH PERCHLORIC ACID

This method is described by Hill et al. (Ref. 9, Sec. 7.4) and presupposes the ignition of the sample at 1000°C (Sec. 10.2).

Quantitatively transfer the sintered mass from the determination of the loss on ignition to a 140 ml, black-glazed porcelain casserole† with a minimum of water and add 25–30 ml of $HClO_4$ (72%). Evaporate to copious fumes, cover and increase the heat until a steady but gentle re-

---

* If fluorine is known to be present in significant concentration, the methods given in Sec. 9.3.1 or Sec. 9.3.2 should be used.

† The black color facilitates the quantitative transfer of the silica.

fluxing begins, and heat for at least 20 min. Allow to cool for 10 min or more.

To the cool solution add 75 ml of boiling water, rinsing the cover at the same time. Stir and allow to cool to at least 60°C.

Filter through a Whatman No. 40 paper and wash the silica 5 times with very hot dilute HCl (5:95) and then at least 10 times with very hot water. The filtrate may be reserved for further determinations.

Transfer the paper and precipitate to the original crucible. Cover the paper with aqueous ammonia and let stand at room temperature overnight. Evaporate to dryness and then ignite slowly, starting with a cold muffle and not allowing the temperature to rise above 400°C until combustion is complete*; finally, ignite at 1200°C for 30 min, cool and weigh as usual. Repeat the ignition at 1200°C until constant weight is obtained.

Treat the ignited residue with HF and H_2SO_4 as described in Sec. 9.2.1 to obtain the weight of SiO_2 and calculate the % SiO_2.

10.5. Determination of the R_2O_3 group

It is unlikely that sufficient platinum will have been introduced into the filtrate from the SiO_2 determination to warrant a separation with H_2S (Sec. 6.3) but many carbonate rocks may have Cu, Pb and Zn present as sulfides and a preliminary spectrographic analysis will indicate whether or not some of these elements should be removed at this stage (Sec. 9.5). If the separation is made, the filtrate must be evaporated and treated with bromine to expel the excess H_2S and to reoxidize the iron to the Fe(III) state as given in Sec. 9.5, before continuing with the precipitation of the R_2O_3 group.

To the filtrate from Sec. 10.4.1 or Sec. 10.4.2, or from the above H_2S separation, in a 250 ml beaker and having a volume of about 100 ml, add sufficient ammonium chloride to bring its total concentration to about 5 g when the acid solution is neutralized with aqueous ammonia. Add 2–3 drops of bromcresol purple indicator (0.04% aqueous solution), or 0.2% methyl red (60% alcoholic solution) and add pure aqueous ammonia slowly until a precipitate forms and just redissolves, or the indicator color shows signs of changing to bluish-purple (or yellow).

Heat to boiling and continue with the precipitation of the mixed oxides as described in Sec. 9.6; if the precipitate is small, add a pea size

---

* The addition of aqueous ammonia and the slow ignition is a precaution against the retention of sufficient $HClO_4$ by the paper to cause explosive ignition at about 300°C.

ball of filter paper pulp to aid its coagulation and filtration. Reduce the size of filter paper accordingly. Prior to the second precipitation of the R_2O_3 group, add 3 g NH_4Cl; if the precipitate is bulky, a third precipitation should be done, again after the addition of 3 g NH_4Cl. Destroy the accumulated ammonium salts as described but omit the subsequent "clean up" precipitation.

Ignite the precipitate as described, in the platinum crucible used for the HF treatment of the silica, and calculate the % R_2O_3.

A recovery of silica from the R_2O_3 can be made as usual by fusion of the ignited residue with $K_2S_2O_7$, solution in H_2SO_4 and evaporation to fumes (Sec. 9.6.1).

A calculation of the % Al_2O_3 by difference is made as usual.

10.6. Determination of calcium and magnesium

The manganese content of most carbonates is usually too low to warrant separation of the manganese prior to the precipitation of calcium. The ammonium persulfate method (Sec. 9.9.2) is fully applicable; it may be more advantageous, however, to employ the ammonium sulfide method (Sec. 9.9.1) if appreciable zinc is also present, and thus separate it at the same time.

The precipitation of calcium as oxalate (Sec. 9.10.1) is made on the filtrate from the separation of the R_2O_3 group (Sec. 10.5), or on that from either Sec. 9.9.1 or Sec 9.9.2 (above). If only a small amount of calcium is present and magnesium predominates, omit the oxalate separation and precipitate the Ca plus Mg with dibasic ammonium phosphate (Sec. 9.12.1); recover the calcium as the sulfate from the solution of the ignited pyrophosphates (Sec. 9.10.2).

The calcium may also be determined titrimetrically with EDTA, as described in Sec. 9.11.2, using an aliquot of the solution B prepared for rapid rock analysis, or with $KMnO_4$, as described in Sec. 9.11.1.

The filtrate from the separation of calcium as the oxalate, or that from subsequent treatment for the removal of Sr and Ba, is treated as described in Sec. 9.12.1 for the gravimetric separation and determination of magnesium with dibasic ammonium phosphate. An aliquot of solution B may be used for the EDTA titration of calcium and magnesium (Sec. 9.13.2); when the magnesium content is very high, as in magnesites, the modifications suggested by Bennett and Hawley (Ref. 2, Sec. 7.6) should be considered.

Conversion Factors

$$CaO \underset{0.5603}{\overset{1.7848}{\rightleftharpoons}} CaCO_3$$

$$MgO \underset{0.4781}{\overset{2.0915}{\rightleftharpoons}} MgCO_3$$

10.7. Determination of titanium and manganese

Titanium, usually present in very low concentration in carbonates, can be determined in the sulfuric acid solution of the pyrosulfate fusion of the ignited mixed oxides (Sec. 10.5) with hydrogen peroxide as described in Sec. 9.15.1, or in an aliquot of solution B (a larger aliquot than that recommended for silicates may be necessary) with Tiron (Sec. 9.15.2).

The content of manganese in most carbonates will be similarly low and a colorimetric determination as the permanganate, using KIO_4 as oxidant, is to be preferred. This can be done using a separate sample (there is usually not sufficient manganese present to warrant its separation and determination prior to the separation of calcium), as in Sec. 9.17a, or an aliquot of solution B (Sec. 9.17b), again increasing the size of the aliquot taken if necessary. For samples having a high manganese content the titrimetric methods (Sec. 9.16) are readily applicable.

10.8. Determination of phosphorus

10.8.1. GRAVIMETRIC

The gravimetric method, involving the preliminary separation of phosphorus as ammonium phosphomolybdate, dissolving of the precipitate in aqueous ammonia and final precipitation as magnesium ammonium phosphate, has been described fully in Sec. 9.18. A nitric acid solution of the sample is a prerequisite and this may be obtained by the decomposition procedure used for silicates (Sec. 9.16.1), if the sample is rather siliceous, or by treating it with dilute nitric acid, fusing the residue with Na_2CO_3 and combining its nitric acid solution with the first filtrate, if the SiO_2 content is very low. A concentration procedure has been previously mentioned in Sec. 7.8.

10.8.2. COLORIMETRIC

The colorimetric determination of phosphorus as the yellow molybdovanadophosphoric acid complex, for P_2O_5 contents of $<2\%$, has been

described in Secs. 9.19.1 and 9.19.2, and either procedure is fully applicable to carbonates. A rapid procedure utilizing the heteropoly blue
complex is given in Sec. 9.19.3.

10.9. Determination of the alkali metals

See Sec. 9.20 for a description of the gravimetric–flame photometric
method, following decomposition of the sample by the J. Lawrence
Smith sinter procedure. Flame photometric methods, employing either
a radiation buffer or a "neutral leach" step are given in Sec. 9.21.
These methods for silicates are equally applicable to carbonates with
modification (see Sec. 7.9). If desired, the Ca and Mg may be omitted
from the radiation buffer added to the sample solutions, but should be
included in that added to the standard and blank solutions.

10.10. Determination of iron

10.10.1. FERROUS IRON

The three methods given in Sec. 9.22 for silicates are applicable without modification to carbonates, although a larger sample weight may be
desirable.*

For pure, or nearly pure, carbonates a simpler approach is possible.
Transfer a sample of one to several grams, depending on the expected
iron content, to a 250–300 ml erlenmeyer flask (the type provided with a
ground glass stopper is a convenient form), add 10 ml H_2O and boil
vigorously to expel air from the flask. While boiling continues, cautiously add dilute (1:1) H_2SO_4 (or dilute HCl (1:1), if $K_2Cr_2O_7$ is to be
used) until effervescence ceases, and then add 5 ml in excess ($CaSO_4$
may precipitate but this will not affect the ferrous iron determination).
Stopper the flask and cool rapidly in a cold water bath; when cold, carefully remove the stopper to equalize pressure,† add 100 ml of cold distilled water and titrate at once with $KMnO_4$ or $K_2Cr_2O_7$ (see Sec. 9.22.1).

---

* The addition of a "spike" consisting of 5 ml of ferrous ammonium sulfate solution
(0.5 g in 500 ml H_2O containing a few ml H_2SO_4) to the sample solution to insure a
definitive end point change for samples containing very small amounts of FeO has
been recommended (Chapter 7, Ref. 16, p. A48).

† Hillebrand *et al.* (Chapter 7, Ref. 10, p. 975) suggest inserting a rubber stopper
through which passes a small funnel fitted with a stopcock. Cold water is placed in
the funnel, the stopcock is opened and the water is drawn into the flask as it cools,
more water being added to the funnel as needed, to a volume of 100–150 ml.

If the sample contains appreciable siliceous material, it may be desirable to determine the ferrous iron in both the acid-soluble and acid-insoluble portions. This can be done by first proceeding as above but keeping the decomposition time and degree of heating with acid to a minimum (limestones can be decomposed in cold acid but dolomites will require some heating). At the end of the titration, filter the solution through a 9 cm Whatman No. 40 filter paper into a 400 ml beaker, wash the residue and paper thoroughly with hot water, and then rinse the residue into a 45 ml platinum crucible used for the determination of ferrous iron by the modified Pratt method. Evaporate most of the water and proceed as in Sec. 9.22.1.

In the presence of carbonaceous matter the value obtained for ferrous iron will be in doubt (see comments in Sec. 7.10). If the higher oxides of manganese are present the use of HCl and $K_2Cr_2O_7$ should be avoided.

Conversion Factors

$$FeO \underset{0.6201}{\overset{1.6126}{\rightleftharpoons}} FeCO_3$$

10.10.2. TOTAL IRON

The total iron content of the acid-soluble and acid-insoluble portions, or of the whole sample, may be determined by the methods described in Sec. 9.23. It may be also determined colorimetrically, in an aliquot of solution B (if the Rapid Method procedure is used) or on a separate portion, as described in Sec. 9.24.

10.11. Determination of combined water

The determination of total water in carbonate is made precisely as described for silicates in Sec. 9.25 (see also Sec. 7.11 for additional comment).

Subtract the value obtained for moisture (Sec. 10.1) from that for total water to give that for combined water.

10.12. Determination of "carbonate" and "carbonaceous" carbon

An acid evolution-gravimetric method for the determination of "carbonate" carbon in silicates is given in Sec. 9.26.1, and a volumetric one in Sec. 9.26.2; both are applicable to carbonates precisely as given. For the determination of carbonaceous matter, see the method given in Sec. 9.27; prior removal of the "carbonate" carbon is recommended (see Sec. 9.27, footnote and Sec. 7.12.2).

10.13. Determination of sulfur, chlorine and fluorine

Sulfur. For "sulfide" sulfur, see Sec. 9.28. For acid-soluble "sulfate" sulfur, see Sec. 9.29. For total sulfur, see Sec. 9.30.

Chlorine. For acid-soluble chlorine, see Sec. 9.31.1. For total chlorine, see Sec. 9.31.2.

Fluorine. See Sec. 9.32.

10.14.   Determination of strontium and barium

See Sec. 9.33

PART IV

Methods of Analysis—Special

CHAPTER 11

X-RAY EMISSION SPECTROGRAPHY

11.1. General

The methods and techniques described in the preceding chapters have been largely "wet" chemical in nature and the instrumentation associated with these is, for the most part, within the budget of most laboratories engaged in rock and mineral analysis. Although frequent mention has been made of the valuable assistance to be had from an optical emission spectrograph, particularly for the determination of those important elements so often present only as trace constituents, no detailed procedures have been given; this in no way detracts from the importance of the optical emission spectrographic method in rock and mineral analysis but recognizes that the special nature and complexity of the method require a detailed treatment beyond the scope of this text. Use of the two instrumental techniques which are discussed in the following pages, particularly X-ray emission spectrography, requires a major investment in capital equipment but the relative newness of these methods, and the impact that they have had upon rock analysis in particular, warrant a brief treatment at least.

X-ray emission spectrography was, for several years, used chiefly for the determination of the heavier elements, particularly those such as niobium, tantalum and the rare earths whose determination by chemical or other methods was a difficult and unreliable venture. The state of the art did not permit it to be applied to the analysis of silicates; air-path spectrometers allowed absorption of the emitted radiation characteristic of the rock-forming elements (2–10 Å) and the early counters were not sensitive at these longer wavelengths. With the development of vacuum and helium-path spectrometers, and of the gas-flow proportional counter, the way was opened to determine the lighter elements as well. Claisse[10] proposed the flux dilution and polished glass bead or disc technique; other analysts investigated the use of unfused powders that were pressed into holders or briquetted. Further advances in instrumentation extended the method to the determination of sodium, and to the very light elements such as fluorine and oxygen. The application of the X-ray emission spectrographic method to rock analysis has been the most significant and revolutionary advance towards the attainment of a universal and truly rapid method since the introduction of the flame photometer into the determination of the alkali metals. Further studies, particularly in the field of computerized corrections for mass absorption effects, should serve to make it one of the most versatile tools in the hands of the rock and mineral analyst.

There is already an extensive literature on this subject, of which only a brief and incomplete survey will be presented here. The writer's experience with the method has been largely confined to the installation and use of automated equipment and to the development of a rapid and routine analytical method to replace the "rapid" colorimetric and titrimetric methods then in use; the methods to be given are modifications of those already described in the literature, adapted to the needs and facilities of the Geological Survey of Canada. The analyst who is considering the application of the X-ray emission spectrographic method to rock analysis should consult the references cited at the end of this chapter, in particular the authoritative texts by Liebhafsky et al.[24] and Adler.[1]

11.2. Some theoretical considerations

When matter (the target) is bombarded by a stream of electrons in an evacuated tube, the electrons are stopped by the target atoms and their kinetic energy is transformed, by a process involving the inner electrons in the atoms, into X-radiation. This X-radiation (having wavelengths approximately 100–0.1 Å) is generated at all voltages applied to the X-ray tube and has a continuous spectrum (continuum), the shape of which is independent of the atomic number of the target element; above certain critical voltages, however, certain sharp lines (or peaks, if the spectra are plotted) occur which are characteristic of that particular element. X-ray emission is a broad term for the various methods of exciting X-ray spectra; the method of interest here is that in which secondary fluorescent X-rays are excited by irradiation with a beam of primary X-rays produced as described above. The spectra of these fluorescent X-rays are identical in wavelength with, and the relative intensities are similar to, those excited when the specimen is bombarded with electrons; the former method has the advantage that any sample, whether solid, powder or liquid, can be irradiated with a permanent sealed-off source of primary X-radiation and the secondary X-ray fluorescence analyzed in a spectrometer or spectrograph.

In an X-ray tube the electrons from the heated filament strike a positively charged metal target, such as tungsten, molybdenum or chromium (elements having a high atomic number have certain advantages, although aluminum targets are now being used to yield "soft" or longer wavelength radiation), and produce the primary radiation mentioned previously. The irradiation of a sample by this primary beam of X-radiation results in the photoelectric absorption of X-ray photons having sufficient energy to eject electrons from the inner (K and L) shells of the constituent atoms, with the resultant appearance of the characteristic spectra of these atoms as other electrons move to fill the vacated positions and emit specific X-ray quanta in the process. As already stated, the wavelengths of the emitted X-rays are independent of the primary radiation, depending only upon the nature of the excited atoms. Because there are only 2 electrons in the K shell, and not more than 8 in the L shell, the resulting X-ray spectra consist of only a few strong lines in contrast to the complexity of optical spectra. As in the latter, the X-ray spectral lines are independent of oxidation state and structural relationship.

It is necessary to separate the various wavelengths of the X-radiation in order to identify the elements responsible for it and to measure the corresponding line intensities. This is done by means of a dispersive optics system utilizing diffraction by analyzer crystals, either flat or curved, which serve as three-dimensional gratings according to Bragg's law which states that if X-rays of a certain wavelength (λ) pass through a crystal having a distance d between two planes of atoms, interference or reinforcement will occur at a certain angle of incidence (θ) expressed by the relationship

$$n\lambda = 2d \sin \theta.$$

Thus, for any crystal having a given interplanar spacing, a particular wavelength (or multiple of it) will diffract at one angle only. The chemical composition of the analyzing crystal is important because any secondary X-rays excited from it must not interfere with the X-rays being measured.

Because the X-ray fluorescence spectroscopic method is fast, nondestructive and readily applied, it is particularly well-suited to qualitative analysis; the simplicity of the X-ray spectrum makes the identification of all elements in a given range a simple process. Quantitative analysis requires the measurement of the intensities of specific X-ray lines characteristic of the element and the reduction of these measurements, after certain necessary and routine corrections, to mass concentrations, usually by a comparative procedure utilizing working curves prepared from appropriate standard materials. The intensity of an analytical line depends not only upon the concentration of the element in question but upon the matrix in which the element is dispersed and, particularly for those elements of low atomic number, upon the homogeneity of the active sample layer undergoing irradiation. The latter is not too great a problem but the effect of varying matrix may be positive or negative and becomes increasingly serious as the composition of the unknown diverges from that of the standards.

Brief mention was made at the beginning of this chapter of the reasons for the relatively late entry of X-ray emission spectrography into the field of silicate analysis–absorption by air of the longer, soft wavelengths of the characteristic radiation, the prohibitively long time required for photographic recording of the intensity of the radiation, and the inability of early counting equipment to detect these radiations. The instrumental developments which have revolutionized the method and have given it almost universal application can be summarized as follows:

(a) the design and construction of X-ray tubes with extremely thin windows which produce high-intensity primary X-rays, thus giving amplified intensity of excited fluorescent radiation;

(b) the development of Geiger, proportional and scintillation counters of sufficiently high sensitivity to function efficiently at these longer wavelengths;

(c) the selection and application of large crystals having a high diffracting power for the dispersive optics system, such as ethylene diamine ditartrate (EDDT) and ammonium dihydrogen phosphate (ADP);

(d) the development of helium and vacuum path spectrometers to minimize the absorption of the X-radiation by air, and thus extend the range covered to the lighter (and silicate rock-forming) elements.

11.3. Preparation of the sample

11.3.1. GENERAL

Because the active portion of a sample undergoing X-ray fluorescence is confined to a very thin surface layer, it follows that the nature of the sample surface presented to the beam of X-rays will exert a major influence on the analytical results to be obtained, particularly on their reproducibility.   This is particularly important in the case of the light elements such as Si, Al, Mg and Na* because the intensity of the radiation decreases sharply as the atomic number decreases; the "soft" radiation emitted by these elements is easily absorbed by the sample matrix and their effective radiation comes only from those atoms which are at or near the surface of the sample. For these elements the slightest inhomogeneity of the sample surface, whether it be an inhomogeneity introduced by differences in the sizes of the particles making up the surface, or in their compositions, can introduce serious errors into the results. The area of the surface to be irradiated, and the distance of the surface from the window of the X-ray tube, are also of vital importance but these are conditions that are easily kept constant.

Among the factors which influence the nature of the sample surface are some which are obvious, and others which are less so.[26] It is obvious, for example, that inhomogeneity of the sample itself will markedly affect the accuracy of the final result but it may not be so obvious that the manner in which an element is distributed throughout a rock, i.e., its mineralogical distribution, can have a similarly undesirable effect; certain minerals such as biotite have a tendency to occupy a disproportionate amount of surface area through parallel orientation of flakes, and this may lead to the enrichment or impoverishment of a given element in the surface layer (the "biotite effect" found by Volborth[2,36,37] in rocks containing about 3 to 7% biotite). Again, it is recognized that the composition of the matrix in which a given element occurs will affect the intensity of the radiation emitted by the element; mineralogical differences in two samples having the same bulk chemical composition can, however, result in two surface layers having chemical differences of sufficient magnitude to affect the degree of matrix absorption that will occur in each.

Various ways have been suggested to overcome these sources of error. Grinding of the sample to the point at which the particle size distribution remains constant is an obvious approach; it has been found that the intensity of radiation rises rapidly and irregularly with decreasing particle size until the latter is approximately 325-mesh, at which point it tends to level off. However, in addition to this being a lengthy and tedious operation, it results in the introduction of extraneous material;

---

* The extension of the X-ray fluorescence method to the very light elements has added a new dimension to methods of rock and mineral analysis. The direct determination of oxygen by means of a special "soft" X-ray source[3] now makes possible a truly complete analysis of a silicate. A neutron activation method, such as that of Volborth[35,38] is not as much within the capability of the average rock and mineral analysis laboratory as is the X-ray fluorescence method.

this can be circumvented by preparing two samples for analysis using different grinding media (e.g., both ceramic and steel plates) and analyzing both samples to obtain the necessary corrections for the material thus introduced ("average" corrections based upon generalities have been found *not* feasible).[37,41]

The use of untreated ground sample powder has the advantages of speed of preparation, simplicity of operation and non-destruction of the sample. The results obtained with powders packed into holders have not been satisfactory, however, and it is preferable to press the powder into a briquette, with or without the addition of an inert binder to give better cohesion. This procedure will be discussed in Sec. 11.3.3.

The direct use of the ground sample powder requires that corrections be applied to the measured intensities to compensate for absorption by the matrix, or that only standards which closely resemble the unknowns in both chemical and mineralogical composition[8,9] be used for comparison in converting measured intensities to weight per cent (because of the uncertainties inherent in most geochemical standards, some analysts prefer to use synthetic standards only[21]). Claisse[10] sought to avoid these problems by fusing the sample with a suitable flux so that not only would the physical properties of the matrix be destroyed but the dilution of the sample by the flux would be sufficient to minimize the effect of matrix changes from one sample to another. This is the popular alternative to direct pelletization of the sample powder and is much used in a variety of ways. The resulting glass bead, pellet or disk can be used without further treatment, except perhaps the polishing of one side; if the homogeneity of the glass is still suspect, it can be crushed, powdered and pressed into a briquet with a backing of inert material to give the necessary strength. Against the disadvantages of the destruction of the sample, the dilution of element concentrations to the point at which their radiation intensities are unduly weakened and the increased complexity of the operations required, one must consider the advantage of a common matrix (Czamanske *et al.*[12] have warned about the need to still consider the possible interelement enhancement or absorption in fused samples). Fusion techniques are discussed in Sec. 11.3.2.

It should be recognized that no one method can be cited as the proper one for all purposes and one should select that which, in terms of time and effort, will give the best return. Dumesnil and Perrault,[13] for example, have recently described a rapid analytical scheme involving a combination of X-ray fluorescence, flame photometric, atomic absorption and wet chemical methods for 15 elements (including some duplication); the quantitative determination of those elements having concentrations $>1\%$ is done by X-ray fluorescence spectroscopy on a fused and polished bead, whereas both the finely ground sample powder and a briquet prepared from it are used for the quantitative determination of those elements having concentrations $<1\%$.

A general description of various methods for sample preparation has been given.[7,20]

11.3.2. FUSION WITH FLUX

Fusion of a sample has as its chief object the elimination of problems caused by matrix differences and sample heterogeneity (including the heterogeneous physical properties of its constituent minerals). There is a difference of opinion as to the effectiveness of the fusion in completely eliminating the matrix problem,[12] but there is no question about the latter one, particularly as it applies to the light elements when they are distributed among minerals of very different physical properties. The total relative deviations are less for fusions than for ground powders, and lower and less varying standard deviations are obtained. Against these advantages must be listed the disadvantages of dilution of the sample (some methods call for extreme dilutions in order to reduce adsorption errors to negligible proportions), and the more complicated preparation procedure required.

The sample is commonly fused at 1000–1050°C with either sodium tetraborate (borax), lithium tetraborate or a mixture of sodium tetraborate and lithium carbonate[4,40]; the lithium salts are generally preferred to the sodium salts. After crushing the fused bead and grinding it to about 325-mesh, the powder is pressed into a disk with a backing of inert material such as boric acid, corn starch or bakelite (some analysts prefer also to rim the sample with the backing material) to give increased handling strength, at pressures ranging from 30,000 to 70,000 psi. The inner surface of the mold cap of the press must be kept scrupulously clean and polished if it is to be used to form the surface of the disk; Volborth[35] introduced the use of a glass disk placed beneath the sample powder in the mold to give a very smooth and uniform surface to the briquet. Welday et al.[40] have investigated various dilutions and recommend a moderate-dilution sample to flux ratio of 35 to 65 (2.8 g sample $+$ 5.2 g $Li_2B_4O_7$); the mixture is fused in a graphite crucible at 1050°C for 20 min. Palau crucibles are said to be less subject to wetting by the fusion, and the film of graphite which may cover the surface of the bead is also thereby avoided. Some analysts prefer to add a binder, such as boric acid or bakelite, to the ground powder before briquetting it and this is often added in an amount sufficient to compensate for the loss in weight of the sample and flux mixture resulting from the escape of volatiles during fusion. Longobucco[25] prefers to calcine all samples at 1350°C first in order to eliminate this variable. When bakelite is used as a binder it is common practice to harden the briquet by "curing" it at 110°C for 15 min.[18]

In order to ensure that matrix absorption effects will be negligible it is necessary to dilute the sample sufficiently with the flux, which may result in a considerable reduction in the radiation intensities, especially for the light elements. Claisse[10] suggested the addition of a heavy absorber to the sample–flux mixture, for the same reason that a radiation buffer is added to solutions prepared for flame photometry (Sec. 9.21), that is, to smooth out compositional differences between samples and thus achieve linearity in the intensity measurements. The dilution also used by Claisse (100 mg sample plus 10 g borax) was extreme, and others, particularly Rose and his colleagues,[27,28,29] reduced the degree of dilution to about 1:10 (Wilson et al.[42] use 1:4) and placed more reliance upon the leveling effect of the heavy absorber. Lanthanum oxide, La_2O_3, is most commonly used for this purpose but the L line of lanthanum interferes with the $K\alpha$ line of magnesium, resulting in a very considerable loss of intensity for the magnesium radiation. Cerium oxide, CeO_2, has been used with less interference with magnesium but the fusion must then be done at not less than 1100°C and a double fusion may be necessary to ensure the complete solution of CeO_2 in the glass. The procedure using La_2O_3 has been applied to as little as 50 mg of sample.[29]

Claisse[10] poured the fused mass into a retaining ring set on a hot plate, and obtained a thin glass disk which was allowed to cool slowly; intensity measurements were made on the flat surface thus produced without further treatment. Variations on this technique have been described by others. Longobucco[25] poured the fusion on an aluminum slab maintained at 260°C, cooled the bead slowly to room temperature and then polished one side of it. Kodama et al.[21] fused a sample with $Li_2B_4O_7 + Li_2CO_3$ (in ratio 1:8:2) in a platinum crucible, quenched the melt, crushed and ground the glass and transferred it to a disk-shaped graphite mold which was then heated to 1000°C for 1 hr. The glass disk was cooled slowly and one side of it was ground and polished. Dumesnils and Perrault[13] used $Li_2B_4O_7 + La_2O_3$ as the fusion mixture in a platinum crucible and poured the fused mass into a graphite mold to produce a disk, one side of which was then polished. Norrish and Chappell[43] have used a somewhat different approach: the sample (0.28 g) is mixed with 1.5 g of a prefused mixture of $Li_2B_4O_7$–Li_2CO_3–La_2O_3 and 0.02 g $NaNO_3$, and fused in a Palau crucible at 980°C; the melt is poured into a heated graphite disk-shaped mold surrounded by a brass retaining ring and an aluminum plunger is immediately inserted to flatten the solidifying disk. The latter is then annealed between asbestos sheets at 200°C and allowed to cool slowly. Measurements are made on the thin glass disk without further treatment of it; it is necessary to use particular care in the preparation to prevent cracking of the disk on cooling.

According to Rose et al.,[28] X-ray microprobe analyses of glass disks, prepared as previously described, indicate that the casting process can result in heterogeneities that will cause serious errors in the determination of the light elements.

11.3.3. DIRECT PELLETIZATION (BRIQUETTING)

As in the fusion method, there are several approaches to the direct pelletization or briquetting of finely ground powders, such as the use of the powder alone, the addition of a binder and/or a heavy absorber, and the use of an internal standard.

The advantage of briquetting over the simple packing of the finely ground sample into a holder is that the former gives more reproducible intensity readings. Maximum advantage in speed of preparation is obtained when the ground sample, which should be about 325-mesh in particle size, is briquetted without further treatment,[5,35,41,43] but some analysts prefer to add a binder to give increased mechanical strength, such as boric acid or even just a few drops of water.[35] Volborth's use of a thin glass disk between the bottom of the mold and the sample powder has already been mentioned as a way of producing surfaces having a mirror-like finish; plastic disks, although much less subject to breakage, proved unsatisfactory because the sharp mineral grains tended to imbed themselves in the soft plastic. The sample powder used by Volborth was mostly finer than 400-mesh and required extremely long periods of grinding (about 1 hr per gram of rock powder).

Some analysts have used inert material as both a binder and a diluent. Webber,[39] in his early studies, mixed 0.1 g of sample with 1 g NaCl; Gunn[15] used a 1:1 mixture of Li_2CO_3 and starch with a sample to give a dilution of about 20:1. Hornung[19] found that, while undiluted powder backed by cellulose gave maximum intensity measurements, a briquet prepared from a cellulose-sample mixture in the ratio 4:1 had a reduced background and was more robust. Ball[6] considers the direct pelletization method to be more suitable to semi-skilled routine preparation than is the fusion method and recommends the use of a sample-cellulose mixture (1:2) to which is added Bi_2O_3 (10% by weight) as a heavy absorber.

Hornung[19] also investigated the use of an added internal standard (vanadium, added as Na_3VO_4) and found it to be unnecessary.

11.3.4. MASS ABSORPTION CORRECTIONS

It has been stated that the measured intensity of a characteristic X-ray line of an element depends not only upon the mass concentration of that element but also upon the nature and abundance of the other constituent elements. This variable would appear to be easily controllable by confining the use of a particular set of working curves to only those samples which are compositionally similar to the standards used to prepare the curves, thus giving rise to "analysis by type" and the use of a "family" of working curves ranging from granitic to ultrabasic in character. This is too much of a simplification, however, for the routine analysis of geological samples; field names are often intended to be approximate only and it is clearly wasteful of time and effort to make a preliminary analysis in order to be able to classify the sample compositionally. In addition, mention has already been made of the need to have samples that are *mineralogically* similar to the reference samples, as well as being *compositionally* similar, because the differing physical properties of the various minerals may result in a heterogeneous distribution of certain elements in the final briquet.

Fusion of the sample with a flux was proposed to overcome these difficulties. The fusion process breaks down the constituent minerals and all samples end up as a similar glass; the dilution of the sample by the flux also serves to minimize changes in composition from one sample to another. As a further precaution against the effect of matrix changes, a heavy absorber is also added to smooth out these compositional variations. The advantages and disadvantages of this method have been discussed in Sec. 11.3.2; it has been shown,[12,14] however, that matrix effects must still be considered even when a moderate-dilution method (admittedly without the addition of a heavy absorber) is used and that for accurate results a correction must be applied to compensate for them.

Interelemental effects are generally considered to be capable of delineation as follows: the enhancement of one element by another may be regarded as the reverse of absorption and is governed by the laws of absorption, and each element exerts an effect on another element which, from a practical point of view, can be considered as being linearly proportional to the concentration of each of the elements. These can be expressed as an equation which can be solved by the application of experimentally determined coefficients. These latter are obtained by measuring the intensities of selected elements in a series of standard samples, correcting the measurements for the effect of background and such other variables as may be necessary, and then making the necessary calculations, usually with the aid of a computer. The mathematical procedure followed is an iterative one in which the correction coefficients are calculated and applied until a set is obtained which gives the best fit between the given and found values for the standards. These are then used to correct the measurements obtained for unknowns which are run under similar conditions. Such correction schemes have been described in detail.[14,16,17,21,43]

Much work has been done on this problem at the Geological Survey of Canada by G. R. Lachance and R. J. Traill.[22,23,33,34] Based upon their hypothesis that the relative intensity of a constituent is directly proportional to its weight fraction and inversely proportional to 1 plus the sum of the products of the weight fractions of the remaining constituents times their respective alpha constants, the relative intensity of a characteristic line emitted by an element in a multicomponent system can be expressed by the following:

$$R_A = \frac{I_A}{I(A)} = \frac{C_A}{I + C_B\alpha_{AB} + C_C\alpha_{AC} + \ldots C_n\alpha_{An}}$$

where I_A = the net measured intensity of X-ray line A from a sample; $I(A)$ = the net measured intensity of X-ray line A from pure element A; C_A, C_B, C_C, $\ldots C_n$ = wt. fractions of elements A, B, C, $\ldots n$, respectively, in the sample; α_{AB}, α_{AC}, $\ldots \alpha_{An}$ = constants that account separately for the effect of element B, C, $\ldots n$, respectively, on the intensity emitted by element A.

For a system of n elements a series of n linear equations must be written, and are used to convert relative intensities to weight fractions. A detailed description of the computer program written for this purpose is available,[23] as well as alternate methods for manual calculation if the services of a computer are not available. Claisse and Quintin[11] have suggested a modification of the calculations to account for the polychromaticity of the excitation radiation and thus improve the accuracy of an analysis.

11.3.5. Other techniques

Schnetzler and Pinson[31] simplified the preparation of tektite samples for X-ray emission spectrographic analysis by using a bubble-free, cut and polished surface for measurements; previous attempts to use a powdered sample had proved to be unsatisfactory.

Rose *et al.*[29] reduced the sample weight of tektite needed for fusion to 50 mg or less. A further reduction of the sample weight to only a few milligrams, and applicable to silicate samples in general, has been a-chieved by using a combination solution-absorption technique to prepare the sample for analysis.[30] The sample is dissolved and the solution is absorbed on powdered cellulose which is then dried at 80°C and briquetted. Standards are easily prepared from reagent grade chemicals. The determination may be made at the microgram level and detection of major constituents is possible at still lower levels of concentration. The possibilities of extending this to such techniques as concentration and precipitation on carriers, to selective solvent extraction and microelectrolytic separations are obvious.

11.4. Methods used at the Geological Survey of Canada

Among the references listed at the end of this chapter are several which give detailed outlines of procedures for both the fusion and direct pelletization methods of X-ray emission spectrographic analysis. The methods now in routine use at the Geological Survey of Canada are the result of the adaptation and modification of published methods, particularly those of Rose, Adler and Flanagan.[27,28] These are used, however, to provide X-ray emission spectrographic data for eight major and minor constituents at the rate of approximately 2000 rapid rock analyses each year (the rapid chemical methods used to determine Na, P, Fe(II), total H_2O and CO_2 have been described in Part III) and, as such, may serve as a guide to others faced with a similar analytical demand.

The methods and techniques to be described are the result of the combined efforts of a number of analysts during the last five years. Special recognition must be given, however, to the contributions made by Serge Courville and G. R. Lachance.

11.4.1. INSTRUMENTATION

The instrument used in the Analytical Chemistry Section of the Geological Survey is a multichannel polychromator designed to provide the simultaneous determination of up to 9 elements. These 9 channels,together with an external standard channel, are arranged radially about a beryllium end-window X-ray tube having a tungsten target and capable of continuous operation at 50 kV and 50 mA; 2 channels, those for Fe and Mn, are airpath types, the others (Ti, K, Ca, Si, Al and Mg) are vacuum monochromators. The sample, in the form of a 1⅛ in. diameter, disk-shaped briquet, backed by boric acid, is placed in the loading chamber which is then revolved to a position below the beam of primary X-radiation. A portion of the secondary radiation emitted from the disk surface is directed towards each of the 9 detectors and the monitoring external standard, all of which are so arranged as to give optimum intensity at the proper wavelength. Each detector is connected to an integrating capacitor which is charged during the integration period by pulses from the detector. When the monitor channel reaches a preset charge, the integration period ends and the charge in each element capacitor is recorded as an integrated intensity ratio of the element line to the external standard.

The detectors are of two types: a *multitron*, a sealed gas-filled (Ne or Ar) tube which operates in the Geiger region, is used for measurement of the intensities of Fe, Mn, Ti, K and Ca; a *minitron*, or gas-flow proportional counter, is used for the light elements Si, Al and Mg (sealed proportional counters, called *exatrons*, are also available). The gas used for the minitrons is a mixture of 50% neon, 49% helium and 1% butane. A summary of the types of crystals and detectors used is given in Table 11.1.

TABLE 11.1

Types of crystals and detectors used for rapid rock analysis by
X-ray fluorescence spectrography

Element	Analyzing crystal	Path	Detector
Fe, Mn, Ti, Ca, K	LiF	air	sealed (multitron)
Si	EDT	vacuum	sealed (exatron)
Al	EDT	vacuum	sealed (exatron)
Mg	ADP	vacuum	flow (minitron)

The recording of intensity ratios was originally done by pen chart, but this has been replaced by a digital voltmeter and paper tape recorder which give the measurement in millivolts.

Two disks may be in use at one time, one of them in stand-by position. Shapiro and Massoni[32] have designed an automatic sample changer and controller for a similar X-ray quantometer which permits the exposure of 32 samples in sequence under vacuum conditions, and which controls the various steps up to and including the final print-out.

The instrument is installed in a laboratory having controlled temperature and humidity, and is left continuously in operating condition in order to minimize instrumental fluctuation.

11.4.2. Fusion with Lithium Tetraborate and Lanthanum Oxide

This method was originally used at the Survey in 1962 for all samples submitted for "rapid" analysis. At the start it involved the preparation of only one disk but the reproducibility, and consequently the accuracy, of the data proved to be unsatisfactory. The reliability of the data could be improved by making duplicate disks for every sample; this would not only be time consuming, however, but would be unnecessary for many samples of a strictly reconnaissance nature. The fusion method, in which duplicate, or more frequently, triplicate disks are prepared for each unknown, is now reserved for those samples requiring data of a quality which more nearly approaches that obtainable by chemical methods. The direct pelletization method, which gives results of an accuracy comparable to those obtained by the single disk fusion procedure but which has a decided advantage in both speed of preparation and reproducibility of results, is used for the majority of the rapid method samples.

The samples are received in the laboratory as -150 mesh powder and two variations of the fusion procedure are followed, the choice usually being governed by the amount of sample available for analysis.

Procedure A (Flux to sample ratio $\cong 3.6:1$)

A 0.500 g sample is quantitatively transferred to a 1 in. i.d. graphite crucible* containing a preweighed, premixed flux charge (1.000 g $Li_2B_4O_7$ + 0.125 g La_2O_3) and intimately mixed. Eight such loaded crucibles are placed in a numbered silica tray (6 in. $\times$ $3\frac{1}{2}$ in. $\times$ 1 in.) and the latter is slowly introduced into an electric muffle furnace maintained at 1100°C, and heated at this temperature for 20 min. It is then removed and placed upon a metal air-cooling support (6 in. $\times$ 5 in. $\times$ 4 in., with 6 rows of 6 holes, 1/8 in. diameter, in the top) through which air is blown to cool the tray and its contents rapidly to room temperature. Figures 11.1 and 11.2 show the equipment used in these steps.

The bead, with any smaller satellite beads, is carefully transferred to a weighing dish and the weight adjusted to 1.800 g with boric acid. The contents of the dish are then transferred quantitatively to a 15 cc hardened steel grinding container ($2\frac{1}{2}$ in. o.d., 2 in. i.d.) in which the bead is crushed by means of a hand-pressure punch operating through a hole in a closely fitting bakelite container cover. The bakelite cover is re-

---

* Crucibles with a cone-shaped internal base are best for gathering the fused material into a bead that is easily recoverable and will minimize loss of fused material during the recovery step.

Fig. 11.1. Equipment and materials for fusion method. From left to right are shown the graphite crucibles with added flux charge (a vial of premixed flux is shown above them, next to an empty crucible), four small grinding containers (one is open to show the hardened steel puck) and the modified container holder, a tray of fused beads, a finished sample disk (the vial next to it contains approximately 5 g of boric acid for backing material) and its labelled envelope, and a spring-loaded hand punch for breaking the fused bead in the grinding container (this has since been replaced by a simple hand pressure punch which is not shown). Photograph by Photography Section, Geological Survey.

moved and any fragments adhering to it and to the punch are brushed into the container. A $1\frac{1}{2}$ in. × 7/8 in. hardened steel puck is added, the steel container cover is inserted, and four of these grinding vials are placed in the special holder shown in Fig. 11.1. This is in turn placed in the grinding mill shown in Fig. 11.3* and the contents of the containers are ground for 4 min (automatic timer) to yield a powder that is -325 mesh. The contents of each container, from which the puck is removed, is then transferred to a piece of glazed paper (quantitative recovery of the powder is not necessary).

* This mill is designed to take one large grinding dish. It was found preferable, both from the standpoints of recovery of sample powder and of speed of preparation, to modify the mill to accomodate four small grinding containers simultaneously.

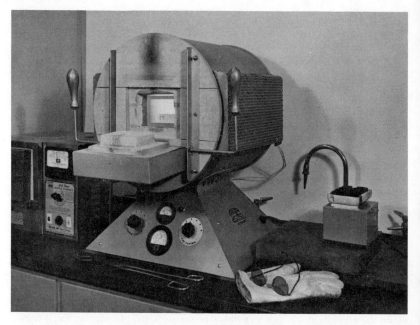

Fig. 11.2.   Fusion.   A silica tray containing 8 crucibles is shown in position in the electric muffle furnace and a second tray is shown cooling rapidly to room temperature on the metal air cooler.   Photograph by Photography Section, Geological Survey.

A vial of boric acid (approximately 5 g) is emptied into the mold cavity of the hydraulic pellet press shown in Fig. 11.3 and is lightly tamped down with a bakelite plunger.   The sample powder is then carefully sprinkled over the boric acid base, the mold cap is inserted (the inside surface of the cap is maintained in a mirror-like state by daily polishing on a cloth lap with alumina (0.05 micron) paste) and the contents are pressed into a 1⅜ in. × 1/8 in. disk (the sample layer is about 1/64 in. thick) for 30 seconds at 70,000 psi.   The disk is removed and stored in a labelled envelope.

The hardened steel containers and pucks are an obvious source of contamination, chiefly of iron, and it is preferable to use liners and pucks of tungsten carbide instead.   Such material is extremely costly, however, and a compromise of sorts is possible by subjecting the standard samples used for the preparation of the working curves to a similar grinding period, in the reasonable hope that the contamination of both unknowns and standards will be the same.   The greater number of steps involved in the

Fig. 11.3. Preparation of disk. To the left is the grinding mill with the four small containers clamped in position. The hydraulic pellet press is shown to the right and the contents of a grinding container are being transferred to a piece of glazed paper prior to being poured into the mold cavity. Photograph by Photography Section, Geological Survey.

preparation of the disk tends to decrease the reproducibility of results obtained by the fusion method. It is possible to improve the reproducibility by replication but only at the expense of speed. The average rate of production of disks is 24, which can be made up of 4 standards and 4 samples, each in triplicate, or of 4 standards and 8 samples, each in duplicate, depending on the degree of accuracy being sought.

Procedure B (Flux to sample ratio $\cong$ 10:1)

A 0.125 g sample is transferred to a graphite crucible containing a pre-weighed, premixed charge of $Li_2B_4O_7$ and La_2O_3 as described in *Procedure*

A and the fusion and preparation of the disk continued as before, except that the weight of the fused bead is adjusted to 1.300 g with boric acid.

The greater dilution of the sample by this procedure reduces possible matrix absorption effects but also results in a considerable reduction in the sensitivity of the magnesium radiation.

11.4.3. DIRECT PELLETIZATION

Because of the fewer steps required to prepare a disk, the average production of disks per operator is 32, of which 4 are secondary internal laboratory standards covering a wide range of composition, 2 are reference samples included as a performance check, and the remaining 26 are samples.   It requires a full working day to analyze these sample disks with the automatic spectrometer, and the production of analyses is thus 13 per operator per day.

A 1.70 g sample of −150 mesh powder is weighed on a top-loading balance and 0.30 g boric acid added to it.   The mixture is ground to −325 mesh as previously described and then pressed into a disk as before, with a boric acid backing.   The prepared disk is stored in a labelled envelope.

11.4.4. PRECISION AND ACCURACY

If by *precision* we mean the extent to which one can reproduce results from one disk to the next, then the X-ray emission spectrographic method is capable of the highest precision of determination of an element, subject only to the skill of the operator in preparing the disk.   From a study of the precision of the direct pelletization procedure (Sec. 11.4.3) it was concluded that the preparation of a duplicate disk for each sample was not warranted, based upon a realistic evaluation made of duplicate disks prepared for 102 analyzed samples.   The usual method of determining the precision of replication is to determine the average standard deviation for intensities measured on a series of disks prepared for one or two samples, a situation closely approaching that of the ideal and something never encountered in the working analytical laboratory.   By using instead the difference between the intensity measurements for the first and second disks of each of the 102 samples, one obtains the reproducibility that is to be expected under normal working conditions for samples having a wide range of composition and processed by different operators over a long period of time.   The precision of determination of each of the eight elements, in terms of one standard deviation, are summarized in Table 11.2 for both the fusion and the direct pelletization procedures. In terms of actual percentage differences between the measured intensities of the 2 disks by both procedures, the direct pelletization procedure has better reproducibility, particularly for iron and aluminum.* With additional care, of course, the reproducibility can be considerably improved over that shown here, but only at the expense of production.

TABLE 11.2

Precision and Accuracy of the fusion (F) and direct pelletization (DP) procedures

Con-stituent	Range %	Precision		Accuracy	
		F SD[a]	DP SD[a]	F SD[a]	DP SD[a]
Fe_2O_3	0–20	0.20	0.06	0.25	0.4
MnO	0–1	0.004	0.005	0.01	0.03
TiO_2	0–5	0.012	0.015	0.07	0.05
CaO	0–20	0.08	0.10	0.23	0.4
K_2O	0–6	0.02	0.03	0.09	0.1
SiO_2	35–75	0.78	0.80	1.0	1.2
Al_2O_3	0–20	0.74	0.30	1.0	1.0
MgO	0–40	0.42	0.40	0.56	0.5

[a] SD = standard deviation.

The data were obtained as follows: for the *precision* measurements, *duplicate* disks were prepared for 102 samples having a wide range of composition and the differences between the intensity measurements for each pair of disks used to calculate the standard deviation; for the measurement of the *accuracy* of the procedures, *one* disk only was prepared for each of 26 chemically analyzed samples used as standards and the differences between the values determined by X-ray emission spectrography and the "accepted" values were used to calculate the standard deviation.

Accuracy is a much more difficult subject to define but in its simplest form it can be considered to be the extent to which a result obtained by a particular method deviates from the accepted value. Reproducibility or precision can, for the most part, be controlled by proper care in the preparation of the disk but the accuracy of a determination is subject to the influence of many factors, of which precision is only one. These include matrix absorption effects, particle size, mineralogical composition, instrumental behavior and the quality of the standards used to prepare

* It must be emphasized that all data for the 102 samples are included in the calculations of the standard deviations (SD) given in Table 11.2. Although the SD of the direct pelletization method is given as 0.80% for SiO_2, 57 of the 102 samples had a difference of $< 0.3\%$ between the intensity measurements for the first and second disks; for Al_2O_3, 64 showed a difference of $< 0.2\%$.

the analytical working curves from which the intensity measurements are converted to weight per cent. This latter is probably the most important factor which influences the accuracy of the determination. The X-ray emission spectrographic method is a comparative one and the results can be no better than permitted by the standards used, not only because of the uncertain accuracy of the "accepted" values, but also because of dissimilarities between the elemental *and mineralogical* compositions of the standards and the unknowns. Unless one is to have a room full of analyzed reference material with which to closely bracket every unknown (providing one knows the composition of the latter!), a compromise is necessary which will be governed by the conditions under which the sample was taken, and the degree of accuracy required for the results.

The data given for the accuracy of the determination of the eight elements by the two procedures in Table 11.2 are based upon a direct comparison of the results obtained with a *single* disk only. 26 previously analyzed samples of igneous rocks were used for the evaluation. Better accuracy is obtained by the fusion procedure than by direct pelletization; the accuracy of the fusion procedure can be considerably improved by using duplicate or triplicate disks prepared with extreme care. The accuracy of the direct pelletization data for Mn, Ti and K is satisfactory without further refinement but that of the five lighter elements is suitable only for chemical reconnaissance work; the accuracy of these latter can be improved considerably by using a narrower compositional range of standards for preparation of the working curve, or by applying a correction to compensate for the matrix absorption effects.

References

1. Adler, I. *X-Ray Emission Spectrography in Geology*, 1966. Amsterdam: Elsevier, 258 pp.
2. Baird, A. K., Copeland, D. A., McIntyre, D. B. and Welday, E. E. *Am. Mineral.* **50**, 792–795 (1965).
3. Baird, A. K., and Henke, B. L. Oxygen determinations in silicates and total major element analysis of rocks by soft X-ray spectrometry. *Anal. Chem.* **37**, 727–729 (1965).
4. Baird, A. K., MacColl, R. S., and McIntyre, D. B. A test of the precision and sources of error in quantitative analysis of light major elements in granitic rocks by X-ray spectrography. *Advances in X-Ray Analysis*, Vol. 5, 1962. New York: Plenum Press, pp. 412–422.
5. Baird, A. K., McIntyre, D. D., and Welday, E. E. Sodium and magnesium fluorescence analysis—Part II: Application to silicates. *Advances in X-Ray Analysis*, Vol. 6, 1963. New York: Plenum Press, pp. 377–388.

6. Ball, D. F. Rapid analysis for some major elements in powdered rock by X-ray fluorescence spectrography. *Analyst* **90**, 258–265 (1965).
7. Bertin, E. P., and Longobucco, R. Sample preparation methods for X-ray fluorescence emission spectrometry. *Norelco Reporter* **9**, 31–43 (1962).
8. Chodos, A. A., and Engle, C. G. Fluorescent X-ray spectrographic analyses of amphibolite rocks. *Am. Mineral.* **46**, 120–133 (1961).
9. Chodos, A. A., and Engle, C. G. Fluorescent X-ray spectrographic analyses of amphibolite rocks and constituent hornblendes. *Advances in X-Ray Analysis*, Vol. 4, 1961. New York: Plenum Press, pp. 401–413.
10. Claisse, F. Accurate X-ray fluorescence analysis without internal standards. *Quebec Dept. Mines Prelim. Rept.* **32**, 1956, 16 pp.
11. Claisse, F., and Quintin, M. Generalization of the Lachance-Traill method for the correction of the matrix effect in X-ray fluorescence analysis. *Can. Spectr.* **12**, 129–133, 146 (1967).
12. Czamanske, G. K., Hower, J., and Millard, R. C. Non-proportional, non-linear results from X-ray emission techniques involving moderate-dilution rock fusion. *Geochim. Cosmochim. Acta* **30**, 745–756 (1966).
13. Dumesnil, J. C., and Perrault, G. La part de la fluorescence X dans le dosage rapide des éléments des roches. *Can. Spectr.* **12**, 3–11 (1967).
14. Gunn, B. M. Matrix corrections for X-ray fluorescence spectrometry by digital computer. *Can. Spectr.* **12**, 41–46, 64 (1967).
15. Gunn, E. L. Fluorescent X-ray spectral analysis of powdered solids by matrix dilution. *Anal. Chem.* **29**, 184–189 (1957).
16. Holland, J. G., and Brindle, D. W. A self-consistent mass absorption correction for silicate analysis by X-ray fluorescence. *Spectrochim. Acta* **22**, 2083–2093 (1966).
17. Holland, J. G., and Hamilton, E. I. Mass absorption corrections in X-ray fluorescence analysis of natural igneous rocks and their metamorphosed equivalents. *Spectrochim. Acta* **21**, 206–208 (1965).
18. Hooper, P. R. Rapid analysis of rocks by X-ray fluorescence. *Anal. Chem.* **36**, 1271–1276 (1964).
19. Hornung, G. The analysis of basalts and granites by X-ray spectrography. *Seventh Ann. Rept. Univ. Leeds Res. Inst. African Geol.* **1961–62**, 59–63 (1963).
20. Houseknecht, T. M., and Patterson, W. Sample preparation for X-ray analysis. *Spectrographer's ARL News Letter* **17**, 2–3, 6 (1964).
21. Kodama, H., Brydon, J. E., and Stone, B. C. X-ray spectrochemical analysis of silicates using synthetic standards with a correction for interelement effects by a computer method. *Geochim. Cosmochim. Acta* **31**, 649–659 (1967).
22. Lachance, G. R. A simple X-ray method for converting measured X-ray intensities into mass compositions. *Geol. Surv. Can. Paper* **64–50**, 1964, 8 pp.
23. Lachance, G. R., and Traill, R. J. A practical solution to the matrix problem in X-ray analysis. I. Method. *Can. Spectr.* **11**, 43–48 (1966).
24. Liebhafsky, H. A., Pfeiffer, H. G., Winslow, E. H., and Zemany, P. D. *X-Ray Absorption and Emission in Analytical Chemistry*, 1960. New York: John Wiley and Sons, 355 pp.
25. Longobucco, R. J. Determination of major and minor constituents in ceramic materials by X-ray spectrometry. *Anal. Chem.* **34**, 1263–1267 (1962).
26. Madlem, K. W. Matrix and particle size effects in analyses of light elements,

zinc through oxygen, by soft X-ray spectrometry. *Advances in X-Ray Analysis*, Vol. 9, 1966. New York: Plenum Press, pp. 441–455.

27. Rose, H. J., Jr., Adler, I., and Flanagan, F. J.   Use of La_2O_3 as a heavy absorber in the X-ray fluorescence analysis of silicate rocks. *U.S. Geol. Surv. Profess. Papers* **450–B**, B80–B82 (1962).

28. Rose, H. J., Jr., Adler, I., and Flanagan, F. J.   X-ray fluorescence analysis of the light elements in rocks and minerals. *Appl. Spectr.* **17**, 81–85 (1963).

29. Rose, H. J., Jr., Cuttitta, F., Carron, M. K., and Brown, R.   Semimicro X-ray fluorescence analysis of tektites, using 50-mg samples. *U.S. Geol. Surv. Profess. Papers* **475–D**, D171–D173 (1964).

30. Rose, H. J., Jr., Cuttitta, F., and Larsen, R. R.   Use of X-ray fluorescence in determination of selected major constituents in silicates. *U.S. Geol. Surv. Profess. Papers* **525–B**, B155–B159 (1965).

31. Schnetzler, C. C., and Pinson, W. J., Jr.   A report on some recent major element analyses of tektites. *Geochim. Cosmochim. Acta* **28**, 793–806 (1964).

32. Shapiro, L., and Massoni, C.   Automatic sample changer and controller for an X-ray quantometer. *U.S. Geol. Surv. Profess. Papers* **525–D**, D178–D183 (1965).

33. Traill, R. J., and Lachance, G. R.   A new approach to X-ray spectrochemical analysis. *Geol. Surv. Can. Paper* **64–57**, 1965, 22 pp.

34. Traill, R. J., and Lachance, G. R.   A practical solution to the matrix problem in X-ray analysis. II.   Application to a multicomponent alloy system. *Can. Spectr.* **11**, 63–71 (1966).

35. Volborth, A.   Total instrumental analysis of rocks. Part A.   X-ray spectrographic determination of all major oxides in igneous rocks and precision and accuracy of a direct pelletizing method. *Nevada Bur. Mines Rept.* **6**, 1963, 72 pp.

36. Volborth, A.   Biotite mica effect in X-ray spectrographic analysis of pressed rock powders. *Am. Mineral.* **49**, 634–643 (1964).

37. Volborth, A.   Dual grinding and X-ray analysis of all major oxides in rocks to obtain true composition. *Appl. Spectr.* **19**, 1–7 (1965).

38. Volborth, A., and Banta, H. E.   Oxygen determination in rocks, minerals and water by neutron activation. *Anal. Chem.* **35**, 2203–2205 (1963).

39. Webber, G. R.   Applications of X-ray emission spectrometry to rock and ore analysis. *Can. Min. Met. Bull., Trans.* **60**, 138–143 (1957).

40. Welday, E. E., Baird, A. K., McIntyre, D. B., and Madlem, K. W.   Silicate sample preparation for light-element analyses by X-ray spectrography. *Am. Mineral.* **49**, 889–903 (1964).

41. Weyler, P. A.   Silicate analysis by X-ray spectrography. *Nevada Bur. Mines Rept.*, **13**, Part B, 99–105 (1966).

42. Wilson, H. D. B., Andrews, P., Moxham, R. L., and Ramlal, K.   Archean volcanism in the Canadian Shield. *Can. J. Earth Sciences*, **2**, 161–175 (1965).

43. Zussman, J., (Ed.).   *Physical Methods in Determinative Mineralogy*, 1967. "X-Ray Fluorescence Analysis," by K. Norrish and B. W. Chappell.   London: Academic Press, pp. 161–214.

Supplementary References

Guillemaut, A., and Mohadjer, K. Analyse des roches silicatées par spectrometrie de fluorescence X. *Rapport CEA-R3286*, Centre d'Etudes Nucléaires de Fontenoy-Aux-Roses, 1967, 14 pp. The sample is fused with borax and Co_3O_4 is added as an internal standard.

Nicholas, J., Quintin, M., and Douillet, Ph. Utilization de la méthode des poudres, sans fusion ni pastillage, pour le dosage rapide de K, Ca, Ti, Mn, and Fe dans les roches silicatées à l'aide de la fluorescence des rayons X. *Bull. soc. franc. Minér. Crist.* **89**, 469–476 (1966).

Price, N. B., and Angell, G. R. Determination of minor elements in rocks by thin film X-ray fluorescence techniques. *Anal. Chem.* **40**, 660–663 (1968). A few mg of 200-mesh powder are placed on an adhesive strip and mounted in a sample holder.

Strasheim, A., and Brandt, M. P. A quantitative X-ray fluorescence method of analysis for geological samples using a correction technique for the matrix effects. *Spectrochim. Acta* **23B,** 183–196 (1967). Disks are prepared by fusion both with $Na_2B_4O_7$ and $Li_2B_4O_7$. Results for Mg, Al and Si are averaged from measurements made on both disks, but Na is determined only on $Li_2B_4O_7$ disk. Corrections are applied to measurements for Ca, Fe, Mn, Ni and Ti.

CHAPTER 12

ATOMIC ABSORPTION SPECTROSCOPY

12.1. General

As far as rock and mineral analysis is concerned, atomic absorption spectroscopy* is unquestionably the newest analytical technique to receive enthusiastic acceptance from analysts. It was introduced by Walsh[10] in 1955 and, as has been so often the case, passed through a period of quiescence until commercially available instruments were produced which in turn sparked a surge of interest in, and experimentation with, the technique. The challenge that existed for the interested analyst to fabricate his own necessary instrumentation has now been replaced by the problem of deciding which one instrument, among the abundance of these now on the market, will do the job best for him.

The compelling reason for adding atomic absorption spectroscopy to the analytical facilities of the Geological Survey of Canada was to supplement the combination X-ray emission spectrographic and chemical rapid methods of rock analysis with a reliable procedure for determining low concentrations of magnesium ($<2\%$), a range of concentration over which the X-ray emission spectrographic procedure was generally unsatisfactory. Dumesnil and Perrault[5] have similarly incorporated this step into their published procedure. The determination of magnesium has always been particularly well served by atomic absorption spectroscopy and numerous references are available on the subject.

Belt[4] decomposed a 0.5 g sample of silicate rock with HF, HNO_3 and $HClO_4$, with final solution of the residue in HCl and a dilution to 50 ml; Na, K, Mn and Fe are determined in this solution, and Ca and Mg in an aliquot to which is added lanthanum as a releasing agent to suppress the interference of aluminum and phosphorus with these determinations. Shapiro[9] has proposed a single-solution method, based upon initial fusion of the sample with lithium metaborate (see the reference to the work of Ingamells and Suhr in Sec. 1.1) and solution of the cooled bead in dilute

---

* It perhaps would be more consistent to refer to this analytical technique as "flame absorption photometry" in order to indicate its close kinship with "flame emission photometry," usually referred to simply as "flame photometry." The more general term "atomic absorption spectroscopy" does, however, allow for the future use of an excitation medium other than a flame.

HCl; a concentration of 100 mg of sample per liter has been found best for the atomic absorption spectroscopic determination of Ca, Mg, Na, K and Mn (lanthanum is used as a buffer) and the spectrophotometric determination of Si, Al, Fe, Ti and P. Althaus[3] extended the method to include Na, K, Ca, Sr, Fe, Ni, Mn and Cr, using a solution obtained by decomposing the sample (100 mg) with HF–H$_2$SO$_4$; a large excess of cesium is added to suppress the mutual ionization interference of Na and K, and the interference of aluminum is eliminated by extracting the latter as hydroxyquinolate.

For detailed discussions of the technique the informative textbooks by Elwell and Gidley,[6] and Robinson[8] should be consulted, as should the general reviews by Allan,[2] Prugger[7] and West,[11] to mention only a few. No attempt has been made to include an extensive bibliography in this chapter. The publications cited will cumulatively contain references to a large portion of the published literature of atomic absorption spectroscopy.

The procedures to be described, and the supporting discussions, are based for the most part upon the work of Sydney Abbey of the Geological Survey of Canada, described by him in a Survey Paper.[1] The three methods are now in routine use, and those for several additional elements are being studied.

12.2. Some theoretical considerations

It is useful to consider the principles of atomic absorption spectroscopy in comparison with those of flame (emission) photometry and absorption spectrophotometry because certain features are shared by the three techniques. The consideration will be brief, however, because of the detailed discussions which are already available in the published literature.

Both the atomic absorption and flame photometric methods involve the spraying of a solution into a flame, in which a portion of the solute is reduced to free atoms of its constituent elements. In flame (emission) photometry some of these atoms, promoted out of their ground-state levels into higher excited states, emit characteristic radiation as they drop back to lower levels, and this is used to measure the concentrations of the constituent elements; a direct measurement is made of the intensity of light of a particular characteristic wavelength emitted by a specific atom species. Such emission is, however, normally received from only a relatively few such excited atoms and the bulk of the atoms remain in their unexcited ground state in the flame. These free atoms can also be raised to an excited state by the absorption of light but only by light of the energy corresponding to the difference in energy between the ground and excited states, that is, by incident energy at certain discrete wavelengths known as "resonant wavelengths." Thus, if light of a resonant wavelength of a particular element is passed through a flame into which a solution of a compound of that element is being sprayed, a fraction of that light will be absorbed by the ground-state atoms of the element and the extent of the absorption (or degree to which the incident light is diminished on passage through the flame) is a measure of the concentration of the absorbing atom species. The flame thus serves as an atom reservoir which can be likened to the cuvet or cell used in absorption spectrophotometry which provides a reservoir of molecular species; incident light, passing through the cell, is absorbed to a varying degree by the molecules.

The difference between absorption spectrophotometry and atomic absorption spectroscopy lies in the nature of the incident energy and absorption mechanism. Molecules have a complex electronic structure and their absorption spectra are characterized by broad molecular bands, rather than the narrow absorption bands (ideally, single lines) in the spectra of free atoms, a reflection of the relatively simple composition of atoms. Thus, while the light source for molecular absorption measurements is a continuous spectrum (commonly obtained from tungsten filament or hydrogen discharge lamps), that for atomic absorption spectroscopy consists of sharply defined spectrum lines and is restricted as much as possible to the resonant wavelengths of the particular atom species being measured. In practice, this incident light is generally produced by a "hollow cathode" lamp designed to emit the characteristic spectrum lines, including the resonant wavelengths, of the element in question. These latter, reduced in intensity by passage through the flame, are isolated by means of a grating or prism monochromator and their intensities measured by a photomultiplier tube, as is done in both absorption spectrophotometry and flame (emission) photometry.

The means by which selectivity of the absorbing species is obtained sharply differentiates between the three techniques that are being considered. In molecular absorption spectrophotometry it is necessary to insure by preliminary chemical treatment the absence of other molecular species which will also absorb some of the incident light, and it may require isolation of the desired species to achieve this. In both flame (emission) photometry and atomic absorption spectroscopy the flame is a reservoir of all the constituent atom species; the selectivity, in the case of flame photometry, is based upon the ability of a monochromator to discriminate between the spectrum lines of all the elements in the flame but, in atomic absorption, it is the narrowness of the band of incident energy provided by the source hollow cathode lamp. An interrupter, or pulsed power supply, in the source is attuned to an alternating current detector circuit in the receiver and the effect of continuous radiation from the flame is thus eliminated.

12.3. The determination of magnesium, lithium and zinc in silicates

The three applications of the atomic absorption method to be described in detail cover a range of analytical problems, including those of de-composition of the silicates and the treatment of insoluble residues.

The instrument used by Abbey[1] for the atomic absorption measurements is a Techtron Model AA-3 Atomic Absorption Spectrophotometer equipped with Atomic Spectral Lamps Pty., Ltd., hollow cathode lamps, with a flat-topped 10 cm slot laminar flow burner, an R-136 photomultiplier and a scale expander unit. A Varian G-10 Recorder was used for some measurements. Acetylene, mixed with air, was used as the burner fuel for all measurements; the air pressure was maintained at 18 lb/sq in. and the flow rate of the acetylene was varied to attain the desired flame conditions.

Among the other commercially available instruments there are differences in the types of source lamps used, the design of the burner and the sample aspirating system, the optical arrangement (including monochromator design and wavelength range) and the read-out system. The procedures to be described, although designed for use with the Techtron instrument, can probably be adapted for use with other instruments as well, with little or no modification. In the following pages no attempt will be made to give details of the operating parameters for the Techtron instrument used in the development and testing of these methods. These should be obtained from the original description of the methods.[1]

12.3.1. MAGNESIUM

The determination of magnesium is one of the most common applications of atomic absorption spectroscopy and is used at the Geological Survey of Canada to supplement the combined X-ray fluorescence–chemical rapid methods scheme for those samples expected to have magnesium concentrations $<2\%$. An aliquot of the sample solution prepared for the rapid spectrophotometric determination of phosphorus and the flame photometric determination of sodium (Sec. 9.19.3a) is used for this purpose. For the precise determination of magnesium a 100 mg sample is decomposed by treatment with hydrofluoric, nitric and perchloric acids.

The amount of absorption by magnesium is substantially reduced by "chemical" interference from aluminate, silicate, phosphate and sulfate anions, possibly as a result of the formation of compounds which are not completely decomposed in the flame and which thus reduce the concentration of free Mg atoms. Silica is eliminated by volatilization with hydrofluoric acid during preparation of the sample solution, and the concentrations of phosphate and sulfate are usually too low to be of significance (provided that H_2SO_4 is not used in the preparation of the sample solution); the only interference encountered was that caused by aluminum.

This interference may be eliminated by the use of the higher temperature flame obtained from a mixture of acetylene and nitrous oxides, or by the addition of a "releasing agent," a much higher concentration of another cation which preferentially combines with the anion causing the interference, thus releasing the cation whose concentration is being determined. Among such releasing agents for magnesium are lanthanum, calcium plus 8-hydroxyquinoline, and strontium. Preliminary tests indicated certain advantages in the use of strontium for this purpose; 500 ppm Sr in $0.12N$ HCl are sufficient to release up to 2 ppm Mg from the effect of as much as 20 ppm Al (up to 3.3% MgO in the presence of as much as 38% Al_2O_3, for a 10 mg sample).

In this determination, as in those which follow, the measurement steps consist of making a rough preliminary assessment of the concentration of Mg in the sample in order to determine the concentration of the standard solution to be used for comparison, and then taking at least two readings on the sample, each being preceded and followed by a reading on the selected standard solution. "Background" or "baseline" readings are taken, while water is aspirated into the flame, between all other readings.

Beer's law was found to apply in all measurements made of Mg concentration in the range 0–2 ppm in solution.

Reagents

Standard magnesium solution (10 ppm Mg): dilute a more concentrated solution, prepared from pure Mg metal, to a final concentration of 10 ppm Mg. The final solution should contain about 10 ml of free HCl per liter.

Strontium solution (5000 ppm Sr): weigh 8.43 g $SrCO_3$ into a 2-liter beaker, cover with water and slowly add 110 ml concentrated HCl. Boil to remove CO_2, cool and dilute to 1 liter in a volumetric flask.

Procedure

Preparation of the sample solution

(a) *Rapid method.* Decompose a 100 mg sample (−150 mesh) in a Teflon dish with HNO_3, HF and HCl as described in Sec. 9.19.3a for the preparation of a (100 ml) solution for the rapid colorimetric and flame photometric determinations of phosphorus and sodium respectively. For samples containing up to 15% MgO, transfer a 10 ml aliquot (corresponding to 10 mg of sample) to another 100 ml volumetric flask (it will be necessary, at the higher MgO concentrations, to reduce the sensitivity of measurement by some means, such as rotating the burner to decrease the length of the flame path). Add 10 ml of strontium solution and dilute to volume.

(b) *Precise method* (see also Abbey, Supp. Ref. Sec. 6.5.3). Weigh 1.000 g sample (−150 mesh) into a 100 ml platinum dish. Moisten with water and add 10 ml 15M HNO_3 and 5 ml HF (47%). Cover and heat on a steam bath until decomposition appears to be complete; rinse and remove the cover and evaporate the contents of the dish to dryness.

Add 5 ml concentrated HNO_3, rinse down the walls of the dish with a small amount of water and repeat the evaporation to dryness.

Add 2 ml $HClO_4$ (72%), rinse down the walls of the dish and swirl the contents to dissolve the soluble salts. Place the dish on a medium temperature sand bath and evaporate the contents to fumes of perchloric acid. Rinse down the walls of the dish and repeat the evaporation, continuing until no fumes are evolved.

Add 20 ml HCl (1:19), cover and warm on a steam bath to dissolve the salts. Filter through a 7 cm Whatman No. 42 paper into a 100 ml volumetric flask and wash with warm HCl (1:19), then twice with water. Use an aliquot of this solution for the appropriate preparation of the final solution as described in (a), *Rapid method*.

Calibration

Into five 100 ml volumetric flasks, measure 0, 5, 10, 15 and 20 ml aliquots of the standard magnesium solution (10 ppm). Add 10 ml strontium solution to each flask and dilute to volume. These solutions will contain 0.0, 0.5, 1.0, 1.5 and 2.0 ppm magnesium.

Adjust the operating parameters of the atomic absorption spectrophotometer being used, including the Mg hollow cathode lamp, acetylene–air fuel mixture and burner head, to give a convenient reading on the dial or chart for a standard Mg solution containing 1 ppm.

Aspirate these solutions into the flame and record the readings; a duplicate run should be made. Take a reading, with water aspirating, before and after each reading on a standard. Convert each reading to absorbance.

Calculate the net absorbance from each absorbance reading by subtracting from the latter the mean of the water absorbance readings immediately following and preceding it. Calculate the average of the two net absorbances obtained for each standard solution. Plot mean net absorbance against ppm magnesium; these points should fall on a straight line.

Because it is not possible to guarantee that operating parameters are always identical, it is not advisable to use this calibration curve for accurate analysis. It is drawn to indicate that Beer's law is obeyed and used to obtain a first approximation of the Mg content of unknown samples. The final determination is done by comparing the readings obtained for the sample solution with that for a closely similar standard.

Sample analysis

Adjust the instrument as previously described (Calibration) and, using the 1.0 ppm standard solution for comparison, take a rough reading for each sample. Select a standard as close as possible in Mg content to each sample and continue taking readings for the samples and standards; a suggested sequence is

$$0\text{-}s\text{-}0\text{-}x\text{-}0\text{-}s\text{-}0\text{-}x\text{-}0\text{-}s\text{-}0$$

where 0 is water, s the standard and x is a sample. Before going from one sample to another, check the wavelength for the desired response and readjust the zero setting if necessary.

Convert all readings to absorbance and calculate the net absorbance for each as described under **Calibration.** Assuming that concentration is directly proportional to absorbance, calculate the apparent concentration for each net absorbance reading for each sample, using the standard readings preceding and following each sample reading. Determine the mean of the four apparent concentrations thus obtained by using the above sequence of readings. If the four values are not in close agreement, further readings should be taken.

For a *10* mg sample in *100* ml of solution;

% Mg in the sample = ppm Mg (in the solution)

%MgO = % Mg × 1.658

12.3.2. LITHIUM

Lithium is an easily excited element and in a flame into which a lithium solution is aspirated there will exist an equilibrium which governs the relative proportions of the lithium atoms that are in the ground state and at the several possible levels of excitation. The presence of another easily excited element in the same flame could upset this equilibrium by redistribution of the excitation energy between the two elements. An increase in the population of ground state atoms of lithium arising out of this redistribution would enhance the absorption of lithium radiation and result in a positive error in the absorbance reading; this is the "ionization effect," a common interference in both flame photometry and atomic absorption spectroscopy.

It might have been expected that ionization interference would result from the relatively high concentrations of sodium and potassium found in most silicate rocks. Preliminary tests showed, however, that up to 5% of sodium or potassium in the sample had no effect upon the absorbance of 50 ppm lithium in the same sample.

The lithium absorbance reading was found to be markedly affected by the concentration of free acid in the sample solution. The free acid concentration selected for routine work was $0.5N$ in HCl; an increase in acidity to $1.0N$ reduced the absorbance by about 10%, to $3.0N$ by 50%.

An important result of preliminary studies was the recognition of the apparent depression of the lithium absorption signal by the total salt concentration of the solution.[1] There appeared to be a relationship between the magnitude of the negative bias and the overall composition of the solution. Measurements of the lithium absorbances in synthetic sample solutions of varying lithium content, and having a gross composition ranging from that equivalent to a granite to one equivalent to a dunite, revealed that the lithium absorption signal was weakened with increasing total salt content of the sample solution but that the depression was not directly proportional to the salt concentration. For samples of unknown composition it is thus necessary to establish the magnitude of the matrix depressant effect by "standard addition" tests; where a large number of samples of similar composition are involved, a correction factor can be established for a few samples, and then applied to the remainder.

The absorption sensitivity of lithium is significantly affected by changes in air- and fuel-flow rates in the flame. The instrumental readings for lithium tend to be less stable than those for magnesium; Beer's law is obeyed over the range 0.0–2.5 ppm Li, corresponding to 0–125 ppm Li in the original sample.

Reagents

Standard lithium solution (10 ppm Li): dilute an aliquot of a more concentrated solution, prepared from reagent grade Li_2CO_3 and HCl, to give a solution containing 10 ppm lithium.

Procedure

Preparation of the sample solution

Decompose a 0.5 g sample of -150 mesh material as described in Sec. 12.3.1, Precise Method, but catch the filtrate in a 100 ml beaker instead of a volumetric flask. The possibility of some lithium being retained in any insoluble residue from this acid treatment must be considered and it may be necessary to recover it by fusing the insoluble residue in a platinum crucible with a small quantity (0.5 g) of sodium carbonate, leaching the melt with water (plus a few drops of 95% ethanol), filtering and diluting the filtrate to 25 ml in a volumetric flask; the residue on the filter paper should also be dissolved and examined for lithium. A blank solution should be made at the same time by dissolving 0.5 g Na_2CO_3 in water plus alcohol in a 20 ml beaker.

Calibration

Into six 100 ml volumetric flasks measure 0, 5, 10, 15, 20 and 25 ml aliquots of the standard lithium solution (10 ppm). Add 4 ml concentrated HCl to each flask, and dilute to volume. This gives a series of standard solutions containing 0.0, 0.5, 1.0, 1.5, 2.0 and 2.5 ppm lithium.

Adjust the operating parameters of the atomic absorption spectrometer being used, including the lithium hollow cathode lamp, acetylene–air fuel mixture and burner head, to give a convenient reading on the dial or chart for a standard Li solution containing 1 ppm. Obtain readings, calculate the net absorbances and plot the mean net absorbances for these standard solutions against ppm lithium as in the determination of magnesium. The plotted points should join to give a straight line. As for magnesium, the use of this working curve should be limited to the determination of the approximate lithium content of the sample; the effect of changes in flame conditions is even more marked for lithium.

Sample analysis

The further treatment of the sample solution obtained as previously described will depend upon the nature of the sample.

(*a*) If the matrix correction factor is unknown, the standard addition method must be used.

(*b*) If the correction factor has been established, use the direct method and apply the correction.

(*c*) If a fusion of unattacked residue was necessary, the filtered leachate requires special treatment.

(*a*) Standard addition method

Measure the volume of the filtrate obtained from the acid decomposition of the 0.5 g sample in a graduated cylinder and dilute to the next multiple of 5 ml. Mix and measure two aliquots, each 2/5 of the final volume, into 50 ml beakers. Add 1.00 ml of the lithium standard solution (10 ppm) to *one* of the beakers and evaporate the contents of both beakers to dryness.

Warm the dried residue with 8 ml of HCl (1:19) to dissolve it, transfer the solution quantitatively to a 10 ml volumetric flask and dilute to volume. The two solutions thus obtained represent the sample solution and the sample solution with 1 ppm Li added.

Using the same operating conditions for the instrument as were used in the calibration step, obtain a first approximation of the lithium content of the sample solution by comparison with any standard Li solution. Select standard solutions that compare closely with the sample solution and with the sample solution plus 1 ppm lithium, and continue taking readings for the samples and standards; a suggested sequence is

$$O\text{-}Sx\text{-}O\text{-}X\text{-}O\text{-}Sx\text{-}O\text{-}X\text{-}O\text{-}Sx\text{-}O\text{-}Sy\text{-}O\text{-}Y\text{-}O\text{-}Sy\text{-}O\text{-}Y\text{-}O\text{-}Sy$$

where O is water, X is the sample solution, Y the sample solution plus 1 ppm Li, and Sx and Sy are respectively the standard solutions similar in composition to X and Y.

Calculate net absorbances for X and Y, as described in Sec. 12.3.1, (**Calibration**) and from these obtain the "apparent concentrations" for each unknown. The "true concentration," Tx, of the sample solution is given by

$$Tx = (Ax/Ay - Ax)$$

where Ax = apparent concentration of sample solution and Ay = apparent concentration of sample solution + 1 ppm Li. Multiply Tx by the dilution factor to obtain ppm Li in the original sample (for 2/5 of 0.5 g in a final volume of 10 ml, the dilution factor is 50).

(b) Direct method

Evaporate the filtrate obtained from the acid decomposition of the 0.5 g sample to dryness and redissolve the residue by warming it with 20 ml HCl (1:19). Transfer the solution quantitatively to a 25-ml volumetric flask and dilute to volume.

Determine the apparent concentration of lithium in the sample solution as described in the preceding standard addition method. Calculate the true concentration (i.e., correct for the matrix depressant effect) by multiplying the apparent concentration by a factor

$$F = (1/Ay - Ax)$$

where Ay and Ax have the same meaning as in the standard addition method. The value for F that is used should be the mean of those obtained in analyzing a number of samples of similar composition, using the standard addition method.

(c) Fusion method

Measure the volume of the blank solution (see **Preparation of the Sample Solution**) in a graduated cylinder and dilute to the next multiple of 5 ml. Measure two portions, each 2/5 of the volume, into 50 ml beakers and add 1.00 ml of the standard lithium solution (10 ppm) to one of them. Evaporate the solutions to about 5 ml, transfer quantitatively to 10 ml volumetric flasks and dilute to volume.

Determine the apparent concentration of lithium in the filtrate from the water leach of the Na_2CO_3 fusion and in the blank and "spiked" blank solutions, as described in the preceding standard addition method. Calculate the true concentrations by multiplying apparent concentrations by a factor

$$G = (1/By - Bx)$$

where Bx = apparent concentration of the blank, and By = apparent concentration of the blank to which 1 ppm Li has been added. If Bx is not zero, it must be subtracted from the concentration of the sample solution as a blank correction.

Dissolve the residue from the leaching of the fusion cake on the filter paper by passing through it repeated portions of HCl (1:19) which have been warmed in the platinum crucible in which the fusion was done, catching the solution in a 50 ml beaker. Wash the paper several times with additional warm HCl (1:19). Evaporate the contents of the beaker to dryness.

Dissolve the dried salts in 20 ml HCl (1:19), warming if necessary, transfer the solution quantitatively to a 25 ml volumetric flask and dilute to volume. Determine the lithium concentration as for the leach solution; no correction factor is necessary because the salt concentration is relatively low.

12.3.3. ZINC

Like magnesium, zinc has received much attention from users of the atomic absorption spectroscopic method because of the high sensitivity of determination that is possible; the numerous studies that have been made also indicate that little, if any, interference will result from the major constituents present in the sample. The tests made at the Geological Survey showed that the determination of zinc by this method is a remarkably simple matter.[1]

"Standard addition" runs were made, as previously described in Sec. 12.3.2 (Lithium), with synthetic mixtures approximating sample compositions ranging from granite to dunite. Plots of the "apparent zinc found" against known zinc content were straight lines in all cases, with slopes that departed from unity by no more than 3%. No additional manipulation is required to correct for matrix effect.

Varying quantities of HCl were added to these synthetic mixtures also and no noticeable difference in the zinc absorption values occurred for HCl concentrations of 0.25–1.5N. It is thus not necessary to evaporate the filtrate from the acid treatment of the sample and redissolve it in a controlled amount of acid.

Using the acid decomposition procedure described in Sec. 12.3.1 (*Precise method*), it was found that, for the range of silicate rock samples tested, over 95% of the zinc was obtained in the filtrate from the acid treatment.

Reagents

Standard zinc solution (10 ppm Zn): dilute a more concentrated solution, prepared from reagent-grade zinc metal and HCl, to a final concentration of 10 ppm Zn.

Procedure

Preparation of the sample solution

Decompose a 0.5 g sample as described in Sec. 12.3.1 (*Precise method*), but catch the filtrate from the acid treatment in a 100 ml beaker. Evaporate the solution to about 40 ml, if necessary, then transfer it quantitatively to a 50 ml volumetric flask and dilute to volume.

The residue remaining on the filter paper may usually be discarded without further treatment. If the presence of zinc in the residue is suspected, however, then the fusion and subsequent steps should be followed.

Calibration

Into six 100 ml volumetric flasks, measure 0, 5, 10, 15, 20 and 25 ml aliquots of the standard zinc solution. Add 2 ml concentrated HCl to each and dilute to volume. This gives a series of standards containing 0.0, 0.5, 1.0, 1.5, 2.0 and 2.5 ppm Zn. Adjust the operating parameters of the atomic absorption instrument being used, including the zinc hollow cathode lamp, acetylene–air fuel mixture and burner head to give a convenient reading on the dial or chart for a standard Zn solution containing 1 ppm. Obtain readings, calculate the net absorbances and plot the mean net absorbances for these standard solutions against ppm zinc, as described in Sec. 12.3.1 (**Calibration**). The calibration plot should be a straight line over more than half its length and may be assumed to be straight for the purpose of obtaining a first approximation of the zinc content of unknown samples. In the final determination, if the sample and the standard solution used for comparison are sufficiently similar in zinc content, very little error is introduced into the determination by assuming that the calibration plot is linear.

Sample analysis

Adjust the instrumental operating conditions as described under **Calibration** and, using the 1 ppm Zn standard for comparison, take a rough reading for each sample. Select a standard solution as close as possible in zinc content to the unknown, and obtain readings and calculate the results as described for the determination of magnesium (Sec. 12.3.1).

References

1. Abbey, Sydney. Analysis of rocks and minerals by atomic absorption spectroscopy. Part I. Determination of magnesium, lithium, zinc and iron. *Geol. Surv. Canada Paper* **67-37**, 1967, 35 pp.
2. Allan, J. E. A review of recent work in atomic absorption spectroscopy. *Spectrochim. Acta* **18**, 605–614 (1962).
3. Althaus, E. Die Atom-Absorptions-Spektralphotometrie—ein neues Hilfsmittel zur Mineralanalyse. *Neues Jahrb. Mineral. Monatsh.* **9**, 259–280 (1966).
4. Belt, C. B. Partial analysis of silicate rocks by atomic absorption. *Anal. Chem.* **39**, 676–678 (1967).

5. Dumesnil, J. C., and Perrault, G.   La part de la fluorescence X dans le dosage rapide des éléments des roches. *Canadian Spectr.* **12**, 3–11 (1967).
6. Elwell, W. T., and Gidley, J. A. F.   *Atomic Absorption Spectrophotometry*, 2nd ed., rev., 1966. Oxford: Pergamon Press, 138 pp.
7. Prugger, H.   The development of absorption flame photometry to a chemical laboratory technique. *Carl Zeiss Information* **No. 56**, 54–59 (1966).
8. Robinson, J. W.   *Atomic Absorption Spectroscopy*, 1966. New York: Marcel Dekker, 204 pp.
9. Shapiro, L.   Rapid analysis of rocks and minerals by a single-solution method. *U.S. Geol. Surv. Profess. Papers* **575–B**, B187–B191 (1967).
10. Walsh, A.   The application of atomic absorption spectra to chemical analysis. *Spectrochim. Acta* **7**, 108–117 (1955).
11. West, T. S.   Atomic analysis in flames. *Endeavor* **26**, 44–49 (1967).

Supplementary References

Abbey, Sydney.   Analyses of rocks and minerals by atomic absorption spectroscopy. Part 2. Determination of total iron, magnesium, calcium, sodium and potassium. *Geol. Surv. Canada Paper* **68–20, 1968**, 21 pp.

Galle, O. K., and Angino, E. E.   Determination of calcium and magnesium in carbonate and silicate rocks by atomic absorption. *Bull. Geol. Surv. Kansas* **187**, Part 1, 9–11 (1967).

Katz, A.   The direct and rapid determination of alumina and silica in silicate rocks and minerals by atomic absorption spectroscopy. *Am. Mineral.* **53**, 283–289 (1968).

Shapiro, L., and Massoni, C. J.   Automatic sample changer for atomic absorption spectrophotometry. *U.S. Geol. Surv. Profess. Papers* **600–B**, B126–B129 (1968).

APPENDIXES

APPENDIX 1

A SCHEME FOR CONVENTIONAL SILICATE ROCK ANALYSIS (FLUORINE <2%)

Dotted lines indicate optional methods

Main Portion

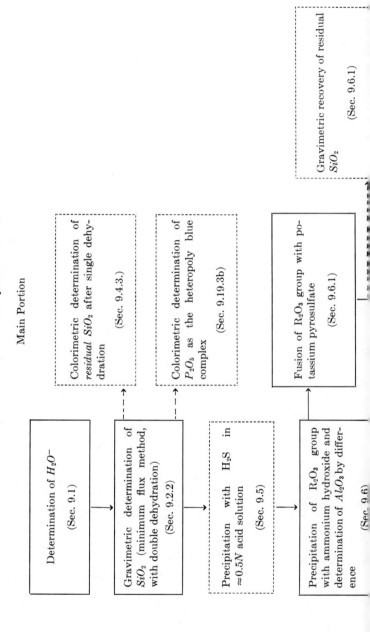

Determination of H_2O^-

(Sec. 9.1)

Gravimetric determination of SiO_2 (minimum flux method, with double dehydration)

(Sec. 9.2.2)

Precipitation with H_2S in $\approx 0.5N$ acid solution

(Sec. 9.5)

Precipitation of R_2O_3 group with ammonium hydroxide and determination of Al_2O_3 by difference

(Sec. 9.6)

Colorimetric determination of *residual* SiO_2 after single dehydration

(Sec. 9.4.3.)

Colorimetric determination of P_2O_5 as the heteropoly blue complex

(Sec. 9.19.3b)

Fusion of R_2O_3 group with potassium pyrosulfate

(Sec. 9.6.1)

Gravimetric recovery of residual SiO_2

(Sec. 9.6.1)

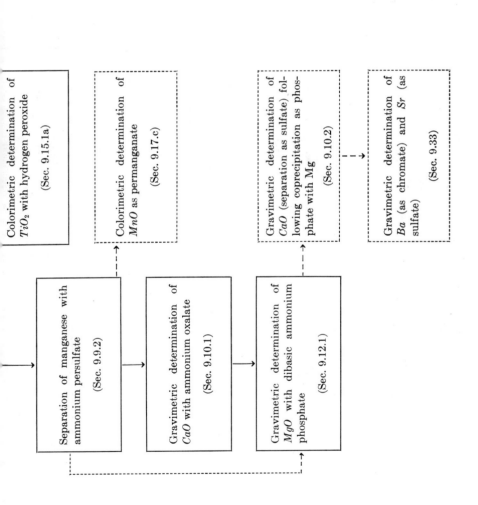

Colorimetric determination of TiO_2 with hydrogen peroxide
(Sec. 9.15.1a)

Colorimetric determination of MnO as permanganate
(Sec. 9.17.e)

Gravimetric determination of CaO (separation as sulfate) following coprecipitation as phosphate with Mg
(Sec. 9.10.2)

Gravimetric determination of Ba (as chromate) and Sr (as sulfate)
(Sec. 9.33)

Separation of manganese with ammonium persulfate
(Sec. 9.9.2)

Gravimetric determination of CaO with ammonium oxalate
(Sec. 9.10.1)

Gravimetric determination of MgO with dibasic ammonium phosphate
(Sec. 9.12.1)

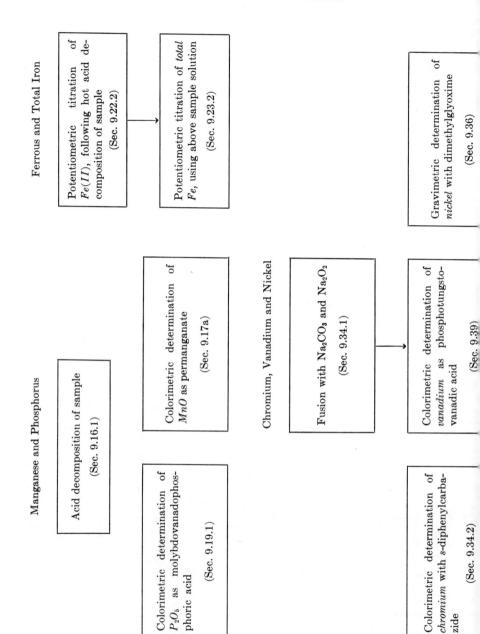

Ferrous and Total Iron

Potentiometric titration of Fe(II), following hot acid decomposition of sample (Sec. 9.22.2)

Potentiometric titration of *total Fe*, using above sample solution (Sec. 9.23.2)

Gravimetric determination of *nickel* with dimethylglyoxime (Sec. 9.36)

Manganese and Phosphorus

Acid decomposition of sample (Sec. 9.16.1)

Colorimetric determination of *MnO* as permanganate (Sec. 9.17a)

Colorimetric determination of P_2O_5 as molybdovanadophosphoric acid (Sec. 9.19.1)

Chromium, Vanadium and Nickel

Fusion with Na_2CO_3 and Na_2O_2 (Sec. 9.34.1)

Colorimetric determination of *vanadium* as phosphotungstovanadic acid (Sec. 9.39)

Colorimetric determination of *chromium* with s-diphenylcarbazide (Sec. 9.34.2)

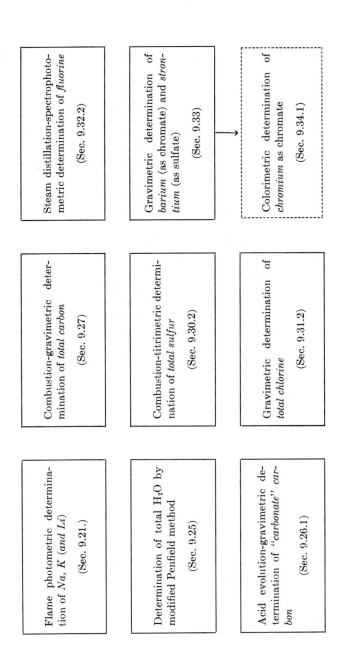

Flame photometric determination of Na, K (and Li)

(Sec. 9.21.)

Combustion-gravimetric determination of *total carbon*

(Sec. 9.27)

Steam distillation-spectrophotometric determination of *fluorine*

(Sec. 9.32.2)

Determination of total H_2O by modified Penfield method

(Sec. 9.25)

Combustion-titrimetric determination of *total sulfur*

(Sec. 9.30.2)

Gravimetric determination of *barium* (as chromate) and *strontium* (as sulfate)

(Sec. 9.33)

Acid evolution-gravimetric determination of *"carbonate"*, *carbon*

(Sec. 9.26.1)

Gravimetric determination of *total chlorine*

(Sec. 9.31.2)

Colorimetric determination of *chromium* as chromate

(Sec. 9.34.1)

APPENDIX 2

A SCHEME FOR THE RAPID CHEMICAL ANALYSIS OF SILICATE ROCKS

Solution A

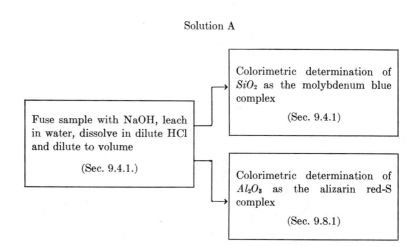

Fuse sample with NaOH, leach in water, dissolve in dilute HCl and dilute to volume

(Sec. 9.4.1.)

Colorimetric determination of SiO_2 as the molybdenum blue complex

(Sec. 9.4.1)

Colorimetric determination of Al_2O_3 as the alizarin red-S complex

(Sec. 9.8.1)

Separate Portions

Titrimetric determination of $Fe(II)$ by cold acid–ammonium metavanadate method

(Sec. 9.22.3)

Determination of *total* H_2O by modified Penfield method

(Sec. 9.25)

Acid evolution–gravimetric determination of *"carbonate"* carbon

(Sec. 9.26.1)

Solution B

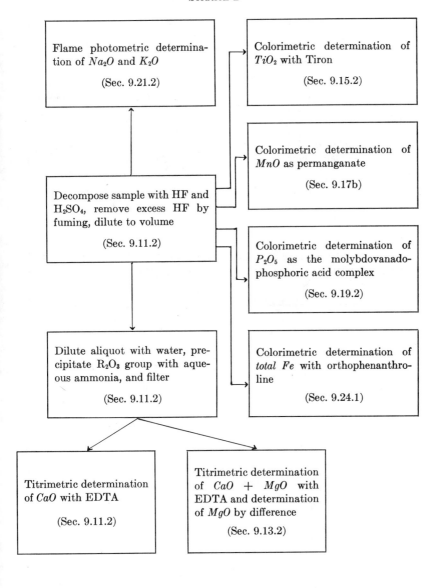

APPENDIX 3

A SCHEME FOR THE COMBINED X-RAY FLUORESCENCE-CHEMICAL RAPID ANALYSIS OF SILICATE ROCKS

X-Ray Fluorescence Spectrography

Fusion (Sec. 11.4.2)   Direct Pelletization (Sec. 11.4.3)

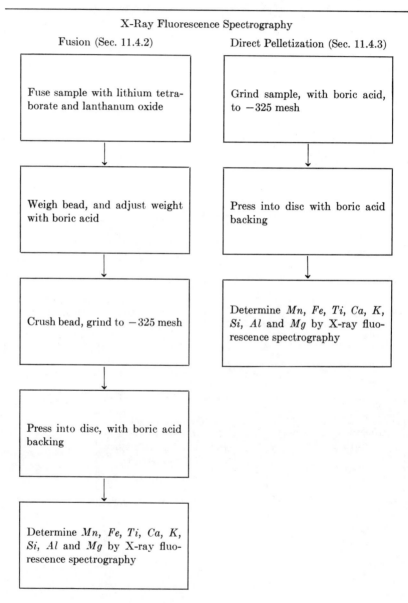

Fuse sample with lithium tetra-borate and lanthanum oxide

Grind sample, with boric acid, to −325 mesh

Weigh bead, and adjust weight with boric acid

Press into disc with boric acid backing

Crush bead, grind to −325 mesh

Determine *Mn*, *Fe*, *Ti*, *Ca*, *K*, *Si*, *Al* and *Mg* by X-ray fluorescence spectrography

Press into disc, with boric acid backing

Determine *Mn*, *Fe*, *Ti*, *Ca*, *K*, *Si*, *Al* and *Mg* by X-ray fluorescence spectrography

Chemical

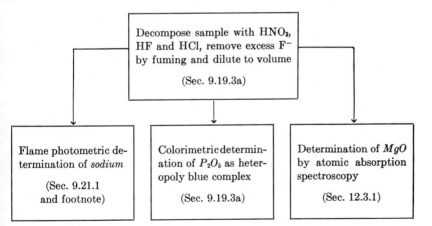

Titrimetric determination of *Fe(II)* by cold acid–ammonium metavanadate method

(Sec. 9.22.3)

Determination of *total water* by modified Penfield method

(Sec. 9.25)

Acid evolution–gravimetric determination of *"carbonate"* carbon

(Sec. 9.26.1)

Decompose sample with HNO₃, HF and HCl, remove excess F⁻ by fuming and dilute to volume

(Sec. 9.19.3a)

Flame photometric determination of *sodium*

(Sec. 9.21.1 and footnote)

Colorimetric determination of *P₂O₅* as heteropoly blue complex

(Sec. 9.19.3a)

Determination of *MgO* by atomic absorption spectroscopy

(Sec. 12.3.1)

AUTHOR INDEX

Numbers in parentheses are reference numbers and indicate that the author's work is referred to although his name is not mentioned in the text. Numbers in italics show the pages on which the complete references are listed.

A

Abbey, S., 93(1), *106*, 160, *162*, 169, 171, *172*, *173*, 198, 199, *200*, *202*, 211, 213, *217*, 290, *291*, 298, *299*, 409, 479, 521, 523, 527(1), 531(1), *532*, *533*
Abdullah, M. I., *186*
Adams, J. A. S., 195(3), *200*, 260(3), *261*, 269(4), *271*
Adler, I., 498, *516*, 504(27,28), 505(28), 508, *518*
Agnihotri, S. K., *217*
Ahlers, P. E., 117(36), 126(36), *130*, 340
Ahrens, L. H., 5, *15*, 22(1), *33*, 39 (15), 45(1), 51(1), 53, *56*, *57*, 79(1), *86*, 95, *107*, 247(1), *247*, 256(1), *256*, 261(1), *261*, 267(1), 269(1), *270*, 281(1), *281*, 283(4), 284(4), 285, 287(1,3), *287*, 293(1), 293, 296, 297(1), *298*, *300*, *301*, 303, *312*
Alimarin, I. P., 77, *86*, *87*, 293(6), *293*
Allan, J. E., 521, *532*
Alon, A., 128, *129*
Althaus, E., 84, *87*, 521, *532*
Amos, M. D., 146, *148*
Anderson, D. H., *35*
Andersson, L. H., 117, *129*, 331
Andrews, P., 504(42), *518*
Angell, G. R., *519*
Angino, E. E., *533*
Antweiler, J. C., 90, *106*
Archer, D. W., 189(1), 190, *193*
Armand, M., 124, *129*
Armstrong, A. R., 177, *180*, 213(44), *216*
Armstrong, G. W., 244(10), *248*
Asklund, A.-M., *16*
Asta'vfev, V. P., 100(4), *106*
Athavale, V. T., 101, 103(5), *106*
Azeem, M., 237, *243*

B

Baadsgaard, H., 191(2), *193*, 392
Babko, A. K., *34*
Badrinas, A., 168(13), *172*
Bagshawe, B., 293(7), *293*
Baird, A. K., *59*, 501(2,3), 503(4,40), 505(5), *516*, *518*
Bakes, J. M., 250, *257*
Ball, D. F., 505, *517*
Ball, T. K., 117(13), *129*
Ballard, J. W., 47, *56*
Bal'yan, K. V., 230, *233*, 438(Sec. 6.13, Ref. 25)
Balyuk, S. T., 103(6), *106*
Bankier, J. D., 37(56), 38, *59*
Banks, C. V., 264(8), 266(8), 267(8), 269(8), *271*
Banta, H. E., 11(23), *16*
Barnard, A. J., Jr., 155(1), *160*
Barnett, P. R., 45, 47(3), *56*, *58*
Baugh, C. A., 155, *160*
Beamish, F. E., 61(17), 66, 76(17), *87*
Beater, B. E., 169(2), *172*
Beeson, M. H., *17*
Béguinot, J., *256*, 303(6), 307(6), *312*
Belcher, R., 77, *86*, 157, *160*
Belt, C. B., 520, *532*
Bennett, H., 121, 124, *129*, *131*, 143, *147*, 157(4), *160*, 167 (Sec. 6.5, Ref. 4), 168 (3), *172*, 198(2), *200*, 236(1), 241(1), *242*, 253, *256*, 260, *261*, 266, *270*, 279, *281*, 295, *298*, 303, 306–308, *312*, 314 (1), *319*, *320*, 332, 333, 354, 399, 490
Berman, S. S., 260(9), *262*, *292*
Bernas, B., 128(1), *129*, 254, *258*
Berry, H., 48(4), 56, *56*
Berthelay, J. C., *148*, *257*
Berthoux, J., 124, *129*

548 AUTHOR INDEX

Edwards, J. W., 211 (22), *215*
Eeckhaut, J., 154, *161*, 259(Sec. 6.5.1a, Ref. 25)
Efros, S. M., *107*
Elinson, S. V., 277, 278(14), 279, 281, *282*, 472(Sec. 6.20, Ref. 14)
Ellestad, R. B., 199(9), *200*, 414
Ellingboe, J. L., 156(8), 157(8), *161*, 230 (3), *232*, 438(Sec. 6.13, Ref. 3)
Elliott, C. J., *111*, 318(12), *319*
Elving, P. J., 19(18), *34*, 43(27), *57*, 81, *87*, 90(34), 104(34), *108*, 117(24), 123 (24), 124(24), 126(24), 128(24), *130*, 136, *149*, 150, *163*, 163(16), 166(16), 168(16), 169(16), *172*, 176(4), 179(4), *180*, 182(2), *186*, 189(6), 191(6), 192 (6), *193*, 194–198(20), *201*, 203(17), 211(17), *215*, 218(12), 223(12), *224*, 236(11), *243*, 244(10), 246(10), *248*, 250–256(19), *257*, 260, 261, *262*, 264 (8), 266(8), 267(8), 269(8), *271*, 273–275(2), *276*, 276–278(8), 279, 280(8), *282*, 283(9), 284(9), 287(9), *287*, 289 (3), *291*, 293(3), *293*, 295(7), 298(7), *298*, *301*, *302*, 384 (Sec. 6.8, Ref. 2), 473
Elwell, W. T., 176(17), 179(17), *180*, 264 (12), 267(12), 269(12), *271*, 521, *533*
Engle, C. G., 502(8,9), *517*
Epshtein, R. Ya., 95, *107*
Erd, R. C., 314(7), *319*
Erlank, A. J., 195, *202*
Eshelman, H. C., 146, *147*
Esikov, A. D., 103, *107*
Esson, J., 293, *293*
Evans, W. H., *148*, *257*
Ewing, G. W., 79(7), *87*
Exley, C. S., 37(13), *57*
Eyring, L., 283(4), 284, *287*

F

Fabregas, R., 168(13), *172*
Fahey, J. J., 205, 207, *215*, *216*, 226, *232*, 314(7,8), *319*
Fairbairn, H. W., 19(4,5), 21, 22(6), 25, 27(4), *33*, 38(15), 54, *57*

Farncomb, F. J., 237(21), 238(21), 241 (21), 242(21), *243*
Farrah, G. H., *149*
Fassel, V. A., 287, *287*
Feldman, C., 197, 199(10), *200*
Ferguson, M., 159(31), *162*
Filby, R. H., 23, 117(13), *129*, 297, *299*
Finn, A. N., 102, *107*
Fisher Scientific Co., 158(11), 159(10), *161*
Fitzgerald, D. M., 237(24), 242(24), *243*
Flanagan, F. J., 31, *35*, 504(27,28), 505 (28), 508, *518*
Flaschka, H., 135, *135*, 155(1), *160*
Fleischer, M., 19(7,8), 22(7), 29, *33*, 284, *287*
Fletcher, M. H., 197(11), *200*
Flinn, D., 41, *57*
Flynn, L. R., *173*
Foner, H. A., 69, *87*
Fox, E. J., 250(7,8), 251, 252, *256*
Fraser, W. E., 160, *163*
Frenkel, C., 128(1), *129*
Frid, B. I., 77, *87*
Friedman, I., 218(4), 222, *223*
Friese, G., *35*
Fritz, J. S., 154(12), *161*, 259(Sec. 6.5.1a, Ref. 12)
Frost, I. C., 230–232, *232*, 438(Sec. 6.13, Ref. 5)
Fuchs, R. J., 123(12), *129*
Fuge, R., *271*, 275, *276*
Fukasawa, T., 96, *107*

G

Galle, O. K., 40, *57*, 226, *232*, 303–305 (9), 307(9), 308(9), *312*, 488(Chap. 9, Ref. 7), *533*
Gaon, M., 164, *172*
Gardner, K., 194(32), *202*
Garralda, B. P., 154(12), *161*, 259(Sec. 6.5.1a, Ref. 12)
Garrett, H. E., 126(14), *129*
Gault, H. R., 305(4), *312*
Geijer, P., 245, 246, *247*, 251, 252, *256*, 447(Sec. 6.15, Ref. 5)
Gidley, J. A. F., 521, *533*

Gilbert, T W., Jr., 295, 298(7), *298*
Gill, H. H., 244(10), *248*
Gillberg, M., 245, 247, *247*, 249, *256*
Ginsberg, G. P., 95, *107*
Ginsburg, L., *298*
Glasö, O. S., 253, 255(11), *256*
Glover, M. J., 316, *319*
Glukhova, L. P., *292*
Goddard, A. P., 145(18), *147*, 355
Goldberg, E. D., 204, 211, *217*
Goldich, S. S., 27(9), *33*, *35*, 120(15), *129*, 156, *161*, 167, 227, *232*, 240, *242*, 303, 305, *312*, 330, 433
Goles, K., *17*
Gordon, C. L., 104, *107*, *110*
Gordon, G. E , *17*
Gordon, L., 153(14), *161*
Gorfinkle, L. G., 38(15), *57*
Goryushina, V. G., *301*
Goto, K., 291, *292*
Govindaraju, K., 6(6), *15–17*, *35*, 195, *202*
Grady, H. R., 273(2), 274(2), 275, *276*
Graff, P. R., 4(11), 6, 7, *15*, 23, 32, 91, 105, *107*, *108*, 205, 207, *215*, 417
Graham, A. L., *35*
Graham, R. P., 177(7,10), 179(2,7,10), *179*, *180*, 273(3), *276*, 471(Sec. 6.19, Ref. 3)
Grant, J. A., 229(9), *233*
Grassman, H., 24, *35*
Gray, G. A., *173*
Greenfield, S., 128, *129*, 189(12), *194*
Greenhalgh, R., 255(12), *256*
Greenland, L., 245, *248*, 256(13), *256*
Gregory, G. R. E. C., 179, *180*
Grillot, H., 38, 44, 46, *57*, 195(12), *200*, *256*, 303, 307, *312*
Grimaldi, F. S., 96, 101(22), 103, *107–109*, 289, *291*, 310(7), *312*
Grimes, M. D., 236(11), *243*
Grimshaw, R. W., 55, *59*
Grout, F. F., 36(19), 40, 44, *57*
Groves, A. W., 3(7), 9(7), 14, *15*, 18, 19, 21, *33*, 37, *57*, 140, 141, *147*, 153, *161*, 196(13), *201*, 204, *215*, 218, 219(5), 220(5), 222(5), *223*, 227(10), 231, *233*, 236–238, 241(7), *242*, 245, 246, *248*,

260(7), *262*, 278(6), *281*, 284(7), *287*, 300, *300*, *301*, 347, 399
Grundulis, V., *16*
Guillemaut, A., *519*
Gunn, B. M., 506(12), 507(12), *517*
Gunn, E. L., 505, *517*
Gurney, J. J., 195, *202*
Guthrie, W. C. A., 222, *223*
Gwyn, M. E., 31, *35*

H

Habashy, M. G., 214, *215*
Haddock, L. A., 236(23), *243*, 244(19), *248*
Hadijoannou, T. P., 158, *162*, 167(Sec. 6.5, Ref. 28).
Hagner, A. F., 208, 210, 211(4), *214*
Hahn, R B., 276(8), 277, 278(8), 280(8), *282*, 472
Hall, R. A., 106, *110*
Hall, R. J., 169, *173*
Hall, W. T., 95(23), *107*
Halstead, R. L., 226(23,24), *233*
Hamilton, E. I., 507(17), *517*
Hampel, J., *16*
Hanna, W. C., 303, *312*
Harden, G., 100, *110*
Hardesty, J. O., *58*
Hardwick, A. J., 296, *299*
Harpum, J. R., 38, *57*
Harris, R. E., 283(9), 284(9), 287, *287*
Hartford, W. H., 264, 266, 267(8), 269 (8), *271*
Hartwig-Bendig, M., 218(7), 219, 220 (7), 221(7), *223*
Harvey, C. O., 222, *223*
Harwood, H. F., 140, *147*
Hawkes, H. E., 37, *57*
Hawley, W. G., 124, *129*, 143, *147*, 157 (4), *160*, 167(Sec. 6.5, Ref. 4), 168(3), *172*, 198(2), *200*, 236(1), 241(1), *242*, 253, *256*, 260, *261*, 266(2), *270*, 279, *281*, 295, *298*, 303, 306–308, *312*, 314 (1), *319*, 332, 333, 399, 490
Hazel, W. M., 124, *129*
Hecht, F., 221(9), *223*
Hedin, R., 126, *129*

558   AUTHOR INDEX

SUBJECT INDEX

A

Acid-insoluble residue in carbonates,
304–305, 487–488
  determination of, 487–488
  discussion of, 304–305
    choice of acids, for digestion, 305
    conditions for determination of,
304
    solubilizing effect of ignition upon
non-carbonate impurities, 304
Alkali metals in carbonates, 308, 492
  determination of, 492
  discussion of, 308
    use of aluminum sulfate in flame
photometric determination of,
308
Alkali metals in silicates, 194–202
  flame emission and absorption methods for, discussion of, 198–200
    lithium, 199
    rubidium and cesium, 199–200
    following precipitation of cesium
on a carrier and organic extraction, 199
    in the potassium chloroplatinate
fraction from J. Lawrence
Smith method, 199
    method of Abbey, 199
    method of Horstman, 199
    method of Ingamells, 199–200
    sodium and potassium, 198
    effect of lithium, rubidium and
cesium on radiation intensities
of, 198
    effect of sodium and potassium
on each other, 198
    interferences in, and removal of,
198
    use of an internal standard or
radiation buffer, 198
  general, 194–195
  gravimetric methods for, discussion
of, 195–197

potassium, 196–197
  with cobaltinitrite, 197
  separation with chloroplatinic
acid, 196
  separation as perchlorate, 197
  with tetraphenylboron, 197
  preliminary decomposition of sample, 195
  by digestion with acids, 195
  by sintering and extraction,
method of J. Lawrence Smith,
195
  separation of lithium, 197
  separation of rubidium and
cesium, 197
  cesium, by adsorption on am-
monium-12-molybdophos-
phate, 197
sodium, 196
  indirectly, with chloroplatinic
acid, 196
  as perchlorate, 196
  with zinc (or magnesium) uranyl
acetate, 196
Aluminum in carbonates, 306
  discussion of, 306
  colorimetric methods for the determination of, 306
  precipitation of R_2O_3 group, 306
  possibility of coprecipitation of
calcium and magnesium, 306
  titrimetric methods for the determination of, 306
Aluminum in silicates, 136–149, 345–
361
  colorimetric determination of, 355–
361
  by methods of Riley and Williams,
357–361
  with complexing of interfering
elements (Procedure I), 357,
358–359
  interfering elements, 357

559

562 SUBJECT INDEX

effective decomposition versus complete decomposition, 88
by fusion, 95–101
  with alkaline fluorides, 101
  advantage of ammonium fluoride, 101
  low temperature fluxes for refractory silicates and oxides, 101
  for zircon, 101
  with ammonium chloride and ammonium nitrate, 101
  with borax (sodium tetraborate), 98–99
  of refractory oxides, 99
  viscous nature of melt, 99
  with boric oxide, 98–99
  for determination of silica in fluorine-bearing samples, 99
  high temperature flux, 99
  with calcium chloride, 103
  with lithium metaborate, 99
  with lithium tetraborate, 99
  with other fluxes, 101
  with potassium bicarbonate and chloride, 97
  with potassium carbonate, 96–97
  disadvantages of, 97
  and sodium carbonate (fusion mixture), 97
  with potassium pyrosulfate, 100–101
  attack on platinum at high temperature, 100
  ·effectiveness with oxides and silicates, 100
  formation of normal sulfate during fusion, 100
  more effective than sodium salt, 100
  in porcelain or silica vessels, 100
  preparation of, from bisulfate, 100
  with sodium carbonate, 95–96
  addition of oxidant, 96
  and current of oxygen, 96
  and potassium chlorate, 96
  and potassium nitrate, 96
  and sodium peroxide, 96

  and sodium tetraborate, 96
  temperature of fusion, 96
  with sodium fluoride–boric acid, 99
  with sodium and potassium hydroxides, 97
  hygroscopic nature of, 97
  use of other than platinum for, 97
  with sodium peroxide, 98
  attack of crucible by, 98
  explosive fusion with carbon, 98
  and sodium hydroxide, 98
  with zinc oxide and sodium carbonate, 101
general, 88–90
loss of constituents by volatilization, 89
methods for, discussion of, 89–90
  by acids, 89
  in bombs, 89
  by chlorination, 89
  by fluxes, 89
  in sealed tubes, 89
  after preliminary treatment, 90
  by roasting, 90
in sealed tubes, 104–105
  glass, 104
  platinum, 104
by sintering, 102–103
  with calcium carbonate and ammonium chloride, 103
  changes in mixture, 103
  J. Lawrence Smith method, 103
  substitution of barium chloride, 103
  with calcium chloride and oxide, and magnesium oxide, 103
  with calcium oxide and potassium nitrate, 103
  of ruby and sapphire, 102
  with sodium carbonate, 102–103
  minimum flux method, 102
  with sodium carbonate and oxalic acid, 103
  with sodium peroxide, 102
  for decomposition of earth-acid and rare-earth minerals, 102

alumina in, 315
calcium in, 315
chromium in, 315
ferrous iron in, 315
magnesium in, 315
manganese in, 315
nickel in, 315
silica in, 315
titanium in, 315
total iron in, 315
vanadium in, 315
method of Malhotra and Prasada
Rao, 315–316
aluminum in, 316
chromium in, 316
nickel in, 316
silicon in, 316
water in, 316
ores, 314
phosphate, 316–318
method of Cruft, Ingamells and
Muysson, 317
alkali metals, in, 317
aluminum in, 317
calcium and magnesium in, 317
lithium in, 317
phosphate in, 317
silica in, 317
thorium in, 317
total rare earths in, 317
method of Wilson, 316
aluminum in, 316
calcium in, 316
iron in, 316
magnesium in, 316
silica in, 316
in monazite and monazite concentrates, 317–318
Methods for the analysis of silicates,
115–302, 323–486
discussion of, 115–302
procedures for, 323–486
Minus water in silicates. See *Water in
silicates, moisture* (H_2O^-)
Moisture in carbonates. See *Water in
carbonates, moisture* (H_2O^-)
Molybdenum in silicates, 296, 485–486
behavior of, during analysis, 296
colorimetric determination of, with

thiocyanate (method of
Sandell), 485–486
conversion factors for, 486
extraction with isopropyl ether,
486
interferences in, 486
general, 296
methods of determination for, discussion of, 296
by atomic absorption spectroscopy,
296
colorimetric, 296

N

Nickel in silicates, 262–264, 267–269,
462–466
behavior of, during analysis, 263
colorimetric determination of, with
dimethylglyoxime (method of
Sandell), 464–466
extraction with chloroform in, 465
treatment of undecomposed residue in, 465
general, 262–264
gravimetric determination of, with
dimethylglyoxime, 462–464
after acid decomposition, 462–464
conversion factors for, 464
after decomposition by fusion, 463
interference of copper and cobalt
in, 463
reprecipitation in, 464
use of tartaric acid during precipitation, 463
methods of determination for, discussion of, 267–269
by atomic absorption spectroscopy,
269
colorimetric, using dimethylglyoxime, 268
choice of pH for, 268
extraction into chloroform, 268
interference in, 268
gravimetric, with dimethylglyoxime, 267–268
interference of palladium in,
267–268
by optical emission spectroscopy,
268